TEXTBOOK OF FLUID DYNAMICS

TEXTBOOK OF FLUID DYNAMICS

FRANK CHORLTON
F.I.M.A., A.F.R.Ae.S.

Senior Lecturer in Applied Mathematics
University of Aston in Birmingham

CBS

CBS Publishers & Distributors Pvt. Ltd.

New Delhi • Bengaluru • Chennai • Kochi • Kolkata • Mumbai
Hyderabad • Nagpur • Patna • Pune • Vijayawada

ISBN: 81-239-0881-4

First Indian Edition: 1985

This edition has been reprinted in India by arrangement with F. Chorlton.

Published by:
Satish Kumar Jain for CBS Publishers & Distributors Pvt. Ltd.,
4819/XI Prahlad Street, 24 Ansari Road, Daryaganj, New Delhi - 110002
delhi@cbspd.com, cbspubs@airtelmail.in • www.cbspd.com
Ph.: 23289259, 23266861, 23266867 • Fax: 011-23243014
Corporate Office: 204 FIE, Industrial Area, Patparganj, Delhi - 110 092
Ph: 49344934 • Fax: 011-49344935
E-mail: publishing@cbspd.com • publicity@cbspd.com
Branches:

- ***Bengaluru:*** 2975, 17th Cross, K.R. Road, Bansankari 2nd Stage, Bengaluru - 70 • Ph: +91-80-26771678/79 • Fax: +91-80-26771680 E-mail: cbsbng@gmail.com, bangalore@cbspd.com
- ***Chennai:*** No. 7, Subbaraya Street, Shenoy Nagar, Chennai - 600030 Ph: +91-44-26681266, 26680620 • Fax: +91-44-42032115 E-mail: chennai@cbspd.com
- ***Kochi:*** Ashana House, 39/1904, A.M. Thomas Road, Valanjambalam, Ernakulum, Kochi • Ph: +91-484-4059061-65 Fax: +91-484-4059065 • E-mail: cochin@cbspd.com
- ***Kolkata:*** 6-B, Ground Floor, Rameshwar Shaw Road, Kolkata - 700014 Ph: +91-33-22891126/7/8 • E-mail: kolkata@cbspd.com
- ***Mumbai:*** 83-C, Dr. E. Moses Road, Worli, Mumbai - 400018 Ph: +91-9833017933, 022-24902340/41 • E-mail: mumbai@cbspd.com

Representatives:
• Hyderabad: 0-9885175004
• Nagpur: 0-9021734563
• Patna: 0-9334159340
• Pune: 0-9623451994
• Vijayawada: 0-9000660880

Printed at:
Neekunj Print Process, Delhi

PREFACE

This textbook gives a modern treatment of incompressible and compressible fluid flows and it is intended as a companion to my *Textbook of Dynamics*. It is suitable for undergraduate students of mathematics in universities and colleges of technology and it should also satisfy the needs of those general science degree students who are offering mathematics as a principal subject in the final part of their course.

Throughout, the subject is based on vectorial methods, the elements of which are built up in the first chapter. Some of the more important results have been summarized in the tensor notation. Although the novice is unlikely to require a full tensor treatment, the introduction of the notation should facilitate further study at the postgraduate level.

The kinematics of a moving fluid is considered in Chapter 2. The basic ideas of fluid velocity, streamlines, vortex motion, rotational and irrotational flows are first introduced and distinctions are drawn between local and particle rates of change. The chapter ends with the general analysis of fluid motion.

A general treatment of fluids in motion follows. An attempt is made to clarify the meanings of many such fundamental terms as pressure and body force. Unless these notions are clearly understood at the outset, the beginner is apt to become confused when he sees that in some derivations the fluid element is taken to be at rest (as in the equation of continuity) whereas in others (such as the Eulerian equations of motion) it is taken to be of fixed mass and moving along with the main body of fluid. The next two chapters are on incompressible fluids and they treat special mathematical methods involving images and complex variables. Images in three dimensions have been derived rather more directly than one usually finds. The development given is based on my paper 'Some Potential Problems Involving Spheres' published in *Mathematical Gazette*. May 1967.

The remainder of the book treats mainly the dynamics of real fluids in which compressibility and viscosity feature. In recent times these topics have appeared in most undergraduate syllabuses and there is obvious need for an undergraduate textbook containing them as well as the inviscid incompressible flow. These subjects are preceded by a chapter on basic thermodynamics, another topic of growing importance in contemporary undergraduate courses.

The book ends with a chapter on magnetohydrodynamics. This up-to-date topic is now being taught in many colleges and universities and the treatment given should be a suitable introduction to more specialized studies in the field. It should also serve to unify many of the principles developed in earlier chapters. Illustrations are drawn from cosmic electrodynamics and from nuclear engineering.

The theory is rigorously treated and profusely illustrated throughout with worked examples. There are also many problems for solution at the end of each chapter. Several of the problems have been drawn from the degree examinations of various universities and colleges. I wish to thank the bodies concerned for kind permission to use them and I hereby expressly assume full responsibility for any solutions that I have given.

I am indebted to my former student Mr. G. J. Briley for having drawn the charts that are used for oblique shock wave analysis and also for some of the problems in this field. My thanks are due to Dr. J. A. Archer-Hall of the Physics Department of this University for his comments and suggestions on the chapter treating thermodynamics and also to my friend and former colleague Dr. H. G. Cuming (now Vice-Principal at the Napier Technical College) for the many discussions we had on fluid dynamics several years ago. Acknowledgement is also made to the Controller of Her Majesty's Stationery Office for kind permission to reproduce Table I from 'The Method of Characteristics in Supersonic Flow' by Professor G. Temple (R and M 2091).

In conclusion I should like to thank Dr. J. H. Wilkinson of the University of Surrey for the help and advice he has given to me in checking the manuscript, Mrs. M. Hasleton of the Mathematics Department for typing the manuscript, and the publishers D. Van Nostrand Company Ltd., for their unstinted help and encouragement.

F. Chorlton
Department of Mathematics
The University of Aston in Birmingham

SOURCES OF EXAMINATION QUESTIONS

The Universities of:

London	Durham
Cambridge	Manchester
Leeds	Liverpool
Aston in Birmingham	Bristol
Reading	Nottingham

also:

The Imperial College of Science and Technology
University of Surrey (formerly the Battersea College of Technology)
The University of Manchester Institute of Science and Technology (formerly the Manchester College of Science and Technology)

KEY TO ABBREVIATIONS

D.C.'s	direction cosines
D.R.'s	direction ratios
K.E.	kinetic energy
MHD	magnetohydrodynamics
P.E.	potential energy
w.r.t.	with respect to
L.H.S.	left-hand side
R.H.S.	right-hand side

CONTENTS

1

VECTOR ANALYSIS

1.1 Vectors and Scalars

Physical measurements are mainly of two kinds: (i) *scalars* which have magnitude but no associated direction, and (ii) *vectors* which have both magnitude and an associated direction. Examples of scalars include mass, volume and energy. On the other hand, displacement, velocity and force—all of which require the stipulation of a direction for their complete specification—are vectorial quantities. Thus if we say that a train is travelling north-east at 60 m.p.h., the statement specifies completely the magnitude and direction of its velocity. If, on the other hand, we merely say it is travelling at 60 m.p.h. without specifying the direction of motion the statement only defines a physical magnitude which we term the speed.

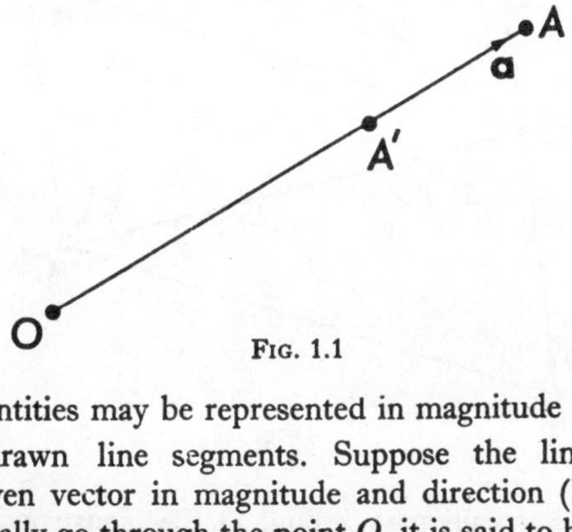

FIG. 1.1

Vectorial quantities may be represented in magnitude and direction by appropriately drawn line segments. Suppose the line segment $\overline{OA}$ represents a given vector in magnitude and direction (Fig. 1.1). If the vector does actually go through the point O, it is said to be *localized* at O: otherwise it is *non-localized*.

It is convenient to have a single algebraic symbol to represent a vector such as $\overline{OA}$. Denote it by **a**. Then we write $\overline{OA} \equiv \mathbf{a}$. In full the statement means that the line segment directed from O to A represents a given vector **a** in magnitude and direction. This representation is incomplete in so far as the position of the vector is unspecified. A further statement needs to be made if we are interested in this.

Now suppose we mark off a segment $\overline{OA'}$ in the direction $\overline{OA}$ and of unit length. Further, suppose $\overline{OA'} \equiv \hat{\mathbf{a}}$. Then $\hat{\mathbf{a}}$ is termed the *unit vector* in the direction of **a**. The *magnitude* of **a** is denoted either by a or by $|\mathbf{a}|$, so that

$$\mathbf{a} = a\hat{\mathbf{a}} = |\mathbf{a}|\hat{\mathbf{a}},$$

since knowledge of a, $\hat{\mathbf{a}}$ define **a** in magnitude and direction.

The *negative* of the vector **a** is denoted by $-\mathbf{a}$ and is defined via the relationship $\overline{AO} \equiv -\mathbf{a}$.

A *zero vector* is one having zero magnitude.

1.2 Addition and Subtraction of Vectors

Suppose we wish to add together two given vectors **a**, **b**. Through any chosen origin O, draw $\overline{OA} \equiv \mathbf{a}$, $\overline{OB} \equiv \mathbf{b}$ (Fig. 1.2). It is immaterial whether

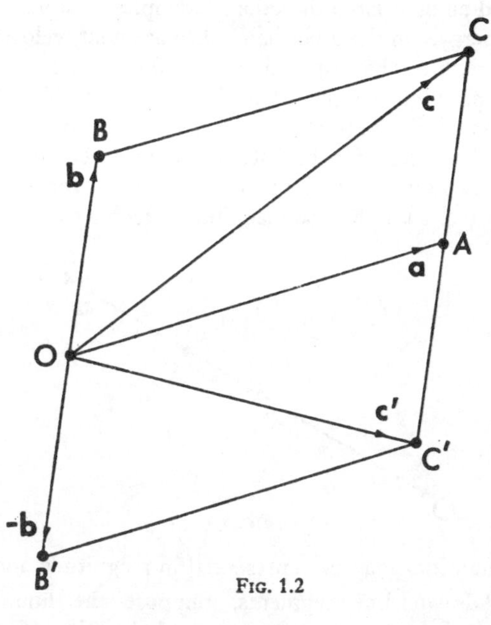

FIG. 1.2

or not the vectors actually pass through O. The construction merely means drawing line segments $\overline{OA}$, $\overline{OB}$ parallel to **a** and **b** and of appropriate lengths.

Now complete the parallelogram $OACB$ and let $\overline{OC} \equiv \mathbf{c}$. Then we say that $\mathbf{c}$ is the sum of $\mathbf{a}$ and $\mathbf{b}$ and we write

$$\mathbf{c} = \mathbf{a} + \mathbf{b}$$

or $\overline{OC} = \overline{OA} + \overline{OB}$.

This is the *Parallelogram Law of Vector Addition*. Since $\overline{AC} \equiv \mathbf{b}$, we could equally write

$$\overline{OC} = \overline{OA} + \overline{AC}$$

which form comprises the *Triangle Law of Addition*. This latter form is very useful in that the form $\overline{OA} + \overline{AC} = \overline{OC}$ shows that the same letter occurring in adjacent positions in summations is annihilated so that the operation $\overline{OA} + \overline{AC}$ may be performed without drawing a diagram.

Now suppose we draw $\overline{OB'} \equiv -\mathbf{b}$ (i.e. $\overline{OB'} = \overline{BO}$). Complete the parallelogram $OAC'B'$ and let $OC' \equiv \mathbf{c}'$. Then

$$\mathbf{c}' = \mathbf{a} + (-\mathbf{b}).$$

This last statement is contracted, in accordance with elementary algebra, to the form

$$\mathbf{c}' = \mathbf{a} - \mathbf{b},$$

thereby giving a method for finding the difference between two vectors.

The process of addition may be extended to several vectors. Thus to add the n vectors $\mathbf{a}_1, \mathbf{a}_2, \ldots, \mathbf{a}_n$, first choose an origin O and draw $\overline{OA_1} \equiv \mathbf{a}_1$. Then draw $\overline{A_1A_2} \equiv \mathbf{a}_2$, $\overline{A_2A_3} \equiv \mathbf{a}_3$, ..., $\overline{A_{n-1}A_n} \equiv \mathbf{a}_n$ (Fig. 1.3).

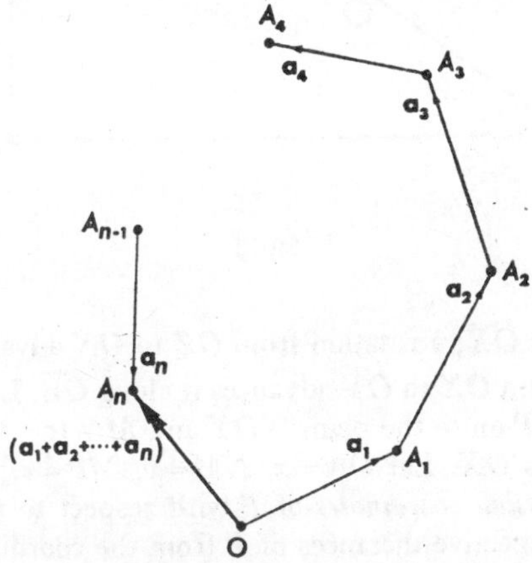

FIG. 1.3

Then

$$\begin{aligned}
&\mathbf{a}_1+\mathbf{a}_2+\mathbf{a}_3+\ldots+\mathbf{a}_n\\
&\equiv\overline{OA_1}+\overline{A_1A_2}+\overline{A_2A_3}+\ldots+\overline{A_{n-1}A_n}\\
&=\overline{OA_2}+\overline{A_2A_3}+\ldots+\overline{A_{n-1}A_n}\\
&=\overline{OA_3}+\overline{A_3A_4}+\ldots+\overline{A_{n-1}A_n}\\
&=\ldots\ldots\ldots\ldots\ldots\\
&=\overline{OA_{n-1}}+\overline{A_{n-1}A_n}\\
&=\overline{OA_n},
\end{aligned}$$

showing that $\overline{OA_n}$ represents the required sum. This is the law of addition by the *polygon of vectors.*

1.3 Use of Coordinates

Suppose $\overline{OP}\equiv\mathbf{r}$ (Fig. 1.4). Through O draw three mutually perpendicular lines OX, OY, OZ to form a right-handed coordinate frame in the sense that the rotation of a right-handed screw from OY to OZ

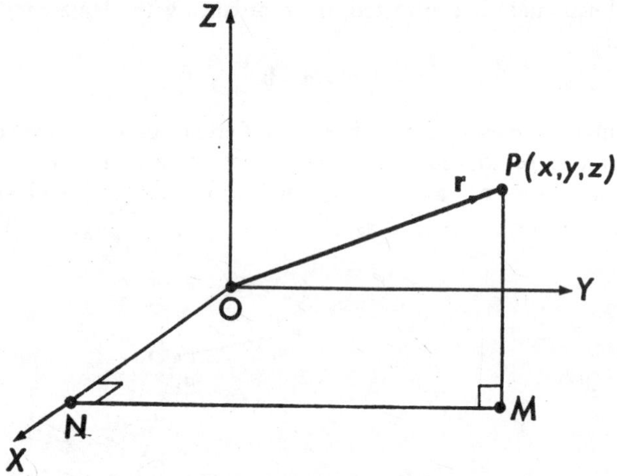

FIG. 1.4

advances it along $\overline{OX}$; a rotation from OZ to OX advances it along $\overline{OY}$; and a rotation from OX to OY advances it along $\overline{OZ}$. Let PM be the perpendicular from P on to the plane XOY and MN the perpendicular from M on to the axis OX. Let $ON=x$, $NM=y$, $MP=z$. Then (x, y, z) are termed the *Cartesian coordinates* of P with respect to the frame. Clearly x, y, z are the respective distances of P from the coordinate planes YOZ, ZOX, XOY.

Now let **i**, **j**, **k** denote unit vectors in the directions $\overline{OX}$, $\overline{OY}$, $\overline{OZ}$. Then

$$\overline{ON} \equiv x\mathbf{i}, \quad \overline{NM} \equiv y\mathbf{j}, \quad \overline{MP} \equiv z\mathbf{k}.$$

Since

$$\overline{OP} = \overline{ON} + \overline{NM} + \overline{MP},$$

$$\overline{OP} \equiv \mathbf{r} = x\mathbf{i} + y\mathbf{j} + z\mathbf{k}. \tag{1}$$

Further,

$$OP^2 = OM^2 + MP^2 = ON^2 + NM^2 + MP^2,$$

$$\therefore\ r^2 = x^2 + y^2 + z^2. \tag{2}$$

The statement (1) will sometimes be more briefly written $\mathbf{r} = [x, y, z]$, meaning that x, y, z are the components of $\mathbf{r}$. The unit vector in the direction of $\mathbf{r}$ is $\hat{\mathbf{r}}$ where

$$\hat{\mathbf{r}} = \mathbf{r}/r = [x/r, y/r, z/r]. \tag{3}$$

1.4 Scalar Product of Two Vectors

In Fig. 1.5, $\overline{OA} \equiv \mathbf{a}$, $\overline{OB} \equiv \mathbf{b}$, $\angle AOB = \theta$. Then the *scalar product of* **a** by **b** is defined to be $ab \cos \theta$ and we write it as $\mathbf{a} \cdot \mathbf{b}$ so that

$$\mathbf{a} \cdot \mathbf{b} = ab \cos \theta.$$

From this definition

$$\mathbf{b} \cdot \mathbf{a} = ba \cos (2\pi - \theta) = ba \cos \theta = \mathbf{a} \cdot \mathbf{b}.$$

Thus scalar multiplication is *commutative*.

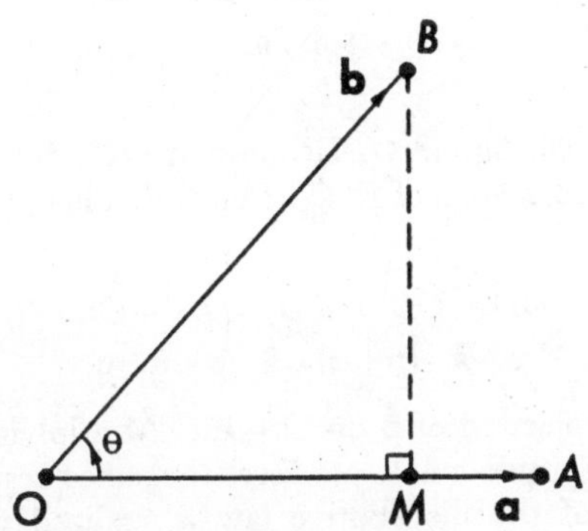

FIG. 1.5

It follows that

$$\hat{\mathbf{a}} \cdot \mathbf{b} = \mathbf{b} \cdot \hat{\mathbf{a}} = b \cos \theta = OM,$$

where OM is the projection of OB on OA.

For any vector **a** we have, writing $\mathbf{a}^2$ for $\mathbf{a} \cdot \mathbf{a}$,

$$\mathbf{a}^2 = \mathbf{a} \cdot \mathbf{a} = (a)(a)(\cos 0) = a^2.$$

The reader will readily see that for unit vectors **i, j, k** in the coordinate ıxes,

$$\mathbf{i}^2 = 1, \quad \mathbf{j}^2 = 1, \quad \mathbf{k}^2 = 1,$$
$$\mathbf{i} \cdot \mathbf{j} = 0, \quad \mathbf{j} \cdot \mathbf{k} = 0, \quad \mathbf{k} \cdot \mathbf{i} = 0.$$

We now prove that for given vectors **a, b, c**, the distributive law for scalar products holds, viz.

$$\mathbf{a} \cdot (\mathbf{b} + \mathbf{c}) = \mathbf{a} \cdot \mathbf{b} + \mathbf{a} \cdot \mathbf{c}.$$

(This result makes the vector method useful in analytical work.)

Proof: In Figure 1.6, let $\overline{OA} \equiv \mathbf{a}$, $\overline{OB} \equiv \mathbf{b}$, $\overline{OC} \equiv \mathbf{c}$.

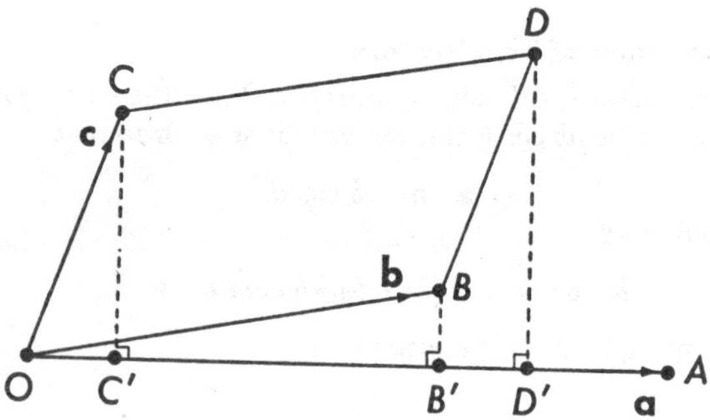

FIG. 1.6

Complete the parallelogram *OBDC* so that $\overline{OD} \equiv \mathbf{b} + \mathbf{c}$. Let B', C', D' be the orthogonal projections of *B*, *C*, *D* on *OA*. Then, from the geometry of the figure,

$$OD' = OB' + B'D' = OB' + OC',$$
$$\therefore \hat{\mathbf{a}} \cdot (\mathbf{b} + \mathbf{c}) = \hat{\mathbf{a}} \cdot \mathbf{b} + \hat{\mathbf{a}} \cdot \mathbf{c},$$

since OB' is the projection of **b** on *OA*, etc. Multiplying this last result through by the magnitude a now produces the stated result.

The usefulness of the distributive law is realized when vectors are expressed in Cartesian components. Thus for given vectors $\mathbf{a} = [a_1, a_2, a_3]$, $\mathbf{b} = [b_1, b_2, b_3]$,

$$\begin{aligned}\mathbf{a} \cdot \mathbf{b} &= (a_1\mathbf{i} + a_2\mathbf{j} + a_3\mathbf{k}) \cdot (b_1\mathbf{i} + b_2\mathbf{j} + b_3\mathbf{k}) \\ &= a_1\mathbf{i} \cdot (b_1\mathbf{i} + b_2\mathbf{j} + b_3\mathbf{k}) + a_2\mathbf{j} \cdot (b_1\mathbf{i} + b_2\mathbf{j} + b_3\mathbf{k}) + a_3\mathbf{k} \cdot (b_1\mathbf{i} + b_2\mathbf{j} + b_3\mathbf{k}) \\ &= \underline{a_1b_1 + a_2b_2 + a_3b_3}.\end{aligned}$$

If $[l_1, m_1, n_1]$, $[l_2, m_2, n_2]$ are the direction cosines of two unit vectors $\overline{OA_1} \equiv \hat{\mathbf{a}}_1$, $\overline{OA_2} \equiv \hat{\mathbf{a}}_2$ where $\angle A_1OA_2 = \theta$, then $\hat{\mathbf{a}}_1 \cdot \hat{\mathbf{a}}_2 = \cos\theta$. But since

$$\hat{\mathbf{a}}_1 = [l_1, m_1, n_1],\ \hat{\mathbf{a}}_2 = [l_2, m_2, n_2],$$
$$\hat{\mathbf{a}}_1 \cdot \hat{\mathbf{a}}_2 = l_1l_2 + m_1m_2 + n_1n_2,$$
$$\therefore\ \underline{\cos\theta = l_1l_2 + m_1m_2 + n_1n_2.}$$

In further illustration of the distributive law, the reader should now show that for any vector **F**,

$$\mathbf{F} = (\mathbf{F}\cdot\mathbf{i})\mathbf{i} + (\mathbf{F}\cdot\mathbf{j})\mathbf{j} + (\mathbf{F}\cdot\mathbf{k})\mathbf{k} = [\mathbf{F}\cdot\mathbf{i},\ \mathbf{F}\cdot\mathbf{j},\ \mathbf{F}\cdot\mathbf{k}].$$

1.5 Vector Product of Two Vectors

In Fig. 1.7, $\overline{OA} \equiv \mathbf{a}$, $\overline{OB} \equiv \mathbf{b}$, $\angle AOB = \theta$ and **n** is the unit vector through O normal to the plane AOB in the sense of a right-handed screw turn from **a** to **b**. Then the *vector product* of **a** by **b** is defined to be the

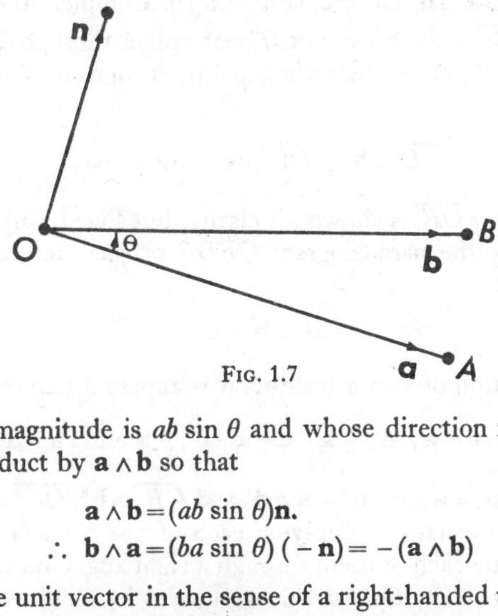

FIG. 1.7

vector whose magnitude is $ab\sin\theta$ and whose direction is **n**. We denote this vector product by $\mathbf{a}\wedge\mathbf{b}$ so that

$$\mathbf{a}\wedge\mathbf{b} = (ab\sin\theta)\mathbf{n}.$$
$$\therefore\ \mathbf{b}\wedge\mathbf{a} = (ba\sin\theta)(-\mathbf{n}) = -(\mathbf{a}\wedge\mathbf{b})$$

since $-\mathbf{n}$ is the unit vector in the sense of a right-handed turn from **b** to **a** through θ. Thus vector multiplication is *non-commutative*.

The reader will readily show that for any vector **a**, $\mathbf{a}\wedge\mathbf{a} = \mathbf{0}$. Further, for the unit vectors **i, j, k**,

$$\mathbf{i}\wedge\mathbf{i} = \mathbf{j}\wedge\mathbf{j} = \mathbf{k}\wedge\mathbf{k} = \mathbf{0};$$
$$\mathbf{i}\wedge\mathbf{j} = \mathbf{k},\ \ \mathbf{j}\wedge\mathbf{k} = \mathbf{i},\ \ \mathbf{k}\wedge\mathbf{i} = \mathbf{j};$$
$$\mathbf{j}\wedge\mathbf{i} = -\mathbf{k},\ \ \mathbf{k}\wedge\mathbf{j} = -\mathbf{i},\ \ \mathbf{i}\wedge\mathbf{k} = -\mathbf{j}.$$

We now establish the distributive law for vector products in the form

$$\mathbf{a}\wedge(\mathbf{b}+\mathbf{c}) = \mathbf{a}\wedge\mathbf{b} + \mathbf{a}\wedge\mathbf{c}.$$

Proof:

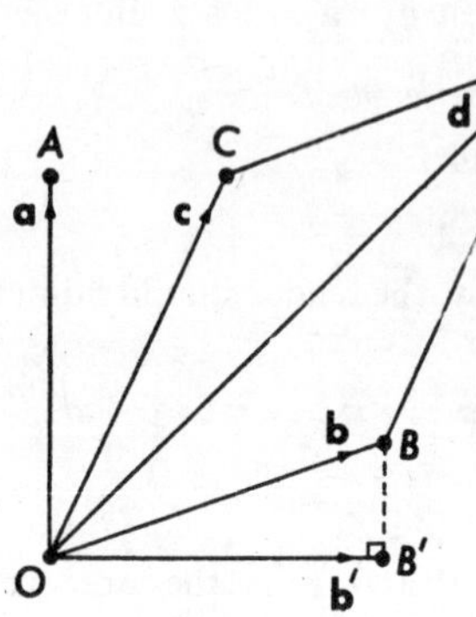

FIG. 1.8(i)

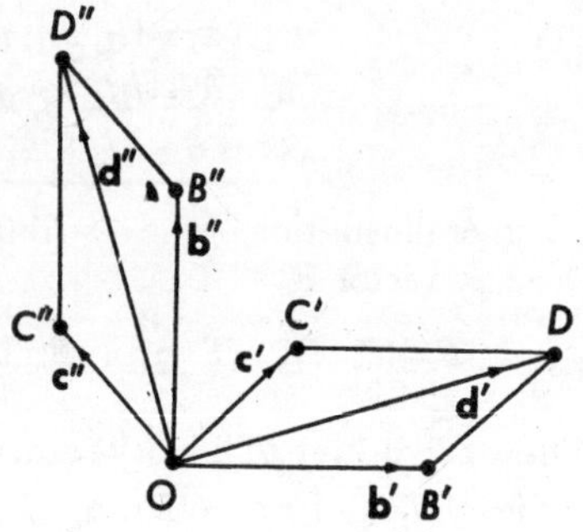

FIG. 1.8(ii)

Let $\overline{OA} \equiv \mathbf{a}$, $\overline{OB} \equiv \mathbf{b}$, $\overline{OC} \equiv \mathbf{c}$ (Fig. 1.8(i)). Complete the parallelogram $OBDC$ so that $\overline{OD} \equiv \mathbf{d} = \mathbf{b} + \mathbf{c}$. Let Π be the plane through O perpendicular to $\mathbf{a}$ and let B', C', D' be the orthogonal projections of B, C, D on Π and write

$$\overline{OB'} \equiv \mathbf{b}', \quad \overline{OC'} \equiv \mathbf{c}', \quad \overline{OD'} \equiv \mathbf{d}'.$$

In Fig. 1.8(i), only $\overline{OB'}$ is shown for clarity, but Fig. 1.8(ii) is a plan of the plane Π. Clearly the parallelogram $OBDC$ projects into a parallelogram $OB'D'C'$ so that

$$\mathbf{d}' = \mathbf{b}' + \mathbf{c}'. \tag{1}$$

From the definition of vector product it is apparent that

$$\mathbf{a} \wedge \mathbf{b} = \mathbf{a} \wedge \mathbf{b}', \quad \mathbf{a} \wedge \mathbf{c} = \mathbf{a} \wedge \mathbf{c}', \quad \mathbf{a} \wedge \mathbf{d} = \mathbf{a} \wedge \mathbf{d}'. \tag{2}$$

Let $\mathbf{b}'' = \mathbf{a} \wedge \mathbf{b}'$, $\mathbf{c}'' = \mathbf{a} \wedge \mathbf{c}'$, $\mathbf{d}'' = \mathbf{a} \wedge \mathbf{d}'$ and $\overline{OB''} \equiv \mathbf{b}''$, $\overline{OC''} \equiv \mathbf{c}''$, $\overline{OD''} \equiv \mathbf{d}''$. The effect of pre-vector multiplying each of the coplanar vectors $\mathbf{b}'$, $\mathbf{c}'$, $\mathbf{d}'$ by $\mathbf{a}$ is to rotate each of them through a right angle in the same sense in the plane Π and to multiply their magnitudes by a. Thus $OB''D''C''$ is a parallelogram and so

$$\mathbf{d}'' = \mathbf{b}'' + \mathbf{c}''. \tag{3}$$

From (3),

$$\mathbf{a} \wedge \mathbf{d}' = \mathbf{a} \wedge \mathbf{b}' + \mathbf{a} \wedge \mathbf{c}',$$

i.e.

$$\mathbf{a} \wedge \mathbf{d} = \mathbf{a} \wedge \mathbf{b} + \mathbf{a} \wedge \mathbf{c}$$

or

$$\underline{\mathbf{a} \wedge (\mathbf{b} + \mathbf{c}) = \mathbf{a} \wedge \mathbf{b} + \mathbf{a} \wedge \mathbf{c}.}$$

This result enables vector products to be evaluated in terms of components. Thus if $\mathbf{a}=[a_1, a_2, a_3]$, $\mathbf{b}=[b_1, b_2, b_3]$,

$$\begin{aligned}\mathbf{a}\wedge\mathbf{b}&=(a_1\mathbf{i}+a_2\mathbf{j}+a_3\mathbf{k})\wedge(b_1\mathbf{i}+b_2\mathbf{j}+b_3\mathbf{k})\\&=a_1\mathbf{i}\wedge(b_1\mathbf{i}+b_2\mathbf{j}+b_3\mathbf{k})+a_2\mathbf{j}\wedge(b_1\mathbf{i}+b_2\mathbf{j}+b_3\mathbf{k})+a_3\mathbf{k}\wedge(b_1\mathbf{i}+b_2\mathbf{j}+b_3\mathbf{k})\\&=a_1b_2\mathbf{k}-a_1b_3\mathbf{j}-a_2b_1\mathbf{k}+a_2b_3\mathbf{i}+a_3b_1\mathbf{j}-a_3b_2\mathbf{i}\\&=(a_2b_3-a_3b_2)\mathbf{i}+(a_3b_1-a_1b_3)\mathbf{j}+(a_1b_2-a_2b_1)\mathbf{k}.\end{aligned}$$

This result may be more convenientlv expressed in the determinantal forms

$$\mathbf{a}\wedge\mathbf{b}=\begin{vmatrix}\mathbf{i}&\mathbf{j}&\mathbf{k}\\a_1&a_2&a_3\\b_1&b_2&b_3\end{vmatrix}=\left\|\begin{matrix}a_1&a_2&a_3\\b_1&b_2&b_3\end{matrix}\right\|.$$

Thus for two unit vectors $\overline{OA_1}\equiv\hat{\mathbf{a}}_1$, $\overline{OA_2}\equiv\hat{\mathbf{a}}_2$, if $\angle A_1OA_2=\theta$, and $\hat{\mathbf{a}}_1=[l_1, m_1, n_1]$, $\hat{\mathbf{a}}_2=[l_2, m_2, n_2]$, then

$$\begin{aligned}\hat{\mathbf{a}}_1\wedge\hat{\mathbf{a}}_2&=\left\|\begin{matrix}l_1&m_1&n_1\\l_2&m_2&n_2\end{matrix}\right\|\\&=[(m_1n_2-m_2n_1), (n_1l_2-n_2l_1), (l_1m_2-l_2m_1)];\end{aligned}$$

$$\therefore\ \sin\theta=|\hat{\mathbf{a}}_1\wedge\hat{\mathbf{a}}_2|=\sqrt{\{(m_1n_2-m_2n_1)^2+(n_1l_2-n_2l_1)^2+(l_1m_2-l_2m_1)^2\}}.$$

1.6 Triple Products

The quantity $\mathbf{a}\cdot(\mathbf{b}\wedge\mathbf{c})$ is the scalar product of the vectors **a** and $\mathbf{b}\wedge\mathbf{c}$: it is called a *scalar triple product* and is denoted by $[\mathbf{a}, \mathbf{b}, \mathbf{c}]$. If we denote the components of **a** by $[a_1, a_2, a_3]$, etc., the reader will easily show that

$$[\mathbf{a}, \mathbf{b}, \mathbf{c}]=\begin{vmatrix}a_1&a_2&a_3\\b_1&b_2&b_3\\c_1&c_2&c_3\end{vmatrix}$$

From elementary determinantal properties it readily follows that

$$[\mathbf{a}, \mathbf{b}, \mathbf{c}]=[\mathbf{b}, \mathbf{c}, \mathbf{a}]=[\mathbf{c}, \mathbf{a}, \mathbf{b}],$$

but $[\mathbf{c}, \mathbf{b}, \mathbf{a}]=-[\mathbf{a}, \mathbf{b}, \mathbf{c}]$, etc. Thus cyclic permuting of the letters leaves the scalar triple product unchanged, but anti-cyclic permuting changes its sign.

The quantity $\mathbf{a}\wedge(\mathbf{b}\wedge\mathbf{c})$, being the vector product of **a** with $(\mathbf{b}\wedge\mathbf{c})$, is vectorial and is called a *vector triple product*. We now prove the result

$$\mathbf{a}\wedge(\mathbf{b}\wedge\mathbf{c})=(\mathbf{a}\cdot\mathbf{c})\mathbf{b}-(\mathbf{a}\cdot\mathbf{b})\mathbf{c}.$$

Proof: With the previous notation,

$$\mathbf{b}\wedge\mathbf{c}=\left\|\begin{matrix}b_1&b_2&b_3\\c_1&c_2&c_3\end{matrix}\right\|=\sum(b_2c_3-b_3c_2)\mathbf{i},$$

the summation being taken cyclically over **i**, **j**, **k**; 1, 2, 3.

$$\therefore\ \mathbf{a}\wedge(\mathbf{b}\wedge\mathbf{c})=\left\|\begin{matrix} a_1 & a_2 & a_3 \\ (b_2c_3-b_3c_2) & (b_3c_1-b_1c_3) & (b_1c_2-b_2c_1)\end{matrix}\right\|$$

$$=\sum\{a_2(b_1c_2-b_2c_1)-a_3(b_3c_1-b_1c_3)\}\mathbf{i}$$

$$=\sum(a_2c_2+a_3c_3)b_1\mathbf{i}-\sum(a_2b_2+a_3b_3)c_1\mathbf{i}$$

$$=\sum(a_1c_1+a_2c_2+a_3c_3)b_1\mathbf{i}-\sum(a_1b_1+a_2b_2+a_3b_3)c_1\mathbf{i}$$

$$=(a_1c_1+a_2c_2+a_3c_3)\sum b_1\mathbf{i}-(a_1b_1+a_2b_2+a_3b_3)\sum c_1\mathbf{i}$$

$$=\underline{(\mathbf{a}\cdot\mathbf{c})\mathbf{b}-(\mathbf{a}\cdot\mathbf{b})\mathbf{c}.}$$

1.7 Vector Moment about a Point and Scalar Moment about an Axis

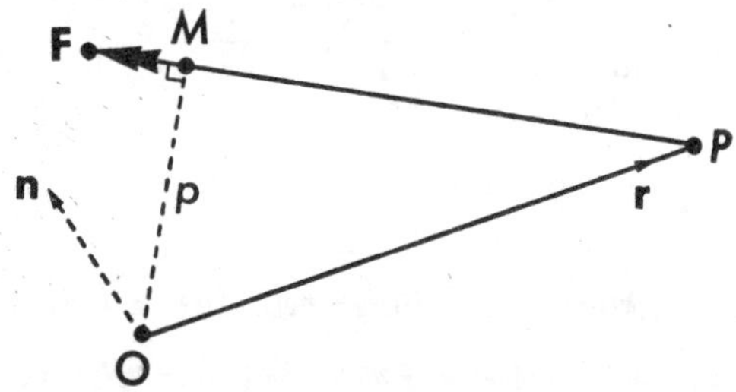

FIG. 1.9

Let **F** be a vector quantity acting at a point P whose position vector with respect to a chosen origin O is given by $\overline{OP}\equiv\mathbf{r}$. Then the *vector moment* **G** of **F** about O is defined to be

$$\mathbf{G}=\mathbf{r}\wedge\mathbf{F}.$$

If $p=OM$ be the perpendicular from O to **F**, then denoting $\theta=\angle OPM$ from the definition of vector product,

$$\mathbf{G}=(rF\sin\theta)\mathbf{n}=(pF)\mathbf{n},$$

where **n** is the unit normal to the plane of OPM in the sense of a right-handed rotation from **r** to **F**.

If we identify **F** with a force, the magnitude of this moment $G=pF$ is clearly the moment of the force about the axis **n** through O. Thinking of OP as the arm of a vice fixed to O by means of a right-handed screw thread, it will readily be appreciated that the effect of this turning moment is to cause O to advance in the direction **n**. Thus we see that associated with the turning effect of the moment of this force about O is a tendency to advance O in the specific direction **n**. In other words, the moment of a force is strictly vectorial in kind and we shall speak of its (vector) moment about O as being $\mathbf{r}\wedge\mathbf{F}$. The magnitude pF will be referred to as the physical moment about the axis **n** through O.

Since moments are vectorial, they may be combined as vectors. Thus if we have n forces $\mathbf{F}_k$ acting at n points P_k with position vectors $\overline{OP_k}=\mathbf{r}_k$ $(k=1, 2, \ldots, n)$, then the total vector moment about O will be

$$\mathbf{G}=\sum_{k=1}^{n}\mathbf{r}_k \wedge \mathbf{F}_k.$$

We now investigate the physical moment of a force about any axis. Let $\mathbf{F}$ be a force acting at a point P distant p from an axis specified by a unit vector $\hat{\mathbf{a}}$ (Fig. 1.10).

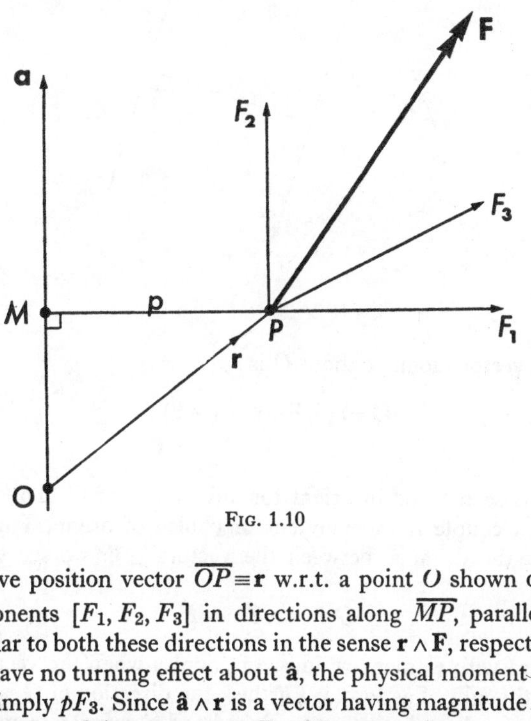

FIG. 1.10

Let P have position vector $\overline{OP}\equiv\mathbf{r}$ w.r.t. a point O shown on $\hat{\mathbf{a}}$. Let $\mathbf{F}$ have components $[F_1, F_2, F_3]$ in directions along $\overline{MP}$, parallel to $\hat{\mathbf{a}}$ and perpendicular to both these directions in the sense $\mathbf{r}\wedge\mathbf{F}$, respectively. Since F_1 and F_2 have no turning effect about $\hat{\mathbf{a}}$, the physical moment of the force about $\hat{\mathbf{a}}$ is simply pF_3. Since $\hat{\mathbf{a}}\wedge\mathbf{r}$ is a vector having magnitude $r\sin\theta$ or p, and direction that of the component F_3 of $\mathbf{F}$, we see that

$$pF_3=(r\sin\theta)F_3=(\hat{\mathbf{a}}\wedge\mathbf{r})\cdot\mathbf{F}=[\hat{\mathbf{a}}, \mathbf{r}, \mathbf{F}].$$

Thus the physical moment of $\mathbf{F}$ about the axis $\hat{\mathbf{a}}$ is given by the scalar triple product $[\hat{\mathbf{a}}, \mathbf{r}, \mathbf{F}]$, where $\mathbf{r}$ is the position vector of $\mathbf{F}$ w.r.t. any chosen point O on $\hat{\mathbf{a}}$. We shall generalize this result to cover cases in which $\mathbf{F}$ is any vector localized at P, force or otherwise.

1.8 Vector and Scalar Couples

Two equal and opposite vectors $\pm\mathbf{F}$ not acting in the same straight line are said to comprise a *couple*. Let P_1, P_2 be any points chosen on $\mathbf{F}$ and $-\mathbf{F}$

and let their position vectors w.r.t. any chosen origin O be given by $\overline{OP_1} \equiv \mathbf{r}_1$, $\overline{OP_2} \equiv \mathbf{r}_2$, and let $\overline{P_2P_1} \equiv \mathbf{s}$ (Fig. 1.11).

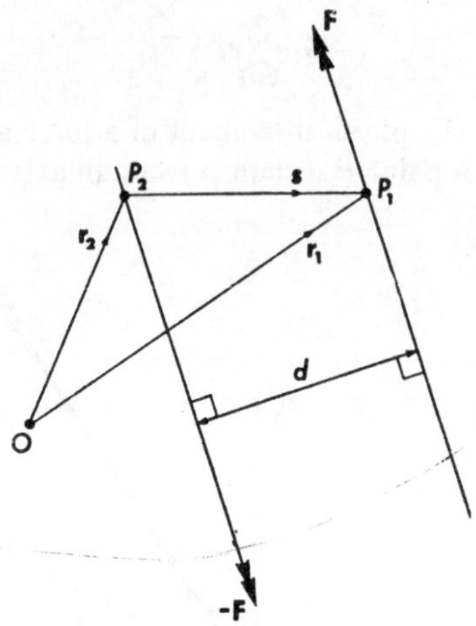

FIG. 1.11

The total vector moment about O is

$$\begin{aligned}\mathbf{G} &= \mathbf{r}_1 \wedge \mathbf{F} + \mathbf{r}_2 \wedge (-\mathbf{F}) \\ &= (\mathbf{r}_1 - \mathbf{r}_2) \wedge \mathbf{F} = \mathbf{s} \wedge \mathbf{F}.\end{aligned}$$

This latter is seen to be invariant for any choice of O. Thus the vector moment of a couple is independent of choice of origin. Further, if we denote by d the distance between the vectors $\pm\mathbf{F}$, we see the physical moment is

$$G = |\mathbf{s} \wedge \mathbf{F}| = dF,$$

in accordance with elementary ideas of couples when the vectors $\pm\mathbf{F}$ are specifically thought of as forces. Further, the direction of $\mathbf{G}$ is the normal to the plane of $\pm\mathbf{F}$ in the sense of a right-handed rotation from $\mathbf{s}$ to $+\mathbf{F}$.

1.9 Centroids

Suppose we have scalar magnitudes m_k associated with n points P_k of space having position vectors $\overline{OP_k} \equiv \mathbf{r}_k$ w.r.t. any chosen origin O $(k = 1, \ldots, n)$. Often in mechanics the quantities m_k will be point masses at the P_k. Then, provided $m_1 + m_2 + \ldots + m_n \neq 0$, the *centroid* or *mean centre* of the scalar magnitudes is defined to be the point G having position vector $\bar{\mathbf{r}}$ where

$$\overline{OG} \equiv \bar{\mathbf{r}} = \left(\sum_{k=1}^{n} m_k \mathbf{r}_k\right) \Big/ \left(\sum_{k=1}^{n} m_k\right).$$

Writing $\mathbf{r}_k=[x_k, y_k, z_k](k=1, \ldots, n)$, $\bar{\mathbf{r}}=[\bar{x}, \bar{y}, \bar{z}]$, we see that the Cartesian coordinates of G are given by

$$\bar{x}=\left(\sum_{k=1}^{n} m_k x_k\right)\Big/\left(\sum_{k=1}^{n} m_k\right), \text{ etc.,} \tag{1}$$

and that, for continuous systems, these summations in limiting form lead to the definite integral expressions

$$\bar{x}=(\int x\,dm)/\int dm, \text{ etc.,} \tag{2}$$

where the integrations extend throughout the entire system.

(2) is derived as follows. If we have point masses δm at coordinate distances x from the plane $x=0$, then in virtue of (1),

$$\bar{x}=(\sum x\delta m)/(\sum \delta m),$$

the summations being throughout the entire distribution. On taking the limit as $\delta m \to 0$, the result (2) follows.

We shall assume the reader is conversant with the methods for determining centroid positions of standard distributions.

Suppose now we take a second origin O' where $\overline{O'O} \equiv \mathbf{s}$ and $\overline{O'P_k} \equiv \mathbf{r}'_k$ $(k=1, \ldots, n)$. Then $\mathbf{s}+\mathbf{r}_k=\mathbf{r}'_k$ $(k=1, \ldots, n)$. If we denote the centroid of the same system for O' by G', then

$$\overline{O'G'} \equiv \left(\sum_{k=1}^{n} m_k \mathbf{r}'_k\right)\Big/\left(\sum_{k=1}^{n} m_k\right)$$

$$= \mathbf{s}+\left(\sum_{k=1}^{n} m_k \mathbf{r}_k\right)\Big/\left(\sum_{k=1}^{n} m_k\right)$$

$$\therefore\ \overline{O'G'}=\overline{O'O}+\overline{OG}=\overline{O'G}.$$

This shows G', G coincide, so that the centroid of the system is a *unique point.*

1.10 Differentiation of Vectors w.r.t. Scalars

Just as in scalar calculus a dependent variable may be a function of an independent variable, having one or more differential coefficients with respect to the independent variable, so a vector quantity $\mathbf{r}$, say, may be a function of a scalar t. Such a functional dependence may be denoted by $\mathbf{r}(t)$. If, corresponding to the value $t+\delta t$ of the independent variable the vector function assumes the value $\mathbf{r}(t+\delta t)=\mathbf{r}+\delta\mathbf{r}$, then the differential coefficient, if it exists, is denoted by $d\mathbf{r}/dt$ and is defined to be

$$\frac{d\mathbf{r}}{dt}=\lim_{\delta t\to 0}\frac{\delta\mathbf{r}}{\delta t}.$$

If $\mathbf{r}$, $\mathbf{s}$ are functions of the scalar t, then

$$\frac{d}{dt}(\mathbf{r}\cdot\mathbf{s})=\lim_{\delta t\to 0}\frac{[(\mathbf{r}+\delta\mathbf{r})\cdot(\mathbf{s}+\delta\mathbf{s})-\mathbf{r}\cdot\mathbf{s}]}{\delta t}$$

$$=\lim_{\delta t\to 0}\frac{\mathbf{r}\cdot\delta\mathbf{s}+\mathbf{s}\cdot\delta\mathbf{r}+\delta\mathbf{r}\cdot\delta\mathbf{s}}{\delta t}$$

$$=\mathbf{r}\cdot\left(\frac{d\mathbf{s}}{dt}\right)+\mathbf{s}\cdot\left(\frac{d\mathbf{r}}{dt}\right).$$

The reader should be able to confirm that

$$\frac{d}{dt}(\mathbf{r}\wedge\mathbf{s})=\mathbf{r}\wedge\left(\frac{d\mathbf{s}}{dt}\right)+\left(\frac{d\mathbf{r}}{dt}\right)\wedge\mathbf{s},$$

where the stated order of vector multiplication must be strictly preserved.

1.11 Notion of Scalar and Vector Fields

If at each point $P(x, y, z)$ of a region $\mathscr{R}$ of space a scalar function $\varphi(x, y, z)$ is defined, then the totality of these values of φ in $\mathscr{R}$ constitutes a *scalar field.* If, in addition, φ is unique at each point P of $\mathscr{R}$, then it is called a *uniform function.* For such a uniform function φ we may construct surfaces on each of which φ is constant: these are called *iso-φ surfaces* or *level surfaces.* Through each point P of $\mathscr{R}$ there will be one and only one level surface when φ is uniform and, moreover, no two level surfaces can intersect; otherwise φ would have more than one value at the points of intersection. An example of a uniform scalar field is afforded by the totality of temperatures at each point of a large room which is not uniformly heated. In this case the isothermals are the level surfaces. Electrostatic potential at points within a region of space containing electric charges affords another example of a uniform scalar field.

In the same way at each point $P(x, y, z)$ of a region $\mathscr{R}$ a unique vector function $\mathbf{F}(x, y, z)$ may be defined, the totality of all such functions comprising a *uniform vector field.* These functions may often be conveniently expressed in terms of Cartesian components:

$$\mathbf{F}=F_1\mathbf{i}+F_2\mathbf{j}+F_3\mathbf{k}=[F_1, F_2, F_3],$$

where $F_1=F_1(x, y, z)$, etc. Examples of vector fields include the velocity of air flowing over an aerofoil and the magnetic lines of force obtained by plotting the field of a bar magnet.

1.12 The Vector Gradient and Direction Differentiator

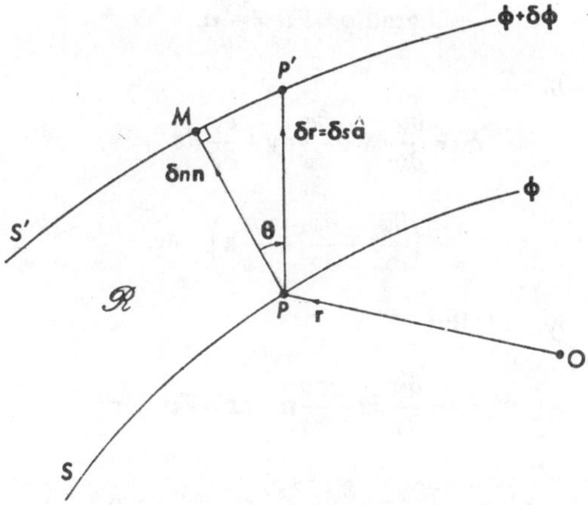

FIG. 1.12

Suppose the iso-φ surfaces of the function $\varphi(x, y, z)$, uniform in a region $\mathscr{R}$, are plotted so that S, S' are two neighbouring surfaces on which the scalar function assumes the values φ, $\varphi+\delta\varphi$ respectively (Fig. 1.12). Let P, P' be neighbouring points on S, S' such that

$$\overline{OP}\equiv\mathbf{r}=[x, y, z],$$
$$\overline{OP'}\equiv\mathbf{r}+\delta\mathbf{r}=[x+\delta x, y+\delta y, z+\delta z],$$

so that $\overline{PP'}\equiv\delta\mathbf{r}=\delta s\hat{\mathbf{a}}$, where $\delta s=|\delta\mathbf{r}|=PP'$. Further let PM be drawn perpendicularly to S' to meet it at M and such that

$$\overline{PM}\equiv\delta n\mathbf{n}, \text{ where } PM=\delta n.$$

When S, S' are ultimately taken to coincide, $\mathbf{n}$ defines the unit normal at P to S in the direction from S to S'. Suppose $\angle MPP'=\theta$ so that $\delta n=\delta s\cos\theta$. Then the rate of change of φ w.r.t. s as we go from P and S to P' on S' is

$$\frac{\delta\varphi}{\delta s}=\frac{\delta\varphi}{\delta n}\frac{\delta n}{\delta s}=\cos\theta\,\frac{\delta\varphi}{\delta n}.$$

Hence as $P'\to P$ along $\overline{P'P}$,

$$\frac{\partial\varphi}{\partial s}=\cos\theta\,\frac{\partial\varphi}{\partial n}$$

and so max. $|\partial\varphi/\partial s|=|\partial\varphi/\partial n|$, at a chosen point P on S. This maximum rate of change takes place when $\theta=0$, i.e. along the direction of $\mathbf{n}$. The vectorial quantity $(\partial\varphi/\partial n)\mathbf{n}$ completely specifies this maximum rate of change together with its direction and is called the (*vector*) *gradient* of φ at P. It is denoted by either grad φ or $\nabla\varphi$ so that

$$\text{grad}\ \varphi = \nabla\varphi = \frac{\partial\varphi}{\partial n}\mathbf{n}. \tag{1}$$

As $\varphi = \varphi(x, y, z)$,

$$\delta\varphi = \frac{\partial\varphi}{\partial x}\,\delta x + \frac{\partial\varphi}{\partial y}\,\delta y + \frac{\partial\varphi}{\partial z}\,\delta z$$

$$= \left(\frac{\partial\varphi}{\partial x}\mathbf{i} + \frac{\partial\varphi}{\partial y}\mathbf{j} + \frac{\partial\varphi}{\partial z}\mathbf{k}\right) \cdot \delta\mathbf{r},$$

since $\delta\mathbf{r} = [\delta x, \delta y, \delta z]$. But

$$\delta\varphi = \frac{\partial\varphi}{\partial n}\,\delta n = \frac{\partial\varphi}{\partial n}\mathbf{n} \cdot \delta\mathbf{r} = \nabla\varphi \cdot \delta\mathbf{r}.$$

Hence

$$\left\{\nabla\varphi - \left(\frac{\partial\varphi}{\partial x}\mathbf{i} + \frac{\partial\varphi}{\partial y}\mathbf{j} + \frac{\partial\varphi}{\partial z}\mathbf{k}\right)\right\} \cdot \delta\mathbf{r} = 0.$$

This last result implies either

(i) $\nabla\varphi - \left(\frac{\partial\varphi}{\partial x}\mathbf{i} + \frac{\partial\varphi}{\partial y}\mathbf{j} + \frac{\partial\varphi}{\partial z}\mathbf{k}\right) = 0;$

(ii) $\delta\mathbf{r} = 0;$

or

(iii) the two vectors are perpendicular.

Since $\delta\mathbf{r}$ is a small non-zero vector of arbitrarily chosen direction, (ii) and (iii) are excluded and (i) holds. Hence we have established

$$\text{grad}\ \varphi = \nabla\varphi = \frac{\partial\varphi}{\partial x}\mathbf{i} + \frac{\partial\varphi}{\partial y}\mathbf{j} + \frac{\partial\varphi}{\partial z}\mathbf{k}. \tag{2}$$

Equations (1) and (2) give alternative methods for evaluating $\nabla\varphi$. Usually (2) is the more convenient.

From

$$\delta\varphi = \frac{\partial\varphi}{\partial x}\delta x + \frac{\partial\varphi}{\partial y}\delta y + \frac{\partial\varphi}{\partial z}\delta z,$$

we have

$$\frac{\partial\varphi}{\partial s} = \frac{\partial\varphi}{\partial x}\,\frac{\partial x}{\partial s} + \frac{\partial\varphi}{\partial y}\,\frac{\partial y}{\partial s} + \frac{\partial\varphi}{\partial z}\,\frac{\partial z}{\partial s}$$

$$= \left(\frac{\partial\varphi}{\partial x}\mathbf{i} + \frac{\partial\varphi}{\partial y}\mathbf{j} + \frac{\partial\varphi}{\partial z}\mathbf{k}\right) \cdot \left(\frac{\partial x}{\partial s}\mathbf{i} + \frac{\partial y}{\partial s}\mathbf{i} + \frac{\partial z}{\partial s}\mathbf{k}\right)$$

$$= \nabla\varphi \cdot \hat{\mathbf{a}},$$

since

$$\hat{\mathbf{a}}=\frac{\partial x}{\partial s}\mathbf{i}+\frac{\partial y}{\partial s}\mathbf{j}+\frac{\partial z}{\partial s}\mathbf{k}.$$

Hence

$$\frac{\partial \varphi}{\partial s}=\hat{\mathbf{a}}\cdot \nabla\varphi.$$

This last result establishes the operational equivalence

$$\partial/\partial s \equiv \hat{\mathbf{a}}\cdot\nabla,$$

since either applied to a scalar function produces the same result. The quantity $\hat{\mathbf{a}}\cdot\nabla$ is called a *direction differentiator* since it differentiates φ in the direction of $\hat{\mathbf{a}}$. Observe that in particular $\partial/\partial n \equiv \mathbf{n}\cdot\nabla$.

EXAMPLE 1

Find the gradient of the function $\varphi(x, y, z)=2x^2-3xy+9z-2$ at the point (1, 0, 0) and differentiate it in the direction specified by [2, −2, 1].

$$\nabla\varphi=\frac{\partial \varphi}{\partial x}\mathbf{i}+\frac{\partial \varphi}{\partial y}\mathbf{j}+\frac{\partial \varphi}{\partial z}\mathbf{k}$$

$$=(4x-3y)\mathbf{i}-3x\mathbf{j}+9\mathbf{k}$$

$$=4\mathbf{i}-3\mathbf{j}+9\mathbf{k} \text{ at } (1, 0, 0).$$

The unit vector in the direction [2, −2, 1] is

$$\hat{\mathbf{a}}=\tfrac{2}{3}\mathbf{i}-\tfrac{2}{3}\mathbf{j}+\tfrac{1}{3}\mathbf{k}.$$

Hence the required directional derivative at (1, 0, 0) is

$$\hat{\mathbf{a}}\cdot\nabla\varphi=(\tfrac{2}{3}\mathbf{i}-\tfrac{2}{3}\mathbf{j}+\tfrac{1}{3}\mathbf{k})\cdot(4\mathbf{i}-3\mathbf{j}+9\mathbf{k})=\tfrac{23}{3}.$$

Alternatively we could use

$$\hat{\mathbf{a}}\cdot\nabla\equiv\frac{2}{3}\frac{\partial}{\partial x}-\frac{2}{3}\frac{\partial}{\partial y}+\frac{1}{3}\frac{\partial}{\partial z}$$

so that $\hat{\mathbf{a}}\cdot\nabla\varphi=\tfrac{2}{3}(4x-3y)-\tfrac{2}{3}(-3x)+\tfrac{1}{3}(9)=\tfrac{23}{3}$, at (1, 0, 0).

EXAMPLE 2

If $\mathbf{r}=[x, y, z]$, find $\nabla(1/r)$.

Method I: Use of level surfaces

Taking $\varphi=1/r$, the iso-φ surfaces are the concentric spheres

$$r=\text{const., centred on } O.$$

Let P be a point on tne sphere S centre O and radius r and let $\overline{OP}\equiv\mathbf{r}$. Then, in the notation of the text,

$$\nabla\varphi=\frac{\partial\varphi}{\partial n}\mathbf{n}.$$

Now

$$\mathbf{n}=\hat{\mathbf{r}}=\frac{\mathbf{r}}{r}.$$

Also, as neighbouring surfaces S, S' are concentric spheres of radii r, $r+\delta r$, and $\varphi=\varphi(r)=1/r$,

$$\frac{\partial\varphi}{\partial n}=\frac{d\varphi}{dr}=-\frac{1}{r^2}$$

$$\therefore\ \nabla\varphi=-\frac{1}{r^2}\hat{\mathbf{r}}=-\frac{1}{r^3}\mathbf{r}.$$

Method II: Use of Cartesians.

$$\varphi=1/r=(x^2+y^2+z^2)^{-1/2},$$

$$\therefore\ \frac{\partial\varphi}{\partial x}=-x(x^2+y^2+z^2)^{-3/2}=-\frac{x}{r^3}.$$

Similarly

$$\frac{\partial\varphi}{\partial y}=-\frac{y}{r^3},\ \frac{\partial\varphi}{\partial z}=-\frac{z}{r^3}.$$

$$\therefore\ \nabla\varphi=-\frac{x}{r^3}\mathbf{i}-\frac{y}{r^3}\mathbf{j}-\frac{z}{r^3}\mathbf{k}=-\frac{\mathbf{r}}{r^3}.$$

Example 3

If $\varphi=\varphi(r)$ where $r=\sqrt{(x^2+y^2+z^2)}$ find $\nabla\varphi$.

The level surfaces are $\varphi=$const. or $r=$const. There are concentric spheres centre O. The reader can now readily show using the arguments of Example 2 that

$$\nabla\varphi=\frac{d\varphi}{dr}\hat{\mathbf{r}}=\frac{1}{r}\frac{d\varphi}{dr}\mathbf{r}.$$

Example 4

Prove $\nabla\int f(u)du=f(u)\nabla u$.

In contracted notation,

$$\nabla\int f(u)du=\sum\left\{\mathbf{i}\frac{\partial}{\partial x}\int f(u)du\right\}.$$

Now

$$\frac{\partial}{\partial x}\int f(u)du=\left\{\frac{d}{du}\int f(u)du\right\}\frac{\partial u}{\partial x}=f(u)\frac{\partial u}{\partial x},$$

$$\therefore\ \nabla\int f(u)du=\sum\left\{\mathbf{i}f(u)\frac{\partial u}{\partial x}\right\}=f(u)\sum\mathbf{i}\frac{\partial u}{\partial x}=\underline{f(u)\nabla u.}$$

We now prove two elementary theorems.

THEOREM I: $$\nabla(\varphi_1\pm\varphi_2)=\nabla\varphi_1\pm\nabla\varphi_2.$$

Proof:

$$\nabla(\varphi_1\pm\varphi_2)=\sum\left\{\mathbf{i}\frac{\partial}{\partial x}(\varphi_1\pm\varphi_2)\right\}$$

$$=\sum\mathbf{i}\frac{\partial\varphi_1}{\partial x}\pm\sum\mathbf{i}\frac{\partial\varphi_2}{\partial x}$$

$$=\underline{\nabla\varphi_1\pm\nabla\varphi_2.}$$

THEOREM II: $$\nabla(\varphi_1\varphi_2)=\varphi_1\nabla\varphi_2+\varphi_2\nabla\varphi_1.$$

Proof:

$$\nabla(\varphi_1\varphi_2)=\sum\left\{\mathbf{i}\frac{\partial}{\partial x}(\varphi_1\varphi_2)\right\}$$

$$=\sum\mathbf{i}\left\{\varphi_1\frac{\partial\varphi_2}{\partial x}+\varphi_2\frac{\partial\varphi_1}{\partial x}\right\}$$

$$=\varphi_1\sum\mathbf{i}\frac{\partial\varphi_2}{\partial x}+\varphi_2\sum\mathbf{i}\frac{\partial\varphi_1}{\partial x}$$

$$=\underline{\varphi_1\nabla\varphi_2+\varphi_2\nabla\varphi_1.}$$

In the above treatment of the direction differentiator, with $\hat{\mathbf{a}}=[l, m, n]$, the method of Example 1 shows

$$\hat{\mathbf{a}}\cdot\nabla\varphi=l\frac{\partial\varphi}{\partial x}+m\frac{\partial\varphi}{\partial y}+n\frac{\partial\varphi}{\partial z},$$

so that

$$\hat{\mathbf{a}}\cdot\nabla\equiv l\frac{\partial}{\partial x}+m\frac{\partial}{\partial y}+n\frac{\partial}{\partial z}.$$

If $\mathbf{F}=[F_1, F_2, F_3]$, which is not necessarily a unit vector, then following the form for $\hat{\mathbf{a}}\cdot\nabla$, we define the operator $\mathbf{F}\cdot\nabla$ to mean $F_1(\partial/\partial x)+F_2(\partial/\partial y)+F_3(\partial/\partial z)$. Moreover such an operator can act on either a scalar function φ or a vector function $\mathbf{V}$ so that we define

$$\mathbf{F}\cdot\nabla\varphi = F_1\frac{\partial\varphi}{\partial x}+F_2\frac{\partial\varphi}{\partial y}+F_3\frac{\partial\varphi}{\partial z},$$

$$\mathbf{F}\cdot\nabla\mathbf{V} = F_1\frac{\partial\mathbf{V}}{\partial x}+F_2\frac{\partial\mathbf{V}}{\partial y}+F_3\frac{\partial\mathbf{V}}{\partial z}.$$

The last relationship requires the evaluation of $\partial\mathbf{V}/\partial x$, $\partial\mathbf{V}/\partial y$, $\partial\mathbf{V}/\partial z$. If $\mathbf{V}=[\mathbf{V}_1,\mathbf{V}_2,\mathbf{V}_3]$ in Cartesians, then

$$\frac{\partial\mathbf{V}}{\partial x}=\lim_{\delta x\to 0}\left[\frac{1}{\delta x}\left\{V_1(x+\delta x,y,z)\mathbf{i}+V_2(x+\delta x,y,z)\mathbf{j}+V_3(x+\delta x,y,z)\mathbf{k} - V_1(x,y,z)\mathbf{i}-V_2(x,y,z)\mathbf{j}-V_3(x,y,z)k\right\}\right]$$

$$=\frac{\partial V_1}{\partial x}\mathbf{i}+\frac{\partial V_2}{\partial x}\mathbf{j}+\frac{\partial V_3}{\partial x}\mathbf{k}.$$

EXAMPLE 5

Prove $\mathbf{F}\cdot\nabla\mathbf{r}\equiv\mathbf{F}$.

With

$$\mathbf{F}=[F_1,F_2,F_3],\quad \mathbf{r}=[x,y,z],$$

$$\mathbf{F}\cdot\nabla = F_1\frac{\partial}{\partial x}+F_2\frac{\partial}{\partial y}+F_3\frac{\partial}{\partial z}.$$

$$\therefore\ \mathbf{F}\cdot\nabla\mathbf{r}=\left(F_1\frac{\partial}{\partial x}+F_2\frac{\partial}{\partial y}+F_3\frac{\partial}{\partial z}\right)(x\mathbf{i}+y\mathbf{j}+z\mathbf{k})$$

$$=F_1\mathbf{i}+F_2\mathbf{j}+F_3\mathbf{k}=\mathbf{F}.$$

1.13 Normal Flux of a Vector over a Surface; Divergence of a Vector

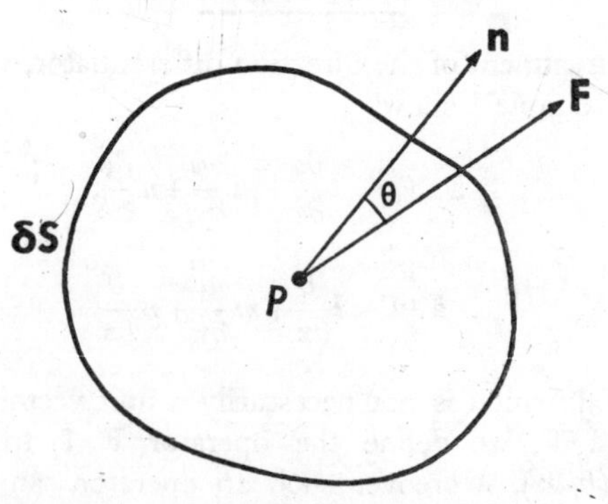

FIG. 1.13

Let $\mathbf{F}$ be a uniform vector function uniquely defined at each point P of a region $\mathscr{R}$ and let δS be a surface element at P having a unit normal specified by $\mathbf{n}$ (Fig. 1.13). Then $\delta\mathbf{S}=\delta S\mathbf{n}$ is called the *vector area* of the element associated with the direction $\mathbf{n}$. If $\mathbf{F}$ at P makes an angle θ with $\mathbf{n}$, then we define *the normal flux of* $\mathbf{F}$ *across* δS *in the sense of* $\mathbf{n}$ to be

$$(F\cos\theta)\delta S=\mathbf{F}\cdot\mathbf{n}\delta S=\mathbf{F}\cdot\delta\mathbf{S}.$$

Over a finite surface S (open or closed) the total normal flux would be

$$\int_S F\cos\theta\, dS=\int_S \mathbf{n}\cdot\mathbf{F}\, dS=\int_S \mathbf{F}\cdot d\mathbf{S},$$

where each integral represents a limited summation such as

$$\lim_{\delta S\to 0}\sum_S (F\cos\theta\delta S).$$

Now suppose that Δv denotes an element of volume of space containing a point P and enclosed by a closed surface ΔS. Let $\mathbf{n}$ be the unit normal at any surface element δS of $\Delta S(\delta S \ll \Delta S)$ *drawn outwards* from Δv. Then the total outward-directed normal flux of $\mathbf{F}$ over ΔS is

$$\int_{\Delta S}\mathbf{n}\cdot\mathbf{F}\, dS=\int_{\Delta S}\mathbf{F}\cdot d\mathbf{S}.$$

Thus the mean outward normal flux of $\mathbf{F}$ per unit volume is

$$\frac{1}{\Delta v}\int_{\Delta S}\mathbf{n}\cdot\mathbf{F}\, dS.$$

Keeping P fixed, let $\Delta v\to 0$ in such a way that ΔS shrinks to the point P. Then if the limit

$$\lim_{\Delta v\to 0}\left\{\frac{1}{\Delta v}\int_{\Delta S}\mathbf{n}\cdot\mathbf{F}\, dS\right\}$$

exists uniquely it measures the outward flux of $\mathbf{F}$ per unit volume at P. It is a scalar quantity and it can be shown that its value is independent of the shape of Δv. (See Section 1.16.1.) Such a limit is termed the *divergence of* $\mathbf{F}$ *at* P and is written div $\mathbf{F}$.

To develop div $\mathbf{F}$ in Cartesians, let $P(x, y, z)$ be the centre of an elemental cuboid $\delta x\times\delta y\times\delta z$, so that $\Delta v=\delta x\delta y\delta z$ (Fig. 1.14). If $\mathbf{F}=[F_1, F_2, F_3]$ then the normal flux of $\mathbf{F}$ across the vector area element $\delta y\delta z\mathbf{i}$ through P (not shown in the diagram) is

$$\mathbf{F}\cdot(\delta y\delta z\,\mathbf{i})=F_1\delta y\delta z.$$

Hence the normal flux in the same direction across the parallel plane centre

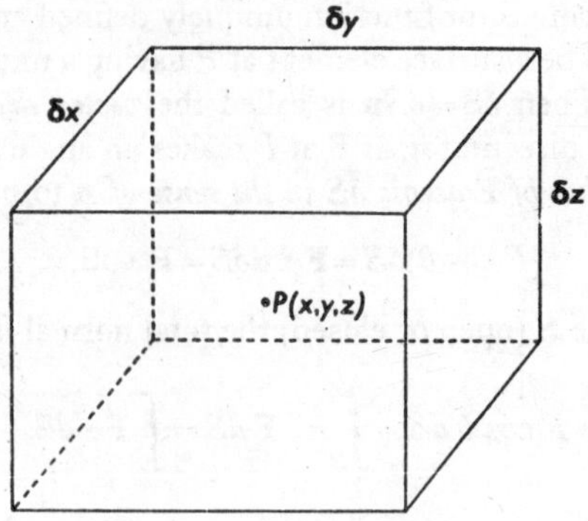

FIG. 1.14

$(x+\frac{1}{2}\delta x, y, z)$ is $F_1\delta y\delta z+\frac{1}{2}\delta x\partial(F_1\delta y\delta z)/\partial x$ to the first order, and that across the parallel plane through $(x-\frac{1}{2}\delta x, y, z)$ is $F_1\delta y\delta z-\frac{1}{2}\delta x\partial(F_1\delta y\delta z)/\partial x$ to the first order. Expressed in terms of the normal directions measured *outwards from the element* (viz. $\mathbf{i}$ and $-\mathbf{i}$ respectively) these quantities become $F_1\delta y\delta z+\frac{1}{2}\delta x\delta y\delta z\ \partial F_1/\partial x$ and $-F_1\delta y\delta z+\frac{1}{2}\delta x\delta y\delta z\ \partial F_1/\partial x$. Thus the total contribution from the two faces perpendicular to the x-direction is $\delta x\delta y\delta z\ \partial F_1/\partial x$. Similarly the total outward flux across the faces perpendicular to the y-direction is $\delta x\delta y\delta z\ \partial F_2/\partial y$ and that across the remaining two is $\delta x\delta y\delta z\ \partial F_3\ \partial z$. Thus, adding contributions together

$$\int_{\Delta S}\mathbf{F}\cdot d\mathbf{S}=\delta x\delta y\delta z\left(\frac{\partial F_1}{\partial x}+\frac{\partial F_2}{\partial y}+\frac{\partial F_3}{\partial z}\right).$$

Hence

$$\text{div }\mathbf{F}=\lim_{\Delta v\to 0}\frac{1}{\Delta v}\int_{\Delta S}\mathbf{F}\cdot d\mathbf{S}=\underline{\frac{\partial F_1}{\partial x}+\frac{\partial F_2}{\partial y}+\frac{\partial F_3}{\partial z}}.$$

The last result may also be written

$$\text{div }\mathbf{F}=\left(\frac{\mathbf{i}\partial}{\partial x}+\frac{\mathbf{j}\partial}{\partial y}+\frac{\mathbf{k}\partial}{\partial z}\right)\cdot(\mathbf{i}F_1+\mathbf{j}F_2+\mathbf{k}F_3)$$

$$=\nabla\cdot\mathbf{F},$$

defining

$$\nabla\equiv\mathbf{i}\frac{\partial}{\partial x}+\mathbf{j}\frac{\partial}{\partial y}+\mathbf{k}\frac{\partial}{\partial z}.$$

The Cartesian form for $\nabla\cdot\mathbf{F}$ is especially useful for evaluation of divergences of vector functions. Thus taking the simple but important case of $\mathbf{r}=[x, y, z]$,

$$\nabla\cdot\mathbf{r}=\frac{\partial x}{\partial x}+\frac{\partial y}{\partial y}+\frac{\partial z}{\partial z}=3.$$

The following are important theorems concerning the divergence.

THEOREM I: For uniform vector functions **F**, **G**,

$$\nabla \cdot (\mathbf{F}+\mathbf{G}) \equiv \nabla \cdot \mathbf{F} + \nabla \cdot \mathbf{G}.$$

Proof: If

$$\mathbf{F}=[F_1, F_2, F_3], \quad \mathbf{G}=[G_1, G_2, G_3],$$

then

$$\mathbf{F}+\mathbf{G}=[(F_1+G_1), (F_2+G_2), (F_3+G_3)]$$

and so

$$\nabla \cdot (\mathbf{F}+\mathbf{G}) = \frac{\partial}{\partial x}(F_1+G_1) + \frac{\partial}{\partial y}(F_2+G_2) + \frac{\partial}{\partial z}(F_3+G_3)$$

$$= \left(\frac{\partial F_1}{\partial x} + \frac{\partial F_2}{\partial y} + \frac{\partial F_3}{\partial z}\right) + \left(\frac{\partial G_1}{\partial x} + \frac{\partial G_2}{\partial y} + \frac{\partial G_3}{\partial z}\right)$$

$$= \underline{\nabla \cdot \mathbf{F} + \nabla \cdot \mathbf{G}.}$$

THEOREM II: For a uniform scalar function φ and a uniform vector function **F**,

$$\nabla \cdot (\varphi \mathbf{F}) \equiv \varphi \nabla \cdot \mathbf{F} + (\nabla \varphi) \cdot \mathbf{F}.$$

Proof: Taking $\mathbf{F}=[F_1, F_2, F_3]$,

$$\nabla \cdot (\varphi \mathbf{F}) = \sum \left\{\frac{\partial}{\partial x}(\varphi F_1)\right\}$$

$$= \sum \left(\varphi \frac{\partial F_1}{\partial x} + \frac{\partial \varphi}{\partial x} F_1\right)$$

$$= \varphi \sum \frac{\partial F_1}{\partial x} + \left(\frac{\partial \varphi}{\partial x}\mathbf{i} + \frac{\partial \varphi}{\partial y}\mathbf{j} + \frac{\partial \varphi}{\partial z}\mathbf{k}\right) \cdot \mathbf{F}$$

$$= \underline{\varphi \nabla \cdot \mathbf{F} + (\nabla \varphi) \cdot \mathbf{F}.}$$

If we can find a uniform scalar function φ such that $\mathbf{F}=\nabla\varphi=(\partial\varphi/\partial x)\mathbf{i} + (\partial\varphi/\partial y)\mathbf{j} + (\partial\varphi/\partial z)\mathbf{k}$, then

$$\nabla \cdot \mathbf{F} = \frac{\partial^2 \varphi}{\partial x^2} + \frac{\partial^2 \varphi}{\partial y^2} + \frac{\partial^2 \varphi}{\partial z^2} = \nabla^2 \varphi,$$

where $\nabla^2 \equiv \nabla \cdot \nabla \equiv \partial^2/\partial x^2 + \partial^2/\partial y^2 + \partial^2/\partial z^2$ and this operator is called the *Laplacian*. If, in particular,

$$\nabla^2 \varphi = 0,$$

then φ is said to satisfy *Laplace's equation* and it is termed a *harmonic function*.

EXAMPLE

Prove that $1/r$ is harmonic.

From Example 2 of Section 1.12,

$$\nabla\varphi = -(1/r^3)\mathbf{r}.$$
$$\therefore\ \nabla^2\varphi = -\nabla\cdot\{(1/r^3)\mathbf{r}\}$$
$$= -(1/r^3)\nabla\cdot\mathbf{r} - \{\nabla(1/r^3)\}\cdot\mathbf{r}.$$

Now $\nabla\cdot\mathbf{r}=3$.

Also since the level surfaces for $1/r^3$ are concentric spheres centred on O,

$$\nabla\frac{1}{r^3} = \left(\frac{d}{dr}\frac{1}{r^3}\right)\hat{\mathbf{r}} = -\frac{3}{r^4}\hat{\mathbf{r}}.$$

$$\therefore\ \nabla^2\varphi = -(1/r^3)3 + (3/r^4)\hat{\mathbf{r}}\cdot\mathbf{r} = 0, \quad \text{since } \hat{\mathbf{r}}\cdot\mathbf{r} = r.$$

As an alternative, by taking $1/r=(x^2+y^2+z^2)^{-1/2}$, one can show by differentiation that

$$\left(\frac{\partial^2}{\partial x^2}+\frac{\partial^2}{\partial y^2}+\frac{\partial^2}{\partial z^2}\right)(x^2+y^2+z^2)^{-1/2}=0.$$

The operator ∇^2 may also be applied to a vector function. Thus if $\mathbf{F}=[F_1, F_2, F_3]$ in Cartesians, then we define $\nabla^2\mathbf{F}$ to mean

$$\nabla^2\mathbf{F} = \left(\frac{\partial^2}{\partial x^2}+\frac{\partial^2}{\partial y^2}+\frac{\partial^2}{\partial z^2}\right)(F_1\mathbf{i}+F_2\mathbf{j}+F_3\mathbf{k})$$
$$=(\nabla^2F_1)\mathbf{i}+(\nabla^2F_2)\mathbf{j}+(\nabla^2F_3)\mathbf{k}.$$

1.14 Line Integrals; Curl of a Vector Function

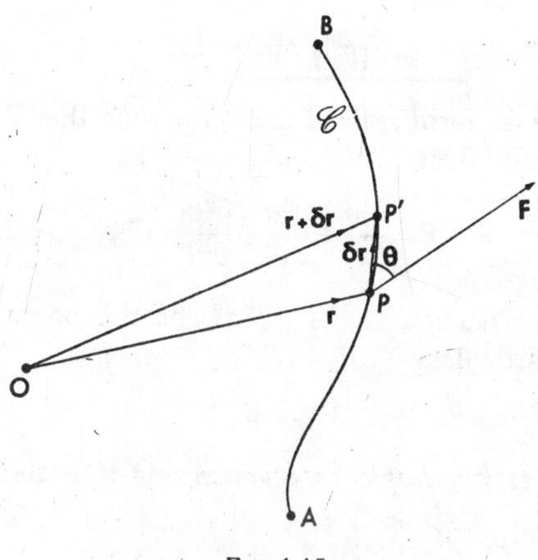

FIG. 1.15

Let $\mathscr{C}$ be an arc of a curve joining points A, B and suppose P, P' are neighbouring points on this arc where $\overline{OP} \equiv \mathbf{r}$, $\overline{OP'} \equiv \mathbf{r} + \delta\mathbf{r}$, so that $\overline{PP'} \equiv \delta\mathbf{r}$ (Fig. 1.15). Let $\mathbf{F}$ be a uniform vector function defined at all points P of $\mathscr{C}$ and suppose $\mathbf{F}$ at P makes angle θ with the tangent at P directed towards P'. If arc $AP = s$, arc $AP' = s + \delta s$, then arc $PP' = \delta s = |\delta\mathbf{r}|$, to the first order. We write $\overline{PP'} \equiv \delta\mathbf{r} = \delta\mathbf{s}$.

The *line integral* of $\mathbf{F}$ along $\overline{PP'}$ is defined to be

$$(F \cos \theta)\delta s = \mathbf{F} \cdot \delta\mathbf{s} = \mathbf{F} \cdot \delta\mathbf{r}$$

and the total line integral of $\mathbf{F}$ from A to B along $\mathscr{C}$ is

$$\int_A^B \mathbf{F} \cos \theta \, ds = \int_A^B \mathbf{F} \cdot d\mathbf{s} = \int_A^B \mathbf{F} \cdot d\mathbf{r}.$$

To give the definition concrete meaning, the reader will readily see that if $\mathbf{F}$ is a force, these line integral expressions each represent the work done by the force in moving a particle from A to B along $\mathscr{C}$ since $F \cos \theta$ is the tangential component of the force at P.

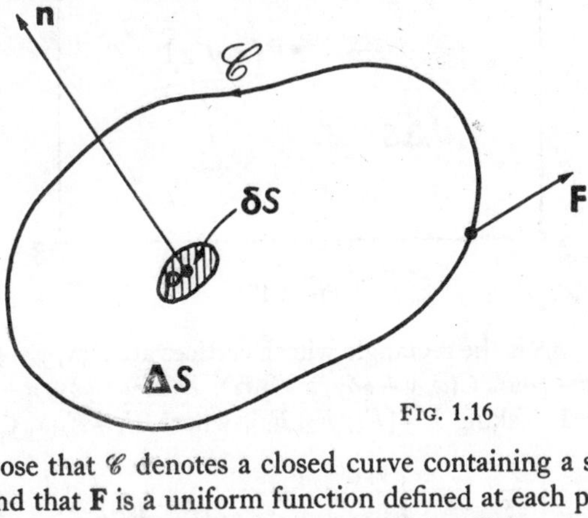

FIG. 1.16

Now suppose that $\mathscr{C}$ denotes a closed curve containing a small element of area ΔS and that $\mathbf{F}$ is a uniform function defined at each point of $\mathscr{C}$ and of ΔS. Let P be a point in ΔS and suppose δS ($\ll \Delta S$) is a small element of ΔS and containing P. Further suppose the sense of description of $\mathscr{C}$ is as indicated (Fig. 1.16). At P draw the unit normal $\mathbf{n}$ to δS in the sense that a positive turn around $\mathbf{n}$ is in the same sense as the description of $\mathscr{C}$. Then the total line integral or *circulation* of $\mathbf{F}$ round $\mathscr{C}$ is denoted by

$$\oint_{\mathscr{C}} \mathbf{F} \cdot d\mathbf{s} = \oint_{\mathscr{C}} \mathbf{F} \cdot d\mathbf{r}.$$

Hence the mean circulation per unit area of ΔS is

$$\frac{1}{\Delta S} \oint_{\mathscr{C}} \mathbf{F} \cdot d\mathbf{s}.$$

Keeping P fixed, let $\Delta S \to 0$ in such a way that the curve $\mathscr{C}$ approaches the point P. Then the limiting value of the mean circulation per unit area at P, if it exists, can be shown to be independent of the shape of $\mathscr{C}$ for a given normal direction $\mathbf{n}$. This limit is a scalar but depends on the normal $\mathbf{n}$ to S. Let us try to find a vector whose scalar product with $\mathbf{n}$ produces the required limit. Let the vector be denoted by curl $\mathbf{F}$ so that

$$\mathbf{n} \cdot \text{curl } \mathbf{F} = \lim_{\Delta S \to 0} \left\{ \frac{1}{\Delta S} \oint_{\mathscr{C}} \mathbf{F} \cdot d\mathbf{s} \right\}.$$

This relationship permits of the development of the vector curl $\mathbf{F}$ in Cartesians.

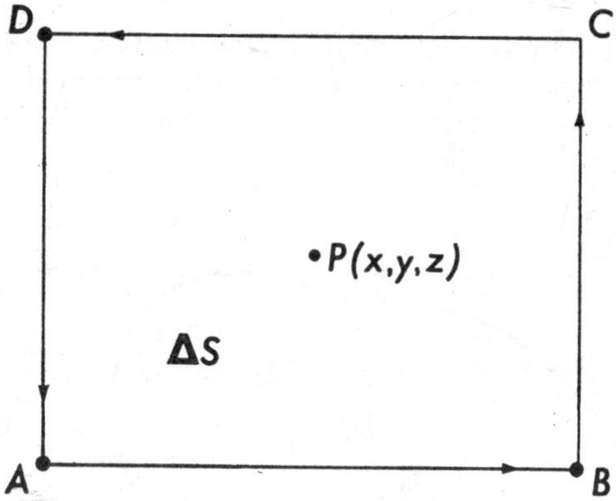

FIG. 1.17

In Fig. 1.17, ΔS is the rectangle whose vertices are $A(x, y-\frac{1}{2}\delta y, z-\frac{1}{2}\delta z)$, $B(x, y+\frac{1}{2}\delta y, z-\frac{1}{2}\delta z)$, $C(x, y+\frac{1}{2}\delta y, z+\frac{1}{2}\delta z)$, $D(x, y-\frac{1}{2}\delta y, z+\frac{1}{2}\delta z)$ so that $\Delta S = \delta y \delta z$, $n = \mathbf{i}$. Taking $\mathbf{F} = [F_1, F_2, F_3]$, where $F_1 = F_1(x, y, z)$, etc.,

$$\int_A^B \mathbf{F} \cdot d\mathbf{s} = \int_{y-(1/2)\delta y}^{y+(1/2)\delta y} F_2(x, y, z-\tfrac{1}{2}\delta z)\, dy,$$

$$\int_B^C \mathbf{F} \cdot d\mathbf{s} = \int_{z-(1/2)\delta z}^{z+(1/2)\delta z} F_3(x, y+\tfrac{1}{2}\delta y, z)\, dz,$$

$$\int_C^D \mathbf{F} \cdot d\mathbf{s} = -\int_{y-(1/2)\delta y}^{y+(1/2)\delta y} F_2(x, y, z+\tfrac{1}{2}\delta z)\, dy,$$

$$\int_D^A \mathbf{F} \cdot d\mathbf{s} = -\int_{z-(1/2)\delta z}^{z+(1/2)\delta z} F_3(x, y-\tfrac{1}{2}\delta y, z)\, dz,$$

each integral being evaluated along the appropriate line segment. Thus the total line integral round the closed rectangle is

$$\oint_{\mathscr{C}} \mathbf{F}\cdot d\mathbf{s}=\int_{z-(1/2)\delta z}^{z+(1/2)\delta z}\{F_3(x,y+\tfrac{1}{2}\delta y,z)-F_3(x,y-\tfrac{1}{2}\delta y,z)\}\,dz,$$

$$-\int_{y-(1/2)\delta y}^{y+(1/2)\delta y}\{F_2(x,y,z+\tfrac{1}{2}\delta z)-F_2(x,y,z-\tfrac{1}{2}\delta z)\}\,dy$$

$$=\int_{z-(1/2)\delta z}^{z+(1/2)\delta z}\left\{\delta y\frac{\partial F_3(x,y,z)}{\partial y}+\ldots\right\}dz-\int_{y-(1/2)\delta y}^{y+(1/2)\delta y}\left\{\delta z\frac{\partial F_2(x,y,z)}{\partial z}+\ldots\right\}dy$$

$$\simeq \delta y\,\delta z\left\{\frac{\partial F_3}{\partial y}-\frac{\partial F_2}{\partial z}\right\},$$

to the second order.

$$\therefore \lim_{\Delta S\to 0}\left\{\frac{1}{\Delta S}\oint_{\mathscr{C}}\mathbf{F}\cdot d\mathbf{s}\right\}=\frac{\partial F_3(x,y,z)}{\partial y}-\frac{\partial F_2(x,y,z)}{\partial z},$$

i.e.

$$\mathbf{i}\cdot \text{curl}\,\mathbf{F}=\frac{\partial F_3}{\partial y}-\frac{\partial F_2}{\partial z}.$$

Similarly

$$\mathbf{j}\cdot \text{curl}\,\mathbf{F}=\frac{\partial F_1}{\partial z}-\frac{\partial F_3}{\partial x},$$

$$\mathbf{k}\cdot \text{curl}\,\mathbf{F}=\frac{\partial F_2}{\partial x}-\frac{\partial F_1}{\partial y}.$$

Hence

$$\text{curl}\,\mathbf{F}=\left[\frac{\partial F_3}{\partial y}-\frac{\partial F_2}{\partial z},\ \frac{\partial F_1}{\partial z}-\frac{\partial F_3}{\partial x},\ \frac{\partial F_2}{\partial x}-\frac{\partial F_1}{\partial y}\right].$$

For convenience this may be remembered in the form

$$\text{curl}\,\mathbf{F}=\left\|\begin{matrix}\dfrac{\partial}{\partial x} & \dfrac{\partial}{\partial y} & \dfrac{\partial}{\partial z}\\ F_1 & F_2 & F_3\end{matrix}\right\|$$

so that we may write

$$\text{curl}\,\mathbf{F}=\nabla\wedge\mathbf{F},$$

where

$$\nabla\equiv\mathbf{i}\frac{\partial}{\partial x}+\mathbf{j}\frac{\partial}{\partial y}+\mathbf{k}\frac{\partial}{\partial z}.$$

1.15 Some Vector Identities

We now prove some important identities.

(i) If φ is any uniform differentiable scalar function, then

$$\text{curl grad}\,\varphi\equiv\mathbf{0}.$$

Proof:
$$\text{grad}\,\varphi=\left[\frac{\partial\varphi}{\partial x},\ \frac{\partial\varphi}{\partial y},\ \frac{\partial\varphi}{\partial z}\right]$$

$$\therefore\ \text{curl grad}\,\varphi=\begin{Vmatrix}\dfrac{\partial}{\partial x} & \dfrac{\partial}{\partial y} & \dfrac{\partial}{\partial z}\\ \dfrac{\partial\varphi}{\partial x} & \dfrac{\partial\varphi}{\partial y} & \dfrac{\partial\varphi}{\partial z}\end{Vmatrix}$$

$$=\left[\frac{\partial^2\varphi}{\partial y\partial z}-\frac{\partial^2\varphi}{\partial z\partial y},\ \frac{\partial^2\varphi}{\partial z\partial x}-\frac{\partial^2\varphi}{\partial x\partial z},\ \frac{\partial^2\varphi}{\partial x\partial y}-\frac{\partial^2\varphi}{\partial y\partial z}\right]$$

$$=[0,0,0],\quad \text{since } \partial^2\varphi/\partial y\partial z=\partial^2\varphi/\partial z\partial y, \text{ etc.}$$

(ii) If **F** is any uniform differentiable vector function, then

$$\text{div curl}\,\mathbf{F}\equiv 0.$$

Proof: With $\mathbf{F}=[F_1, F_2, F_3]$ in Cartesians,

$$\text{curl}\,\mathbf{F}=\begin{Vmatrix}\dfrac{\partial}{\partial x} & \dfrac{\partial}{\partial y} & \dfrac{\partial}{\partial z}\\ F_1 & F_2 & F_3\end{Vmatrix}$$

$$=\left[\frac{\partial F_3}{\partial y}-\frac{\partial F_2}{\partial z},\ \frac{\partial F_1}{\partial z}-\frac{\partial F_3}{\partial x},\ \frac{\partial F_2}{\partial x}-\frac{\partial F_1}{\partial y}\right]$$

and so

$$\text{div curl}\,\mathbf{F}=\frac{\partial}{\partial x}\left(\frac{\partial F_3}{\partial y}-\frac{\partial F_2}{\partial z}\right)+\frac{\partial}{\partial y}\left(\frac{\partial F_1}{\partial z}-\frac{\partial F_3}{\partial x}\right)+\frac{\partial}{\partial z}\left(\frac{\partial F_2}{\partial x}-\frac{\partial F_1}{\partial y}\right)$$

$$=\left(\frac{\partial^2 F_1}{\partial y\partial z}-\frac{\partial^2 F_1}{\partial z\partial y}\right)+\left(\frac{\partial^2 F_2}{\partial z\partial x}-\frac{\partial^2 F_2}{\partial x\partial z}\right)+\left(\frac{\partial^2 F_3}{\partial x\partial y}-\frac{\partial^2 F_3}{\partial y\partial x}\right)$$

$$=0.$$

(*N.B.* In symbolic form,

$$\text{div curl}\,\mathbf{F}=\nabla\cdot(\nabla\wedge\mathbf{F})=[\nabla,\nabla,\mathbf{F}].$$

The two equal elements in the scalar triple product form suggest its value is zero. It should be remembered, however, that ∇ is a vector operator and not a vector, so that this approach is not a strict proof of the identity as we are not entitled to expect the laws of vector algebra to be applicable to vector operators.)

(iii) With the function **F** as in (ii),

$$\nabla\wedge(\nabla\wedge\mathbf{F})\equiv\nabla(\nabla\cdot\mathbf{F})-\nabla^2\mathbf{F},$$

or

$$\text{curl curl}\,\mathbf{F}\equiv\text{grad div}\,\mathbf{F}-\nabla^2\mathbf{F}.$$

Proof: Taking $\mathbf{F}=[F_1, F_2, F_3]$,

$$\nabla\wedge\mathbf{F}=\begin{Vmatrix}\frac{\partial}{\partial x} & \frac{\partial}{\partial y} & \frac{\partial}{\partial z}\\ F_1 & F_2 & F_3\end{Vmatrix}$$

$$=\left[\frac{\partial F_3}{\partial y}-\frac{\partial F_2}{\partial z},\ \frac{\partial F_1}{\partial z}-\frac{\partial F_3}{\partial x},\ \frac{\partial F_2}{\partial x}-\frac{\partial F_1}{\partial y}\right],$$

and so

$$\nabla\wedge(\nabla\wedge\mathbf{F})=\begin{Vmatrix}\frac{\partial}{\partial x} & \frac{\partial}{\partial y} & \frac{\partial}{\partial z}\\ \frac{\partial F_3}{\partial y}-\frac{\partial F_2}{\partial z} & \frac{\partial F_1}{\partial z}-\frac{\partial F_3}{\partial x} & \frac{\partial F_2}{\partial x}-\frac{\partial F_1}{\partial y}\end{Vmatrix}$$

$$=\sum\mathbf{i}\left\{\frac{\partial}{\partial y}\left(\frac{\partial F_2}{\partial x}-\frac{\partial F_1}{\partial y}\right)-\frac{\partial}{\partial z}\left(\frac{\partial F_1}{\partial z}-\frac{\partial F_3}{\partial x}\right)\right\}$$

$$=\sum\mathbf{i}\frac{\partial}{\partial x}\left(\frac{\partial F_2}{\partial y}+\frac{\partial F_3}{\partial z}\right)-\sum\mathbf{i}\left(\frac{\partial^2 F_1}{\partial y^2}+\frac{\partial^2 F_1}{\partial z^2}\right)$$

$$=\sum\mathbf{i}\frac{\partial}{\partial x}\left(\frac{\partial F_1}{\partial x}+\frac{\partial F_2}{\partial y}+\frac{\partial F_3}{\partial z}\right)-\sum\mathbf{i}\left(\frac{\partial^2 F_1}{\partial x^2}+\frac{\partial^2 F_1}{\partial y^2}+\frac{\partial^2 F_1}{\partial z^2}\right)$$

$$=\sum\left\{\mathbf{i}\frac{\partial}{\partial x}(\nabla\cdot\mathbf{F})\right\}-\sum\mathbf{i}\nabla^2 F_1$$

$$=\sum\mathbf{i}\frac{\partial}{\partial x}(\nabla\cdot\mathbf{F})-\nabla^2\sum F_1\mathbf{i}$$

$$=\underline{\nabla(\nabla\cdot\mathbf{F})-\nabla^2\mathbf{F}}.$$

(*N.B.* Application of the vector product formula

$$\mathbf{A}\wedge(\mathbf{B}\wedge\mathbf{C})=(\mathbf{A}\cdot\mathbf{C})\mathbf{B}-(\mathbf{A}\cdot\mathbf{B})\mathbf{C}$$

to $\nabla\wedge(\nabla\wedge\mathbf{F})$ with A, B, C replaced by ∇, ∇, $\mathbf{F}$, gives the desired result so long as the second operator ∇ is brought in front of the 'scalar product' $\nabla\cdot\mathbf{F}$. This affords a convenient mode of recovery of the result but it is not a substitute for the proof given above.)

(iv) For uniform vector functions $\mathbf{F}$, $\mathbf{G}$,

$$\operatorname{div}(\mathbf{F}\wedge\mathbf{G})\equiv\mathbf{G}\cdot\operatorname{curl}\mathbf{F}-\mathbf{F}\cdot\operatorname{curl}\mathbf{G}.$$

Proof:

$$\nabla\cdot(\mathbf{F}\wedge\mathbf{G})=\sum\left\{\mathbf{i}\cdot\frac{\partial}{\partial x}(\mathbf{F}\wedge\mathbf{G})\right\}$$

$$=\sum\left\{\mathbf{i}\cdot\left[\mathbf{F}\wedge\frac{\partial\mathbf{G}}{\partial x}+\frac{\partial\mathbf{F}}{\partial x}\wedge\mathbf{G}\right]\right\}$$

$$=\sum\left[\mathbf{i},\mathbf{F},\frac{\partial\mathbf{G}}{\partial x}\right]+\sum\left[\mathbf{i},\frac{\partial\mathbf{F}}{\partial x},\mathbf{G}\right]$$

$$=-\sum\left[\mathbf{F},\mathbf{i},\frac{\partial\mathbf{G}}{\partial x}\right]+\sum\left[\left(\mathbf{i}\wedge\frac{\partial\mathbf{F}}{\partial x}\right)\cdot\mathbf{G}\right]$$

$$=-\mathbf{F}\cdot\sum\left(\mathbf{i}\wedge\frac{\partial\mathbf{G}}{\partial x}\right)+\left\{\sum\left(\mathbf{i}\wedge\frac{\partial\mathbf{F}}{\partial x}\right)\right\}\cdot\mathbf{G}$$

$$=\underline{-\mathbf{F}\cdot\operatorname{curl}\mathbf{G}+\mathbf{G}\cdot\operatorname{curl}\mathbf{F}}.$$

(v) curl $(\mathbf{F}\wedge\mathbf{G})\equiv(\mathbf{G}\cdot\nabla)\mathbf{F}-(\mathbf{F}\cdot\nabla)\mathbf{G}+\mathbf{F}\operatorname{div}\mathbf{G}-\mathbf{G}\operatorname{div}\mathbf{F}$.

Proof:

$$\nabla\wedge(\mathbf{F}\wedge\mathbf{G})=\sum\left\{\mathbf{i}\wedge\frac{\partial}{\partial x}(\mathbf{F}\wedge\mathbf{G})\right\}$$

$$=\sum\left\{\mathbf{i}\wedge\left(\mathbf{F}\wedge\frac{\partial\mathbf{G}}{\partial x}+\frac{\partial\mathbf{F}}{\partial x}\wedge\mathbf{G}\right)\right\}$$

$$=\sum\left\{\mathbf{i}\wedge\left(\mathbf{F}\wedge\frac{\partial\mathbf{G}}{\partial x}\right)\right\}+\sum\left\{\mathbf{i}\wedge\left(\frac{\partial\mathbf{F}}{\partial x}\wedge\mathbf{G}\right)\right\}$$

$$=\sum\left\{\left(\mathbf{i}\cdot\frac{\partial\mathbf{G}}{\partial x}\right)\mathbf{F}-(\mathbf{i}\cdot\mathbf{F})\frac{\partial\mathbf{G}}{\partial x}\right\}+\sum\left((\mathbf{i}\cdot\mathbf{G})\frac{\partial\mathbf{F}}{\partial x}-\left(\mathbf{i}\cdot\frac{\partial\mathbf{F}}{\partial x}\right)\mathbf{G}\right\}$$

$$=\left(\sum\mathbf{i}\cdot\frac{\partial\mathbf{G}}{\partial x}\right)\mathbf{F}-\mathbf{F}\cdot\left(\sum\mathbf{i}\frac{\partial}{\partial x}\right)\mathbf{G}+\mathbf{G}\cdot\left(\sum\mathbf{i}\frac{\partial}{\partial x}\right)\mathbf{F}-\left(\sum\mathbf{i}\cdot\frac{\partial\mathbf{F}}{\partial x}\right)\mathbf{G}$$

$$=\underline{(\nabla\cdot\mathbf{G})\mathbf{F}-(\mathbf{F}\cdot\nabla)\mathbf{G}+(\mathbf{G}\cdot\nabla)\mathbf{F}-(\nabla\cdot\mathbf{F})\mathbf{G}}.$$

(*N.B.* In this example naïve application of the vector product rule would give the wrong result:

$$\nabla \wedge (\mathbf{F} \wedge \mathbf{G}) = (\nabla \cdot \mathbf{G})\mathbf{F} - (\nabla \cdot \mathbf{F})\mathbf{G}.$$

This shows the fallacy of tacitly assuming that the laws of vector algebra also apply without modification to vector operators.)

(vi) $\nabla(\mathbf{F} \cdot \mathbf{G}) \equiv (\mathbf{G} \cdot \nabla)\mathbf{F} + (\mathbf{F} \cdot \nabla)\mathbf{G} + \mathbf{G} \wedge \operatorname{curl} \mathbf{F} + \mathbf{F} \wedge \operatorname{curl} \mathbf{G}$.

Proof:

$$\begin{aligned}\nabla(\mathbf{F} \cdot \mathbf{G}) &= \sum \left\{ \mathbf{i} \frac{\partial}{\partial x} (\mathbf{F} \cdot \mathbf{G}) \right\} \\ &= \sum \mathbf{i} \left(\mathbf{F} \cdot \frac{\partial \mathbf{G}}{\partial x} + \frac{\partial \mathbf{F}}{\partial x} \cdot \mathbf{G} \right) \\ &= \sum \left(\mathbf{F} \cdot \frac{\partial \mathbf{G}}{\partial x} \mathbf{i} \right) + \sum \left(\mathbf{G} \cdot \frac{\partial \mathbf{F}}{\partial x} \mathbf{i} \right).\end{aligned}$$

From

$$\mathbf{A} \wedge (\mathbf{B} \wedge \mathbf{C}) = (\mathbf{A} \cdot \mathbf{C})\mathbf{B} - (\mathbf{A} \cdot \mathbf{B})\mathbf{C},$$

we have

$$(\mathbf{A} \cdot \mathbf{B})\mathbf{C} = (\mathbf{A} \cdot \mathbf{C})\mathbf{B} - \mathbf{A} \wedge (\mathbf{B} \wedge \mathbf{C}),$$

so that

$$\left(\mathbf{F} \cdot \frac{\partial \mathbf{G}}{\partial x} \right) \mathbf{i} = (\mathbf{F} \cdot \mathbf{i}) \frac{\partial \mathbf{G}}{\partial x} - \mathbf{F} \wedge \left(\frac{\partial G}{\partial x} \wedge \mathbf{i} \right).$$

Thus

$$\begin{aligned}\sum \left(\mathbf{F} \cdot \frac{\partial \mathbf{G}}{\partial x} \mathbf{i} \right) &= \mathbf{F} \cdot \left(\sum \mathbf{i} \frac{\partial \mathbf{G}}{\partial x} \right) + \mathbf{F} \wedge \left(\sum \mathbf{i} \wedge \frac{\partial \mathbf{G}}{\partial x} \right) \\ &= (\mathbf{F} \cdot \nabla)\mathbf{G} + \mathbf{F} \wedge (\nabla \wedge \mathbf{G}).\end{aligned}$$

Similarly

$$\sum \left(\mathbf{G} \cdot \frac{\partial \mathbf{F}}{\partial x} \mathbf{i} \right) = (\mathbf{G} \cdot \nabla)\mathbf{F} + \mathbf{G} \wedge (\nabla \wedge \mathbf{F}).$$

The result now follows by addition.

The following device enables certain of the vector identities to be established more rapidly though perhaps less rigorously. If $u = u(x)$, $v = v(x)$, $D \equiv d/dx$, then

$$\begin{aligned}D(uv) &= D(u_0 v) + D(u v_0) = u_0 Dv + (Du)v_0 \\ &= uDv + (Du)v,\end{aligned}$$

the suffix zero indicating that the corresponding quantity is not varied when the operator is applied. The zero is dropped in the last line. Thus, to establish (v) above

$$\begin{aligned}\nabla \wedge (\mathbf{F} \wedge \mathbf{G}) &= \nabla \wedge (\mathbf{F}_0 \wedge \mathbf{G}) + \nabla \wedge (\mathbf{F} \wedge \mathbf{G}_0) \\ &= (\nabla \cdot \mathbf{G})\mathbf{F}_0 - (\mathbf{F}_0 \cdot \nabla)\mathbf{G} + (\mathbf{G}_0 \cdot \nabla)\mathbf{F} - (\nabla \cdot \mathbf{F})\mathbf{G}_0 \\ &= (\nabla \cdot \mathbf{G})\mathbf{F} - (\mathbf{F} \cdot \nabla)\mathbf{G} + (\mathbf{G} \cdot \nabla)\mathbf{F} - (\nabla \cdot \mathbf{F})\mathbf{G},\end{aligned}$$

on dropping the suffix zero.

Worked Example

A magnetic dipole of constant vector moment **M** is at the origin O and the magnetostatic potential U at P, where $\overline{OP}\equiv\mathbf{r}$, is known to be given by $U=(\mathbf{M}\cdot\mathbf{r})/r^3$. If the magnetostatic field intensity at P is $\mathbf{H}=-\operatorname{grad} U$, show that

$$\mathbf{H}=3r^{-5}(\mathbf{M}\cdot\mathbf{r})\mathbf{r}-r^{-3}\mathbf{M}.$$

$$\begin{aligned}\mathbf{H}&=-\nabla\{(\mathbf{M}\cdot\mathbf{r})r^{-3}\}\\&=-(\mathbf{M}\cdot\mathbf{r})(\nabla r^{-3})-r^{-3}\nabla(\mathbf{M}\cdot\mathbf{r}).\end{aligned}$$

The level surfaces of r^{-3} are concentric spheres centred on O. Thus $\nabla r^{-3}=\{(d/dr)(r^{-3})\}\hat{\mathbf{r}}=-3r^{-4}\hat{\mathbf{r}}=-3r^{-5}\mathbf{r}$.

Also,

$$\begin{aligned}\nabla(\mathbf{M}\cdot\mathbf{r})&=\sum\left\{\mathbf{i}\frac{\partial}{\partial x}(\mathbf{M}\cdot\mathbf{r})\right\}\\&=\sum\mathbf{i}\left(\mathbf{M}\cdot\frac{\partial\mathbf{r}}{\partial x}\right)\\&=\sum\mathbf{i}(\mathbf{M}\cdot\mathbf{i})\\&=(\mathbf{M}\cdot\mathbf{i})\mathbf{i}+(\mathbf{M}\cdot\mathbf{j})\mathbf{j}+(\mathbf{M}\cdot\mathbf{k})\mathbf{k}=\mathbf{M}.\end{aligned}$$

Hence the result follows.

1.16 Relations between Surface and Volume Integrals

In the following, S denotes a closed surface containing a volume v. P is a point either within v or on S.

First suppose the volume to be dissected into many elemental volumes δv by three families of planes parallel to the fixed Cartesian coordinate planes. Then δv will be either a cubical element $\delta x\delta y\delta z$ within v or a fraction of such an element cut off by the surface element δS of S (Fig. 1.18).

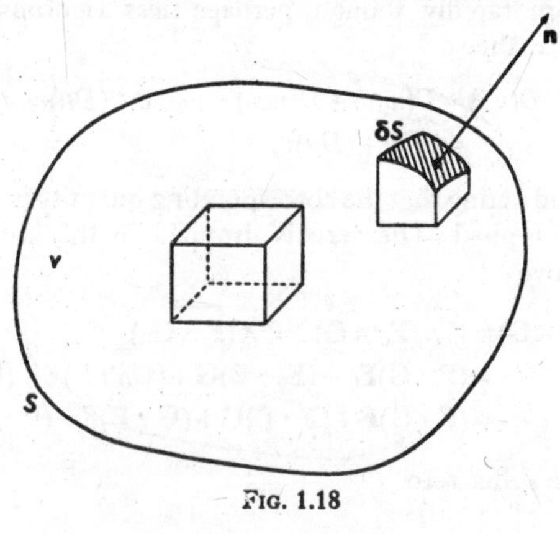

Fig. 1.18

If **F** is a uniform differentiable vector function defined at all points of v and S, then the outward normal flux of **F** across the surfaces of δv is (div **F**)δv. Hence the total outward normal flux across the surfaces of all such elements is $\int_v \operatorname{div} \mathbf{F}\, dv$. But to the first order

$$(\operatorname{div} \mathbf{F})\delta v = \int_{\Delta S} \mathbf{n} \cdot \mathbf{F}\, dS,$$

where ΔS is the total surface containing δv and **n** is drawn outwards from each δv. Thus

$$\int_v \operatorname{div} \mathbf{F}\, dv = \lim \sum \int_{\Delta S} \mathbf{n} \cdot \mathbf{F}\, dS,$$

where the summation on the R.H.S. is over all surfaces of all elements δv. This summation comprises two kinds: one over complete cuboidal elements for which $\delta v = \delta x \delta y \delta z$ and one over those elements δv which are bounded by surface elements δS. Denoting such sums by $\sum\limits_1$ and $\sum\limits_2$ respectively, it follows that

$$\sum \int_{\Delta S} \mathbf{n} \cdot \mathbf{F}\, dS = \sum_1 \int_{\Delta S} \mathbf{n} \cdot \mathbf{F}\, dS + \sum_2 \int_{\Delta S} \mathbf{n} \cdot \mathbf{F}\, dS.$$

The first sum on the R.H.S. vanishes owing to there being equal and opposite contributions across a common interface of two contiguous cuboids. Further,

$$\lim \sum_2 \int_{\Delta S} \mathbf{n} \cdot \mathbf{F}\, dS = \int_S \mathbf{n} \cdot \mathbf{F}\, dS.$$

Thus we have finally

$$\int_v \operatorname{div} \mathbf{F}\, dv = \int_S \mathbf{n} \cdot \mathbf{F}\, dS. \tag{1}$$

This result is known as the *Divergence Theorem* and is due to Gauss. If in (1) we take in turn the values $\varphi\mathbf{i}$, $\varphi\mathbf{j}$, $\varphi\mathbf{k}$ for **F** we obtain

$$\int_v \frac{\partial \varphi}{\partial x}\, dv = \int_S \varphi \mathbf{n} \cdot \mathbf{i}\, dS,$$

$$\int_v \frac{\partial \varphi}{\partial y}\, dv = \int_S \varphi \mathbf{n} \cdot \mathbf{j}\, dS,$$

$$\int_v \frac{\partial \varphi}{\partial z}\, dv = \int_S \varphi \mathbf{n} \cdot \mathbf{k}\, dS.$$

Multiplying these equations through by **i**, **j**, **k** respectively and adding gives

$$\int_v \operatorname{grad}\varphi \, dv = \int_S \varphi \mathbf{n} \, dS = \int_S \varphi \, d\mathbf{S}. \tag{2}$$

Returning to (1), write $\mathbf{F} = \mathbf{G} \wedge \mathbf{d}$ where $\mathbf{G} = \mathbf{G}(x, y, z)$ and $\mathbf{d}$ is an arbitrary constant non-zero vector. Then

$$\operatorname{div} \mathbf{F} = \sum \left\{ \mathbf{i} \cdot \frac{\partial}{\partial x}(\mathbf{G} \wedge \mathbf{d}) \right\} = \sum \mathbf{i} \cdot \left(\frac{\partial \mathbf{G}}{\partial x} \wedge \mathbf{d} \right)$$

$$= \sum \left(\mathbf{i} \wedge \frac{\partial \mathbf{G}}{\partial x} \right) \cdot \mathbf{d} = (\operatorname{curl} \mathbf{G}) \cdot \mathbf{d}.$$

Also, $\mathbf{n} \cdot \mathbf{F} = \mathbf{n} \cdot (\mathbf{G} \wedge \mathbf{d}) = (\mathbf{n} \wedge \mathbf{G}) \cdot \mathbf{d}$. Thus (1) becomes

$$\mathbf{d} \cdot \int_v \operatorname{curl} \mathbf{G} \, dv = \mathbf{d} \cdot \int_S (\mathbf{n} \wedge \mathbf{G}) \, dS.$$

Since **d** is arbitrary, this gives finally

$$\int_v \operatorname{curl} \mathbf{G} \, dv = \int_S (\mathbf{n} \wedge \mathbf{G}) \, dS = \int_S (d\mathbf{S} \wedge \mathbf{G}). \tag{3}$$

The relations (1), (2), (3) may be more conveniently remembered in the following equivalent forms:

$$\int_v \nabla \cdot \mathbf{F} \, dv = \int_S \mathbf{n} \cdot \mathbf{F} \, dS,$$

$$\int_v \nabla \varphi \, dv = \int_S \mathbf{n} \varphi \, dS,$$

$$\int_v \nabla \wedge \mathbf{G} \, dv = \int_S \mathbf{n} \wedge \mathbf{G} \, dS.$$

We next discuss some ramifications of the divergence theorem.

1.16.1 Cartesian Form of the Divergence Theorem

With the above notation we obtained the divergence theorem in the form

$$\int_v \nabla \cdot \mathbf{F} \, dv = \int_S \mathbf{n} \cdot \mathbf{F} \, dS.$$

Now suppose $\mathbf{F} = [P, Q, R]$ and that $\mathbf{n} = [l, m, n]$ in Cartesians so that

$[l, m, n]$ are the D.C.'s of the outward-directed normal to the surface element dS. Then the divergence theorem assumes the form

$$\int_v \left(\frac{\partial P}{\partial x}+\frac{\partial Q}{\partial y}+\frac{\partial R}{\partial z}\right) dv = \int_S (lP+mQ+nR)\, dS.$$

The integral on the L.H.S. may be expressed in the alternative form

$$\iiint_v \left(\frac{\partial P}{\partial x}+\frac{\partial Q}{\partial y}+\frac{\partial R}{\partial z}\right) dx\, dy\, dz,$$

which is particularly useful for evaluation purposes. We derive an equivalent Cartesian form for the R.H.S.

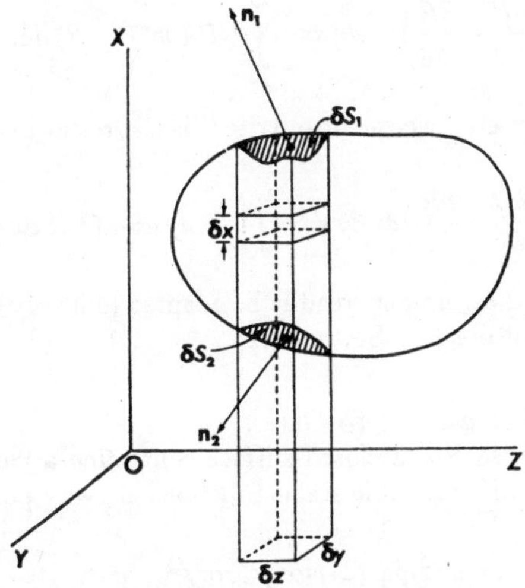

FIG. 1.19

Consider Fig. 1.19 in which the prism perpendicular to the rectangular section having the four vertices $(0, y\pm\frac{1}{2}\delta y, z\pm\frac{1}{2}\delta z)$ cuts the surface S in surface elements $\delta S_1, \delta S_2$ which have unit outward-drawn normals $\mathbf{n}_1=[l_1, m_1, n_1]$, $\mathbf{n}_2=[l_2, m_2, n_2]$. Then, using the conventions of the diagram in which the rectangular prism is shown intersecting the surface S but twice, we have $l_1\delta S_1=\delta y\delta z=-l_2\delta S_2$. Take δv to be the rectangular element $\delta x\times\delta y\times\delta z$ within the prism and between $\delta S_1, \delta S_2$. Integrate $\partial P/\partial x$ through the prism. Then

$$\iiint_{\text{prism}} \frac{\partial P}{\partial x} dx\, dy\, dz = \left(\int \frac{\partial P}{\partial x} dx\right)\delta y\delta z$$
$$=(P_1-P_2)\delta y\delta z = l_1P_1\delta S_1 + l_2P_2\delta S_2.$$

The reader should obtain similar results for a prism giving 4, 6, 8, . . . surface intersections. Thus

$$\iiint_v \frac{\partial P}{\partial x}\,dx\,dy\,dz=\int(l_1P_1\,dS_1+l_2P_2\,dS_2)=\int_S lP\,dS.$$

Similarly,

$$\iiint_v \frac{\partial Q}{\partial y}\,dx\,dy\,dz=\int_S mQ\,dS;$$

$$\iiint_v \frac{\partial R}{\partial z}\,dx\,dy\,dz=\int_S nR\,dS.$$

Hence

$$\iiint_v\left(\frac{\partial P}{\partial x}+\frac{\partial P}{\partial y}+\frac{\partial R}{\partial z}\right)dx\,dy\,dz=\int_S(lP+mQ+nR)\,dS.$$

Since $l\,dS=dy\,dz$, etc., we can also write this last result in the form

$$\iiint_v\left(\frac{\partial P}{\partial x}+\frac{\partial Q}{\partial y}+\frac{\partial R}{\partial z}\right)dx\,dy\,dz=\iint_S(P\,dy\,dz+Q\,dz\,dx+R\,dx\,dy).$$

This form of the theorem may readily be adapted to justify the uniqueness of the definition of div **F** in Section 1.13.

1.16.2 Some Theorems due to Green

As before suppose S is a closed surface containing a volume v and let U, V be uniform differentiable scalar functions at every point of S and v. Then since

$$\nabla\cdot(U\nabla V)=U\nabla^2V+(\nabla U)\cdot(\nabla V),$$

we have, in virtue of the divergence theorem,

$$\int_S U\mathbf{n}\cdot\nabla V\,dS=\int_S\mathbf{n}\cdot(U\nabla V)\,dS=\int_v\nabla\cdot(U\nabla V)\,dv,$$

so that

$$\int_S U\frac{\partial V}{\partial n}\,dS=\int_v(U\nabla^2V+\nabla U\cdot\nabla V)\,dv, \tag{1}$$

using $\mathbf{n}\cdot\nabla\equiv\partial/\partial n$. Equation (1) expresses a theorem due to Green.

Reversing the roles of U, V in (1) gives

$$\int_S V\frac{\partial U}{\partial n}\,dS=\int_v(V\nabla^2U+\nabla V\cdot\nabla U)\,dv. \tag{2}$$

Subtracting (2) from (1) gives Green's theorem in symmetrical form:

$$\int_S \left(U\frac{\partial V}{\partial n} - V\frac{\partial U}{\partial n} \right) dS = \int_v (U\nabla^2 V - V\nabla^2 U)\, dv. \tag{3}$$

In (3), take $U = 1/r$, where r is the *distance from a fixed point P outside S* to a variable point in v or on S. Then, since $\nabla^2(1/r) = 0$, we have

$$\int_S \left\{ V\frac{\partial}{\partial n}\left(\frac{1}{r}\right) - \frac{1}{r}\frac{\partial V}{\partial n} \right\} dS + \int_v \frac{1}{r}\nabla^2 V\, dv = 0. \tag{4}$$

Suppose, however, that P is taken within S so that $1/r$ is infinite at P. With centre P and radius ε describe a small sphere Σ within v (Fig. 1.20). Let v' be the volume of the region between S, Σ and let S' denote the surface $S + \Sigma$ bounding v'.

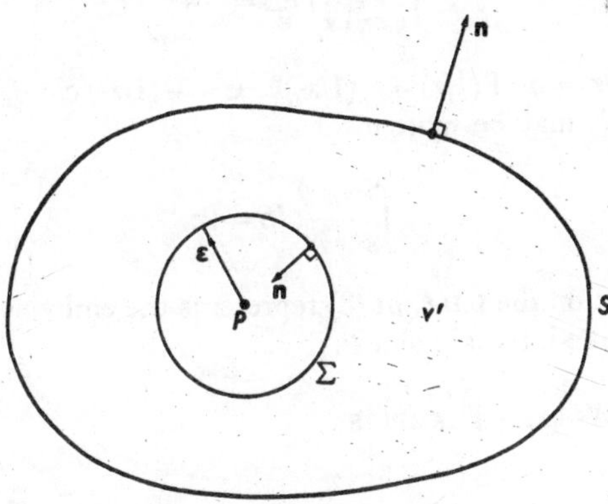

FIG. 1.20

If $\mathbf{n}$ denotes the unit normal to any element δS of S', drawn outwards from v', then (4) shows that since P' lies outside S',

$$\int_{S'} \left\{ V\frac{\partial}{\partial n}\left(\frac{1}{r}\right) - \frac{1}{r}\frac{\partial V}{\partial n} \right\} dS + \int_{v'} \frac{1}{r}\nabla^2 V\, dv = 0,$$

or

$$\left(\int_S + \int_\Sigma\right) \left\{ V\frac{\partial}{\partial n}\left(\frac{1}{r}\right) - \frac{1}{r}\frac{\partial V}{\partial n} \right\} dS + \int_{v'} \frac{1}{r}\nabla^2 V\, dv = 0. \tag{5}$$

On Σ the unit normal outwards from v' goes into the interior of Σ. Thus on Σ, $r = \varepsilon$ and

$$\left\{ \left(\frac{\partial}{\partial n}\right)\left(\frac{1}{r}\right) \right\}_{r=\varepsilon} = \left\{ -\left(\frac{\partial}{\partial r}\right)\left(\frac{1}{r}\right) \right\}_{r=\varepsilon} = 1/\varepsilon^2.$$

Hence

$$\int_\Sigma \left\{ V\frac{\partial}{\partial n}\left(\frac{1}{r}\right) - \frac{1}{r}\frac{\partial V}{\partial n} \right\} dS \simeq \left(\frac{1}{\varepsilon^2}V - \frac{1}{\varepsilon}\frac{\partial V}{\partial n} \right) \cdot 4\pi\varepsilon^2 \to 4\pi V_P,$$

as $\varepsilon \to 0$. Thus in the limit as $\varepsilon \to 0$, (5) gives

$$4\pi V_P = \int_S \left\{ \frac{1}{r}\frac{\partial V}{\partial n} - V\frac{\partial}{\partial n}\left(\frac{1}{r}\right) \right\} dS - \int_v \frac{1}{r}\nabla^2 V dv. \tag{6}$$

Equation (6) is often called *Green's formula*: it expresses the value of V at an internal point P of the region v bounded by S. If in particular V is harmonic, as is often the case in potential theory, then $\nabla^2 V = 0$, resulting in the evanescence of the volume integral on the R.H.S. of (6).

If we put $V = 1$ in v and on S, then (6) gives

$$\int_S \frac{\partial}{\partial n}\left(\frac{1}{r}\right) dS = -4\pi. \tag{7}$$

Now $(\partial/\partial n)(1/r) = \mathbf{n} \cdot \nabla(1/r) = -(1/r^2)\hat{\mathbf{r}} \cdot \mathbf{n} = -(1/r^2)\cos\theta$, where θ = angle $(\mathbf{n}, \hat{\mathbf{r}})$. Thus (7) may be written

$$\int_S \frac{\cos\theta}{r^2} dS = 4\pi. \tag{8}$$

The expression on the L.H.S. of (8) represents the entire solid angle which S subtends at the internal point P.

1.16.3 Some Worked Examples

Example 1

If v is the volume contained within a closed surface S, prove

$$\int_v \mathbf{V} \cdot \operatorname{curl} \mathbf{U} dv = \int_S (\mathbf{U} \wedge \mathbf{V}) \cdot d\mathbf{S} + \int_v \mathbf{U} \cdot \operatorname{curl} \mathbf{V} dv.$$

$$\int_S (\mathbf{U} \wedge \mathbf{V}) \cdot d\mathbf{S} = \int_S \mathbf{n} \cdot (\mathbf{U} \wedge \mathbf{V}) dS = \int_v \nabla \cdot (\mathbf{U} \wedge \mathbf{V}) dv.$$

Using the methods of Example (iv) Section 1.15,

$$\nabla \cdot (\mathbf{U} \wedge \mathbf{V}) = \mathbf{V} \cdot \operatorname{curl} \mathbf{U} - \mathbf{U} \cdot \operatorname{curl} \mathbf{V},$$

so that

$$\int_S (\mathbf{U} \wedge \mathbf{V}) \cdot d\mathbf{S} = \int_v \mathbf{V} \cdot \operatorname{curl} \mathbf{U}\, dv - \int_v \mathbf{U} \cdot \operatorname{curl} \mathbf{V}\, dv,$$

whence the result follows on transposition.

EXAMPLE 2

Prove the Gauss divergence theorem and deduce Green's theorem

$$\int_\tau (\varphi\nabla^2\psi - \psi\nabla^2\varphi)\, d\tau = \int_S (\varphi\nabla\psi - \psi\nabla\varphi)\cdot d\mathbf{S},$$

where φ, ψ are functions of position defined in a region τ and on its boundary S. Show that $(\cos\lambda r)/r$ satisfies the equation

$$(\nabla^2+\lambda^2)\varphi=0,$$

where $r=\sqrt{(x^2+y^2+z^2)}$, in any region excluding the origin. If $\varphi=\Phi$ is any solution of the above equation which has no singularity on or inside S, show that the value of Φ at the origin, assumed to be inside S, is given by

$$-\frac{1}{4\pi}\int_S\left(\Phi\frac{\partial}{\partial n}\frac{\cos\lambda r}{r}-\frac{\cos\lambda r}{r}\frac{\partial\Phi}{\partial n}\right)dS,$$

where $\partial/\partial n$ denotes differentiation along the outward normal to S.

(London Univ., B.Sc. Spec. Hons., 1960)

The first part is standard bookwork.

The level surfaces of $(\cos\lambda r)/r$ are concentric spheres centred on $r=0$. Hence

$$\nabla\left(\frac{1}{r}\cos\lambda r\right)=\hat{\mathbf{r}}\frac{d}{dr}\left(\frac{1}{r}\cos\lambda r\right)=-\mathbf{r}\left(\frac{1}{r^3}\cos\lambda r+\frac{\lambda}{r^2}\sin\lambda r\right).$$

$$\therefore\ \nabla^2\left(\frac{1}{r}\cos\lambda r\right)=-(\nabla\cdot\mathbf{r})\left(\frac{1}{r^3}\cos\lambda r+\frac{\lambda}{r^2}\sin\lambda r\right)-\mathbf{r}\cdot\nabla\left(\frac{1}{r^3}\cos\lambda r+\frac{\lambda}{r^2}\sin\lambda r\right)$$

$$=-3\left(\frac{1}{r^3}\cos\lambda r+\frac{\lambda}{r^2}\sin\lambda r\right)-r\frac{d}{dr}\left(\frac{1}{r^3}\cos\lambda r+\frac{\lambda}{r^2}\sin\lambda r\right)$$

$$=-\frac{\lambda^2}{r}\cos\lambda r.$$

This shows $\varphi=(1/r)\cos\lambda r$ satisfies $(\nabla^2+\lambda^2)\varphi=0$.

Taking $\varphi=(1/r)\cos\lambda r$, $\psi=\Phi$, the symmetrical form of Green's theorem shows that

$$\int_{\tau'}\left[\left(\frac{\cos\lambda r}{r}\right)\nabla^2\Phi-\Phi\nabla^2\left(\frac{\cos\lambda r}{r}\right)\right]d\tau$$

$$=\int_{S'}\left[\left(\frac{\cos\lambda r}{r}\right)\frac{\partial\Phi}{\partial n}-\Phi\frac{\partial}{\partial n}\left(\frac{\cos\lambda r}{r}\right)\right]dS, \quad (1)$$

where S' is the entire boundary $S+\Sigma$ containing volume τ between S, Σ, and Σ is the surface of the small sphere centre P and radius ε, as in Fig. 1.20.

Now $\nabla^2\{(1/r)\cos\lambda r\} = -(\lambda^2/r)\cos\lambda r$, so that the L.H.S. of (1) becomes

$$\int_{\tau'}\left(\frac{\cos\lambda r}{r}\right)(\nabla^2\Phi+\lambda^2\Phi)\,d\tau=0, \quad \text{since } \nabla^2\Phi+\lambda^2\Phi=0.$$

Thus (1) gives

$$\left(\int_S+\int_\Sigma\right)\left\{\left(\frac{\cos\lambda r}{r}\right)\frac{\partial\Phi}{\partial n}-\Phi\frac{\partial}{\partial n}\left(\frac{\cos\lambda r}{r}\right)\right\}dS=0. \qquad (2)$$

On Σ, $\partial/\partial n \equiv -d/dr$ and so

$$\frac{\partial}{\partial n}\left(\frac{\cos\lambda r}{r}\right)=-\frac{d}{dr}\left(\frac{\cos\lambda r}{r}\right)=\frac{\lambda\sin\lambda r}{r}+\frac{\cos\lambda r}{r}.$$

Thus

$$\int_\Sigma\{\ \}dS=\int_\Sigma\left\{\left(\frac{\cos\lambda\varepsilon}{\varepsilon}\right)\frac{\partial\Phi}{\partial n}-\Phi\left(\frac{\lambda\sin\lambda\varepsilon}{\varepsilon}+\frac{\cos\lambda\varepsilon}{\varepsilon^2}\right)\right\}dS$$

$$\simeq 4\pi\varepsilon^2\left\{\frac{\cos\lambda\varepsilon}{\varepsilon}\frac{\partial\Phi}{\partial n}-\Phi\left(\frac{\lambda\sin\lambda\varepsilon}{\varepsilon}+\frac{\cos\lambda\varepsilon}{\varepsilon^2}\right)\right\}$$

$$\to -4\pi\Phi_P \text{ as } \varepsilon\to 0.$$

Hence, as $\varepsilon\to 0$, (2) gives

$$\int_S\left\{\left(\frac{\cos\lambda r}{r}\right)\frac{\partial\Phi}{\partial n}-\Phi\frac{\partial}{\partial n}\left(\frac{\cos\lambda r}{r}\right)\right\}dS-4\pi\Phi_P=0$$

which yields the stated result.

1.17 Relations between Line and Surface Integrals

If $\mathscr{C}$ is a closed curve, not necessarily plane, containing an area S and if $\mathbf{F}$ is a uniform differentiable vector function defined at all points on $\mathscr{C}$ and in S, then we prove that

$$\oint_{\mathscr{C}}\mathbf{F}\cdot d\mathbf{s}=\int_S\mathbf{n}\cdot\operatorname{curl}\mathbf{F}\,dS.$$

This result is known as *Stokes's theorem.*

Proof: On S superpose a fine mesh as in Fig. 1.21(i). Then the mesh dissects S into a number of cells such as γ shown enlarged in Fig. 1.21(ii). Let the boundaries of all the cells be described in the manner indicated in the

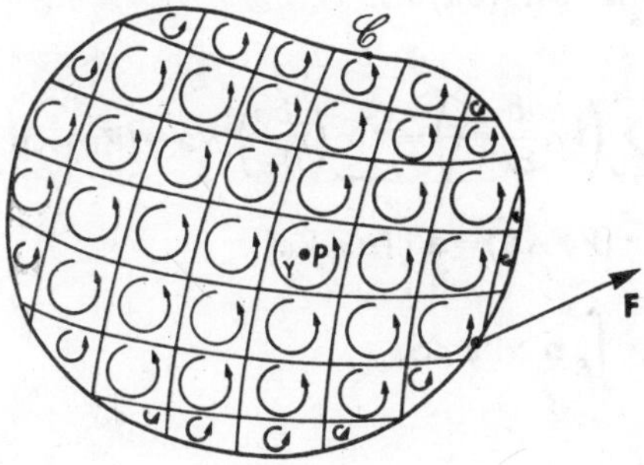

FIG. 1.21(i)

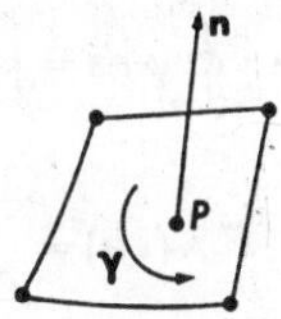

FIG. 1.21(ii)

diagram. Then the contributions to the total line integral arising from common sides of contiguous cells cancel out so that

$$\oint_{\mathscr{C}} \mathbf{F}\cdot d\mathbf{s} = \sum_{\text{all }\gamma} \oint_{\gamma} \mathbf{F}\cdot d\mathbf{s}.$$

Let δS be the area of the cell γ (appreciably flat) and let P be an internal point. Also let $\mathbf{n}$ be the unit normal at P to δS so that positive rotation about $\mathbf{n}$ is in the sense of description of γ. Then, by definition,

$$(\mathbf{n}\cdot \text{curl }\mathbf{F})\delta S = \oint_{\gamma} \mathbf{F}\cdot d\mathbf{s}.$$

Thus

$$\oint_{\mathscr{C}} \mathbf{F}\cdot d\mathbf{s} = \sum_{\text{all }\gamma} (\mathbf{n}\cdot \text{curl }\mathbf{F})\delta S$$

$$\rightarrow \int_S (\mathbf{n}\cdot \text{curl }\mathbf{F})\, dS,$$

as the network is made finer. This establishes the required result.

EXAMPLE

With the same notation as in the text, prove that for a uniform differentiable scalar function φ,

$$\oint_{\mathscr{C}} \varphi\, d\mathbf{s} = \int_S d\mathbf{S} \wedge \text{grad }\varphi.$$

In Stokes's theorem, take $\mathbf{F} = \varphi\mathbf{a}$ where $\mathbf{a}$ is an arbitrary constant non-zero vector. Then

$$\oint_{\mathscr{C}} \varphi\mathbf{a}\cdot d\mathbf{s} = \int_S \mathbf{n}\cdot \text{curl }(\varphi\mathbf{a})\, dS,$$

or

$$\mathbf{a}\cdot\oint_{\mathscr{C}}\varphi\,d\mathbf{s}=\int_S \mathbf{n}\cdot\operatorname{curl}(\varphi\mathbf{a})\,dS.$$

Now

$$\nabla\wedge(\varphi\mathbf{a})=\sum\left(\mathbf{i}\wedge\frac{\partial}{\partial x}\varphi\mathbf{a}\right)=\sum\left(\mathbf{i}\frac{\partial\varphi}{\partial s}\right)\wedge\mathbf{a}=(\nabla\varphi)\wedge\mathbf{a}.$$

$$\therefore\ \mathbf{n}\cdot\nabla\wedge(\varphi\mathbf{a})=\mathbf{n}\cdot(\nabla\varphi\wedge\mathbf{a})=\mathbf{a}\cdot(\mathbf{n}\wedge\nabla\varphi).$$

$$\therefore\ \mathbf{a}\cdot\oint_{\mathscr{C}}\varphi\,d\mathbf{s}=\mathbf{a}\cdot\int_S\mathbf{n}\wedge\nabla\varphi\,dS,$$

$$\mathbf{a}\cdot\left\{\oint_{\mathscr{C}}\varphi\,d\mathbf{s}-\int_S\mathbf{n}\wedge\nabla\varphi\,dS\right\}=0.$$

Now **a** is a non-zero vector. Moreover, as it is otherwise arbitrary, it is, in general, not perpendicular to the braced vectorial expression. Hence this expression is zero, which establishes the desired result.

1.18* Conservative Vector Fields

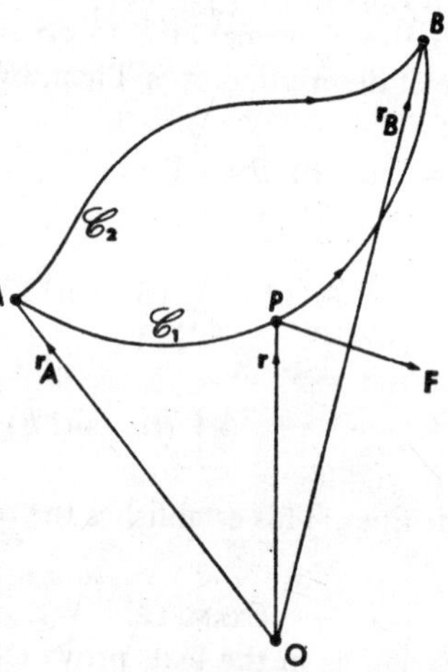

FIG. 1.22

Let **F** be a uniform vector field within a region $\mathscr{R}$ and let A and B be any two points of $\mathscr{R}$. Then the line integral of **F** from A to B is denoted by the forms

$$\int_A^B \mathbf{F}\cdot d\mathbf{r}=\int_A^B\mathbf{F}\cdot d\mathbf{s}$$

* See also F. CHORLTON, *Textbook of Dynamics*, Van Nostrand, Section 3.3.

where $\mathbf{r} \equiv \overline{OP}$, the position vector of any point P in $\mathscr{R}$ with reference to a fixed origin and $d\mathbf{r} = d\mathbf{s}$. In general the value of either integral depends on the positions of both A and B and also on the shape and length of the path joining them. Thus if A and B are fixed, the value of either integral will not, in general, be unique. If, however, the value of the line integral depends solely on the positions of A and B and not at all on the path joining them for all pairs of points A, B in $\mathscr{R}$, then the totality of values of $\mathbf{F}$ in $\mathscr{R}$ is said to constitute a *conservative* or *irrotational* field. (The second term is preferred in hydrodynamics.)

In the special case when $\mathbf{F}$ is conservative, it follows that if we compute its line integral round any closed curve $\mathscr{C}$ in $\mathscr{R}$ the value obtained is zero. This case is realized when A and B above are brought into coincidence. Mathematically this result is expressed by

$$\oint_{\mathscr{C}} \mathbf{F} \cdot d\mathbf{s} = 0.$$

Conservative fields are of special importance in most branches of potential theory, including classical hydrodynamics. We therefore establish necessary and sufficient conditions for a given vector field to be conservative.

THEOREM: *If* $\mathbf{F}$ *is a uniform vector function throughout a region* $\mathscr{R}$ *then a necessary and sufficient condition for the field of* $\mathbf{F}$ *to be conservative in* $\mathscr{R}$ *is that there exists a uniform differentiable scalar function* φ *such that* $\mathbf{F} = \nabla \varphi$ *at each point of* $\mathscr{R}$.

Proof: (i) First suppose that $\varphi = \varphi(x, y, z)$ exists at each point (x, y, z) of $\mathscr{R}$ such that $\mathbf{F} = \nabla \varphi$. Then we show that the field of $\mathbf{F}$ is conservative. Let A and B be any two points within $\mathscr{R}$. Then along any chosen path from A to B in $\mathscr{R}$,

$$\int_A^B \mathbf{F} \cdot d\mathbf{r} = \int_A^B \nabla \varphi \cdot d\mathbf{r}.$$

Now

$$d\mathbf{r} \cdot \nabla \varphi = (dx\mathbf{i} + dy\mathbf{j} + dz\mathbf{k}) \cdot \left(\mathbf{i}\frac{\partial \varphi}{\partial x} + \mathbf{j}\frac{\partial \varphi}{\partial y} + \mathbf{k}\frac{\partial \varphi}{\partial z}\right)$$

$$= \frac{\partial \varphi}{\partial x} dx + \frac{\partial \varphi}{\partial y} dy + \frac{\partial \varphi}{\partial z} dz = d\varphi,$$

$$\therefore \int_A^B \mathbf{F} \cdot d\mathbf{r} = \int_A^B d\varphi = \varphi_B - \varphi_A.$$

This latter expression depends only on the positions of A and B and not at all on the nature of the path joining them. Thus the field of $\mathbf{F}$ is conservative.

(ii) Now suppose the field of $\mathbf{F}$ is conservative. Then we have to show that there exists a uniform scalar field φ for which $\mathbf{F} = \nabla \varphi$.

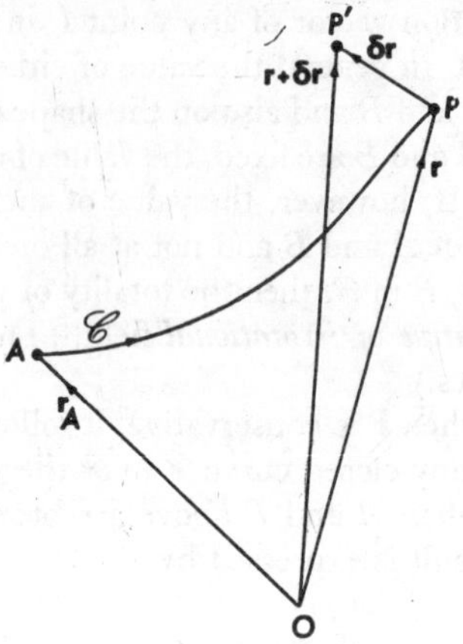

FIG. 1.23

Let $\mathscr{C}$ be an arc of a curve joining A to P where $\overline{OA}\equiv \mathbf{r}_A$, $\overline{OP}\equiv\mathbf{r}$. Also let P' be a point near to P such that $\overline{OP'}\equiv\mathbf{r}+\delta\mathbf{r}$ and $\overline{PP'}\equiv\delta\mathbf{r}$. P, P' have coordinates (x, y, z), $(x+\delta x, y+\delta y, z+\delta z)$. Write

$$I=\int_A^P \mathbf{F}\cdot d\mathbf{r}.$$

The value of I is dependent only on the positions of A and P and not at all on the path $\mathscr{C}$. Since A is fixed and P has variable coordinates (x, y, z) $I=I(x, y, z)$ and so at P', $I=I(x+\delta x, y+\delta y, z+\delta z)$. Thus

$$\delta I=I(x+\delta x, y+\delta y, z+\delta z)-I(x, y, z)$$

$$=\int_P^{P'}\mathbf{F}\cdot d\mathbf{r}\simeq\mathbf{F}\cdot\delta\mathbf{r},\quad\text{to the first order.}$$

Since $I=I(x, y, z)$, $\delta I=\delta\mathbf{r}.\nabla I$. Hence $\delta\mathbf{r}\cdot\nabla I=\mathbf{F}\cdot\delta\mathbf{r}$, to the first order, i.e. $\delta\mathbf{r}\cdot(\nabla I-\mathbf{F})=0$.

If now $\delta\mathbf{r}$ is any small, arbitrary, non-zero displacement from P, then $\delta\mathbf{r}\neq\mathbf{0}$ and, on account of the direction of $\delta\mathbf{r}$ being arbitrary, $\delta\mathbf{r}$ is not in general perpendicular to $\nabla I-\mathbf{F}$. Thus we infer $\nabla I-\mathbf{F}=\mathbf{0}$, which, on taking $I=\varphi$, gives the required result.

COROLLARY: With $\mathbf{F}$ defined as above, a necessary and sufficient condition for the field of $\mathbf{F}$ to be conservative is that curl $\mathbf{F}=0$ at each point of $\mathscr{R}$.

Proof: (i) If the field of $\mathbf{F}$ is conservative, then we can find a scalar field φ in $\mathscr{R}$ such that

$$\mathbf{F}=\nabla\varphi.$$

Then

$$\underline{\text{curl }\mathbf{F}=\nabla\wedge\nabla\varphi\equiv\mathbf{0}.}$$

Alternatively, without using the previous theorem, if $\mathscr{C}$ denotes a closed curve containing a small area $\triangle S$ with unit normal $\mathbf{n}$ then

$$0=\oint_{\mathscr{C}} \mathbf{F} \cdot d\mathbf{s}=\int_{\Delta S} \mathbf{n} \cdot \operatorname{curl} \mathbf{F}\, dS.$$

Thus to the first order

$$(\mathbf{n} \cdot \operatorname{curl} \mathbf{F})\Delta S=0.$$

Since ΔS is arbitrary in size and since its normal $\mathbf{n}$ is of arbitrary direction, it follows that

$$\underline{\operatorname{curl} \mathbf{F}=0.}$$

(ii) If curl $\mathbf{F}=0$ at each point P of $\mathscr{R}$, then for all circuits $\mathscr{C}$ in $\mathscr{R}$,

$$\oint_{\mathscr{C}} \mathbf{F} \cdot d\mathbf{s}=\int_{\Delta S} \mathbf{n} \cdot \operatorname{curl} \mathbf{F}\, dS=0.$$

This shows that the field of $\mathbf{F}$ is conservative.

1.18.1 Conservative Fields of Force

Suppose that $\mathbf{F}$ now denotes a vector field of force, known to be conservative in a region $\mathscr{R}$. Then we can find a uniform differentiable scalar function φ in $\mathscr{R}$ such that $\mathbf{F}=-\nabla\varphi$. In these circumstances φ is called a *potential function.* If we move $\mathbf{F}$ from A to B along any path in $\mathscr{R}$ then the work done by $\mathbf{F}$ is unambiguously equal to

$$\int_A^B \mathbf{F} \cdot d\mathbf{r}=-\int_A^B \nabla\varphi \cdot d\mathbf{r}.$$

Thus the *work done against* $\mathbf{F}$ is

$$\int_A^B \nabla\varphi \cdot d\mathbf{r}=W, \text{ say.}$$

Thus

$$W=\int_A^B d\varphi=\varphi_B-\varphi_A.$$

Now regard A as fixed and B as varying. Then the last result leads to the following definition of a potential function for a conservative field of force: *for a conservative field of force, the work done against the field in moving the point of application of the force from some standard position A to a variable position B is the potential of the field at B.* Often A is taken as the point at infinity. Thus in the case of the earth's gravitational field, the force on unit mass at any point P is $\mathbf{g}$, the acceleration due to gravity. It is also equal to the work done against the earth's gravitational field in moving a particle of unit mass from infinity to the position P in the field.

Example

A *central force* is one such that the force at P is $F(r)\hat{\mathbf{r}}$, where $\overline{OP}\equiv\mathbf{r}$ and O is fixed. Prove that the field of such a force is conservative.

The work done by such a force in moving its point of application from A to B is

$$W=\int_A^B F(r)\hat{\mathbf{r}}\cdot d\mathbf{r}.$$

Now $\mathbf{r}^2=r^2$, so that $2\mathbf{r}\cdot d\mathbf{r}=2rdr$.

$$\therefore\ \hat{\mathbf{r}}\cdot d\mathbf{r}=dr.$$

$$\therefore\ W=\int_A^B F(r)\,dr.$$

This integral has a value dependent only on the positions of A, B and not at all on the path joining them.

The reader can also confirm that

$$\text{curl}\,\{F(r)\hat{\mathbf{r}}\}=0,\quad \text{by using } F(r)\hat{\mathbf{r}}=\{F(r)/r\}\mathbf{r}.$$

1.19 General Orthogonal Curvilinear Coordinates

Suppose that at each point $P(x, y, z)$ of space, referred to a system of tri-rectangular coordinate axes, there exist three uniform scalar functions $\lambda(x, y, z)$, $\mu(x, y, z)$, $\nu(x, y, z)$ such that the surfaces

$$\lambda(x, y, z)=\text{const.},\ \mu(x, y, z)=\text{const.},\ \nu(x, y, z)=\text{const.} \tag{1}$$

are mutually orthogonal at their intersections, i.e. the tangents to the three curves of intersection at P are mutually at right angles. By assigning specific values to λ, μ, ν we can evaluate x, y, z and conversely. It follows that (λ, μ, ν) is an alternative and equivalent coordinate system for defining P in space. Here we assume that each coordinate in either system is a single-valued function of the three coordinates in the other system. The system (x, y, z) is the Cartesian system. The system of coordinates (λ, μ, ν) is called a *general orthogonal curvilinear coordinate system* because in general the three surfaces of equations (1) intersect along curved lines which are mutually orthogonal at each point P.

The Cartesian coordinate system (x, y, z) is the simplest of all orthogonal coordinate systems because the three planes

$$x=\text{const.},\ y=\text{const.},\ z=\text{const.}$$

through any point P of space intersect along three lines which are mutually orthogonal. In many problems involving vector field theory it is, however, sometimes convenient to work with other systems of orthogonal coordinates. The two most common are those afforded by *cylindrical polar coordinates* and *spherical polar coordinates.*

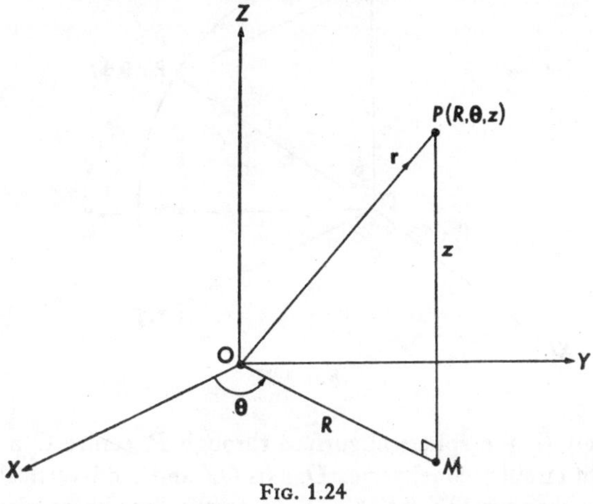

FIG. 1.24

Figure 1.24 shows a point P in a Cartesian tri-rectangular frame specified by axes OX, OY, OZ. PM is drawn perpendicularly to the coordinate plane XOY. Let $XOM=\theta$, $OM=$R, $PM=z$. Then (R, θ, z) are called the cylindrical polar coordinates of P. Knowledge of these elements clearly enables P to be determined uniquely. The three surfaces through P

$$R=\text{const.},\quad \theta=\text{const.},\quad z=\text{const.}$$

are respectively (i) the right circular cylinder through P, axis OX and radius R; (ii) the plane through OZ set at angle θ to the coordinate plane ZOX; (iii) the plane through P parallel to the coordinate plane XOY and distant z from it. Clearly these three surfaces intersect mutually orthogonally, so that the system of coordinates is an orthogonal one. We observe that if (x, y, z) are the Cartesian coordinates of P, then

$$x=R\cos\theta,\quad y=R\sin\theta,\quad z=z;$$
$$R=\sqrt{(x^2+y^2)},\quad \theta=\tan^{-1}(y/x),\quad z=z.$$

The first system of equations expresses the Cartesian coordinates in terms of the cylindrical polars: the second system gives the cylindrical polar coordinates in terms of the Cartesians.

In Fig. 1.25, P is the point chosen in a Cartesian tri-rectangular frame of axes OX, OY, OZ, such that $OP=r$, $\angle ZOP=\theta$, $\varphi=$angle between planes ZOX, ZOP. Then (r, θ, φ) are called the spherical polar coordinates of P. Knowledge of these elements clearly enables P to be determined uniquely. The three surfaces through P

$$r=\text{const.},\quad \theta=\text{const.},\quad \varphi=\text{const.}$$

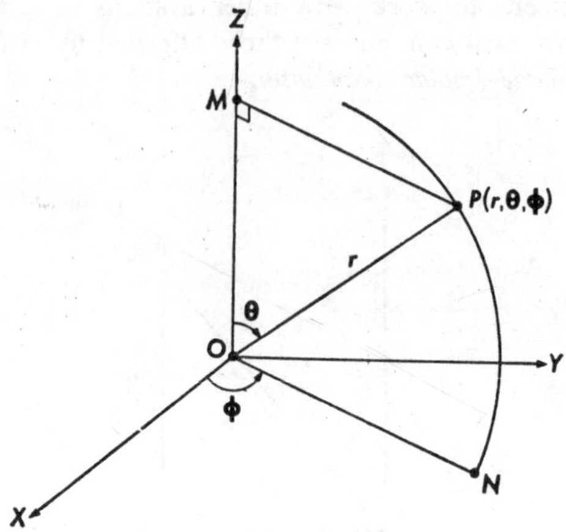

Fig. 1.25

are respectively (i) the spherical surface through P, centre O and radius r; (ii) the right circular cone, vertex O, axis OZ and semi-vertical angle θ; (iii) the plane through OZ set at angle φ to the coordinate plane ZOX. Clearly these three surfaces intersect mutually orthogonally, so that the system of coordinates is an orthogonal one. If (x, y, z) are the Cartesian coordinates of P, then

$$x = r\sin\theta\cos\varphi, \quad y = r\sin\theta\sin\varphi, \quad z = r\cos\theta;$$
$$r = \sqrt{(x^2+y^2+z^2)}, \quad \theta = \cos^{-1}\{z(x^2+y^2+z^2)\}, \quad \varphi = \tan^{-1}(y/x).$$

1.19.1 Arc Length in General Orthogonal Curvilinear Coordinates

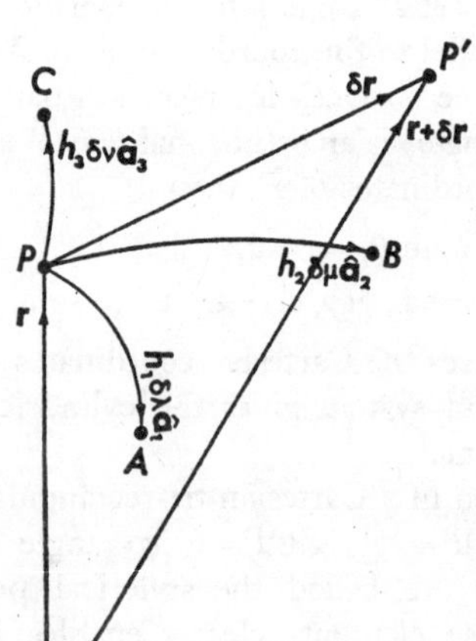

Fig. 1.26(i)

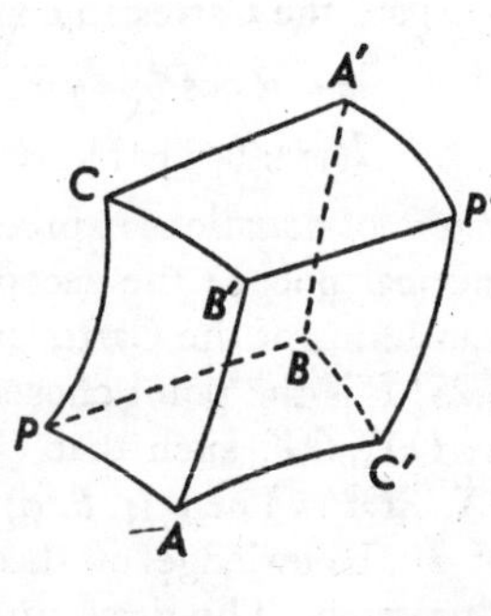

Fig. 1.26(ii)

Let P, P' be neighbouring points (Fig. 1.26(i)) such that

$$\overline{OP} \equiv \mathbf{r}, \quad \overline{OP'} \equiv \mathbf{r} + \delta\mathbf{r}, \quad \overline{PP'} \equiv \delta\mathbf{r} = \delta\mathbf{s}.$$

Also let (λ, μ, ν), $(\lambda+\delta\lambda, \mu+\delta\mu, \nu+\delta\nu)$ denote the orthogonal coordinates of P, P'. Through P we can draw a curved line PA marking the intersection of the two surfaces $\mu=$const., $\nu=$const. Along this line, λ alone varies and so we shall call PA the λ-line through P. Likewise we can draw PB, PC, the μ- and ν-lines through P. These three lines are mutually orthogonal at P. Let the points A, B, C on the λ-, μ-, ν-lines be so chosen that the orthogonal curvilinear coordinates of A, B, C are respectively $(\lambda+\delta\lambda, \mu, \nu)$, $(\lambda, \mu+\delta\mu, \nu)$, $(\lambda, \mu, \nu+\delta\nu)$. Then A, P' are on the surface for which the first orthogonal coordinate has the constant value $(\lambda+\delta\lambda)$. Similarly B, P' are on that surface for which the second orthogonal coordinate has the constant value $(\mu+\delta\mu)$ and C, P' are on that surface for which the third orthogonal coordinate has the constant value $(\nu+\delta\nu)$. To a first order of approximation, PA, PB, PC, PP' are straight-line segments. Since the first three lines are mutually orthogonal they form adjacent edges of three rectangles $PAC'BP$, $PBA'CP$, $PAB'CP$. Complete these rectangles and join P' to A', P' to B', P' to C'. In this way the small rectangular parallelepiped of Fig. 1.26(ii) is generated.

In Fig. 1.26(ii) we have at once

$$\overline{PP'} = \overline{PA} + \overline{PB} + \overline{PC}. \tag{1}$$

Let $\hat{\mathbf{a}}_1$, $\hat{\mathbf{a}}_2$, $\hat{\mathbf{a}}_3$ be the unit vectors in the edges $\overline{PA}$, $\overline{PB}$, $\overline{PC}$, and suppose they are so ordered that $\hat{\mathbf{a}}_1 = \hat{\mathbf{a}}_2 \wedge \hat{\mathbf{a}}_3$, etc. Since P is (λ, μ, ν) and A is $(\lambda+\delta\lambda, \mu, \nu)$, arc length along the line PA is a function of λ only and the arc length of PA is proportional to $\delta\lambda$. We suppose $\overline{PA} \equiv h_1 (\lambda, \mu, \nu)\delta\lambda\hat{\mathbf{a}}_1$, the factor of proportionality h_1 being a function of the coordinates (λ, μ, ν) of P. Similarly let $\overline{PB} \equiv h_2 (\lambda, \mu, \nu)\delta\mu\hat{\mathbf{a}}_2$, $\overline{PC} \equiv h_3(\lambda, \mu, \nu)\delta\nu\hat{\mathbf{a}}_3$. Then (1) gives

$$\delta\mathbf{r} = \delta\mathbf{s} = h_1\delta\lambda\hat{\mathbf{a}}_1 + h_2\delta\mu\hat{\mathbf{a}}_2 + h_3\delta\nu\hat{\mathbf{a}}_3. \tag{2}$$

Thus

$$\delta s = |\delta\mathbf{r}| = \sqrt{\{(h_1\delta\lambda)^2 + (h_2\delta\mu)^2 + (h_3\delta\nu)^2\}}. \tag{3}$$

In the case of Cartesian coordinates (x, y, z),

$$\lambda = x, \quad \mu = y, \quad \nu = z; \quad \hat{\mathbf{a}}_1 = \mathbf{i}, \quad \hat{\mathbf{a}}_2 = \mathbf{j}, \quad \hat{\mathbf{a}}_3 = \mathbf{k}$$

and

$$\delta\mathbf{s} = \delta x\mathbf{i} + \delta y\mathbf{j} + \delta z\mathbf{k}. \tag{4}$$

Comparison of (2), (4) shows $h_1 = 1$, $h_2 = 1$, $h_3 = 1$.

To obtain the arc differential $\delta\mathbf{s}$ in terms of cylindrical polar coordinates and the values of h_1, h_2, h_3, proceed in the following way with reference to Fig. 1.27. Keeping θ, z fixed, increase R to $R+\delta R$ so that P travels out along $\overline{NP}$ to A where $PA = \delta R$. Denote $\overline{PA} \equiv \delta R\hat{\mathbf{R}}$ to define the unit

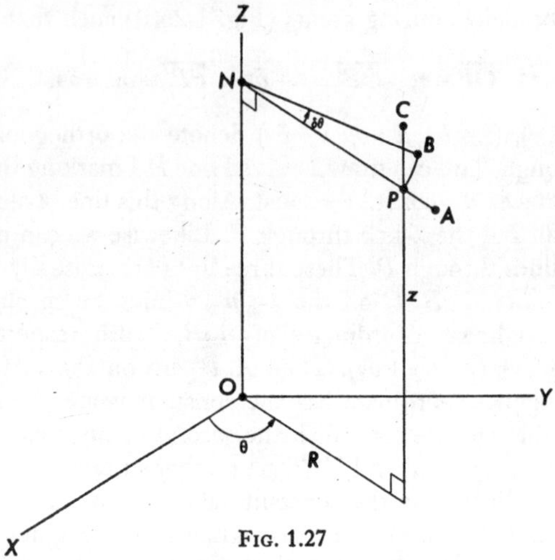

FIG. 1.27

vector $\hat{\mathbf{R}}$ at P in the direction of R increasing. Reverting to P, now keep z, R fixed but let θ increase to $(\theta+\delta\theta)$. Then P travels along a circular arc PB to B, centre N and radius R and such that $\angle PNB=\delta\theta$. Then $PB=R\delta\theta$. Denote $\overline{PB}\equiv R\delta\theta\hat{\boldsymbol{\theta}}$ to define the unit vector $\hat{\boldsymbol{\theta}}$ at P in the direction of θ increasing. Reverting to P again, now increase to $z+\delta z$, keeping R, θ fixed. Then P travels to C along MP where $PC=\delta z$. We write $\overline{PC}\equiv\delta z\mathbf{k}$ since $\mathbf{k}$ is the unit vector in the fixed axis $\overline{OZ}$. Thus, since in the notation of Fig. 1.26(i)

$$\overline{PP}=\overline{PA}+\overline{PB}+\overline{PC},$$
$$\delta\mathbf{s}=\delta R\hat{\mathbf{R}}+R\delta\theta\hat{\boldsymbol{\theta}}+\delta z\mathbf{k}.$$

$\hat{\mathbf{R}}$, $\hat{\boldsymbol{\theta}}$ are variable in direction from point to point, but $\mathbf{k}$ is in a fixed direction. This difference will usually be emphasized by the two different notations that are adopted for variable and fixed unit vectors. Taking $\lambda=R$, $\mu=\theta$, $\nu=z$; $\hat{\mathbf{a}}_1=\hat{\mathbf{R}}$, $\hat{\mathbf{a}}_2=\hat{\boldsymbol{\theta}}$, $\hat{\mathbf{a}}_3=\mathbf{k}$ and comparing the last equation with

$$\delta\mathbf{s}=h_1\delta\lambda\hat{\mathbf{a}}_1+h_2\delta\mu\hat{\mathbf{a}}_2+h_3\delta\nu\hat{\mathbf{a}}_3,$$

we see that for cylindrical polars,

$$h_1=1, \quad h_2=R, \quad h_3=1.$$

Also,

$$\delta s=\sqrt{\{(\delta R)^2+(R\delta\theta)^2+(\delta z)^2\}}.$$

In Fig. 1.28, P has spherical polar coordinates (r, θ, φ). Keeping θ, φ constant, allow r to increase to $(r+\delta r)$ so that P travels out along $\overline{OP}$ to A where $PA=\delta r$. Let $\overline{PA}\equiv\delta r\hat{\mathbf{r}}$ so as to define the unit vector $\hat{\mathbf{r}}$ at P in the

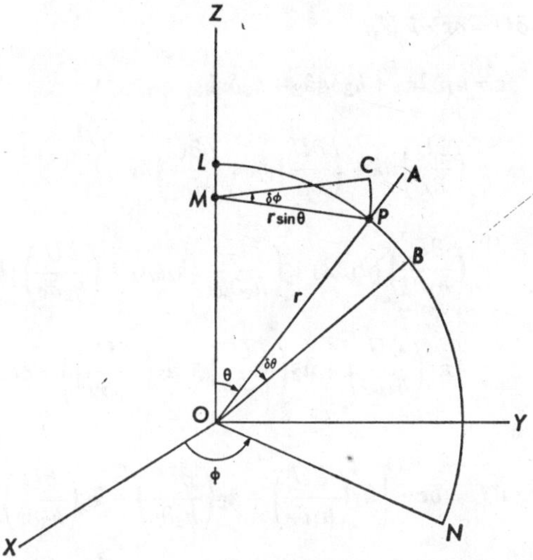

Fig. 1.28

direction of r increasing. Reverting to P and keeping r, φ constant, allow θ to increase to $(\theta+\delta\theta)$. Then P moves along the circular arc LPN to PB where $\angle POB=\theta$ and arc $PB=r\delta\theta$. Write $\overline{PB}\equiv r\delta\theta\hat{\boldsymbol{\theta}}$: this defines the unit vector $\hat{\boldsymbol{\theta}}$ at P in the direction of θ increasing. Reverting again to P, keeping r, θ fixed, increase φ to $(\varphi+\delta\varphi)$. Then P travels out to C along a circular arc through P centre M and radius $r\sin\theta=PM$. Also $\angle PMC=\delta\varphi$ so that arc $PC=r\sin\theta\delta\varphi$. Let $\overline{PC}\equiv r\sin\theta\delta\varphi\hat{\boldsymbol{\varphi}}$ so as to define the unit vector $\hat{\boldsymbol{\varphi}}$ at P in the direction of φ increasing. Then $\delta\mathbf{s}=\delta r\hat{\mathbf{r}}+r\delta\theta\hat{\boldsymbol{\theta}}+r\sin\theta\delta\varphi\hat{\boldsymbol{\varphi}}$. Take $\lambda=r$, $\mu=\theta$, $\nu=\varphi$; $\hat{\mathbf{a}}_1=\hat{\mathbf{r}}$, $\hat{\mathbf{a}}_2=\hat{\boldsymbol{\theta}}$, $\hat{\mathbf{a}}_3=\hat{\boldsymbol{\varphi}}$ and compare this last result with

$$\delta\mathbf{s}=h_1\delta\lambda\hat{\mathbf{a}}_1+h_2\delta\mu\hat{\mathbf{a}}_2+h_3\delta\nu\hat{\mathbf{a}}_3.$$

Then

$$h_1=1,\quad h_2=r,\quad h_3=r\sin\theta.$$

Also,

$$\delta s=\sqrt{\{(\delta r)^2+(r\delta\theta)^2+(r\sin\theta\delta\varphi)^2\}}.$$

1.19.2 Gradient in Orthogonal Coordinates

Let $U(\lambda,\mu,\nu)$ be a uniform differentiable scalar function defined at each point $P(\lambda,\mu,\nu)$ of a region $\mathscr{R}$. Let P' be the neighbouring point $(\lambda+\delta\lambda, \mu+\delta\mu,\nu+\delta\nu)$, as in Fig. 1.26(i), (ii), so that $\overline{PP'}\equiv\delta\mathbf{r}$. If $U+\delta U$ denotes the value of the scalar at P',

$$\delta U=\frac{\partial U}{\partial\lambda}\delta\lambda+\frac{\partial U}{\partial\mu}\delta\mu+\frac{\partial U}{\partial\nu}\delta\nu.$$

But

$$\delta U = \delta \mathbf{r} \cdot \nabla U.$$

Also

$$\delta \mathbf{r} = h_1 \delta\lambda \hat{\mathbf{a}}_1 + h_2 \delta\mu \hat{\mathbf{a}}_2 + h_3 \delta\nu \hat{\mathbf{a}}_3,$$

so that

$$\left(\frac{\partial U}{\partial \lambda}\right)\delta\lambda + \left(\frac{\partial U}{\partial \mu}\right)\delta\mu + \left(\frac{\partial U}{\partial \nu}\right)\delta\nu$$

$$= \left(\frac{\partial U}{h_1 \partial \lambda}\right)(h_1\delta\lambda) + \left(\frac{\partial U}{h_2 \partial \mu}\right)(h_2\delta\mu) + \left(\frac{\partial U}{h_3 \partial \nu}\right)(h_3\delta\nu)$$

$$= \left\{\hat{\mathbf{a}}_1\left(\frac{\partial U}{h_1 \partial \lambda}\right) + \hat{\mathbf{a}}_2\left(\frac{\partial U}{h_2 \partial \mu}\right) + \hat{\mathbf{a}}_3\left(\frac{\partial U}{h_3 \partial \nu}\right)\right\} \cdot \delta\mathbf{r}.$$

Thus we have

$$\delta\mathbf{r} \cdot \nabla U = \delta\mathbf{r} \cdot \left\{\hat{\mathbf{a}}_1\left(\frac{\partial U}{h_1 \partial \lambda}\right) + \hat{\mathbf{a}}_2\left(\frac{\partial U}{h_2 \partial \mu}\right) + \hat{\mathbf{a}}_3\left(\frac{\partial U}{h_3 \partial \nu}\right)\right\}.$$

If $\delta\mathbf{r}$ is an arbitrary small non-zero displacement, this shows that

$$\nabla U = \left(\frac{\partial U}{h_1 \partial \lambda}\right)\hat{\mathbf{a}}_1 + \left(\frac{\partial U}{h_2 \partial \mu}\right)\hat{\mathbf{a}}_2 + \left(\frac{\partial U}{h_3 \partial \nu}\right)\hat{\mathbf{a}}_3.$$

In cylindrical polar coordinates (R, θ, z)

$$\delta\mathbf{s} = \delta R \hat{\mathbf{R}} + R\delta\theta\hat{\boldsymbol{\theta}} + \delta z \mathbf{k}$$

so that

$$\nabla U = \left(\frac{\partial U}{\partial R}\right)\hat{\mathbf{R}} + \left(\frac{\partial U}{R\partial \theta}\right)\hat{\boldsymbol{\theta}} + \left(\frac{\partial U}{\partial z}\right)\mathbf{k}.$$

In spherical polars (r, θ, φ),

$$\delta\mathbf{s} = \delta r\hat{\mathbf{r}} + (r\delta\theta)\hat{\boldsymbol{\theta}} + (r \sin\theta\delta\varphi)\hat{\boldsymbol{\phi}},$$

so that

$$\nabla U = \left(\frac{\partial U}{\partial r}\right)\hat{\mathbf{r}} + \left(\frac{\partial U}{r\partial \theta}\right)\hat{\boldsymbol{\theta}} + \left(\frac{\partial U}{r \sin\theta\partial \varphi}\right)\hat{\boldsymbol{\phi}}.$$

1.19.3 Divergence in Orthogonal Coordinates

Let $\mathbf{F}(\lambda, \mu, \nu) = F_1\hat{\mathbf{a}}_1 + F_2\hat{\mathbf{a}}_2 + F_3\hat{\mathbf{a}}_3$, where $F_1 = F_1(\lambda, \mu, \nu)$, etc., be a uniform differentiable vector function defined at each point $P(\lambda, \mu, \nu)$ of a region $\mathscr{R}$. Then, with reference to Fig. 1.26(ii), since area $PAB'CP = h_1h_3\delta\lambda\delta\nu$, the normal flux of $\mathbf{F}$ across this face in the direction of $\hat{\mathbf{a}}_2$ is $(\mathbf{F} \cdot \hat{\mathbf{a}}_2)h_1h_3\delta\lambda\delta\nu = F_2h_1h_3\delta\lambda\delta\nu$. The normal flux of $\mathbf{F}$ in the same direction across the parallel face $P'A'BC'P'$ is

$$F_2h_1h_3\delta\lambda\delta\nu + \delta\mu(\partial/\partial\mu)(F_2h_1h_3\delta\lambda\delta\nu).$$

Thus the two faces give a total contribution to the normal flux of **F** measured outwards from the box of amount

$$\delta\mu(\partial/\partial\mu)\,(F_2 h_1 h_3 \delta\lambda\delta\nu) = \delta\lambda\delta\mu\delta\nu(\partial/\partial\mu)\,(h_3 h_1 F_2).$$

Similar contributions arise from the two other pairs of parallel faces. Thus the total outward normal flux of **F** from all six faces of the box is

$$\delta\lambda\delta\mu\delta\nu\left\{\left(\frac{\partial}{\partial\lambda}\right)(h_2h_3F_1)+\left(\frac{\partial}{\partial\mu}\right)(h_3h_1F_2)+\left(\frac{\partial}{\partial\nu}\right)(h_1h_2F_3)\right\}.$$

Since the volume of the box is $\Delta v = h_1h_2h_3\delta\lambda\delta\mu\delta\nu$, the divergence of **F** at P is given by

$$\operatorname{div}\mathbf{F} = \lim_{\Delta v\to 0}\left\{\frac{\delta\lambda\delta\mu\delta\nu}{\Delta v}\left[\frac{\partial}{\partial\lambda}(h_2h_3F_1)+\frac{\partial}{\partial\mu}(h_3h_1F_2)+\frac{\partial}{\partial v}(h_1h_2F_3)\right]\right\}$$

$$=\frac{1}{h_1h_2h_3}\left\{\frac{\partial}{\partial\lambda}(h_2h_3F_1)+\frac{\partial}{\partial\mu}(h_3h_1F_2)+\frac{\partial}{\partial\nu}(h_1h_2F_3)\right\}.$$

1.19.4 Laplacian in Orthogonal Coordinates

If $U = U(\lambda, \mu, \nu)$ as in Section 1.19.2, then

$$\nabla U = (\partial U/h_1\partial\lambda)\hat{\mathbf{a}}_1 + (\partial U/h_2\partial\mu)\hat{\mathbf{a}}_2 + (\partial U/h_3\partial\nu)\hat{\mathbf{a}}_3.$$

Taking $\nabla U = \mathbf{F}$, $F_1 = (\partial U/h_1\partial\lambda)$, etc.,

$$\nabla^2 U = \nabla\cdot(\nabla U)$$

$$=\frac{1}{h_1h_2h_3}\left\{\frac{\partial}{\partial\lambda}\left(\frac{h_2h_3}{h_1}\frac{\partial U}{\partial\lambda}\right)+\frac{\partial}{\partial\mu}\left(\frac{h_3h_1}{h_2}\frac{\partial U}{\partial\mu}\right)+\frac{\partial}{\partial\nu}\left(\frac{h_1h_2}{h_3}\frac{\partial U}{\partial\nu}\right)\right\}.$$

In cylindrical polars (R, θ, z),

$$\lambda = R,\ \mu = \theta,\ \nu = z;\quad h_1 = 1,\ h_2 = R,\ h_3 = 1,$$

so that

$$\nabla^2 U = \frac{1}{R}\left\{\frac{\partial}{\partial R}\left(\frac{R\partial U}{\partial R}\right)+\frac{\partial}{\partial\theta}\left(\frac{\partial U}{R\partial\theta}\right)+\frac{\partial}{\partial z}\left(\frac{R\partial U}{\partial z}\right)\right\}$$

$$=\frac{\partial^2 U}{\partial R^2}+\frac{1}{R}\frac{\partial U}{\partial R}+\frac{1}{R^2}\frac{\partial^2 U}{\partial\theta^2}+\frac{\partial^2 U}{\partial z^2}.$$

1.19.5 Curl of a Vector Function in Orthogonal Coordinates

In Fig. 1.29, $\overline{PA} \equiv h_1\delta\lambda\hat{\mathbf{a}}_1$, $\overline{PB} \equiv h_2\delta\mu\hat{\mathbf{a}}_2$, as before. Complete the rectangle $PAC'B$. Then $\hat{\mathbf{a}}_3 = \hat{\mathbf{a}}_1 \wedge \hat{\mathbf{a}}_2$ is the unit normal to this elemental rectangle in the sense that a positive rotation about $\hat{\mathbf{a}}_3$ is in the sense of description $PAC'BP$ of the rectangle. Let $\mathbf{F} = F_1\hat{\mathbf{a}}_1 + F_2\hat{\mathbf{a}}_2 + F_3\hat{\mathbf{a}}_3$, where

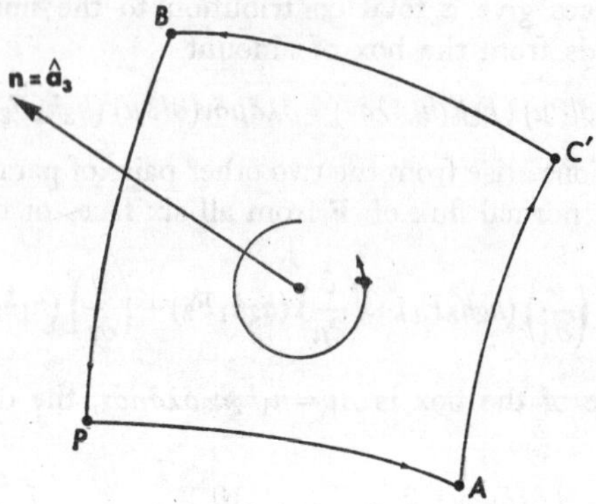

FIG. 1.29

$F_1 = F_1(\lambda, \mu, \nu)$, etc. Then denoting by $\mathscr{C}$ the entire boundary of the rectangle,

$$\oint_{\mathscr{C}} \mathbf{F} \cdot d\mathbf{s} = \int_{PA} \mathbf{F} \cdot d\mathbf{s} + \int_{AC'} \mathbf{F} \cdot d\mathbf{s} - \int_{BC'} \mathbf{F} \cdot d\mathbf{s} - \int_{PB} \mathbf{F} \cdot d\mathbf{s}.$$

Now

$$\int_{PA} \mathbf{F} \cdot d\mathbf{s} \simeq \mathbf{F} \cdot (h_1 \delta\lambda \hat{\mathbf{a}}_1) = h_1 F_1 \delta\lambda,$$

$$\therefore \int_{BC'} \mathbf{F} \cdot d\mathbf{s} \simeq h_1 F_1 \delta\lambda + \delta\mu \frac{\partial}{\partial\mu}(h_1 F_1 \delta\lambda) = h_1 F_1 \delta\lambda + \delta\lambda\delta\mu \frac{\partial}{\partial\mu}(h_1 F_1).$$

Also

$$\int_{PB} \mathbf{F} \cdot d\mathbf{s} \simeq \mathbf{F} \cdot (h_2 \delta\mu \hat{\mathbf{a}}_2) = h_2 F_2 \delta\mu,$$

$$\therefore \int_{AC'} \mathbf{F} \cdot d\mathbf{s} \simeq h_2 F_2 \delta\mu + \delta\lambda \frac{\partial}{\partial\lambda}(h_2 F_2 \delta\mu) = h_2 F_2 \delta\mu + \delta\lambda\delta\mu \frac{\partial}{\partial\lambda}(h_2 F_2).$$

Hence

$$\oint_{\mathscr{C}} \mathbf{F} \cdot d\mathbf{s} = \delta\lambda\delta\mu \left\{ \frac{\partial}{\partial\lambda}(h_2 F_2) - \frac{\partial}{\partial\mu}(h_1 F_1) \right\}.$$

By definition, since $\mathbf{n} = \hat{\mathbf{a}}_3$ in this case, and the area of the rectangle $PAC'BP$ is $\Delta S = h_1 h_2 \delta\lambda\delta\mu$,

$$\hat{\mathbf{a}}_3 \cdot \operatorname{curl} \mathbf{F} = \lim_{\Delta S \to 0} \left[\frac{1}{\Delta S} \delta\lambda\delta\mu \left\{ \frac{\partial}{\partial\lambda}(h_2 F_2) - \frac{\partial}{\partial\mu}(h_1 F_1) \right\} \right]$$

$$= \frac{1}{h_1 h_2} \left\{ \frac{\partial}{\partial\lambda}(h_2 F_2) - \frac{\partial}{\partial\mu}(h_1 F_1) \right\}.$$

Similarly

$$\hat{\mathbf{a}}_1 \cdot \operatorname{curl} \mathbf{F} = \frac{1}{h_2 h_3}\left\{\frac{\partial}{\partial \mu}(h_3 F_3) - \frac{\partial}{\partial \nu}(h_2 F_2)\right\},$$

$$\hat{\mathbf{a}}_2 \cdot \operatorname{curl} \mathbf{F} = \frac{1}{h_3 h_1}\left\{\frac{\partial}{\partial \nu}(h_1 F_1) - \frac{\partial}{\partial \lambda}(h_3 F_3)\right\}.$$

Thus

$$\operatorname{curl} \mathbf{F} = \sum \left\{\frac{\hat{\mathbf{a}}_1}{h_2 h_3}\left[\frac{\partial}{\partial \mu}(h_3 F_3) - \frac{\partial}{\partial \nu}(h_2 F_2)\right]\right\}.$$

This result is more easily remembered in the determinantal form

$$\operatorname{curl} \mathbf{F} = \frac{1}{h_1 h_2 h_3}\begin{vmatrix} h_1 \hat{\mathbf{a}}_1 & h_2 \hat{\mathbf{a}}_2 & h_3 \hat{\mathbf{a}}_3 \\ \dfrac{\partial}{\partial \lambda} & \dfrac{\partial}{\partial \mu} & \dfrac{\partial}{\partial \nu} \\ h_1 F_1 & h_2 F_2 & h_3 F_3 \end{vmatrix}.$$

1.19.6 Worked Examples

Example 1

Parabolic coordinates (ξ, η, φ) are related to Cartesian coordinates (x, y, z) by

$$x = \tfrac{1}{2}(\xi - \eta), \quad y = (\xi\eta)^{1/2} \cos \varphi, \quad z = (\xi\eta)^{1/2} \sin \varphi.$$

Show that this system is an orthogonal curvilinear one and that the line element in it is given by

$$(ds)^2 = \tfrac{1}{4}(\xi + \eta)\left\{\frac{1}{\xi}(d\xi)^2 + \frac{1}{\eta}(d\eta)^2\right\} + \xi\eta(d\varphi)^2.$$

Show further that if $\psi = \psi(\xi, \eta, \varphi)$ is a uniform differentiable scalar function,

$$\nabla^2 \psi = \frac{4}{\xi + \eta}\left\{\frac{\partial}{\partial \xi}\left(\xi \frac{\partial \psi}{\partial \xi}\right) + \frac{\partial}{\partial \eta}\left(\eta \frac{\partial \psi}{\partial \eta}\right)\right\} + \frac{1}{\xi\eta}\frac{\partial^2 \psi}{\partial \varphi^2}.$$

Regarding ξ, η, φ as the independent variables,

$$dx = \frac{\partial x}{\partial \xi} d\xi + \frac{\partial x}{\partial \eta} d\eta + \frac{\partial x}{\partial \varphi} d\varphi = \tfrac{1}{2}(d\xi - d\eta),$$

$$dy = \frac{\partial y}{\partial \xi} d\xi + \frac{\partial y}{\partial \eta} d\eta + \frac{\partial y}{\partial \varphi} d\varphi$$

$$= \frac{1}{2}\left(\frac{\eta}{\xi}\right)^{1/2} \cos \varphi \, d\xi + \frac{1}{2}\left(\frac{\xi}{\eta}\right)^{1/2} \cos \varphi \, d\eta - (\xi\eta)^{1/2} \sin \varphi \, d\varphi,$$

$$dz=\frac{\partial z}{\partial \xi}d\xi+\frac{\partial z}{\partial \eta}d\eta+\frac{\partial z}{\partial \varphi}d\varphi$$

$$=\frac{1}{2}\left(\frac{\eta}{\xi}\right)^{1/2}\sin\varphi\, d\xi+\frac{1}{2}\left(\frac{\xi}{\eta}\right)^{1/2}\sin\varphi\, d\eta+(\xi\eta)^{1/2}\cos\varphi\, d\varphi.$$

Hence

$$(ds)^2=(dx)^2+(dy)^2+(dz)^2$$

$$=\tfrac{1}{4}(\xi+\eta)\left[\frac{1}{\xi}(d\xi)^2+\frac{1}{\eta}(d\eta)^2\right]+\xi\eta(d\varphi)^2.$$

Now let $\hat{\mathbf{a}}_1$, $\hat{\mathbf{a}}_2$, $\hat{\mathbf{a}}_3$ be unit vectors in the sense of ξ, η, φ increasing and write

$$d\mathbf{r}=\tfrac{1}{2}\xi^{-1/2}(\xi+\eta)^{1/2}\,d\xi\hat{\mathbf{a}}_1+\tfrac{1}{2}\eta^{-1/2}(\xi+\eta)^{1/2}\,d\eta\,\hat{\mathbf{a}}_2+(\xi\eta)^{1/2}\,d\varphi\hat{\mathbf{a}}_3.$$

Then $(ds)^2=(d\mathbf{r})^2$ only if $\hat{\mathbf{a}}_1\cdot\hat{\mathbf{a}}_2=\hat{\mathbf{a}}_2\cdot\hat{\mathbf{a}}_3=\hat{\mathbf{a}}_3\cdot\hat{\mathbf{a}}_1=0$, which establishes the orthogonality of the system.

Take

$$d\mathbf{r}=h_1\,d\xi\hat{\mathbf{a}}_1+h_2\,d\eta\hat{\mathbf{a}}_2+h_3\,d\varphi\hat{\mathbf{a}}_3.$$

Then

$$h_1=\tfrac{1}{2}\xi^{-1/2}(\xi+\eta)^{1/2},\; h_2=\tfrac{1}{2}\eta^{-1/2}(\xi+\eta)^{1/2},\; h_3=(\xi\eta)^{1/2}.$$

Thus

$$\nabla^2\psi=\frac{1}{h_1h_2h_3}\left[\frac{\partial}{\partial\xi}\left(\frac{h_2h_3}{h_1}\frac{\partial\psi}{\partial\xi}\right)+\frac{\partial}{\partial\eta}\left(\frac{h_3h_1}{h_2}\frac{\partial\psi}{\partial\eta}\right)+\frac{\partial}{\partial\varphi}\left(\frac{h_1h_2}{h_3}\frac{\partial\psi}{\partial\varphi}\right)\right]$$

$$=\frac{4}{\xi+\eta}\left[\frac{\partial}{\partial\xi}\left(\xi\frac{\partial\psi}{\partial\xi}\right)+\frac{\partial}{\partial\eta}\left(\eta\frac{\partial\psi}{\partial\eta}\right)+\frac{\partial}{\partial\varphi}\left(\frac{\xi+\eta}{4\xi\eta}\right)\frac{\partial\psi}{\partial\varphi}\right],$$

which yields the stated result.

Example 2

The components of a vector **V** with regard to spherical polar coordinates (r, θ, φ) are $[u, v, w]$; write down the expression for the components of curl **V** in terms of u, v, w.

If $u=v=\partial w/\partial\varphi=0$, and curl curl $\mathbf{V}=\mathbf{0}$, show that

$$r^2\frac{\partial^2 w}{\partial r^2}+2r\frac{\partial w}{\partial r}+\frac{\partial}{\partial\theta}\left(\frac{\partial w}{\partial\theta}\, w\cot\theta\right)=0.$$

Show also that there is a solution of this equation of the form $f(r)\sin\theta$, and find the general expression for $f(r)$.

(London Univ., B.Sc. Spec. Hons., 1959)

In orthogonal coordinates (λ, μ, ν) with line element

$$d\mathbf{s} = h_1 d\lambda \hat{\mathbf{a}}_1 + h_2 d\mu \hat{\mathbf{a}}_2 + h_3 d\nu \hat{\mathbf{a}}_3$$

and

$$\mathbf{V} = V_1 \hat{\mathbf{a}}_1 + V_2 \hat{\mathbf{a}}_2 + V_3 \hat{\mathbf{a}}_3,$$

$$\text{curl } \mathbf{V} = \frac{1}{h_1 h_2 h_3} \begin{vmatrix} h_1 \hat{\mathbf{a}}_1 & h_2 \hat{\mathbf{a}}_2 & h_3 \hat{\mathbf{a}}_3 \\ \dfrac{\partial}{\partial \lambda} & \dfrac{\partial}{\lambda \mu} & \dfrac{\partial}{\partial \nu} \\ h_1 V_1 & h_2 V_2 & h_3 V_3 \end{vmatrix}.$$

In spherical polars (r, θ, φ),

$$d\mathbf{s} = dr\hat{\mathbf{r}} + r\, d\theta \hat{\boldsymbol{\theta}} + r \sin\theta\, d\varphi \hat{\boldsymbol{\varphi}},$$

so that if

$$\lambda = r,\ \mu = \theta,\ \nu = \varphi;$$
$$h_1 = 1,\ h_2 = r,\ h_3 = r \sin\theta;$$
$$V_1 = u,\ V_2 = v,\ V_3 = w$$

Hence

$$\text{curl } \mathbf{V} = \frac{1}{r^2 \sin\theta} \begin{vmatrix} \hat{\mathbf{r}} & r\hat{\boldsymbol{\theta}} & r \sin\theta \hat{\boldsymbol{\varphi}} \\ \dfrac{\partial}{\partial r} & \dfrac{\partial}{\partial \theta} & \dfrac{\partial}{\partial \varphi} \\ u & rv & r \sin\theta w \end{vmatrix}$$

$$= \frac{1}{r^2 \sin\theta} \left\{ r\left(w \cos\theta + \sin\theta \frac{\partial w}{\partial \theta} - \frac{\partial v}{\partial \varphi} \right) \hat{\mathbf{r}} \right.$$

$$+ r\left(\frac{\partial u}{\partial \varphi} - w \sin\theta - r \sin\theta \frac{\partial w}{\partial r} \right) \hat{\boldsymbol{\theta}}$$

$$\left. + r \sin\theta \left(v + r \frac{\partial v}{\partial r} - \frac{\partial u}{\partial \theta} \right) \hat{\boldsymbol{\varphi}} \right\}.$$

If $u = v = \partial w / \partial \varphi = 0$, then

$$\text{curl } \mathbf{V} = \left(\frac{1}{r} \right) \left(\cot\theta \cdot w + \frac{\partial w}{\partial \theta} \right) \hat{\mathbf{r}} - \left(\frac{w}{r} + \frac{\partial w}{\partial r} \right) \hat{\boldsymbol{\theta}}$$

and hence curl curl $\mathbf{V}$

$$= \frac{1}{r^2 \sin\theta} \begin{vmatrix} \hat{\mathbf{r}} & r\hat{\boldsymbol{\theta}} & r \sin\theta \hat{\boldsymbol{\varphi}} \\ \dfrac{\partial}{\partial r} & \dfrac{\partial}{\partial \theta} & \dfrac{\partial}{\partial \varphi} \\ \left(\dfrac{1}{r} \right) \left(w \cot\theta + \dfrac{\partial w}{\partial \theta} \right) & -\left(w + r \dfrac{\partial w}{\partial r} \right) & 0 \end{vmatrix}$$

Thus if curl curl $\mathbf{V}=\mathbf{0}$, then

$$0\hat{\mathbf{r}}+0\hat{\boldsymbol{\theta}}+\frac{1}{r}\left\{-\frac{\partial}{\partial r}\left(w+r\frac{\partial w}{\partial r}\right)-\frac{1}{r}\frac{\partial}{\partial \theta}\left(w\cot\theta+\frac{\partial w}{\partial \theta}\right)\right\}\hat{\boldsymbol{\phi}}=\mathbf{0},$$

i.e.

$$2\frac{\partial w}{\partial r}+r\frac{\partial^2 w}{\partial r^2}+\frac{1}{r}\frac{\partial}{\partial \theta}\left(\frac{\partial w}{\partial \theta}+w\cot\theta\right)=0$$

or

$$r^2\frac{\partial^2 w}{\partial r^2}+2r\frac{\partial w}{\partial r}+\frac{\partial}{\partial \theta}\left(\frac{\partial w}{\partial \theta}+w\cot\theta\right)=0.$$

In the last equation, put $w=f(r)\sin\theta$. Then

$$r^2 f''(r)\sin\theta+2rf'(r)\sin\theta+f(r)\,(d/d\theta)\,(2\cos\theta)=0,$$

or

$$r^2 f''(r)+2rf'(r)-2f(r)=0.$$

The substitution $r=\exp t$ reduces this to

$$(D+2)\,(D-1)f=0 \quad (D\equiv d/dt),$$

so that

$$\begin{aligned} f&=A\exp t+B\exp(-2t)\\ &=Ar+Br^{-2}. \end{aligned}$$

Thus the required solution is

$$\underline{w=(Ar+Br^{-2})\sin\theta.}$$

1.20 Some Cartesian Tensor Notation

In modern treatises and papers on fluid dynamics, as well as in other branches of continuum mechanics, the use of tensor notation is becoming more frequent. Essentially this is a shorthand notation the use of which makes the writing of many equations less cumbersome.

We denote the Cartesian coordinates of a point P in three-dimensional space by (x_1, x_2, x_3) or more simply by $x_i (i=1, 2, 3)$, so that $(x_1, x_2, x_3)=(x, y, z)$. The corresponding Cartesian coordinate axes are OX_1, OX_2, OX_3 or OX_i $(i=1, 2, 3)$.

Crucial to the tensor calculus is the contracted notation used for summations. According to this *summation convention* a repeated suffix implies summation w.r.t. that suffix. Thus if, as will be understood once and for all, the letters i, j, k can each take the discrete values 1, 2, 3, then

$$\alpha_{ij}x_j=\sum_{j=1}^{3}\alpha_{ij}x_j \quad (i=1, 2, 3).$$

Written out fully this gives the three summations

$$\begin{aligned} &\alpha_{11}x_1+\alpha_{12}x_2+\alpha_{13}x_3,\\ &\alpha_{21}x_1+\alpha_{22}x_2+\alpha_{23}x_3,\\ &\alpha_{31}x_1+\alpha_{32}x_2+\alpha_{33}x_3, \end{aligned}$$

where i is allowed to range through the values 1, 2, 3. Other examples of the summation convention are:

(i) $$A_{ij}u_iu_j = \sum_{i=1}^{3}\sum_{j=1}^{3} A_{ij}u_iu_j;$$

(ii) $$A_{ijk}u_iu_ju_k = \sum_{i=1}^{3}\sum_{j=1}^{3}\sum_{k=1}^{3} A_{ijk}u_iu_ju_k.$$

We note that

$$\alpha_{ij}x_j = \alpha_{ik}x_k,$$

since each means $\sum_{k=1}^{3}\alpha_{ik}x_k$, both of which are independent of j, k when written out fully. Thus the repeated suffix is a *dummy suffix* and it may be replaced by any other letter. No letter may be repeated more than once.

The *Kronecker delta* or *substitution tensor* δ_{ij} is defined to be

$$\delta_{ij} = 1 \text{ when } i=j \text{ (not summed)};$$
$$= 0 \text{ when } i \neq j.$$

Clearly

$$\delta_{ij} = \delta_{ji}.$$

Note, however, that in accordance with the summation convention

$$\delta_{ii} = \delta_{11} + \delta_{22} + \delta_{33} = 1+1+1 = 3.$$

Further, we observe that

$$\delta_{ij}x_j = \delta_{i1}x_1 + \delta_{i2}x_2 + \delta_{i3}x_3.$$

If $i=1$, the above gives x_1. If $i=2$, it gives x_2. If $i=3$, it gives x_3.

$$\therefore \underline{\delta_{ij}x_j = x_i.}$$

This last result shows that the operation of δ_{ij} on x_j merely substitutes i for j: hence the name 'substitution tensor'.

Partial derivatives are denoted as follows:

$$\partial\varphi/\partial x_i = \varphi_{,i} \quad (i=1, 2, 3);$$
$$\partial u_i/\partial x_j = u_{i,j} \quad (i, j = 1, 2, 3).$$

Thus a comma preceding a suffix i means differentiate partially w.r.t. x_i. Observe that

$$\varphi_{,ii} = \frac{\partial^2\varphi}{\partial x_1^2} + \frac{\partial^2\varphi}{\partial x_2^2} + \frac{\partial^2\varphi}{\partial x_3^2} = \nabla^2\varphi,$$

$$u_{i,i} = \frac{\partial u_1}{\partial x_1} + \frac{\partial u_2}{\partial x_2} + \frac{\partial u_3}{\partial x_3}.$$

The quantities div **A**, grad φ for fixed coordinate axes are expressible in the forms

$$\operatorname{div} \mathbf{A} = \frac{\partial A_1}{\partial x_1} + \frac{\partial A_2}{\partial x_2} + \frac{\partial A_3}{\partial x_3} = A_{i,i};$$

$$\operatorname{grad} \varphi = \frac{\partial \varphi}{\partial x_1} \hat{\mathbf{a}}_1 + \frac{\partial \varphi}{\partial x_2} \hat{\mathbf{a}}_2 + \frac{\partial \varphi}{\partial x_3} \hat{\mathbf{a}}_3 = \varphi_{,i} \hat{\mathbf{a}}_i.$$

Here, $\mathbf{A} = A_i \hat{\mathbf{a}}_i$, where $\hat{\mathbf{a}}_i$ is the unit vector in the positive x_i-axis. Thus the divergence theorem in the vector form

$$\int_V \nabla \cdot \mathbf{A}\, dv = \int_S \mathbf{n} \cdot \mathbf{A}\, dS,$$

where $\mathbf{n} = n_i \hat{\mathbf{a}}_i$, becomes

$$\int_V A_{i,i}\, dv = \int_S n_i A_i\, dS$$

in tensor form.

Now suppose OX_i, OX_i' $(i=1, 2, 3)$ determine two sets of rectangular Cartesian coordinate axes through O. Further, suppose the direction cosines of each axis of the one system w.r.t. the three axes of the other are as shown in the following table.

	OX_1	OX_2	OX_3
OX_1'	l_{11}	l_{12}	l_{13}
OX_2'	l_{21}	l_{22}	l_{23}
OX_3'	l_{31}	l_{32}	l_{33}

The table means that, e.g., $[l_{21}, l_{22}, l_{23}]$ are the D.C.'s of OX_2' w.r.t. the frame OX_i and that $[l_{13}, l_{23}, l_{33}]$ are the D.C.'s of OX_3 w.r.t. the frame OX_i' $(i=1, 2, 3)$. We observe that the following six direction-cosine properties hold:

$$l_{11}^2 + l_{12}^2 + l_{13}^2 = 1, \text{ etc.},$$
$$l_{11}l_{21} + l_{12}l_{22} + l_{13}l_{23} = 0, \text{ etc.}$$

These are conveniently expressed by means of the summation convention in the form

$$l_{ij}l_{kj} = \delta_{ik}.$$

Also we have the six equations

$$l_{11}^2 + l_{21}^2 + l_{31}^2 = 1, \text{ etc.},$$
$$l_{11}l_{12} + l_{21}l_{22} + l_{31}l_{32} = 0, \text{ etc.},$$

epitomized by

$$l_{ij}l_{ik} = \delta_{jk}.$$

If x_i denotes the Cartesian coordinates of a point P in the frame OX_i and x' those in the frame OX'_i ($i=1, 2, 3$), then

$$x'_i = l_{ij}x_j,$$
$$x_i = l_{ji}x_j.$$

These are the required transformation fo mulae. They are easily established by vector methods. Thus taking

$$\mathbf{r} = x_1\hat{\mathbf{a}}_1 + x_2\hat{\mathbf{a}}_2 + x_3\hat{\mathbf{a}}_3 = x'_1\hat{\mathbf{a}}'_1 + x'_2\hat{\mathbf{a}}'_2 + x'_3\hat{\mathbf{a}}'_3,$$

and using the table

$$\begin{aligned}\mathbf{r}\cdot\hat{\mathbf{a}}_1 = x_1 &= x'_1\hat{\mathbf{a}}'_1\cdot\hat{\mathbf{a}}_1 + x'_2\hat{\mathbf{a}}'_2\cdot\hat{\mathbf{a}}_1 + x_3\hat{\mathbf{a}}'_3\cdot\hat{\mathbf{a}}_1\\ &= x'_1 l_{11} + x'_2 l_{21} + x'_3 l_{31}.\end{aligned}$$

The set of three quantities x_i ($i=1, 2, 3$) is called a *vector* or a *tensor of the first order* (or *rank*). The set x'_i comprises another first-order tensor. Single quantities such as mass and distance which remain invariant for all sets of axes are called *scalars* or *tensors of zero order*. Thus in three dimensions,

$$x'_i = l_{ij}x_j = l_{ik}x_k,$$

so that

$$x'_i x'_i = (l_{ij}x_j)(l_{ik}x_k) = (l_{ij}l_{ik})(x_j x_k).$$

Now $l_{ij}l_{ik} = \delta_{jk}$ from the table of D.C.'s. Thus

$$x'_i x'_i = \delta_{jk}x_j x_k = x_k x_k$$

or

$$x'^2_1 + x'^2_2 + x'^2_3 = x^2_1 + x^2_2 + x^2_3.$$

This shows that OP^2 is invariant under transformation of axes and it illustrates the use of the summation convention in manipulation.

The set of nine quantities a_{ij} ($i, j=1, 2, 3$) obeying the transformation law

$$a'_{rs} = l_{ri}l_{sj}a_{ij},$$

the l's being D.C.'s, such that $l_{ri}l_{si} = \delta_{rs} = l_{ir}l_{is}$, constitutes the *components of a tensor of the second order*.

The following examples showing that the quantities

$$a_{ii},\quad a_{ij}a_{ji},\quad a_{ij}a_{jk}a_{ki}$$

are invariant under transformation further illustrate the algebraic manipulation of tensors. These results are of importance in the mathematical theories of elasticity and plasticity.

EXAMPLE 1

Show that

$$a'_{ii}=a_{ii}.$$

We have

$$a'_{rs}=l_{ri}l_{sj}a_{ij},$$

so that

$$a'_{rr}=l_{ri}l_{rj}a_{ij}=\delta_{ij}a_{ij}=a_{jj},$$

i.e.

$$\underline{a'_{ii}=a_{ii}.}$$

EXAMPLE 2

Show that

$$a'_{ij}a'_{ji}=a_{ij}a_{ji}.$$

We have

$$a'_{rs}=l_{ri}l_{sj}a_{ij},$$
$$a'_{sr}=l_{sm}l_{rn}a_{mn}.$$

Thus

$$\begin{aligned}a'_{rs}a'_{sr}&=(l_{ri}l_{sj}a_{ij})\,(l_{sm}l_{rn}a_{mn})\\&=(l_{ri}l_{rn})\,(l_{sj}l_{sm})\,(a_{ij}a_{mn})\\&=\delta_{in}\delta_{jm}a_{ij}a_{mn}\\&=\delta_{in}(a_{im}a_{mn})\\&=a_{nm}a_{mn},\quad\text{as required.}\end{aligned}$$

EXAMPLE 3

Show that

$$a'_{ij}a'_{jk}a'_{ki}=a_{ij}a_{jk}a_{ki}.$$

$$\begin{aligned}a'_{rs}a'_{st}a'_{tr}&=(l_{ri}l_{sj}a_{ij})\,(l_{sm}l_{tn}a_{mn})\,(l_{t\alpha}l_{r\beta}a_{\alpha\beta})\\&=(l_{ri}l_{r\beta})\,(l_{sj}l_{sm})\,(l_{tn}l_{t\alpha})a_{ij}a_{mn}a_{\alpha\beta}\\&=\delta_{i\beta}\delta_{jm}\delta_{n\alpha}a_{ij}a_{mn}a_{\alpha\beta}\\&=a_{\beta m}a_{m\alpha}a_{\alpha\beta},\quad\text{as required.}\end{aligned}$$

The last example is very formidable to establish without using the summation convention.

EXERCISE 1

1. *Coplanar Vectors.* If three vectors **a**, **b**, **c** lie in the same plane prove that in general one can find scalar magnitudes λ, μ such that

$$\mathbf{c}=\lambda\mathbf{a}+\mu\mathbf{b}.$$

If a linear relationship of the above type holds between any three vectors **a**, **b**, **c** show that the vectors are all parallel to a plane. Show that the vectors

$$(2\mathbf{a}-3\mathbf{b}+\mathbf{c}),\ (\mathbf{a}-2\mathbf{b}+4\mathbf{c}),\ (7\mathbf{a}-12\mathbf{b}+14\mathbf{c})$$

are all parallel to the same plane.

2. *Internal Division of a Line Segment.* Let $\overline{OP_1} \equiv \mathbf{r}_1 = [x_1, y_1, z_1]$, $\overline{OP_2} \equiv \mathbf{r}_2 = [x_2, y_2, z_2]$ and suppose P divides P_1P_2 internally in the ratio $P_1P : PP_2 = \lambda : \mu$. Show that the position vector $\mathbf{r}$ of P is given by

$$\overline{OP} \equiv \mathbf{r} = (\lambda \mathbf{r}_2 + \mu \mathbf{r}_1)/(\lambda + \mu)$$

and deduce that the coordinates of P are

$$\left(\frac{\lambda x_2 + \mu x_1}{\lambda + \mu}, \frac{\lambda y_2 + \mu y_1}{\lambda + \mu}, \frac{\lambda z_2 + \mu z_1}{\lambda + \mu} \right).$$

3. *External Division of a Line Segment.* If, in the previous case, the division of P_1P_2 be external and in the ratio $P_1P : PP_2 = \lambda : \mu$, show that

$$\overline{OP} \equiv \mathbf{r} = (\lambda \mathbf{r}_2 - \mu \mathbf{r}_1)/(\lambda - \mu)$$

and find the coordinates of P.

4. Prove the following results:

(i) $(\mathbf{P} \wedge \mathbf{Q}) \wedge \mathbf{R} = (\mathbf{P} \cdot \mathbf{R})\mathbf{Q} - (\mathbf{Q} \cdot \mathbf{R})\mathbf{P}$;

(ii) $(\mathbf{P} \wedge \mathbf{Q}) \cdot (\mathbf{R} \wedge \mathbf{S}) = (\mathbf{P} \cdot \mathbf{R})(\mathbf{Q} \cdot \mathbf{S}) - (\mathbf{P} \cdot \mathbf{S})(\mathbf{Q} \cdot \mathbf{R})$;

(iii) $\mathbf{P} \wedge (\mathbf{Q} \wedge \mathbf{R}) + \mathbf{Q} \wedge (\mathbf{R} \wedge \mathbf{P}) + \mathbf{R} \wedge (\mathbf{P} \wedge \mathbf{Q}) = 0$;

(iv) $(\mathbf{P} \wedge \mathbf{Q}) \wedge (\mathbf{R} \wedge \mathbf{S}) = [\mathbf{P}, \mathbf{Q}, \mathbf{S}]\mathbf{R} - [\mathbf{P}, \mathbf{Q}, \mathbf{R}]\mathbf{S}$
$= [\mathbf{P}, \mathbf{R}, \mathbf{S}]\mathbf{Q} - [\mathbf{Q}, \mathbf{R}, \mathbf{S}]\mathbf{P}$.

5. If φ, φ_1, φ_2 are uniform differentiable scalar functions and $\mathbf{F}$ a uniform differentiable vector function, prove the following results:

(i) $\text{curl}\,(\varphi \mathbf{F}) = \varphi\, \text{curl}\, \mathbf{F} + (\text{grad}\, \varphi) \wedge \mathbf{F}$;

(ii) $\text{div}\,(\varphi_1\, \text{grad}\, \varphi_2) = \varphi_1 \nabla^2 \varphi_2 + (\text{grad}\, \varphi_1) \cdot (\text{grad}\, \varphi_2)$.

6. Define the divergence of a vector function of position in Euclidean three-dimensional space.

Let $\mathbf{r} = [x, y, z]$, and let $\mathbf{a}$ be a fixed vector of unit length. Denote the length of $\mathbf{r}$ by $|\mathbf{r}|$. Evaluate the divergence of each of the following vector functions:

(i) $|\mathbf{r}|^n \mathbf{a}$ $(\mathbf{r} \neq \mathbf{0})$;

(ii) $|\mathbf{r}|^n (\mathbf{a} \wedge \mathbf{r})$ $(\mathbf{r} \neq \mathbf{0})$;

(iii) $(\mathbf{a} \cdot \mathbf{r})\mathbf{a}$;

(iv) $(\mathbf{a} \wedge \mathbf{r}) \wedge \mathbf{a}$.

(Cambridge Univ., M.T. I., 1964)

7. State and prove a necessary and sufficient condition for a vector field $\mathbf{u}$ to be expressible in the form

$$\mathbf{u} = -\text{grad}\, \varphi,$$

where φ is a scalar field.

Determine whether the following vector fields are expressible in this form:

(i) $(\mathbf{a}\cdot\mathbf{r})\mathbf{r}$;
(ii) $(\mathbf{a}\cdot\mathbf{r})\mathbf{a}$;
(iii) $r^{-1}[r^2\mathbf{a}+(\mathbf{a}\cdot\mathbf{r})\mathbf{r}]$;
(iv) $r^{-1}[r^2\mathbf{a}-(\mathbf{a}\cdot\mathbf{r})\mathbf{r}]$;

where $\mathbf{a}$ is a constant non-zero vector. Find suitable scalar fields φ in appropriate cases.

(Cambridge Univ., Prelim. to M.T. II, 1956)

8. If S is any closed surface containing a volume v, prove

(i) $\int_S \mathbf{r}\cdot d\mathbf{S}=3v$ and use the result to find the volume of a right circular cone;

(ii) $\int_S d\mathbf{S}\cdot \operatorname{curl}\mathbf{F}\equiv 0$;

(iii) $\int_S d\mathbf{S}\wedge \nabla\varphi\equiv\mathbf{0}$;

(iv) $\int_v \mathbf{q}\cdot\nabla p\,dv=\int_S p\mathbf{q}\cdot\mathbf{n}\,dS-\int_v p\nabla\cdot\mathbf{q}\,dv.$

9. If $\mathscr{C}$ is a closed circuit containing an area S, prove

(i) $\int_S \varphi\mathbf{n}\cdot\operatorname{curl}\mathbf{F}\,dS=\oint_{\mathscr{C}}\varphi\mathbf{F}\cdot d\mathbf{s}-\int_S(\nabla\varphi\wedge\mathbf{F})\cdot\mathbf{n}\,dS$;

(ii) $\oint_{\mathscr{C}}\varphi\nabla\varphi\cdot d\mathbf{s}=0.$

10. Evaluate the integral $\int \mathbf{n}\cdot\operatorname{curl}\mathbf{F}\,dS$ for the vector function

$$\mathbf{F}=(2y^2+3z^2-x^2)\mathbf{i}+(2z^2+3x^2-y^2)\mathbf{j}+(2x^2+3y^2-z^2)\mathbf{k}$$

over the part of the surface

$$x^2+y^2-2ax+az=0$$

which lies above the plane $z=0$.

(London Univ., B.Sc.(Eng.) III, 1957)

11. Prove Stokes's theorem that if $\mathscr{C}$ is any closed curve spanned by a surface S, and $\mathbf{f}$ a vector function of position, then, under suitable restrictions,

$$\oint_{\mathscr{C}}\mathbf{f}\cdot d\mathbf{l}=\iint_S d\mathbf{S}\cdot(\nabla\wedge\mathbf{f}).$$

Deduce that

$$\oint_{\mathscr{C}}\mathbf{f}\wedge d\mathbf{l}=\iint_S\left\{d\mathbf{S}\nabla\cdot\mathbf{f}-(d\mathbf{S}\cdot\nabla)\mathbf{f}-d\mathbf{S}\wedge(\nabla\wedge\mathbf{f})\right\}.$$

(Cambridge Univ., Prelim. to M.T. II, 1964)

12. The mathematical formulation of Faraday's law of magnetic induction through a closed circuit $\mathscr{C}$ is

$$\oint_{\mathscr{C}} c\,\mathbf{E}\cdot d\mathbf{s} = -\frac{\partial}{\partial t}\int_S \mathbf{B}\cdot d\mathbf{S}$$

where S is the area within $\mathscr{C}$, $\mathbf{E}$ is the electric intensity, $\mathbf{B}$ the magnetic induction, t signifies time, and c is a constant. Prove that

$$\operatorname{curl}\mathbf{E} = -\frac{1}{c}\frac{\partial\mathbf{B}}{\partial t}.$$

[First show $\int_S \mathbf{n}\cdot\left(\operatorname{curl}\mathbf{E} - \frac{1}{c}\frac{\partial\mathbf{B}}{\partial t}\right)dS = 0$. When $\mathscr{C}$ contains a small area ΔS, to the first order this gives

$$\mathbf{n}\cdot\left[\operatorname{curl}\mathbf{E} - \frac{1}{c}\frac{\partial\mathbf{B}}{\partial t}\right]\Delta S = 0.$$

Since ΔS is arbitrary and non-zero, $\mathbf{n}$ is not in general perpendicular to the bracketed vector and so the result follows.]

13. Prove that

$$\begin{aligned}\mathbf{n}\cdot[\mathbf{u}\wedge(\mathbf{v}\wedge\mathbf{t})] &= \mathbf{t}\cdot[(\mathbf{n}\wedge\mathbf{u})\wedge\mathbf{v}]\\ &= \mathbf{t}\cdot[(\mathbf{n}\cdot\mathbf{u})\mathbf{v} + \mathbf{n}\wedge(\mathbf{u}\wedge\mathbf{v}) - \mathbf{n}(\mathbf{u}\cdot\mathbf{v})].\end{aligned}$$

By applying Stokes's theorem to the vector $\mathbf{v}\wedge\mathbf{t}$, where $\mathbf{v}$ is a vector function of position and $\mathbf{t}$ is a constant vector, or otherwise, prove that

$$\begin{aligned}\oint_{\mathscr{C}} d\mathbf{r}\wedge\mathbf{v} &= \int (d\mathbf{S}\wedge\nabla)\wedge\mathbf{v}\\ &= \int (d\mathbf{S}\cdot\nabla)\mathbf{v} + \int d\mathbf{S}\wedge(\nabla\wedge\mathbf{v}) - \int d\mathbf{S}(\nabla\cdot\mathbf{v}),\end{aligned}$$

where $d\mathbf{S}$ is an oriented element of an open surface, $d\mathbf{r}$ is an element of its bounding curve $\mathscr{C}$, ∇ is the gradient operator, and Cartesian components of all vectors are used.

(London Univ., Gen. B.Sc. II, 1962)

14. Three vector fields $\mathbf{E}$, $\mathbf{H}$, and $\mathbf{v}$ satisfy the equations

$$\nabla\cdot\mathbf{E} = \rho, \quad \nabla\cdot\mathbf{H} = 0$$

$$\nabla\wedge\mathbf{E} = -\frac{1}{c}\frac{\partial\mathbf{H}}{\partial t}, \quad \nabla\wedge\mathbf{H} = \frac{1}{c}\frac{\partial\mathbf{E}}{\partial t} + \rho\mathbf{v},$$

ρ being a scalar field and c a constant. Assuming that **E**, **H**, **v** and ρ are at least twice differentiable functions of x, y, z and t, establish the following equations:

$$\nabla^2\mathbf{E}-\frac{1}{c^2}\frac{\partial^2\mathbf{E}}{\partial t^2}=\nabla\rho+\frac{1}{c^2}\frac{\partial}{\partial t}(\rho\mathbf{v}),$$

$$\nabla^2\mathbf{H}-\frac{1}{c^2}\frac{\partial^2\mathbf{H}}{\partial t^2}=-\frac{1}{c}\nabla\wedge(\rho\mathbf{v}),$$

$$\nabla\cdot(\rho\mathbf{v})+\frac{\partial\rho}{\partial t}=0.$$

(London Univ., Gen. B.Sc. II, 1963)

15. (i) Show that

$$\operatorname{div}(\mathbf{u}\wedge\mathbf{v})=\mathbf{v}\cdot\operatorname{curl}\mathbf{u}-\mathbf{u}\cdot\operatorname{curl}\mathbf{v}$$

for any two vector fields **u** and **v**.

If curl $\mathbf{a}=\mathbf{b}$, and curl $\mathbf{b}=\mathbf{a}$, show that

(a) div (curl $\mathbf{a}\wedge$ curl $\mathbf{b})=\mathbf{a}^2-\mathbf{b}^2$,

(b) $\nabla^2\mathbf{a}+\mathbf{a}=\mathbf{0}$, $\nabla^2\mathbf{b}+\mathbf{b}=\mathbf{0}$.

(ii) If θ, φ, ψ are suitable functions of position, prove that

$$\int_V(\psi\nabla\theta\cdot\nabla\varphi)\,dV=\int_S(\varphi\psi\nabla\theta)\cdot d\mathbf{S}-\int_V\left\{\varphi\nabla\cdot(\psi\nabla\theta)\right\}dV$$

$$=\int_S(\theta\psi\nabla\varphi)\cdot d\mathbf{S}-\int_V\left\{\theta\nabla\cdot(\psi\nabla\varphi)\right\}dV,$$

where S is a simple closed surface bounding a volume V.

(Battersea College of Technology, Dip. Tech. I, 1964)

16. Using Green's theorem that

$$\iint_S\left(U\frac{\partial V}{\partial n}-V\frac{\partial U}{\partial n}\right)dS=\iiint_\tau(U\nabla^2V-V\nabla^2U)\,d\tau,$$

show that, if P is any point of a volume τ bounded by a surface S,

$$4\pi V_P=\iint_S\left[\frac{1}{R}\frac{\partial V}{\partial n}-V\frac{\partial}{\partial n}\left(\frac{1}{R}\right)\right]dS-\iiint_\tau\frac{1}{R}\nabla^2V\,d\tau,$$

where V is a scalar function of position satisfying conditions which should be stated, V_P is the value of V at P, and R is the distance from P to a variable point of c. Hence (i) show that

$$4\pi=-\iint_S\frac{\partial}{\partial n}\left(\frac{1}{R}\right)dS,$$

(ii) determine V as a function of position inside a sphere of unit radius, given that $\nabla^2 V = -6$ everywhere inside the sphere and that everywhere on the boundary of the sphere $V=4$, $\partial V/\partial n = -2$.

(London Univ., B.Sc. Spec. Hons., 1954)

17. (i) Assuming the divergence theorem, prove Green's theorem that

$$\iiint_\tau (\nabla U \cdot \nabla V + U\nabla^2 V)\, d\tau = \iint_S U \frac{\partial V}{\partial n} dS.$$

Given two single-valued functions of position, U and V, whose second partial derivatives are continuous in a connected volume τ and on its boundary surface S, show that, if U and V are harmonic in τ, and if $\partial U/\partial n = \partial V/\partial n$ on S, then $(U - V)$ is constant in τ.

(ii) Using spherical polar coordinates, find a harmonic function U, finite at the origin, such that $\partial U/\partial r = 1 - 3\cos^2\theta$ on the spherical surface $r=2$.

(London Univ., B.Sc. Spec. Hons., 1955)

18. A volume τ is bounded externally by a sphere S, of radius a and centre C, and internally by a sphere S', of centre C', which lies wholly inside S. State Green's theorem and deduce that

$$\int_S \left(U\frac{\partial V}{\partial n} - V\frac{\partial U}{\partial n}\right) dS + \int_{S'} \left(U\frac{\partial V}{\partial n} - V\frac{\partial U}{\partial n}\right) dS = 0$$

where U and V are functions of position such that $\nabla^2 U = \nabla^2 V = 0$ at all points of τ and $\partial/\partial n$ denotes differentiation along the normal drawn outwards from τ.

A general point P inside S is at a distance R from C' and at a distance R' from the inverse point of C' with regard to S; if $U = (1/R) - (a/bR')$, where $b = CC'$, prove that $\nabla^2 U = 0$ at all points inside S (except C') and that, if P is on S,

$$U = 0, \quad \frac{\partial U}{\partial n} = -\frac{a^2 - b^2}{aR^3}.$$

Hence show that

$$\frac{a^2 - b^2}{a} \int_S \frac{V dS}{R^3} + \int_{S'} \left(U\frac{\partial V}{\partial n} - V\frac{\partial V}{\partial n}\right) dS = 0,$$

and derive the limiting form of this equation as the radius of S' tends to zero.

(London Univ., B.Sc. Spec. Hons., 1958)

19. If x_s are orthogonal curvilinear coordinates, $h_s dx_s$ the corresponding elements of length, whose directions are given by unit vectors $\mathbf{i}_s$ $(s = 1, 2, 3)$, and if v_s are the corresponding components of a vector $\mathbf{v}$, show that

$$h_1h_2h_3 \text{ curl } \mathbf{v} = \begin{vmatrix} h_1\mathbf{i}_1 & h_2\mathbf{i}_2 & h_3\mathbf{i}_3 \\ \dfrac{\partial}{\partial x_1} & \dfrac{\partial}{\partial x_2} & \dfrac{\partial}{\partial x_3} \\ h_1v_1 & h_2v_2 & h_3v_3 \end{vmatrix}.$$

Hence, or otherwise, show that, in spherical polar coordinates (r, θ, φ),

$$\text{curl } (\cos\theta \text{ grad } \varphi) = \text{grad } (1/r).$$

(Durham Univ., Gen. B.Sc. Final, 1963)

20. A system of coordinates $(\rho_1, \rho_2, \varphi)$ is related to Cartesian coordinates (x, y, z) by the equations

$$x = \rho_1\rho_2^2\sigma \cos\varphi, \quad y = \rho_1\rho_2^2\sigma \sin\varphi, \quad z = \rho_1^2\rho_2\sigma,$$

where $\sigma = (\rho_1^2 + \rho_2^2)^{-1}$.

Examine the nature of the surfaces $\rho_1 = \text{const.}$, $\rho_2 = \text{const.}$, and $\varphi = \text{const.}$, and show that they form a triply-orthogonal system.

Transform the equation

$$\frac{\partial^2 V}{\partial x^2} + \frac{\partial^2 V}{\partial y^2} + \frac{\partial^2 V}{\partial z^2} = 0$$

into one having ρ_1, ρ_2 and φ as independent variables.

(Imperial Coll. B.Sc. Spec. Hons., 1948)

21. When a highly conducting homogeneous fluid moves steadily with velocity $\mathbf{q}$ in a magnetic field of intensity $\mathbf{H}$, it is known from magnetohydrodynamics that $\nabla \wedge (\mathbf{q} \wedge \mathbf{H}) = 0$. For axi-symmetric incompressible flow, with z-axis as axis of symmetry and (R, θ, z) as cylindrical polar coordinates, taking

$$\mathbf{q} = U(R, z)\hat{\mathbf{R}} + V(R, z)\mathbf{k}, \quad \mathbf{H} = \mathbf{H}(R, z)\hat{\boldsymbol{\theta}},$$

show, by forming $\nabla \wedge (\mathbf{q} \wedge \mathbf{H})$, that

$$\frac{\partial}{\partial R}(HU) + \frac{\partial}{\partial z}(HV) = 0.$$

Verify that if there exists a scalar function $\psi = \psi(R, z)$ such that

$$U = -\frac{1}{H}\frac{\partial \psi}{\partial z}, \quad V = \frac{1}{H}\frac{\partial \psi}{\partial R}$$

then the last equation is satisfied and by using the equation of continuity for incompressible flow in the form $\nabla \cdot \mathbf{q} = 0$, show that

$$\frac{\partial(\psi, R/H)}{\partial(R, z)} = 0.$$

(Univ. of Aston in Birmingham, Tech. Maths. II, 1966)

22. Write down the components of the curl of the vector field **A** using orthogonal curvilinear coordinates u, v, w.

A system of orthogonal curvilinear coordinates r, θ, φ is related to rectangular Cartesian coordinates x, y, z by the equations

$$x=(c+r\cos\theta)\cos\varphi, \quad y=(c+r\cos\theta)\sin\varphi, \quad z=r\sin\theta,$$

where c is a positive constant. If the vector **A** has components A_1, A_2, A_3 with respect to the coordinates r, θ, φ; if $A_1=A_2=\partial A_3/\partial\varphi=0$, and curl curl $\mathbf{A}=0$, show that A_3 satisfies the differential equation

$$r\frac{\partial^2 A_3}{\partial r^2}+\frac{1}{r}\frac{\partial^2 A_3}{\partial\theta^2}-\left(\frac{\sin\varphi}{c+r\cos\theta}\right)\frac{\partial A_3}{\partial\theta}+\left(\frac{c+2r\cos\theta}{c+r\cos\theta}\right)\frac{\partial A_3}{\partial r} - \frac{r}{(c+r\cos\theta)^2}A_3=0.$$

(Battersea College of Technology, Dip. Tech. I, 1964)

23. If a_{ij} is a second-order tensor, show that $a_{ij}a_{jk}a_{kl}a_{li}$ is invariant under a rotation of orthogonal axes.

24. If a_{ij} and b_{ij} are both second-order tensors, prove

(i) $a_{ik}b_{kj}$ is a second-order tensor;

(ii) $a_{ij}b_{ji}$ is invariant under a rotation of orthogonal axes.

25. If u_i is a first-order tensor and a_{ijk} a third-order one, show that their contracted product u_ia_{ijk} transforms as a second-order Cartesian tensor under an orthogonal transformation.

2

KINEMATICS OF FLUIDS IN MOTION

2.1 Real Fluids and Ideal Fluids

Fluid dynamics is the science treating the study of fluids in motion. By the term *fluid* is meant a substance that flows: one which does not is termed a *solid*. Fluids may be divided into *two* kinds: (i) *liquids* which are *incompressible*, i.e. their volumes do not change when the pressure changes, and (ii) *gases* which are *compressible* fluids suffering change in volume whenever the pressure changes. There are no sharp distinctions between the three states of matter, however. The term *hydrodynamics* is often applied to the science of moving incompressible fluids.

When matter is subjected to examination on the microscopic or molecular scale, it is found to consist of molecules in random motion and separated from one another by distances which are at least comparable with molecular size. In the case of gases, the separation distances are great: in the case of liquids, they are less great and in the case of solids even less so.

For the purpose of macroscopic analysis, however, the molecular structure of matter is, in general, of no interest. It is thus more convenient to treat the fluid as having continuous structure so that at each point we can prescribe a unique velocity, a unique pressure, a unique density, etc. Moreover, for a continuous or ideal fluid we can define a *fluid particle* as the fluid contained within an infinitesimal volume whose size is so small that it may be regarded as a geometrical point.

We conclude this introductory section by mentioning briefly the natures of the different types of force which are called into play in moving fluids. Suppose two fluid particles, moving at different velocities, have a common boundary. Then across the boundary there will be interchange of momentum. The normal transport of molecules across the boundary will lead to a *direct* or *normal* force. In the case of a *viscous fluid* there is friction between the particles: this will manifest itself in the form of equal and opposite tangential or shearing forces on each particle at the common boundary. In the case of *inviscid fluids*, however, there is no friction and consequently there are no tangential or shearing forces. All real fluids exhibit viscosity but in many cases, such as arise when the rates of variation of fluid velocity with distances are small, viscous effects may be ignored.

2.2 Velocity of a Fluid at a Point

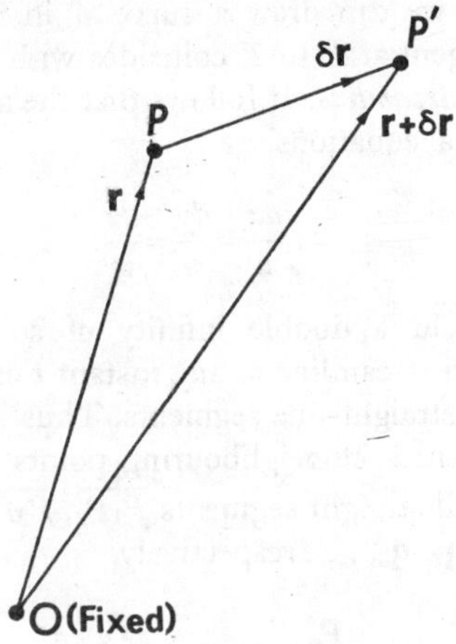

FIG. 2.1

At time t a fluid particle is at P where $\overline{OP}\equiv r$ and at time $(t+\delta t)$ the same particle has reached P' where $\overline{OP'}\equiv \mathbf{r}+\delta\mathbf{r}$. Then in the interval δt the movement of the particle is $\overline{PP'}\equiv\delta\mathbf{r}$ and so the particle velocity $\mathbf{q}$ at P is

$$\mathbf{q}=\lim_{\delta t\to 0}\left(\frac{\delta r}{\delta t}\right)=\frac{dr}{dt}$$

assuming such a limit to exist uniquely. This assumption is reasonable if we postulate that the fluid is continuous. Clearly $\mathbf{q}$ is in general dependent on both $\mathbf{r}$ and t so that we may write

$$\mathbf{q}=\mathbf{q}(\mathbf{r},\, t).$$

Alternatively, if P has Cartesian coordinates $(x,\, y,\, z)$ relative to a fixed tri-rectangular coordinate frame through O, then we may write

$$\mathbf{q}=\mathbf{q}(x,\, y,\, z,\, t).$$

Let us further suppose that $[u,\, v,\, w]$ are the Cartesian components of $\mathbf{q}$ in this frame. Then

$$\mathbf{q}=u\mathbf{i}+v\mathbf{j}+w\mathbf{k}.$$

Moreover, since $\mathbf{r}=x\mathbf{i}+y\mathbf{j}+z\mathbf{k}$,

$$\mathbf{q}=\frac{dx}{dt}\mathbf{i}+\frac{dy}{dt}\mathbf{j}+\frac{dz}{dt}\mathbf{k},$$

so that $u=dx/dt$, $v=dy/dt$, $w=dz/dt$.

2.3 Streamlines and Pathlines; Steady and Unsteady Flows

Suppose that at any time t we know the velocity $\mathbf{q}=[u, v, w]$ at each point $P(x, y, z)$ of the fluid. Further, suppose that at each such point P at any given instant t we can draw a curve $\mathscr{C}$ in the fluid such that the direction of the tangent at P to $\mathscr{C}$ coincides with the direction of $\mathbf{q}$ at P. Then $\mathscr{C}$ is termed a *streamline*. It follows that the streamlines are the solutions of the differential equations

$$\frac{dx}{u}=\frac{dy}{v}=\frac{dz}{w}. \tag{1}$$

These equations yield a double infinity of solutions. Knowing $\mathbf{q}$ at successive points of a streamline at any instant t enables the streamline to be approximated by straight-line segments. Thus in Fig. 2.2, if $\mathbf{q}_1$, $\mathbf{q}_2$, $\mathbf{q}_3$, ... denote the velocities at neighbouring points P_1, P_2, P_3, ... of the streamline, then small straight segments $\overline{P_1P_2}$, $\overline{P_2P_3}$, $\overline{P_3P_4}$, ... are drawn in the directions $\mathbf{q}_1$, $\mathbf{q}_2$, $\mathbf{q}_3$, ... respectively.

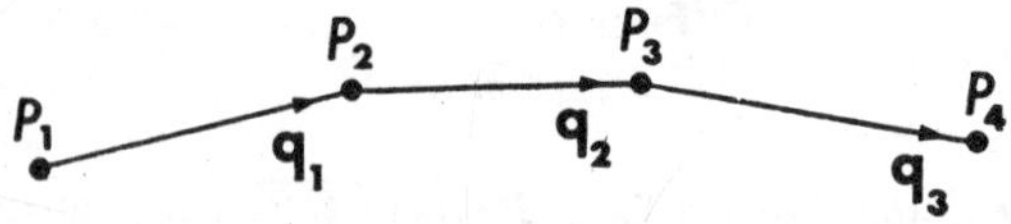

FIG. 2.2

When the motion is *steady* so that the pattern of flow does not vary with time, the paths of the fluid particles coincide with the streamlines. In unsteady motion, however, the flow pattern varies with time and the paths of the particles do not coincide with the streamlines, though the streamline through any point P does touch the *pathline* through P. The pathlines are the solutions of the differential equations

$$\frac{dx}{dt}=u, \quad \frac{dy}{dt}=v, \quad \frac{dz}{dt}=w. \tag{2}$$

These equations have a triply-infinite set of solutions.

A time exposure may be used to photograph the streamlines for steady flow, but a snapshot must be used for unsteady flow. In the case of a liquid the particles are illuminated by a suspension of aluminium dust in the liquid. Alternatively the fluid particles may be rendered visible by means of a few crystals of potassium permanganate. For a gas, smoke streams may be used.

If we draw the streamlines through every point of a closed curve Γ in the fluid we obtain a *stream tube*, as in Fig. 2.3.

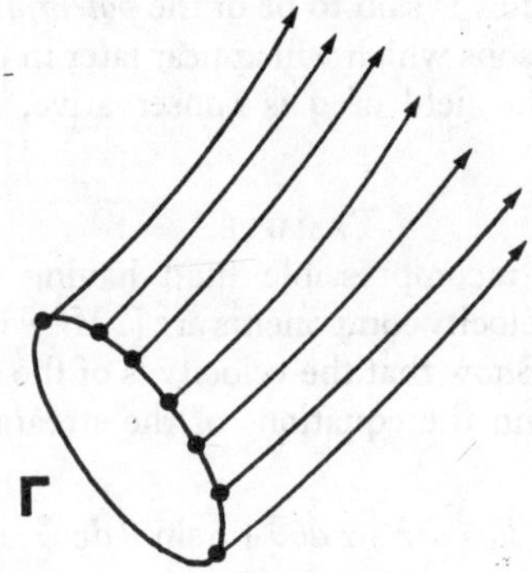

FIG. 2.3

2.4 The Velocity Potential

When the fluid velocity at time t is $\mathbf{q}=[u, v, w]$ in Cartesians, the equations of the streamlines at that instant are

$$\frac{dx}{u}=\frac{dy}{v}=\frac{dz}{w}. \tag{1}$$

These curves cut the surfaces

$$u\,dx+v\,dy+w\,dz=0 \tag{2}$$

orthogonally. Now suppose that at the considered instant t, we can find a scalar function $\varphi(x, y, z, t)$, uniform throughout the entire field of flow and such that

$$-d\varphi=u\,dx+v\,dy+w\,dz. \tag{3}$$

Then the expression on the R.H.S. of (3) is an exact differential and

$$\left.\begin{aligned} u=-\frac{\partial\varphi}{\partial x}, \quad v=-\frac{\partial\varphi}{\partial y}, \quad w=-\frac{\partial\varphi}{\partial z} \\ \text{or} \qquad \mathbf{q}=-\nabla\varphi. \end{aligned}\right\} \tag{4}$$

φ is termed the *velocity potential*. By the elements of Chapter 1 (Section 1.18) the necessary and sufficient condition for (4) to hold is

$$\text{curl}\,\mathbf{q}=\mathbf{0}. \tag{5}$$

The surfaces

$$\varphi(x, y, z, t)=\text{const.} \tag{6}$$

are called *equipotentials*. Equations (1) and (2) show that at all points of the field of flow the *equipotentials are cut orthogonally by the streamlines*.

The negative sign in the equation $\mathbf{q} = -\nabla\varphi$ is a convention. It ensures that flow takes place from the higher to the lower potentials, but some writers adopt the opposite convention.

When (5) holds, the flow is said to be of the *potential* kind. It is also said to be *irrotational* for reasons which will appear later in this chapter (Section 2.11). For such flow the field of $\mathbf{q}$ is conservative, and $\mathbf{q}$ is a *lamellar* vector.

EXAMPLE

At the point in an incompressible fluid having spherical polar coordinates (r, θ, ψ), the velocity components are $[2Mr^{-3}\cos\theta, Mr^{-3}\sin\theta, 0]$, where M is a constant. Show that the velocity is of the potential kind. Find the velocity potential and the equations of the streamlines.

Taking

$$d\mathbf{s} = dr\,\hat{\mathbf{r}} + r\,d\theta\,\hat{\boldsymbol{\theta}} + r\sin\theta\,d\psi\,\hat{\boldsymbol{\psi}}$$

$$\mathbf{q} = 2Mr^{-3}\cos\theta\,\hat{\mathbf{r}} + Mr^{-3}\sin\theta\,\hat{\boldsymbol{\theta}},$$

we obtain

$$\text{curl}\,\mathbf{q} = \frac{1}{r^2\sin\theta}\begin{vmatrix} \hat{\mathbf{r}} & r\hat{\boldsymbol{\theta}} & r\sin\theta\,\hat{\boldsymbol{\psi}} \\ \dfrac{\partial}{\partial r} & \dfrac{\partial}{\partial\theta} & \dfrac{\partial}{\partial\psi} \\ 2Mr^{-3}\cos\theta & Mr^{-2}\sin\theta & 0 \end{vmatrix} = \mathbf{0}.$$

Thus the flow is of the potential kind.

Let $\varphi(r, \theta, \psi)$ be the appropriate velocity potential. Then

$$-\frac{\partial\varphi}{\partial r} = 2Mr^{-3}\cos\theta, \quad -\frac{\partial\varphi}{r\partial\theta} = Mr^{-3}\sin\theta, \quad \frac{\partial\varphi}{r\sin\theta\partial\psi} = 0.$$

$$d\varphi = \frac{\partial\varphi}{\partial r}dr + \frac{\partial\varphi}{\partial\theta}d\theta + \frac{\partial\varphi}{\partial\psi}d\psi$$

$$= -(2Mr^{-3}\cos\theta)\,dr - (Mr^{-2}\sin\theta)\,d\theta + 0\,d\psi.$$

Thus

$$\underline{\varphi = Mr^{-2}\cos\theta.}$$

The streamlines are given by

$$\frac{dr}{2Mr^{-3}\cos\theta} = \frac{r\,d\theta}{Mr^{-3}\sin\theta} = \frac{r\sin\theta\,d\psi}{0},$$

or

$$d\psi = 0$$

$$2\cot\theta\,d\theta = (1/r)\,dr.$$

Integrating, the equations of the streamlines are

$$\begin{cases} \psi = \text{const.} \\ \underline{r = A\sin^2\theta.} \end{cases}$$

The equations $\psi=$const. show that the streamlines lie in planes which pass through the axis of symmetry $\theta=0$.

2.5 The Vorticity Vector

We now consider flows for which curl $\mathbf{q}\neq\mathbf{0}$. The vector

$$\boldsymbol{\zeta}=\nabla\wedge\mathbf{q} \tag{1}$$

is called the *vorticity vector*. The necessary and sufficient condition for potential flow may be expressed by $\boldsymbol{\zeta}=\mathbf{0}$. A *vortex line* is a curve drawn in the fluid such that the tangent to it at every point is in the direction of the vorticity vector $\boldsymbol{\zeta}$. If the Cartesian components of $\boldsymbol{\zeta}$ are $[\zeta_1, \zeta_2, \zeta_3]$, then the equations of the vortex lines are given by

$$\frac{dx}{\zeta_1}=\frac{dy}{\zeta_2}=\frac{dz}{\zeta_3}. \tag{2}$$

In general these do not coincide with the streamlines.

As in Section 2.3, Fig. 2.3, we can draw the vortex lines through all points of a closed curve Γ to form a *vortex tube*. Let δS_1, δS_2 be two sections of a vortex tube and let $\mathbf{n}_1$, $\mathbf{n}_2$ be the unit normals to these sections drawn outwards from the fluid between them. Also, let δS be the curved surface of the vortex tube, $\Delta S=\delta S_1+\delta S+\delta S_2=$total surface area of element, and $\Delta v=$total volume which ΔS contains. Then

$$\int_{\Delta S}\mathbf{n}\cdot\boldsymbol{\zeta}\,dS=\int_{\Delta v}\nabla\cdot\boldsymbol{\zeta}\,dv=0,$$

since $\nabla\cdot\boldsymbol{\zeta}=\nabla\cdot(\nabla\wedge\mathbf{q})\equiv 0$. Thus

$$\int_{\delta S_1}\mathbf{n}\cdot\boldsymbol{\zeta}\,dS+\int_{\delta S}\mathbf{n}\cdot\boldsymbol{\zeta}\,dS+\int_{\delta S_2}\mathbf{n}\cdot\boldsymbol{\zeta}\,dS=0.$$

At each point of δS, $\mathbf{n}\cdot\boldsymbol{\zeta}=0$, since $\boldsymbol{\zeta}$ is tangential to the curved surface. Thus, to the first order, the last equation gives

$$(\mathbf{n}_1\cdot\boldsymbol{\zeta}_1)\delta S_1+(\mathbf{n}_2\cdot\boldsymbol{\zeta}_2)\delta S_2=0. \tag{3}$$

Equation (3) shows that $|\mathbf{n}\cdot\boldsymbol{\zeta}|\,\delta S$ is constant over every section δS of the vortex tube. Its value is called the *strength* of the vortex tube. A vortex tube whose strength is unity is called a *unit vortex tube*.

Now suppose S is any closed surface containing a volume V. Then

$$\int_S\mathbf{n}\cdot\boldsymbol{\zeta}\,dS=\int_V\nabla\cdot\boldsymbol{\zeta}\,dv=0. \tag{4}$$

Equation (4) shows that the total strength of vortex tubes emerging from S is equal to that entering S. This means that *vortex lines and tubes cannot originate or terminate at internal points in a fluid.* They can only form closed curves or terminate on boundaries. In the case of smoke rings, the vortex lines form closed curves. On the other hand, the vortex lines in a whirlpool terminate on the boundary of the fluid.

If $\mathscr{C}$ is a closed curve drawn in a moving fluid and if S denotes an area contained by $\mathscr{C}$, then the *circulation* Γ of the fluid velocity $\mathbf{q}$ is defined to be

$$\Gamma = \oint_{\mathscr{C}} \mathbf{q} \cdot d\mathbf{s} = \int_S \mathbf{n} \cdot \boldsymbol{\zeta}\, dS$$

since $\boldsymbol{\zeta} = \text{curl}\, \mathbf{q}$. For potential flow the circulation round any closed circuit is clearly zero. We shall later obtain wider conditions under which Γ is constant (Section 3.12).

2.6 Local and Particle Rates of Change

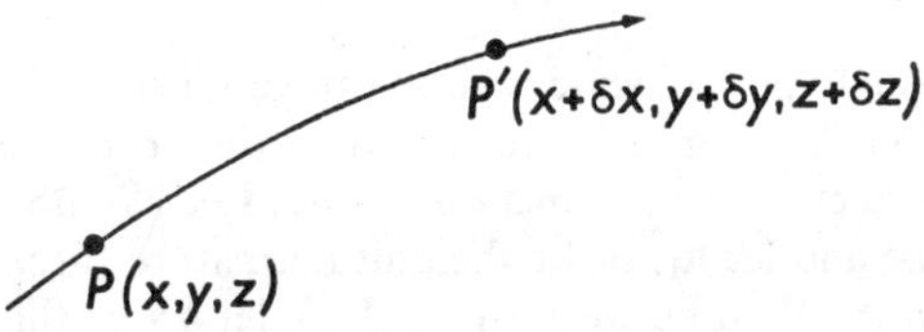

FIG. 2.4

Suppose a particle of fluid moves from $P(x, y, z)$ at time t to $P'(x+\delta x, y+\delta y, z+\delta z)$ at time $t+\delta t$. Let $f(x, y, z, t)$ be a scalar function associated with some property of the fluid (e.g. the density). Then in the motion of the particle from P to P' the total change of f is given by

$$\delta f = \frac{\partial f}{\partial x}\delta x + \frac{\partial f}{\partial y}\delta y + \frac{\partial f}{\partial z}\delta z + \frac{\partial f}{\partial t}\delta t.$$

Thus the total rate of change of f at the point P at time t in the motion of the particle is

$$\frac{df}{dt} = \lim_{\delta t \to 0}\left(\frac{\delta f}{\delta t}\right) = \frac{\partial f}{\partial x}\frac{dx}{dt} + \frac{\partial f}{\partial y}\frac{dy}{dt} + \frac{\partial f}{\partial z}\frac{dz}{dt} + \frac{\partial f}{\partial t},$$

$$= u\frac{\partial f}{\partial x} + v\frac{\partial f}{\partial y} + w\frac{\partial f}{\partial z} + \frac{\partial f}{\partial t},$$

if $\mathbf{q} = [u, v, w]$ is the velocity of the fluid particle at P. Thus we may write

$$\frac{df}{dt} = \mathbf{q} \cdot \nabla f + \frac{\partial f}{\partial t}. \tag{1}$$

Similarly, for a vector function $\mathbf{F}(x, y, z, t)$ associated with some property of the fluid (e.g. its velocity) we can show that

$$\frac{d\mathbf{F}}{dt}=\mathbf{q}\cdot\nabla\mathbf{F}+\frac{\partial\mathbf{F}}{\partial t}. \tag{2}$$

Thus for both scalar and vector functions we have established the operational equivalence

$$\frac{d}{dt}\equiv\mathbf{q}\cdot\nabla+\frac{\partial}{\partial t}, \tag{3}$$

applicable to both scalar and vector functions of position and time, provided that these functions are associated with properties of the moving fluid.

In obtaining (1) and (2) we are considering the total change in f or $\mathbf{F}$ when the fluid particle moves from $P(x, y, z)$ to $P'(x+\delta x, y+\delta y, z+\delta z)$ in time δt. Thus df/dt, $d\mathbf{F}/dt$ are total time differentiations following the fluid particle and are called the *particle rates of change.* On the other hand the partial time derivatives $\partial f/\partial t$, $\partial\mathbf{F}/\partial t$ are only the time rates of change at the point $P(x, y, z)$ considered fixed in space: they are the *local rates of change.* It follows that $\mathbf{q}\cdot\nabla f$ or $\mathbf{q}\cdot\nabla\mathbf{F}$ represents the rate of change due solely to the motion of the particle along its path. This point may also be seen by denoting the arc length of the path by s and PP' by δs. Then if $\overline{PP'}\equiv\delta s\hat{\mathbf{s}}$, $\mathbf{q}=q\delta\hat{\mathbf{s}}$, where $q=|\mathbf{q}|$, and so

$$\mathbf{q}\cdot\nabla f=q\delta\hat{\mathbf{s}}\cdot\nabla f=q\partial f/\partial s,$$

with similar results for the vector function $\mathbf{F}$. Here we are using $\delta\hat{\mathbf{s}}\cdot\nabla\equiv\partial/\partial s$.

2.7 The Equation of Continuity

When a region of a fluid contains neither sources nor sinks, that is to say when there are no inlets or outlets through which fluid can enter or leave the region, the amount of fluid within the region is conserved in accordance with the principle of conservation of matter. We now attempt to formulate this principle mathematically by means of the so-called *equation of continuity.*

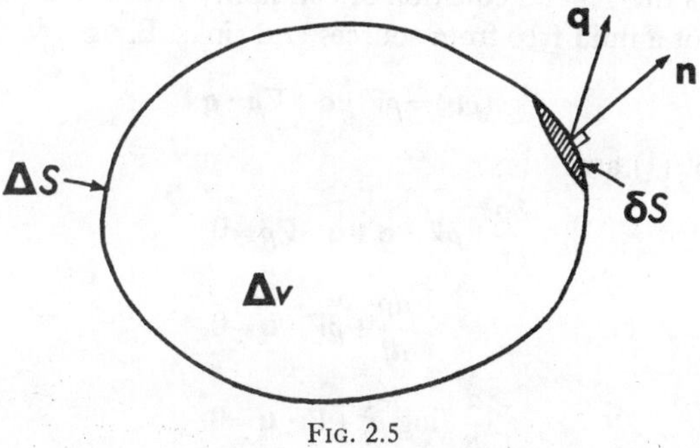

FIG. 2.5

Let ΔS be a closed surface drawn in the fluid and taken *fixed in space.* Suppose it contains a volume Δv of the fluid and let $\rho=\rho(x, y, z, t)$ be the fluid density (mass per unit volume) at any point (x, y, z) of the fluid in Δv at any time t. Suppose $\mathbf{n}$ is the unit outward-drawn normal at any surface element δS of ΔS, where $\delta S \ll \Delta S$. Then if $\mathbf{q}$ is the fluid velocity at the element δS, the normal component of $\mathbf{q}$ measured outwards from Δv is $\mathbf{n}\cdot\mathbf{q}$. Thus

Rate of efflux of fluid mass per unit time across $\delta S=\rho\mathbf{n}\cdot\mathbf{q}\,\delta S$.

Total rate of mass flow out of Δv across $\Delta S=\int_{\Delta S}\rho\mathbf{n}\cdot\mathbf{q}\,dS$.

Total rate of mass flow *into* $\Delta v=-\int_{\Delta S}\mathbf{n}\cdot(\rho\mathbf{q})\,dS=-\int_{\Delta v}\nabla\cdot(\rho\mathbf{q})\,dv$.

At time t, the mass of fluid within the element is $\int_{\Delta v}\rho\,dv$.

Local rate of mass increase within $\Delta v=\dfrac{\partial}{\partial t}\int_{\Delta v}\rho\,dv=\int_{\Delta v}\dfrac{\partial\rho}{\partial t}\,dv$.

In the absence of sources and sinks within Δv, matter is not created or destroyed in this region so that

$$\int_{\Delta v}\frac{\partial\rho}{\partial t}\,dv=-\int_{\Delta v}\nabla\cdot(\rho\mathbf{q})\,dv$$

or

$$\int_{\Delta v}\left\{\frac{\partial\rho}{\partial t}+\nabla\cdot(\rho\mathbf{q})\right\}dv=0.$$

This last relation is true for all volumes Δv if

$$\frac{\partial\rho}{\partial t}+\nabla\cdot(\rho\mathbf{q})=0. \tag{1}$$

Equation (1) is the general equation of continuity which must always hold at any points of a fluid free from sources and sinks. Since

$$\nabla\cdot(\rho\mathbf{q})=\rho\nabla\cdot\mathbf{q}+\nabla\rho\cdot\mathbf{q},$$

other forms of (1) are:

$$\frac{\partial\rho}{\partial t}+\rho\nabla\cdot\mathbf{q}+\mathbf{q}\cdot\nabla\rho=0, \tag{1'}$$

$$\frac{d\rho}{dt}+\rho\nabla\cdot\mathbf{q}=0, \tag{1''}$$

$$\frac{d}{dt}(\log\rho)+\nabla\cdot\mathbf{q}=0. \tag{1'''}$$

In the last two variants, d/dt denotes differentiation following the fluid motion and the operational equivalence $d/dt \equiv \partial/\partial t + \mathbf{q} \cdot \nabla$ has been used.

In the special case of *steady flow* in which the pattern does not vary with time at any location, $\partial \rho/\partial t = 0$ and (1) gives

$$\nabla \cdot (\rho \mathbf{q}) = 0. \tag{2}$$

For an incompressible fluid the density of any particle is invariable with time so that $d\rho/dt = 0$. In such a fluid there could be a variation of ρ from particle to particle as in the case of a non-homogeneous incompressible fluid. For a homogeneous and incompressible fluid ρ is constant throughout the entire fluid. In either case (1″) shows that the equation of continuity is

$$\nabla \cdot \mathbf{q} = 0. \tag{3}$$

From now onwards, unless otherwise stated, the term 'incompressible fluid' will be taken to imply one which is not only incompressible but also homogeneous. If in addition to equation (3), the flow is of the potential kind, then there exists a velocity potential φ such that $\mathbf{q} = -\nabla \varphi$ and (3) becomes

$$\nabla^2 \varphi = 0, \tag{4}$$

which is Laplace's equation.

2.8 Worked Examples

In illustration of the principles so far developed we consider some examples.

Example 1

Test whether the motion specified by

$$\mathbf{q} = \frac{k^2(x\mathbf{j} - y\mathbf{i})}{x^2 + y^2} \quad (k = \text{const.})$$

is a possible motion for an incompressible fluid. If so, determine the equations of the streamlines. Also test whether the motion is of the potential kind and if so determine the velocity potential.

We have

$$\nabla \cdot \mathbf{q} = k^2 \left\{ -\frac{\partial}{\partial x}\left(\frac{y}{x^2 + y^2}\right) + \frac{\partial}{\partial y}\left(\frac{x}{x^2 + y^2}\right) \right\}$$

$$= k^2 \left\{ \frac{2xy}{(x^2 + y^2)^2} - \frac{2xy}{(x^2 + y^2)^2} \right\} = 0.$$

Thus the equation of continuity for an incompressible fluid is satisfied and so such a motion is possible.

In the usual notation, the Cartesian components of velocity are

$$u=-\frac{k^2y}{x^2+y^2},\quad v=\frac{k^2x}{x^2+y^2},\quad w=0$$

and so the equations of the streamlines are

$$\frac{dx}{u}=\frac{dy}{v}=\frac{dz}{w}.$$

On substitution we find

$$\begin{cases} x\,dx+y\,dy=0,\\ dz=0,\end{cases}$$

which are immediately integrable to the forms

$$\begin{cases} x^2+y^2=\text{const.},\\ z=\text{const.}\end{cases}$$

Thus the streamlines are circles whose centres are on the z-axis, their planes being perpendicular to this axis.

Now

$$\nabla\wedge\mathbf{q}=\begin{Vmatrix} \dfrac{\partial}{\partial x} & \dfrac{\partial}{\partial y} & \dfrac{\partial}{\partial z}\\[2ex] \dfrac{-k^2y}{x^2+y^2} & \dfrac{k^2x}{x^2+y^2} & 0\end{Vmatrix}$$

$$=k^2\left\{\frac{y^2-x^2}{(x^2+y^2)^2}+\frac{x^2-y^2}{(x^2+y^2)^2}\right\}\mathbf{k}=\mathbf{0}.$$

Thus the flow is of the potential kind and we can find $\varphi(x, y, z)$ such that $\mathbf{q}=-\nabla\varphi$. We have

$$\frac{\partial\varphi}{\partial x}=-u=\frac{k^2y}{x^2+y^2},$$

$$\frac{\partial\varphi}{\partial y}=-v=-\frac{k^2x}{x^2+y^2},$$

$$\frac{\partial\varphi}{\partial z}=-w=0.$$

The last relation shows that $\varphi=\varphi(x, y)$. The first gives on integration

$$\varphi(x, y)=k^2\tan^{-1}(x/y)+f(y).$$

From this we find

$$\frac{\partial\varphi}{\partial y}=f'(y)-\frac{k^2x}{x^2+y^2}$$

so comparing this with $\partial\varphi/\partial y = -k^2x/(x^2+y^2)$, we see that

$$f'(y)=0$$

or

$$f(y)=\text{const.}$$

As the constant is immaterial, we may take

$$\varphi(x, y)=k^2 \tan^{-1}(x/y).$$

The equipotentials are thus given by the planes

$$x=cy,$$

through the z-axis. They are appropriately intersected orthogonally by the streamlines. Figure 2.6 shows the streamlines and equipotentials.

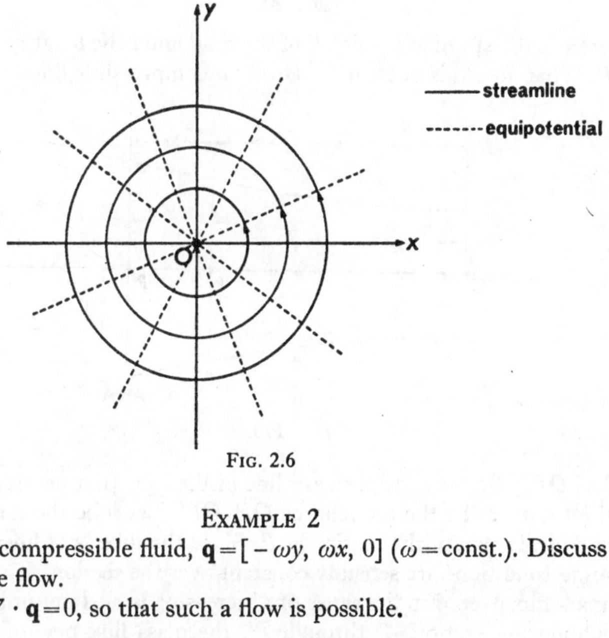

FIG. 2.6

EXAMPLE 2

For an incompressible fluid, $\mathbf{q}=[-\omega y, \omega x, 0]$ ($\omega=$const.). Discuss the nature of the flow.

We find $\nabla \cdot \mathbf{q}=0$, so that such a flow is possible.

Further,

$$\nabla \wedge \mathbf{q} = \begin{Vmatrix} \dfrac{\partial}{\partial x} & \dfrac{\partial}{\partial y} & \dfrac{\partial}{\partial z} \\ -\omega y & \omega x & 0 \end{Vmatrix} = 2\omega\mathbf{k}.$$

Thus the flow is not of the potential kind. It can easily be shown that a rigid body rotating about the z-axis with constant vector angular velocity $\omega\mathbf{k}$ gives the same type of motion. (For the velocity at (x, y, z) in the body is $-\omega y\mathbf{i}+\omega x\mathbf{j}$.)

The equations of the streamlines are

$$\frac{dx}{-\omega y}=\frac{dy}{\omega x}=\frac{dz}{0},$$

i.e. the streamlines are the circles

$$\begin{cases} x^2+y^2=\text{const.}, \\ z=\text{const.} \end{cases}$$

EXAMPLE 3

For a fluid moving in a fine tube of variable section A, prove from first principles that the equation of continuity is

$$A\frac{\partial\rho}{\partial t}+\frac{\partial}{\partial s}(A\rho v)=0$$

where v is the speed at a point P of the fluid and s the length of the tube up to P. What does this become for steady incompressible flow?

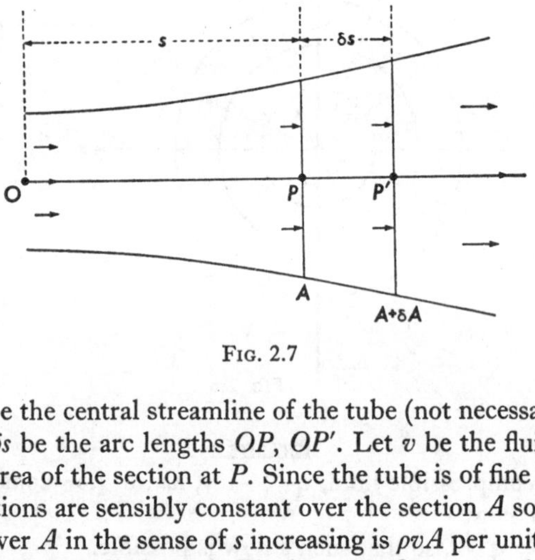

FIG. 2.7

Let OPP' be the central streamline of the tube (not necessarily straight) and let s, $s+\delta s$ be the arc lengths OP, OP'. Let v be the fluid velocity at P and A the area of the section at P. Since the tube is of fine bore we may assume conditions are sensibly constant over the section A so that the rate of mass flux over A in the sense of s increasing is ρvA per unit time. At the neighbouring section A' through P', the mass flux per unit time in the direction of s increasing is therefore

$$\rho vA+\delta s\frac{\partial}{\partial s}(\rho vA)$$

at the same instant of time t. Thus the net rate of flow of mass into the element between the sections A, $A+\delta A$ (considered fixed in space) is $-\delta s(\partial/\partial s)(\rho vA)$. But at time t, the mass between the sections is $\rho A\delta s$ and so its rate of increase is $(\partial/\partial t)(\rho A\delta s)$ or $(\partial\rho/\partial t)A\delta s$. In the absence of sources and sinks, then, we have

$$-\delta s\frac{\partial}{\partial s}(\rho vA)=\frac{\partial \rho}{\partial t}A\delta s,$$

or

$$\frac{A\partial\rho}{\partial t}+\frac{\partial}{\partial s}(\rho vA)=0.$$

For steady incompressible flow, ρ is everywhere constant and the equation reduces to

$$\frac{d}{ds}(vA)=0$$

or $vA=$const. over every section.
This last result means that the volume of fluid crossing every section per unit time is constant.

Example 4

Liquid flows through a pipe whose surface is the surface of revolution of the curve $y=a+kx^2/a$ about the x-axis $(-a\leqslant x\leqslant a)$. If the liquid enters at the end $x=-a$ of the pipe with velocity V, show that the time taken by a liquid particle to traverse the entire length of the pipe from $x=-a$ to $x=+a$ is

$$\{2a/V(1+k)^2\}(1+\tfrac{2}{3}k+\tfrac{1}{5}k^2).$$

[Assume that k is so small that the flow remains appreciably one-dimensional throughout.]

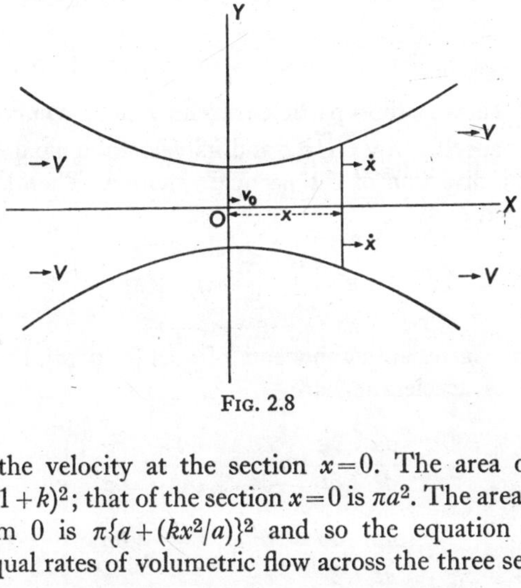

Fig. 2.8

Let v_0 be the velocity at the section $x=0$. The area of the section $x=-a$ is $\pi a^2(1+k)^2$; that of the section $x=0$ is πa^2. The area of the section distant x from 0 is $\pi\{a+(kx^2/a)\}^2$ and so the equation of continuity (expressing equal rates of volumetric flow across the three sections) is

$$\pi a^2(1+k)^2V=\pi a^2v_0=\pi\left\{a+\left(\frac{kx^2}{a}\right)\right\}^2\dot{x}$$

since $\dot{x}$ is the velocity across the plane distant x from 0. Then

$$dt=\left(1+\frac{kx^2}{a^2}\right)^2\frac{dx}{(1+k)^2V}.$$

Thus the required time of flow is

$$\frac{2}{V(1+k)^2}\int_0^a\left(1+\frac{kx^2}{a^2}\right)^2dx,$$

which gives the stated result.

2.9 Acceleration of a Fluid

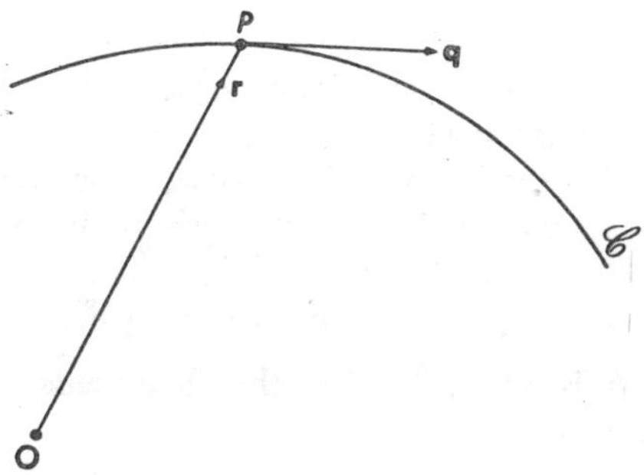

FIG. 2.9

Figure 2.9 shows a fluid particle travelling along a curve $\mathscr{C}$. At time t its position P is specified by $\overline{OP}\equiv\mathbf{r}$ and its velocity $\mathbf{q}$ along the tangent at P to $\mathscr{C}$ is in the direction of the particle's motion. Then the instantaneous acceleration $\mathbf{f}$ at P is

$$\mathbf{f}=\frac{d\mathbf{q}}{dt}=\frac{\partial\mathbf{q}}{\partial t}+(\mathbf{q}\cdot\nabla)\mathbf{q}. \tag{1}$$

Taking the Cartesian components of $\mathbf{q}$ as $[u, v, w]$, this shows that the components of acceleration are

$$\left.\begin{aligned}\frac{du}{dt}&=\frac{\partial u}{\partial t}+u\frac{\partial u}{\partial x}+v\frac{\partial u}{\partial y}+w\frac{\partial u}{\partial z},\\ \frac{dv}{dt}&=\frac{\partial v}{\partial t}+u\frac{\partial v}{\partial x}+v\frac{\partial v}{\partial y}+w\frac{\partial v}{\partial z},\\ \frac{dw}{dt}&=\frac{\partial w}{\partial t}+u\frac{\partial w}{\partial x}+v\frac{\partial w}{\partial y}+w\frac{\partial w}{\partial z}.\end{aligned}\right\} \tag{2}$$

In tensor form, with coordinates x_i and velocity components u_i ($i=1, 2, 3$), the set of equations (2) could be written

$$\frac{du_i}{dt}=\frac{\partial u_i}{\partial t}+u_j u_{i,j}. \tag{3}$$

Reverting to the vectorial form (1), the term $(\mathbf{q}\cdot\nabla)\mathbf{q}$ may be developed in a form more suitable for certain purposes. Thus

$$(\mathbf{q}\cdot\nabla)\mathbf{q}=\sum\left\{(\mathbf{q}\cdot\mathbf{i})\frac{\partial\mathbf{q}}{\partial x}\right\}.$$

Since

$$\mathbf{A}\wedge(\mathbf{B}\wedge\mathbf{C})=(\mathbf{A}\cdot\mathbf{C})\mathbf{B}-(\mathbf{A}\cdot\mathbf{B})\mathbf{C},$$
$$(\mathbf{A}\cdot\mathbf{B})\mathbf{C}=(\mathbf{A}\cdot\mathbf{C})\mathbf{B}-\mathbf{A}\wedge(\mathbf{B}\wedge\mathbf{C}),$$

so that

$$(\mathbf{q}\cdot\mathbf{i})\frac{\partial\mathbf{q}}{\partial x}=\left(\mathbf{q}\cdot\frac{\partial\mathbf{q}}{\partial x}\right)\mathbf{i}-\mathbf{q}\wedge\left(\mathbf{i}\wedge\frac{\partial\mathbf{q}}{\partial x}\right)$$

$$=\mathbf{i}\frac{\partial}{\partial x}\left(\frac{1}{2}\mathbf{q}^2\right)-\mathbf{q}\wedge\left(\mathbf{i}\wedge\frac{\partial\mathbf{q}}{\partial x}\right).$$

$$\therefore\ (\mathbf{q}\cdot\nabla)\mathbf{q}=\sum\mathbf{i}\frac{\partial}{\partial x}\left(\frac{1}{2}\mathbf{q}^2\right)-\mathbf{q}\wedge\sum\left(\mathbf{i}\wedge\frac{\partial\mathbf{q}}{\partial x}\right)$$

$$=\nabla(\tfrac{1}{2}\mathbf{q}^2)-\mathbf{q}\wedge(\nabla\wedge\mathbf{q}).$$

Hence (1) may be developed in the form

$$\mathbf{f}=\partial\mathbf{q}/\partial t+\nabla(\tfrac{1}{2}\mathbf{q}^2)-\mathbf{q}\wedge(\nabla\wedge\mathbf{q}). \tag{4}$$

(4) is especially useful for potential flow for which $\nabla\wedge\mathbf{q}=\mathbf{0}$. Also (4) may be more useful than (1) for general orthogonal curvilinear coordinates.

2.10 Conditions at a Rigid Boundary

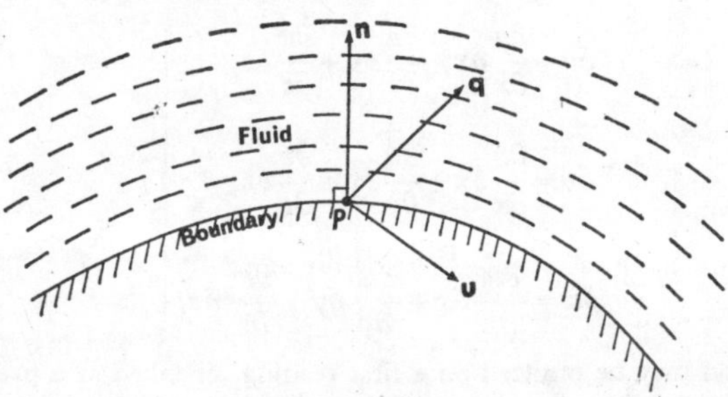

FIG. 2.10

P is a point on the boundary where the fluid velocity is $\mathbf{q}$ and where the boundary has a velocity $\mathbf{U}$. If $\mathbf{n}$ specifies a unit normal direction at P, then since there must be no relative normal velocity at P between boundary and fluid we must have the two normal components equal, i.e.

$$\mathbf{q} \cdot \mathbf{n} = \mathbf{U} \cdot \mathbf{n}.$$

In the case of inviscid fluid this is the only condition. For viscous fluid in which there is no slip, the tangential components must also be equal. In the case when the boundary is at rest (as is often the case), we require $\mathbf{q} \cdot \mathbf{n} = 0$ at every point of the boundary.

2.11* General Analysis of Fluid Motion

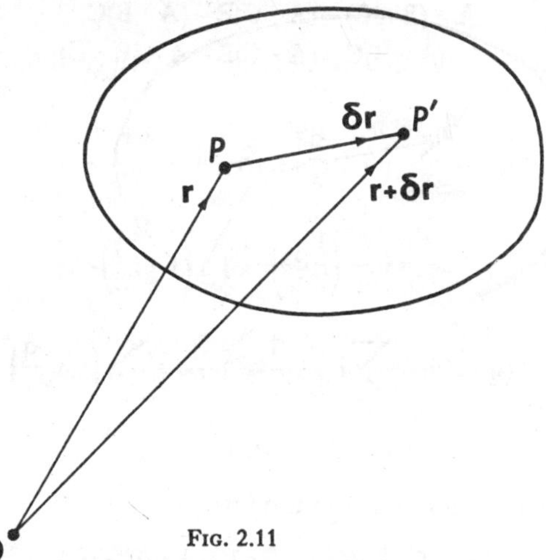

FIG. 2.11

Let P, P' be two neighbouring points in a fluid particle moving along with the fluid so that at any time t, $\overline{OP} \equiv \mathbf{r}$, $\overline{OP'} \equiv \mathbf{r} + \delta\mathbf{r}$ and P, P' have Cartesian coordinates (x, y, z), $(x + \delta x, y + \delta y, z + \delta z)$ respectively (Fig. 2.11). Further, let $\mathbf{q} = [u, v, w]$, $\mathbf{q} + \delta\mathbf{q} = [u + \delta u, v + \delta v, w + \delta w]$ be the velocities of P, P' at the considered instant t. Since $u = u(x, y, z, t)$, etc., and t is not varying here,

$$\left.\begin{aligned} \delta u &= \frac{\partial u}{\partial x}\delta x + \frac{\partial u}{\partial y}\delta y + \frac{\partial u}{\partial z}\delta z, \\ \delta v &= \frac{\partial v}{\partial x}\delta x + \frac{\partial v}{\partial y}\delta y + \frac{\partial v}{\partial z}\delta z, \\ \delta w &= \frac{\partial w}{\partial x}\delta x + \frac{\partial w}{\partial y}\delta y + \frac{\partial w}{\partial z}\delta z. \end{aligned}\right\} \qquad (1)$$

* This section may be omitted on a first reading, or taken as a prelude to the chapter on viscous flow.

These equations may be written in matrix form:

$$\begin{bmatrix} \delta u \\ \delta v \\ \delta w \end{bmatrix} = \begin{bmatrix} \dfrac{\partial u}{\partial x} & \dfrac{\partial u}{\partial y} & \dfrac{\partial u}{\partial z} \\ \dfrac{\partial v}{\partial x} & \dfrac{\partial v}{\partial y} & \dfrac{\partial v}{\partial z} \\ \dfrac{\partial w}{\partial x} & \dfrac{\partial w}{\partial y} & \dfrac{\partial w}{\partial z} \end{bmatrix} \begin{bmatrix} \delta x \\ \delta y \\ \delta z \end{bmatrix}. \qquad (2)$$

Let us now write

$$a=\frac{\partial u}{\partial x}, \qquad b=\frac{\partial v}{\partial y}, \qquad c=\frac{\partial w}{\partial z},$$

$$f=\frac{1}{2}\left(\frac{\partial w}{\partial y}+\frac{\partial v}{\partial z}\right), \quad g=\frac{1}{2}\left(\frac{\partial u}{\partial z}+\frac{\partial w}{\partial x}\right), \quad h=\frac{1}{2}\left(\frac{\partial v}{\partial x}+\frac{\partial u}{\partial y}\right),$$

$$\xi=\frac{1}{2}\left(\frac{\partial w}{\partial y}-\frac{\partial v}{\partial z}\right), \quad \eta=\frac{1}{2}\left(\frac{\partial u}{\partial z}-\frac{\partial w}{\partial x}\right), \quad \zeta=\frac{1}{2}\left(\frac{\partial v}{\partial x}-\frac{\partial u}{\partial y}\right).$$

Then

$$\underset{(=A)}{\begin{bmatrix} \dfrac{\partial u}{\partial x} & \dfrac{\partial u}{\partial y} & \dfrac{\partial u}{\partial z} \\ \dfrac{\partial v}{\partial x} & \dfrac{\partial v}{\partial y} & \dfrac{\partial v}{\partial z} \\ \dfrac{\partial w}{\partial x} & \dfrac{\partial w}{\partial y} & \dfrac{\partial w}{\partial z} \end{bmatrix}} = \underset{(=A_1)}{\begin{bmatrix} a & h & g \\ h & b & f \\ g & f & c \end{bmatrix}} + \underset{(=A_2)}{\begin{bmatrix} 0 & -\zeta & \eta \\ \zeta & 0 & -\xi \\ -\eta & \xi & 0 \end{bmatrix}}.$$

Thus the matrix A is expressed as the sum of a symmetric matrix A_1 and a skew symmetric one A_2. We now investigate the significance of these two matrices.

We first observe that

$$\begin{bmatrix} a & h & g \\ h & b & f \\ g & f & c \end{bmatrix} \begin{bmatrix} \delta x \\ \delta y \\ \delta z \end{bmatrix} = \begin{bmatrix} a\delta x+h\delta y+g\delta z \\ h\delta x+b\delta y+f\delta z \\ g\delta x+f\delta y+c\delta z \end{bmatrix}. \qquad (3)$$

We now consider the quadric surface

$$a(\delta x)^2+b(\delta y)^2+c(\delta z)^2+2f\delta y\delta z+2g\delta z\delta x+2h\delta x\delta y=\text{const.} \qquad (4)$$

in which $(\delta x, \delta y, \delta z)$ are taken as the current coordinates and the coefficients $a, b, \ldots, f, \ldots$ are constants with respect to these coordinates, so that (4) is a quadric whose centre is P. Then partial differentiation of the L.H.S. of (4) w.r.t. δx, δy, δz confirms that the column vector on the R.H.S. of (3) is normal to the quadric surface (4). Writing $\delta\mathbf{q}_1 = u_1\mathbf{i} + v_1\mathbf{j} + w_1\mathbf{k}$ where u_1, v_1, w_1 are the components of (3), this shows that $\delta\mathbf{q}_1$ is everywhere normal to the quadric surface (4) having P as centre. Such a velocity $\delta\mathbf{q}_1$ is ascribable to a motion which is called a *pure strain*. The quadric is called *the rate of strain quadric*. By a suitable rotation of coordinate axes, the equation of the quadric surface (4) may be reduced to the form

$$a'(\delta x')^2 + b'(\delta y')^2 + c'(\delta z')^2 = \text{const.}$$

in which the coordinate axes coincide with the principal axes of the quadric. The velocity components due to pure strain parallel to the axes of the quadric are respectively $[a'\delta x', b'\delta y', c'\delta z']$. Lines drawn parallel to the principal axes are undergoing elongations at uniform rates.

Next, we observe that

$$\begin{bmatrix} 0 & -\zeta & \eta \\ \zeta & 0 & -\xi \\ -\eta & \xi & 0 \end{bmatrix} \begin{bmatrix} \delta x \\ \delta y \\ \delta z \end{bmatrix} = \begin{bmatrix} \eta\delta z - \zeta\delta y \\ \zeta\delta x - \xi\delta z \\ \xi\delta y - \eta\delta x \end{bmatrix}. \tag{5}$$

Writing $\delta\mathbf{q}_2 = u_2\mathbf{i} + v_2\mathbf{j} + w_2\mathbf{k}$, where u_2, v_2, w_2 are the components of the matrix on the R.H.S. of (5), and also $\boldsymbol{\omega} = \xi\mathbf{i} + \eta\mathbf{j} + \zeta\mathbf{k}$, we see that

$$\delta\mathbf{q}_2 = \boldsymbol{\omega} \wedge \delta\mathbf{r}. \tag{6}$$

This shows $\delta\mathbf{q}_2$ is ascribable to a rigid-body rotation* with angular velocity $\boldsymbol{\omega}$. Further substitution for ξ, η, ζ shows that

$$\boldsymbol{\omega} = \tfrac{1}{2}\,\text{curl}\,\mathbf{q} = \tfrac{1}{2}\boldsymbol{\zeta} \tag{7}$$

where $\boldsymbol{\zeta}$ is the vorticity vector. Thus the vortex line through the fluid element is its instantaneous axis of rotation. This analysis justifies the terminology 'irrotational' for those flows for which curl $\mathbf{q} = \mathbf{0}$, leading as we have seen to $\mathbf{q} = -\nabla\varphi$.

We may now summarize the results of the foregoing analysis of the general motion of a fluid particle in the following way:

(i) There is a translational velocity $\mathbf{q}$ of any point P of it.

(ii) Relative to P there is a velocity component $\delta\mathbf{q}_1$ of P' which is ascribable to pure strain in the sense that for all other points P' of the particle the component $\delta\mathbf{q}_1$ is normal to a certain quadric surface centred on P. This type of motion requires that the fluid is deformable.

(iii) Relative to P there is a velocity component $\delta\mathbf{q}_2$ of P' which is ascribable to a rigid-body rotation of the element. This means that if the fluid in the particle were suddenly frozen, then the solid would be instantaneously rotating with vector angular velocity $\tfrac{1}{2}$ curl $\mathbf{q}$.

* See F. CHORLTON, *Textbook of Dynamics*, p. 36. Van Nostrand.

We note that the total velocity of P' relative to P is $\delta\mathbf{q}_1 + \delta\mathbf{q}_2$.

Similar analysis may be obtained more tersely using the methods of Cartesian tensor analysis. Thus if P has coordinates x_i and velocity components u_i and P' coordinates $x_i + \delta x_i$ and velocity components $u_i + \delta u_i$ ($i = 1, 2, 3$), then corresponding to equations (1),

$$\delta u_i = u_{i,j}\delta x_j$$

and

$$u_{i,j} = e_{ij} + \omega_{ij},$$

where

$$e_{ij} = \tfrac{1}{2}(u_{i,j} + u_{j,i}), \quad \omega_{ij} = \tfrac{1}{2}(u_{i,j} - u_{j,i}),$$

so that e_{ij} is a symmetric tensor of the second order and ω_{ij} is a skew-symmetric tensor of the second order. Thus if we write

$$\delta u_i = \delta u_i^{(1)} + \delta u_i^{(2)},$$

where

$$\delta u_i^{(1)} = e_{ij}\delta x_j,$$

$$\delta u_i^{(2)} = \omega_{ij}\delta x_j,$$

then $\delta u_i^{(1)}$ is seen to be normal to the quadric

$$e_{ij}\delta x_i \delta x_j = \text{const.},$$

which is the rate of strain quadric centred on P. Also, we have

$$[\delta u_1^{(2)}, \delta u_2^{(2)}, \delta u_3^{(2)}] = [(\omega_{13}\delta x_3 - \omega_{21}\delta x_2), (\omega_{21}\delta x_1 - \omega_{32}\delta x_3), (\omega_{32}\delta x_2 - \omega_{13}\delta x_1)]$$

$$= \begin{Vmatrix} \omega_{32} & \omega_{13} & \omega_{21} \\ \delta x_1 & \delta x_2 & \delta x_3 \end{Vmatrix},$$

which is the vector product of $[\omega_{32}, \omega_{13}, \omega_{21}]$ and $[\delta x_1, \delta x_2, \delta x_3]$, showing that the component $\delta u_i^{(2)}$ represents a rigid-body rotation. We note further that

$$\frac{\partial}{\partial x_m}(\omega_{ij}) = \frac{1}{2}\frac{\partial}{\partial x_m}\left(\frac{\partial u_i}{\partial x_j} - \frac{\partial u_j}{\partial x_i}\right)$$

$$= \frac{1}{2}\frac{\partial}{\partial x_j}\left(\frac{\partial u_i}{\partial x_m} + \frac{\partial u_m}{\partial x_i}\right) - \frac{1}{2}\frac{\partial}{\partial x_i}\left(\frac{\partial u_j}{\partial x_m} + \frac{\partial u_m}{\partial x_j}\right)$$

$$= \frac{\partial}{\partial x_j}(e_{im}) - \frac{\partial}{\partial x_i}(e_{jm})$$

$$= 0, \text{ if each } e_{ij} = 0.$$

This shows that if each $e_{ij} = 0$ everywhere, then the ω_{ij} are constants, i.e.

the rotation is the same everywhere. The vanishing of the e_{ij} is the necessary and sufficient condition that the fluid moves without strain.

EXERCISE 2

1. Show that a fluid of constant density can have a velocity $\mathbf{q}$ given by

$$\mathbf{q}=\left[-\frac{2xyz}{(x^2+y^2)^2}, \frac{(x^2+y^2)z}{(x^2+y^2)^2}, \frac{y}{x^2+y^2}\right].$$

Find the vorticity vector.

2. $H(\mathbf{r}, t)$ is any scalar or vector property of fluid and dH/dt is the rate of change of H for a fluid element moving with velocity $\mathbf{q}$. Prove that

$$\frac{dH}{dt}=\frac{\partial H}{\partial t}+(\mathbf{q}\cdot\nabla)H.$$

For a fluid of density ρ, establish the equation of continuity,

$$\frac{d\rho}{dt}+\rho\nabla\cdot\mathbf{q}=0.$$

Show that if $\mathbf{q}=m\mathbf{r}/r^3$, where m is a constant, the equation of continuity for an incompressible fluid is satisfied at all points other than the origin O. Show further that, if S is any closed surface not passing through O, the net volume of fluid flowing out of S per unit time is $4\pi m$ if S encloses O and zero if O is outside S.

(London Univ., Gen. B.Sc. II, 1962)

3. If $\int\mathbf{A}\cdot d\mathbf{s}=0$ for all closed curves in a region R show that there exists a scalar function V such that $\mathbf{A}=-\text{grad } V$.

Show that the kinetic energy of uniform incompressible fluid moving irrotationally in the finite region between surfaces S_1, S_2 is given by

$$\tfrac{1}{2}\rho\int_{S_1+S_2}(\mathbf{n}\cdot\text{grad }\varphi)\varphi\, dS,$$

where $\hat{\mathbf{n}}$ is a unit vector directed from the surface normally into the fluid and φ is the velocity potential. Hence or otherwise, show that, if S_1, S_2 are at rest, the fluid cannot move irrotationally.

(London Univ., Gen. B.Sc. II, 1965)

4. Derive the equation of continuity, in Cartesians, for the flow of a compressible fluid.

If the velocity is constant in magnitude everywhere in the irrotational, two-dimensional flow of an incompressible fluid, prove that the flow must be uniform in direction. Prove also the converse of this statement.

(Leeds Univ., B.Sc., 1961)

5. Obtain in the form div $\mathbf{q}=0$ the equation of continuity for a perfect incompressible fluid, $\mathbf{q}$ being the velocity.

In particular, if P is any point in the fluid and M is the foot of the perpendicular from P to the fixed axis Ox and if $\mathbf{q}$ has components U, parallel to Ox, and V, along MP, such that U and V are functions of $x(=OM)$, $R(=MP)$ and the time, show that the equation takes the form

$$\frac{\partial}{\partial x}(UR)+\frac{\partial}{\partial R}(VR)=0.$$

Deduce the existence of a stream function ψ such that

$$U=-\frac{1}{R}\frac{\partial\psi}{\partial R},\quad V=\frac{1}{R}\frac{\partial\psi}{\partial x}.$$

(London Univ., Gen. B.Sc. II, 1958)

6. Explain what is meant in hydrodynamics by a stagnation point.

Referred to Cartesian axes Ox, Oy, Oz, a semi-infinite incompressible inviscid fluid occupies the region $z>0$, and is bounded by a rigid plane at $z=0$. The fluid is in steady irrotational motion with a stagnation point at O. If the velocity potential φ is taken as zero at O, and it is assumed valid to expand φ in a power series of the form

$$\varphi=a_1x+a_2y+a_3z+a_{11}x^2+a_{22}y^2+a_{33}z^2+a_{12}xy+a_{23}yz+a_{31}zx+\ldots,$$

show that, with the axes Ox, Oy suitably orientated, the series is of the form

$$\varphi=Ax^2+By^2-(A+B)z^2+\text{terms of third order and higher.}$$

Discuss the form of the equipotential surfaces near the stagnation point for

(i) motion in the (x, z) plane, and
(ii) motion symmetrical about the axis Oz.

(London Univ., B.Sc. Spec. Hons., 1963)

(*N.B.* A stagnation point is one at which the velocity $\mathbf{q}=\mathbf{0}$. See Chapter 3, Section 3.9.)

7. Show that in the general motion of a fluid, the motion of a particle relative to a neighbouring particle consists of a pure strain compounded with a rigid body rotation.

If the velocity potential at any instant be λxyz, show that the velocity at any point $P'(x+\xi, y+\eta, z+\zeta)$ relative to the fluid at $P(x, y, z)$ is normal to the quadric surface

$$x\eta\zeta+y\zeta\xi+z\xi\eta=\text{const.},$$

with centre P, when ξ, η, ζ are small.

(Univ. of Aston in Birmingham, Tech. Maths. IV, 1963)

3

EQUATIONS OF MOTION OF A FLUID

3.1 Pressure at a Point in a Fluid at Rest

Let us first suppose that a fluid is at rest. Then at any point P within the fluid there is a *hydrostatic pressure* due to the density of the fluid. In order to explain the existence of this pressure, it is necessary to resort to molecular considerations. If at P we insert a small rigid plane area δA, then either side of δA will be repeatedly bombarded with molecules of fluid moving randomly in all directions. As the fluid molecules impinge on one side of the area, they transfer momentum to the area in a specified time, thereby exerting a force δF on the area. An equal and opposite force is exerted on the fluid by this side of the area. If we suppose the linear dimensions of δA to be large in comparison with molecular sizes, but small compared with the macroscopic fluid lengths, then δA is at rest relative to the main body of fluid, the only motion being that on the molecular scale which we have just considered. In these circumstances, irrespective of whether the fluid is viscous or inviscid, δF will act normally to δA. Moreover, the mean force per unit area exerted on the one side of δA will be $\delta F/\delta A$. Now suppose δA be made vanishingly small at P. Then if we were to replace our molecular model by a continuous fluid, it would be reasonable to suppose that at P the value of $\lim_{\delta A \to 0} (\delta F/\delta A)$ would exist uniquely. We shall suppose this to be so; in the last resort, the apparent contradiction between this and the molecular model of a fluid as consisting of discrete randomly-moving molecules must be resolved by experiment. In fact for a medium of very rare density—such as might be created in an attempt to evacuate a closed vessel with a highly efficient pump—the above supposition would probably break down. However, for liquids and for most gases the supposition is believed to be valid. In these circumstances, if p denotes the value of this unique limit, then p is called the *hydrostatic pressure* at P.

Before passing on to the case of pressures in fluids in motion, we state a few of the more important results in hydrostatics. These results may be found in any standard textbook on the subject. They are as follows:

(i) When the fluid is at rest, the pressure at an internal point P is the same in all directions. Thus in the above considerations, the orientation of δA is immaterial.

(ii) The pressure at all points at the same depth below the free horizontal surface of a fluid at rest is the same. At a point P distant h below the free surface, the pressure is $\Pi+\rho gh$, where Π is the atmospheric pressure at the free surface, ρ the fluid density and g the acceleration due to gravity.

(iii) If any pressure is applied to the free surface of a fluid, it is transmitted equally to all parts of the fluid.

These results constitute *Pascal's laws* for static fluids.

3.2 Pressure at a Point in a Moving Fluid

Let P be a point in an inviscid fluid moving locally with the velocity $\mathbf{q}$. As in the last section, insert a small rigid plane area δA into the fluid at P, but instead of having it at rest let it move with the local fluid velocity $\mathbf{q}$ of P. Then the only momentum interchange between fluid and plane is on the molecular scale and, as before, if δF denotes the force exerted on one side of δA by this momentum interchange, then this force is normal to δA and, assuming the limit to exist uniquely, we take

$$\lim_{\delta A\to 0}\frac{\delta F}{\delta A}=p$$

to be the pressure in the fluid at P. We shall show that *p is independent of the orientation of δA*. It is important, however, that *δA is at rest relative to the local fluid*, i.e. that it has the same velocity $\mathbf{q}$ as the fluid at P. Were δA at rest or moving with some velocity $\mathbf{v}$ other than $\mathbf{q}$, the local fluid would have to be brought to rest or to the velocity $\mathbf{v}$, and this requirement would necessitate an additional exchange of momentum between fluid and boundary.

We note that in the case of an inviscid fluid when δA is situated tangentially to the local fluid velocity at P, the only momentum interchange between fluid and surface is of the random molecular type, so that for this special orientation the force δF would be the same, irrespective of whether δA is at rest, moving with the local fluid velocity $\mathbf{q}$, or with some other velocity. Of course, for a viscous fluid different velocities of δA would result in different frictional forces being called into play. Also, for a viscous fluid such friction forces will in general exist, even where there is relative rest between surface and local fluid, provided there is a velocity gradient along the normal. We return to this in Chapter 8.

We now establish that the pressure p at P in a moving inviscid fluid is independent of the orientation of δA. To this end we consider the motion of an elemental tetrahedron $PQRS$ of fluid (Fig. 3.1) whose edges PQ, PR, PS are taken parallel to the coordinate axes OX, OY, OZ and of lengths δx, δy, δz at time t. The tetrahedron is moving with the local velocity $\mathbf{q}$ of the fluid at P.

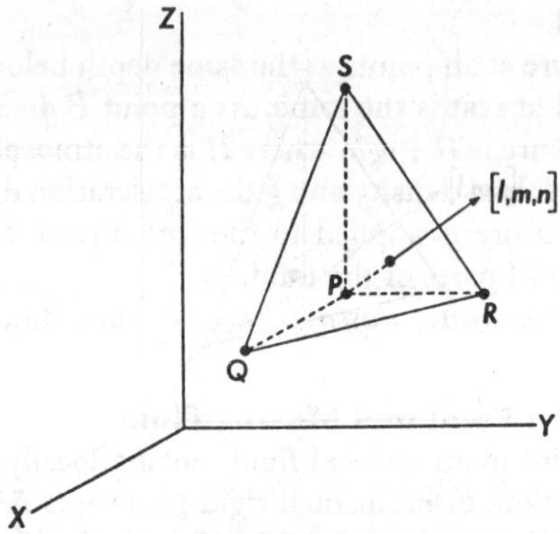

FIG. 3.1

Let p be the pressure on the triangle QRS and δS its area. Further suppose the D.C.'s of the normal to δS drawn outwards from the tetrahedron are $[l, m, n]$. Then the total force exerted by the fluid outside the tetrahedron on the face QRS is

$$-p\delta S(l\mathbf{i}+m\mathbf{j}+n\mathbf{k})=-\tfrac{1}{2}p(\delta y\delta z\mathbf{i}+\delta z\delta x\mathbf{j}+\delta x\delta y\mathbf{k}),$$

since

$$l\delta S=\text{area of face } PRS=\tfrac{1}{2}\delta y\delta z, \text{ etc.}$$

Now suppose p_x, p_y, p_z are the pressures on the planes PRS, PQS, PRQ. Since the areas of these faces are respectively $\frac{1}{2}\delta y\delta z$, $\frac{1}{2}\delta z\delta x$, $\frac{1}{2}\delta x\delta y$, the forces exerted on them by the outside fluid are respectively

$$\tfrac{1}{2}p_x\delta y\delta z\mathbf{i},\quad \tfrac{1}{2}p_y\delta z\delta x\mathbf{j},\quad \tfrac{1}{2}p_z\delta x\delta y\mathbf{k}.$$

Thus the total surface force on the tetrahedron is

$$\tfrac{1}{2}\{(p_x-p)\delta y\delta z\mathbf{i}+(p_y-p)\delta z\delta x\mathbf{j}+(p_z-p)\delta x\delta y\mathbf{k}\}.$$

In addition to the surface forces, the fluid element may be subject to *body forces* which are due to external causes, such as gravity. Let $\mathbf{F}$ be the mean body force per unit mass within the tetrahedron, so that $\frac{1}{6}\rho\mathbf{F}\delta x\delta y\delta z$ is the total body force acting, ρ being the mean fluid density. Then the net force on the element is

$$\tfrac{1}{2}\{(p_x-p)\delta y\delta z\mathbf{i}+(p_y-p)\delta z\delta x\mathbf{j}+(p_z-p)\delta x\delta y\mathbf{k}\}+\tfrac{1}{6}\rho\mathbf{F}\delta x\delta y\delta z.$$

Now let $\mathbf{q}$ be the velocity of P. Then $d\mathbf{q}/dt$ (differentiation following the fluid) is the acceleration of P so that if the mass $\frac{1}{6}\rho\delta x\delta y\delta z$ stays constant (which could be ensured by following a particle of fixed mass) the equation of motion is

$$\tfrac{1}{2}\{(p_x-p)\delta y\delta z\mathbf{i}+(p_y-p)\delta z\delta x\mathbf{j}+(p_z-p)\delta x\delta y\mathbf{k}\}+\tfrac{1}{6}\rho\mathbf{F}\delta x\delta y\delta z$$
$$=\tfrac{1}{6}\rho\delta x\delta y\delta z\, d\mathbf{q}/dt.$$

This equation contains quantities of the second and third orders of small quantities. When we make the edges of the tetrahedron vanishingly small we have, to the second order,

$$\tfrac{1}{2}\{(p_x-p)\delta y\delta z\mathbf{i}+(p_y-p)\delta z\delta x\mathbf{j}+(p_z-p)\delta x\delta y\mathbf{k}\}=0,$$

so that equating coefficients of the unit vectors gives at once

$$p_x=p_y=p_z=p.$$

Since the choice of axes is quite arbitrary, this establishes that *at any point P of a moving inviscid fluid the pressure p is the same in all directions.* This, of course, also recovers the corresponding result for static fluids.

3.3 Conditions at a Boundary of Two Inviscid Immiscible Fluids

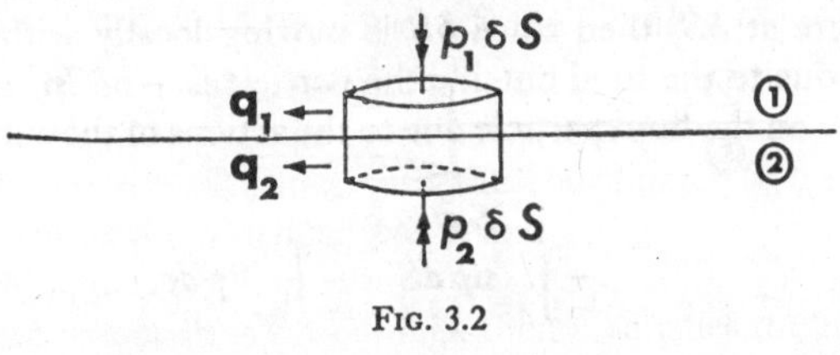

FIG. 3.2

Figure 3.2 shows two fluids separated by a plane boundary, their velocities at P on the boundary being $\mathbf{q}_1$, $\mathbf{q}_2$ respectively. Consider a small cylindrical hat-box shaped element of normal section δS containing P and projecting into both fluids, its generators being normal to the surface. Since there is no fluid transfer across the boundary,

$$p_1\delta S=p_2\delta S$$

or

$$p_1=p_2.$$

In particular for the case of a liquid in contact with the atmosphere, the pressure at the free surface is the same as that of the atmosphere.

When the boundary is curved, however, the above condition has to be modified to account for the effects of surface tension.

3.4 Euler's Equations of Motion

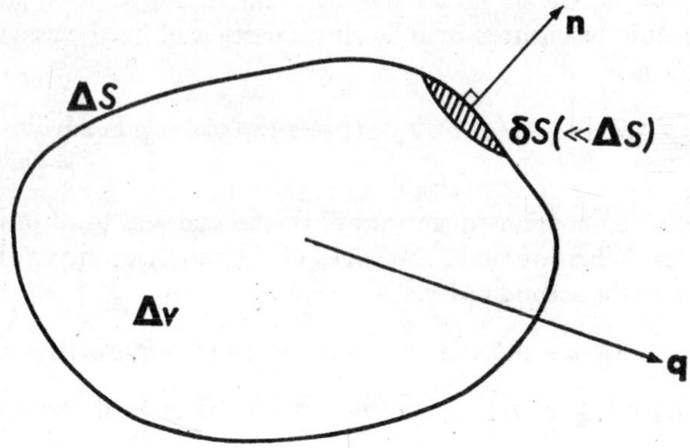

FIG. 3.3

At time t, suppose that ΔS denotes the closed surface of a particle of a moving fluid, the particle *moving along with the fluid* with velocity $\mathbf{q}$. We take the particle to be of *fixed mass* $\rho\Delta v$, ρ being the density and Δv the volume. Ultimately, ΔS and Δv are to be made vanishingly small.

Suppose that δS $(\ll \Delta S)$ is a surface element of ΔS and that $\mathbf{n}$ is the unit normal vector to δS, drawn outwards from the particle. If p denotes the fluid pressure at δS, then since δS is moving locally with the fluid, the force on it due to the fluid outside the particle is $-p\delta S\mathbf{n}$. Hence the total surface force on the fluid particle due to the actions of the surrounding fluid is

$$-\int_{\Delta S} \mathbf{n}p\, dS = -\int_{\Delta v} \nabla p\, dv.$$

Let $\mathbf{F}$ be the body force per unit mass at any volume element δv $(\ll \Delta v)$ of the particle. Then $\rho\mathbf{F}\delta v$ is the body force on the element δv and so the total body force on the entire particle is $\int_{\Delta v} \rho\mathbf{F}\, dv$.

Since the mass of the volume element δv is $\rho\delta v$, that of the entire fluid particle is $\int_{\Delta v} \rho\, dv$. But this mass is constant. Hence the equation of motion of the fluid particle is

$$\int_{\Delta v} (\rho\mathbf{F} - \nabla p)\, dv = \frac{d\mathbf{q}}{dt} \int_{\Delta v} \rho\, dv.$$

The terms on the L.H.S. represent the total force acting. $d\mathbf{q}/dt$ is the acceleration of the fluid particle following its motion. To the first order this equation approximates to

$$(\rho\mathbf{F} - \nabla p)\Delta v = \frac{d\mathbf{q}}{dt}\rho\Delta v,$$

so that in the limit as $\Delta v \to 0$, we obtain

$$\frac{d\mathbf{q}}{dt} = \mathbf{F} - \frac{1}{\rho}\nabla p. \tag{1}$$

Equation (1) is known as *Euler's equation of motion*. Other forms of it, using the results of Section 2.9 are

$$\frac{\partial\mathbf{q}}{\partial t} + (\mathbf{q}\cdot\nabla)\mathbf{q} = \mathbf{F} - \frac{1}{\rho}\nabla p, \tag{2}$$

$$\frac{\partial\mathbf{q}}{\partial t} + \nabla(\tfrac{1}{2}\mathbf{q}^2) - \mathbf{q}\wedge(\nabla\wedge\mathbf{q}) = \mathbf{F} - \frac{1}{\rho}\nabla p. \tag{3}$$

In tensor form, with x_i as coordinates, u_i as velocity components, F_i as components of body force ($i=1, 2, 3$) the equations of motion may be written

$$\frac{\partial u_i}{\partial t} + u_j u_{i,j} = F_i - \frac{1}{\rho}p_{,i}. \tag{4}$$

Worked Example

AB is a tube of small uniform bore forming a quadrantal arc of a circle of radius a and centre O, OA being horizontal and OB vertical with B below O. The tube is full of liquid of density ρ, the end B being closed. If B is suddenly opened, show that initially $du/dt = 2g/\pi$, where $u = u(t)$ is the velocity, and that the pressure at a point whose angular distance from A is θ immediately drops to

$$\rho g a\left(\sin\theta - \frac{2\theta}{\pi}\right)$$

above atmospheric pressure. Prove further that when the liquid remaining in the tube subtends an angle β at the centre,

$$\frac{d^2\beta}{dt^2} = -\frac{2g}{a\beta}\sin^2\left(\frac{\beta}{2}\right).$$

(London Univ., Gen. B.Sc. II, 1965)

In Fig. 3.4, arc $AP = s$, arc $AP' = s + \delta s$ at time t. The equation of motion of the element PP' along the arc in the direction of s increasing is

$$\frac{du}{dt} = g\cos\theta - \frac{1}{\rho}\frac{\partial p}{\partial s}.$$

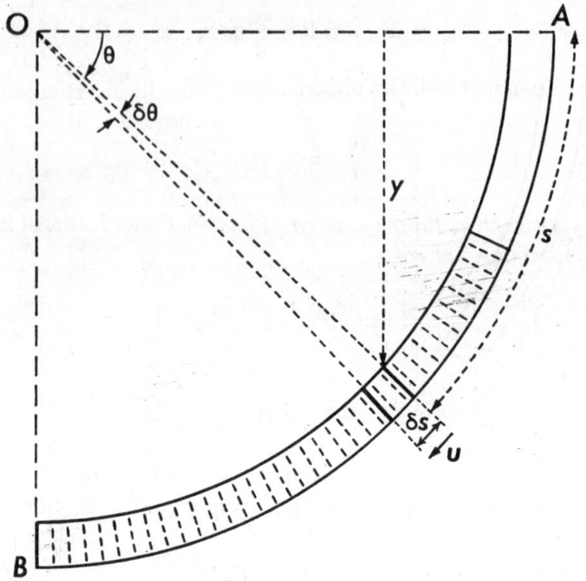

FIG. 3.4

Now

$$\frac{du}{dt}=\frac{\partial u}{\partial t}+\frac{u\partial u}{\partial s}=\frac{\partial u}{\partial t}+\frac{\partial}{\partial s}(\tfrac{1}{2}u^2)\,;\ \cos\ \theta=\frac{dy}{ds}$$

where y is the depth of P below OA. Hence the equation of motion is

$$\frac{\partial u}{\partial t}+\frac{\partial}{\partial s}(\tfrac{1}{2}u^2)=g\frac{dy}{ds}-\frac{1}{\rho}\frac{\partial p}{\partial s}.$$

Integrate both sides of (1) w.r.t. s, keeping t constant. Then

$$s\frac{\partial u}{\partial t}+\tfrac{1}{2}u^2=gy-\frac{p}{\rho}+f(t), \tag{2}$$

since, at any time t, $\partial u/\partial t$ is the same at all points of the liquid. Initially, taking $t=0$, at A we have $s=0$, $u=0$, $p=\Pi$ (the atmospheric pressure), $y=0$, so that (2) gives

$$0=-\frac{\Pi}{\rho}+f(0). \tag{3}$$

Also, when $t=0$, at B we have $s=\tfrac{1}{2}\pi a$, $u=0$, $p=\Pi$, $y=a$, so that (2) gives

$$\tfrac{1}{2}\pi a\frac{\partial u}{\partial t}=ga-\frac{\Pi}{\rho}+f(0). \tag{4}$$

From (3), (4), by subtraction,

$$\left(\frac{\partial u}{\partial t}\right)_{t=0}=\frac{2g}{\pi}. \tag{5}$$

Hence

$$\left(\frac{du}{dt}\right)_{t=0}=\left(\frac{\partial u}{\partial t}+\frac{u\partial u}{\partial s}\right)_{t=0}=\frac{2g}{\pi},$$

since $u=0$ when $t=0$.

In (2), take $s=a\theta$, $u=0$, $t=0$, $y=a\sin\theta$, $\partial u/\partial t=2g/\pi$. Then $a\theta(2g/\pi)=ga\sin\theta-(p/\rho)+f(0)$. Using (3), this gives

$$\underline{p-\Pi=\rho ga(\sin\theta-2\theta/\pi).}$$

In the equation (2), take $s=a(\frac{1}{2}\pi-\beta)$, $y=a\cos\beta$, $p=\Pi$, the conditions at the upper free surface. Then

$$a(\tfrac{1}{2}\pi-\beta)\frac{\partial u}{\partial t}+\tfrac{1}{2}u^2=ga\cos\beta-\frac{\Pi}{\rho}+f(t).$$

Also in (2), take $s=\frac{1}{2}a\pi$, $y=a$, $p=\Pi$, the conditions at the exit. Then

$$(a\cdot\tfrac{1}{2}\pi)\frac{\partial u}{\partial t}+\tfrac{1}{2}u^2=ga-\frac{\Pi}{\rho}+f(t).$$

Subtracting these equations gives

$$a\beta\frac{\partial u}{\partial t}=ga(1-\cos\beta),$$

or

$$\frac{\partial u}{\partial t}=\frac{2g}{\beta}\sin^2\frac{\beta}{2}.$$

Now

$$u=\frac{\partial}{\partial t}\{(\tfrac{1}{2}\pi-\beta)a\}=-a\frac{d\beta}{dt}.$$

Hence

$$\underline{\frac{\partial u}{\partial t}=-\frac{ad^2\beta}{dt^2}=\frac{2g}{\beta}\sin^2\frac{\beta}{2}.}$$

3.5 Bernoulli's Equation

Let us suppose that the body forces are conservative and that the flow is of the potential kind. Then there exist scalar functions Ω, φ such that

$$\mathbf{F}=-\nabla\Omega,\quad \mathbf{q}=-\nabla\varphi.$$

Then $\nabla\wedge\mathbf{q}=\mathbf{0}$ and equation (3) of Section 3.4 becomes

$$-\nabla\left(\frac{\partial\varphi}{\partial t}\right)+\nabla(\tfrac{1}{2}\mathbf{q}^2)=-\nabla\Omega-\frac{1}{\rho}\nabla p.$$

Let $\mathbf{r}$ be the position vector of the fluid particle at time t and let $d\mathbf{r}$ be an instantaneous displacement made in the position of the particle at this instant t. Then scalar multiplying the last equation through by $d\mathbf{r}$ and using $d\mathbf{r} \cdot \nabla\Omega = d\Omega$, etc., we obtain

$$-d\left(\frac{\partial\varphi}{\partial t}\right)+d(\tfrac{1}{2}\mathbf{q}^2)=-d\Omega-\frac{1}{\rho}dp,$$

subject to t being constant. Rearranging and integrating gives

$$\tfrac{1}{2}\mathbf{q}^2+\Omega+\int\frac{dp}{\rho}-\frac{\partial\varphi}{\partial t}=f(t), \tag{1}$$

where $f(t)$ is an arbitrary function of t arising from the integration in which t is being kept constant. (1) is *Bernoulli's equation* in its most general form.

Other forms of Bernoulli's equation, often of considerable use in the solution of problems, may be derived from (1). Thus for steady motion $\partial\varphi/\partial t=0$ and $f(t)$ is constant, so that

$$\tfrac{1}{2}\mathbf{q}^2+\Omega+\int\frac{dp}{\rho}=\text{const.} \tag{2}$$

If, further, the fluid is homogeneous and incompressible so that ρ is constant, (2) becomes

$$\tfrac{1}{2}\mathbf{q}^2+\Omega+(p/\rho)=\text{const.} \tag{3}$$

3.6 Worked Examples

We illustrate the above theory with examples, some of which are of practical importance.

Example 1

The Pitot tube.

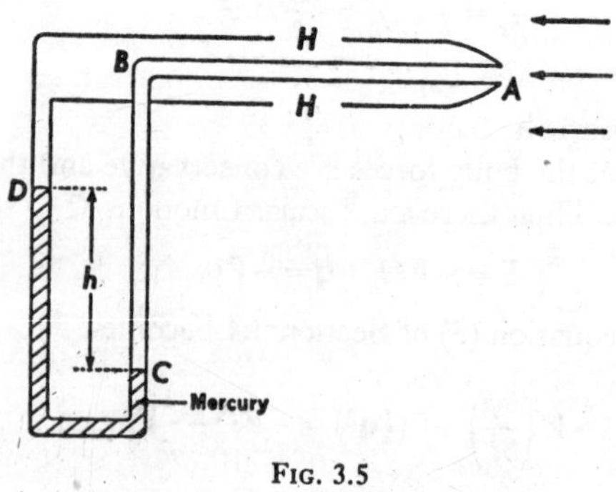

Fig. 3.5

Figure 3.5 shows an instrument, known as a *Pitot tube*, which is used for measuring fluid velocities. Suppose it is desired to measure the velocity q of a stream of water. Then the inner tube BA is placed upstream of the flow. The outer tube contains holes such as H. Let p be the pressure in the stream where the fluid velocity is q. Then p is also the pressure on the inside and outside of the hole and hence also at the meniscus D of the mercury in the U-tube. Considering the stream entering the tube AB, this is brought to rest at the meniscus C. Let p_0 be the pressure here. Then by applying Bernoulli's equation along the streamline extending from infinity, entering the inner tube at A and being finally brought to rest at C, we have

$$\frac{p}{\rho}+\tfrac{1}{2}q^2=\frac{p_0}{\rho},$$

where ρ is the density of the water. Hence

$$q=\sqrt{\left\{\frac{2(p_0-p)}{\rho}\right\}}.$$

The pressure difference p_0-p is measured from the difference in levels of the mercury. In fact if h is this difference and σ the density of the mercury, then

$$p_0-p=\sigma gh.$$

Thus the ambient velocity of the stream of water is measurable.

The device is also very suitable for measuring subsonic airstreams. At low speeds air may be treated as an incompressible fluid for which similar analysis is applied.

EXAMPLE 2

Venturi tube.

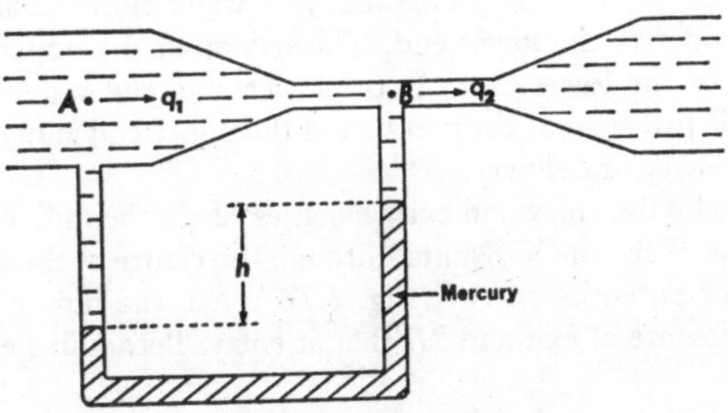

FIG. 3.6

The *Venturi tube* is a device for measuring flow in a pipe. The pipe is steadily constricted from a section S_1 to a much smaller section S_2, a U-tube serving as a mercury manometer being inserted between the broad and narrow sections at A and B (Fig. 3.6). Let q_1, q_2 be the fluid velocities at A, B and p_1, p_2 the pressures. The sectional areas being S_1, S_2 at these locations, the equation of continuity, taking the fluid to be incompressible, is

$$q_1 S_1 = q_2 S_2. \tag{1}$$

Applying Bernoulli's equation along the central streamline from A to B gives

$$\frac{p_1}{\rho} + \tfrac{1}{2}q_1^2 = \frac{p_2}{\rho} + \tfrac{1}{2}q_2^2, \tag{2}$$

ρ being the density of the fluid. Eliminating q_2 from (1), (2), we find

$$q_1 = \sqrt{\left\{\frac{2(p_1 - p_2)S_2^2}{\rho(S_1^2 - S_2^2)}\right\}}. \tag{3}$$

The equation (3) clearly shows that $p_1 > p_2$, i.e. that the fluid pressure is a minimum at a constriction. This result is rather surprising intuitively. Certainly $q_2 > q_1$. The result must mean that when a one-dimensional flow is constricted, the pressure energy of the fluid is mostly converted into kinetic energy.

If h is the difference of meniscus levels in the mercury manometer and σ the density of mercury, then

$$p_1 - p_2 = \sigma g h \tag{4}$$

so that (3) and (4) enable q_1 to be found. The mass of liquid flowing through the unconstricted part of the pipe per unit time is then $\rho q_1 S_1$.

Example 3

A long pipe is of length l and has slowly tapering cross-section. It is inclined at angle α to the horizontal and water flows steadily through it from the upper to the lower end. The section at the upper end has twice the radius of the lower end. At the lower end, the water is delivered at atmospheric pressure. If the pressure at the upper end is twice atmospheric find the velocity of delivery.

Let q_1, q_2 be the entry and exit velocities; $2a$, a the radii of the entry and exit sections. Take the horizontal through the centre of the entry section as zero level of potential energy (Fig. 3.7). When the flow is steady, taking Π as the pressure at exit and 2Π that at entry, Bernoulli's equation is

$$\frac{p}{\rho} + \tfrac{1}{2}q^2 + \Omega = \text{const.},$$

or

$$\frac{2\Pi}{\rho} + \tfrac{1}{2}q_1^2 + 0 = \frac{\Pi}{\rho} + \tfrac{1}{2}q^2 - gl \sin\alpha \tag{1}$$

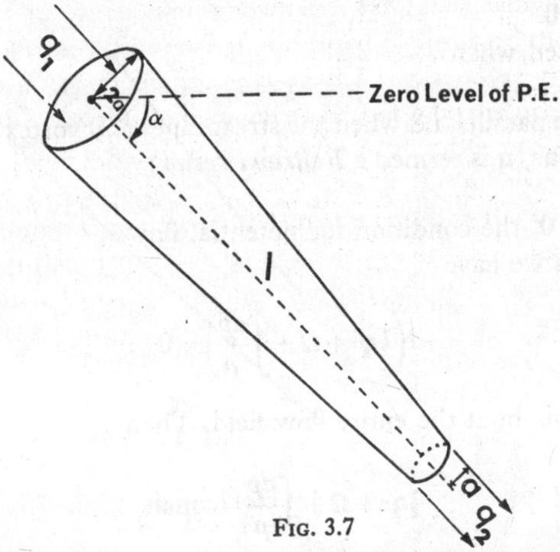

FIG. 3.7

since $-gl \sin \alpha$ is the P.E. per unit mass of the gravitational force at the lower end. Further, the equation of continuity, since the fluid is incompressible, is

$$q_1(4\pi a^2) = q_2(\pi a^2). \tag{2}$$

From (1), (2)

$$\frac{15}{32} q_2^2 = gl \sin \alpha + \frac{\Pi}{\rho},$$

giving the desired exit velocity.

3.7 Discussion of the Case of Steady Motion under Conservative Body Forces

Bernoulli's equation was derived for potential flows under conservative body forces. We now study cases that arise when the flow is no longer of the potential kind but is steady. The body forces are still assumed to be conservative. We shall see that in such conditions equations somewhat akin to Bernoulli's equation arise.

Euler's equation of motion may be written in the form

$$\mathbf{F} - \frac{1}{\rho} \nabla p = \frac{\partial \mathbf{q}}{\partial t} + \nabla(\tfrac{1}{2}\mathbf{q}^2) - \mathbf{q} \wedge \boldsymbol{\zeta},$$

where $\boldsymbol{\zeta} = \text{curl}\, \mathbf{q}$, the vorticity vector. If the forces are conservative, then $\mathbf{F} = -\nabla \Omega$ so that when the flow is steady,

$$\nabla\left(\tfrac{1}{2}\mathbf{q}^2 + \Omega + \int \frac{dp}{\rho}\right) = \mathbf{q} \wedge \boldsymbol{\zeta}.$$

Scalar multiplying this equation through by $d\mathbf{r}$, a time-independent variation in the position vector $\mathbf{r}$ of the fluid particle, gives

$$d\left(\tfrac{1}{2}\mathbf{q}^2 + \Omega + \int \frac{dp}{\rho}\right) = d\mathbf{r} \cdot (\mathbf{q} \wedge \boldsymbol{\zeta}). \tag{1}$$

Case I: $\mathbf{q} \wedge \boldsymbol{\zeta} = \mathbf{0}$.

This is realized when either

(i) $\mathbf{q}$ and $\boldsymbol{\zeta}$ are parallel, i.e. when the streamlines and vortex lines coincide. For such motions, $\mathbf{q}$ is termed a *Beltrami vector*;

or

(ii) when $\boldsymbol{\zeta} = \mathbf{0}$, the condition for potential flow.

In both cases we have

$$d\left(\tfrac{1}{2}\mathbf{q}^2 + \Omega + \int \frac{dp}{\rho}\right) = 0$$

at all times throughout the entire flow field. Then

$$\tfrac{1}{2}\mathbf{q}^2 + \Omega + \int \frac{dp}{\rho} = \text{const.} \tag{2}$$

throughout the entire field of flow. The constant is the same throughout the entire field since the differential $d\mathbf{r}$ in (1) is any arbitrary small variation of position vector $\mathbf{r}$ in the field.

Case II: $\mathbf{q} \wedge \boldsymbol{\zeta} \neq \mathbf{0}$.

Now $\mathbf{q} \wedge \boldsymbol{\zeta}$ is perpendicular to the vectors $\mathbf{q}$, $\boldsymbol{\zeta}$. Hence, if $d\mathbf{r} \neq \mathbf{0}$, then $d\mathbf{r} \cdot (\mathbf{q} \wedge \boldsymbol{\zeta}) = 0$ whenever $d\mathbf{r}$ lies in the plane of $\mathbf{q}$, $\boldsymbol{\zeta}$. Thus if we take the variation $d\mathbf{r}$ in the surface containing both the streamlines and vortex lines, then (1) shows that

$$d\left(\tfrac{1}{2}\mathbf{q}^2 + \Omega + \int \frac{dp}{\rho}\right) = 0$$

over such a surface, or

$$\tfrac{1}{2}\mathbf{q}^2 + \Omega + \int \frac{dp}{\rho} = \text{const.} \tag{3}$$

over a surface containing the streamlines and vortex lines. We note that the constant in (3) is the same everywhere on any one such surface, but that its value varies from one such surface to another. Also (3) holds irrespective of whether the motion is rotational or irrotational.

3.8* Some Potential Theorems

We here prove a number of theorems for steady irrotational incompressible flows for which the velocity potential φ satisfies $\nabla^2\varphi = 0$ at all points within the moving fluid.

THEOREM I: If S is the boundary of a spherical surface lying wholly within the fluid, then the mean value of the velocity potential is equal to its value at the centre of the sphere.

* This section may be omitted on a first reading.

Proof: Denoting by r the radius of S, the mean value of φ on S is $\bar{\phi}$ where

$$\bar{\phi}=\frac{1}{4\pi r^2}\int_S \varphi\, dS.$$

If the element dS subtends a solid angle $d\omega$ at the centre of the sphere, then $dS=r^2 d\omega$ and so

$$\bar{\phi}=\frac{1}{4\pi}\int_S \varphi\, d\omega.$$

$$\therefore\quad \frac{\partial\bar{\phi}}{\partial r}=\frac{1}{4\pi}\int_S \frac{\partial\varphi}{\partial r}\, d\omega=\frac{1}{4\pi r^2}\int_S \frac{\partial\varphi}{\partial r}\, dS.$$

Now

$$\int_S \frac{\partial\varphi}{\partial r}\, dS=\int_S \frac{\partial\varphi}{\partial n}\, dS=\int_S \mathbf{n}\cdot\nabla\varphi\, dS=\int_V \nabla\cdot\nabla\varphi\, dv=0,$$

since $\nabla^2\varphi=0$ in V, the interior of S.

$$\therefore\quad \partial\bar{\phi}/\partial r=0.$$

Hence $\bar{\phi}$ is independent of r, so that the mean value of φ is the same over all spheres having the same centre and therefore is equal to its value at the centre.

THEOREM II: If Σ is the solid boundary of a large spherical surface of radius R, containing fluid in motion and also enclosing one or more closed surfaces, then the mean value of φ on Σ is of the form

$$\bar{\phi}=(M/R)+C,$$

where M, C are constants, provided that the fluid extends to infinity and is at rest there.

Proof: Suppose that the volume of fluid crossing each of the internal surfaces contained within Σ per unit time is a finite quantity $4\pi M$. Then this is the total volume flux per unit time across Σ. Since the fluid velocity at any point of Σ is $-\partial\varphi/\partial R$ radially outwards, the equation of continuity is

$$\int_\Sigma \left(-\frac{\partial\varphi}{\partial R}\right) dS=4\pi M.$$

If dS subtends a solid angle $d\omega$ at the centre of Σ, then $dS=R^2\, d\omega$ and so

$$\frac{1}{4\pi}\int_\Sigma \frac{\partial\varphi}{\partial R}\, d\omega=-\frac{M}{R^2},$$

or

$$\frac{1}{4\pi}\frac{\partial}{\partial R}\int_\Sigma \varphi\, d\omega=-\frac{M}{R^2}.$$

Integrating w.r.t. R gives

$$\frac{1}{4\pi}\int_{\Sigma}\varphi\, d\omega=\frac{M}{R}+C,$$

or

$$\frac{1}{4\pi R^2}\int_{\Sigma}\varphi\, dS=\frac{M}{R}+C=\bar{\phi}.$$

C is here independent of R. To show that it is an absolute constant, we must prove that it is independent of the coordinates of the centre of Σ. To this end, let the sphere be displaced a distance δx in any direction, keeping R constant. Then from the last equation, the attendant change in $\bar{\phi}$ is

$$\frac{\partial\bar{\phi}}{\partial x}\delta x=\frac{1}{4\pi R^2}\int_{\Sigma}\frac{\partial\varphi}{\partial x}\delta x\, dS=\frac{\partial C}{\partial x}\delta x.$$

But $\partial\varphi/\partial x=0$ on Σ when $R\to\infty$, since the liquid is at rest at infinity. Thus for large R, $\partial C/\partial x=0$.

Hence

$$\bar{\phi}=(M/R)+C,$$

where M, C are constants.

We note that in the special case when the closed surfaces within Σ are all rigid, no flow can take place across them so that $M=0$ and $\bar{\phi}=C$.

THEOREM III: With the notation of Theorem II, if φ_P denotes the potential at any point P of the fluid, then $\varphi_P\to C$ as $P\to\infty$.

Proof:

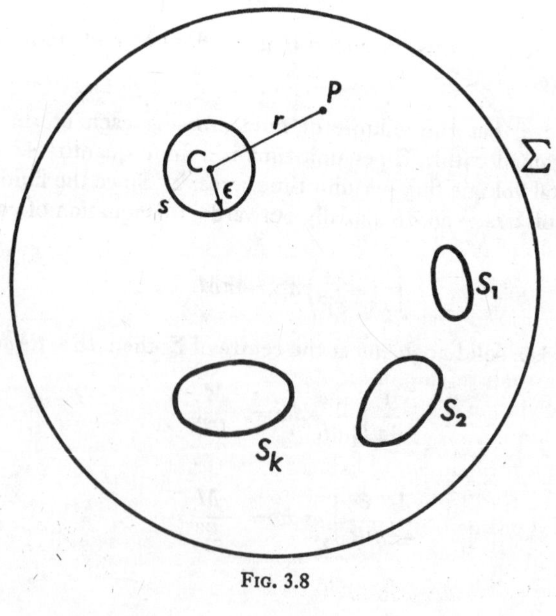

FIG. 3.8

Let $S_1, S_2, \ldots, S_k$ be the surfaces contained within Σ. Also let s be the sphere whose centre is C and radius ε. Then if r is the distance from C to any point P within the fluid,

$$\int_V \left\{\varphi \nabla^2\left(\frac{1}{r}\right) - \frac{1}{r}\nabla^2\varphi\right\} dv = \int_{\text{Total surface}} \left\{\varphi \frac{\partial}{\partial n}\left(\frac{1}{r}\right) - \frac{1}{r}\frac{\partial \varphi}{\partial n}\right\} dS,$$

using the symmetrical form of Green's theorem the unit normal at any point on the boundary being drawn *outwards* from the fluid. But $\nabla^2\varphi = 0$, $\nabla^2(1/r) = 0$, so the last equation gives

$$\int_S \left\{\varphi \frac{\partial}{\partial n}\left(\frac{1}{r}\right) - \frac{1}{r}\frac{\partial \varphi}{\partial n}\right\} dS + \int_\Sigma \left\{\varphi \frac{\partial}{\partial n}\left(\frac{1}{r}\right) \frac{\partial \varphi}{\partial n}\right\} dS$$

$$+ \sum_{m=1}^{k} \int_{S_m} \left\{\varphi \frac{\partial}{\partial n}\left(\frac{1}{r}\right) - \frac{1}{r}\frac{\partial \varphi}{\partial n}\right\} dS = 0.$$

On s,

$$\frac{\partial}{\partial n}\left(\frac{1}{r}\right) = \left\{-\frac{\partial}{\partial r}\left(\frac{1}{r}\right)\right\}_{r=\varepsilon} = +\frac{1}{\varepsilon^2}, \quad \frac{\partial \varphi}{\partial n} = -\frac{\partial \varphi}{\partial r}.$$

$$\int_S \left\{\varphi \frac{\partial}{\partial n}\left(\frac{1}{r}\right) - \frac{1}{r}\frac{\partial \varphi}{\partial n}\right\} dS = \int_S \left\{\varphi\left(\frac{1}{\varepsilon^2}\right) + \frac{1}{\varepsilon}\frac{\partial \varphi}{\partial r}\right\} dS$$

$$= \left\{\varphi\left(\frac{1}{\varepsilon^2}\right) + \frac{1}{\varepsilon}\frac{\partial \varphi}{\partial r}\right\} 4\pi\varepsilon^2, \quad \text{to 1st order.}$$

$$\to 4\pi\varphi_P \text{ as } \varepsilon \to 0.$$

On Σ,

$$\frac{\partial}{\partial n}\left(\frac{1}{r}\right) = \left\{\frac{\partial}{\partial r}\left(\frac{1}{r}\right)\right\}_{r=R} = -\frac{1}{R^2}, \quad \frac{\partial \varphi}{\partial n} = \frac{\partial \varphi}{\partial R}.$$

$$\therefore \int_\Sigma \left\{\varphi \frac{\partial}{\partial n}\left(\frac{1}{r}\right) - \frac{1}{r}\frac{\partial \varphi}{\partial n}\right\} dS = -\frac{1}{R^2}\int_\Sigma \varphi\, dS - \frac{1}{R}\int \frac{\partial \varphi}{\partial R} dS$$

$$= -\frac{1}{R^2}\int_\Sigma \left(\frac{M}{R} + C\right) dS + \frac{1}{R}\int_\Sigma \frac{M}{R^2} dS$$

$$= -\frac{M}{R^3}(4\pi R^2) - \frac{1}{R^2}(4\pi C R^2) + \frac{M}{R^3}(4\pi R^2)$$

$$\to -4\pi C \text{ as } R \to \infty.$$

Thus

$$4\pi(\varphi_P - C) + \sum_{m=1}^{k} \int \left\{\varphi \frac{\partial}{\partial n}\left(\frac{1}{r}\right) - \frac{1}{r}\frac{\partial \varphi}{\partial n}\right\} dS = 0.$$

As $r \to \infty$, $(1/r) \to 0$ and $(\partial/\partial n)(1/r) \to 0$, so the last result shows that

$$\underline{\varphi_P \to C.}$$

THEOREM IV: With the notation of Theorem III, if the fluid is at rest at infinity and if each surface S_m is rigid, then the kinetic energy of the moving fluid is

$$T = \tfrac{1}{2}\rho \int_V \mathbf{q}^2\, dv = \tfrac{1}{2}\rho \sum_{m=1}^{k} \int_{S_m} \varphi \frac{\partial \varphi}{\partial n}\, dS,$$

the normal at each surface element dS being drawn outwards from the fluid.

Proof: The K.E. of a fluid particle of mass $\rho\delta v$ is $\tfrac{1}{2}\rho\delta v \mathbf{q}^2 = \tfrac{1}{2}\rho(\nabla\varphi)^2\, dv$. Hence that of the entire fluid is

$$T = \tfrac{1}{2}\rho \int_V (\nabla\varphi)^2\, dv,$$

V denoting the total volume of fluid. Now

$$(\nabla\varphi)^2 = (\nabla\varphi)\cdot(\nabla\varphi) + \varphi\nabla^2\varphi = \nabla\cdot(\varphi\nabla\varphi).$$

$$\therefore\ T = \tfrac{1}{2}\rho \int_V \nabla\cdot(\varphi\nabla\varphi)\, dv$$

$$= \tfrac{1}{2}\rho \int_{\text{Total surface}} \mathbf{n}\cdot(\varphi\nabla\varphi)\, dS$$

$$= \tfrac{1}{2}\rho \int_\Sigma \varphi\frac{\partial\varphi}{\partial n}\, dS + \tfrac{1}{2}\rho \sum_{m=1}^{k} \int_{S_m} \varphi\frac{\partial\varphi}{\partial n}\, dS. \tag{1}$$

The equation of continuity for the entire region V is

$$\int_\Sigma \frac{\partial\varphi}{\partial n}\, dS + \sum_{m=1}^{k} \int_{S_m} \frac{\partial\varphi}{\partial n}\, dS = 0. \tag{2}$$

Thus from (1), (2),

$$T = \tfrac{1}{2}\rho \int_\Sigma (\varphi - C)\frac{\partial\varphi}{\partial n}\, dS + \tfrac{1}{2}\rho \sum_{m=1}^{k} \int_{S_m} (\varphi - C)\frac{\partial\varphi}{\partial n}\, dS, \tag{3}$$

where C is any constant. Taking C to be the value of φ when $P \to \infty$, we have, on making $R \to \infty$,

$$T = \tfrac{1}{2}\rho \sum_{m=1}^{k} \int_{S_m} (\varphi - C)\frac{\partial \varphi}{\partial n}\, dS. \tag{4}$$

On each rigid surface S_m there is no flow, so that

$$\int_{S_m} \frac{\partial \varphi}{\partial n}\, dS = 0.$$

Thus

$$T = \tfrac{1}{2}\rho \sum_{k=1}^{m} \int_{S_m} \varphi \frac{\partial \varphi}{\partial n}\, dS.$$

THEOREM V: **If the fluid of Theorem IV is at rest at infinity and if either φ or $\partial\varphi/\partial n$ is prescribed on each surface S_m, then φ is determined uniquely throughout V to within an arbitrary constant.**

Proof: From Theorem IV,

$$\int_V (\nabla\varphi)^2\, dv = \sum_{k=1}^{m} \int_{S_m} \varphi \frac{\partial \varphi}{\partial n}\, dS, \tag{1}$$

whilst within V,

$$\nabla^2 \varphi = 0. \tag{2}$$

Suppose $\varphi = \varphi_1$, $\varphi = \varphi_2$ are two solutions of (2), subject to (1). Write $\psi = \varphi_1 - \varphi_2$. Then

$$\nabla^2 \psi = \nabla^2 \varphi_1 - \nabla^2 \varphi_2 = 0 \text{ in } V.$$

Thus ψ is a harmonic function satisfying similar conditions to φ_1, φ_2. Hence

$$\int_V (\nabla\psi)^2\, dv = \sum_{k=1}^{m} \int_{S_m} \psi \frac{\partial \psi}{\partial n}\, dS. \tag{3}$$

On each S_m, either $\varphi_1 = \varphi_2$ or $\partial\varphi_1/\partial n = \partial\varphi_2/\partial n$, i.e. either $\psi = 0$ or $\partial\psi/\partial n = 0$. Thus in all cases, $\psi\, \partial\psi/\partial n = 0$ on each S_m so that the R.H.S. of (3) is zero. Thus

$$\int_V (\nabla\psi)^2\, dv = 0. \tag{4}$$

Since $(\nabla\psi)^2 \geqslant 0$, the condition (4) can hold only if

$$\nabla\psi = \mathbf{0} \text{ in } V. \tag{5}$$

(5) gives

$$\psi = \text{const.}$$

or

$$\varphi_1 - \varphi_2 = \text{const.} \tag{6}$$

The function ψ is continuous, so that if φ is prescribed on any one S_m, the constant is zero. If the velocities are prescribed on each S_m, we cannot evaluate the constant.

THEOREM VI: If the fluid of Theorem V is in uniform motion at infinity and if $\partial\varphi/\partial n$ is prescribed on each surface S_m, then φ is uniquely determined throughout V.

Proof: Let **V** be the velocity of the fluid at infinity. Superpose a velocity $-\mathbf{V}$ on the entire system so as to reduce the velocity of the fluid at infinity to rest. Then the conditions of Theorem V prevail in which the velocities at all points of each S_m are known in a fluid which is at rest at infinity. This leads to a unique value for φ at each point of V.

Theorems V, VI are ***uniqueness theorems*** which are useful in finding solutions of $\nabla^2\varphi = 0$ subject to prescribed boundary conditions.

3.9 Some Flows Involving Axial Symmetry

Let $\varphi(r, \theta, \psi)$ be the velocity potential at any point having spherical polar coordinates (r, θ, ψ) in a field of steady irrotational incompressible flow. Then in this coordinate system, Laplace's equation $\nabla^2\varphi = 0$ becomes

$$\sin\theta \frac{\partial}{\partial r}\left(r^2 \frac{\partial\varphi}{\partial r}\right) + \frac{\partial}{\partial\theta}\left(\sin\theta \frac{\partial\varphi}{\partial\theta}\right) + \frac{1}{\sin\theta}\left(\frac{\partial^2\varphi}{\partial\psi^2}\right) = 0.$$

When there is symmetry about the line $\theta = 0$, $\varphi = \varphi(r, \theta)$ and the equation becomes

$$\sin\theta \frac{\partial}{\partial r}\left(r^2 \frac{\partial\varphi}{\partial r}\right) + \frac{\partial}{\partial\theta}\left(\sin\theta \frac{\partial\varphi}{\partial\theta}\right) = 0. \tag{1}$$

It is easily confirmed by direct substitution that special solutions of (1) are

$$\varphi = r\cos\theta, \quad \varphi = (1/r^2)\cos\theta.$$

Thus a rather more general solution of (1) is

$$\varphi(r, \theta) = (Ar + Br^{-2})\cos\theta. \tag{2}$$

We now investigate a number of flows with axial symmetry which have solutions of the form (2).

EXAMPLE 1

Stationary sphere in a uniform stream.

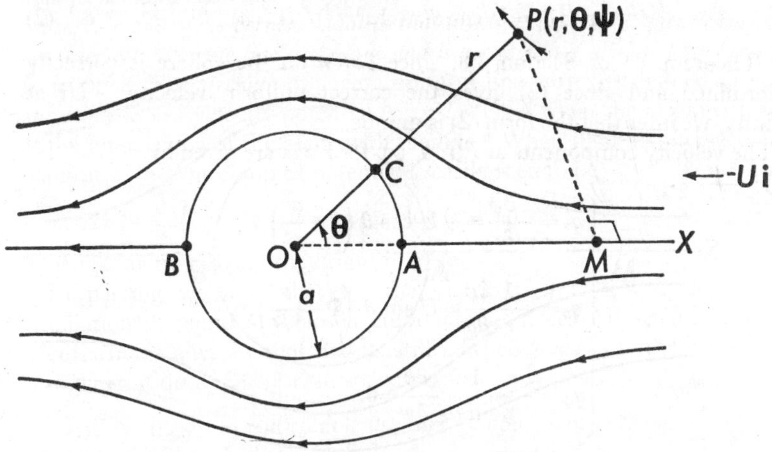

FIG. 3.9

Figure 3.9 shows a solid stationary sphere of radius a placed in a uniform stream of incompressible fluid for which the undisturbed velocity is $-U\mathbf{i}$, where U is a constant and $\mathbf{i}$ a constant unit vector in the direction $\overline{OX}$, the axis of symmetry. PM is the perpendicular from P on OX and $OM=x$. The spherical polar coordinates of P are (r, θ, ψ), referred to OX as the line $\theta=0$. Thus $x=r\cos\theta$.

In the absence of the sphere, the flow would be uniform and parallel to $\overline{XO}$, the velocity everywhere being $-U\mathbf{i}$ which is clearly $-\nabla(Ux)$. Hence the velocity potential due to such uniform flow would be

$$Ux=Ur\cos\theta.$$

When the sphere is inserted as shown, it will produce local perturbation of the stream. The diagram shows how the streamlines are disposed around the sphere. Consequently the undisturbed potential $Ur\cos\theta$ of the parallel flow has to be modified by a 'perturbation potential' due to the presence of the sphere. This perturbation must have the following properties: (i) it must satisfy Laplace's equation for the case of axial symmetry, (ii) it must tend to zero at large distances from the sphere, so as to recover the conditions of uniform parallel flow at infinity. A harmonic function satisfying these requirements is $\cos\theta/r^2$. Let us, then, try the function

$$\varphi(r, \theta)=Ur\cos\theta+Ar^{-2}\cos\theta \ (r\geqslant a) \tag{1}$$

for the velocity potential at $P(r, \theta, \psi)$ when the sphere is inserted in the stream. The constant A in (1) is determinable from the fact that there is no flow normal to the surface $r=a$. This means that $(\partial\varphi/\partial r)_{r=a}=0$, so that

$$U\cos\theta-(2A/a^3)\cos\theta=0.$$

Thus (1) becomes

$$\varphi(r,\theta)=U\cos\theta(r+\tfrac{1}{2}a^3r^{-2})\quad(r\geqslant a). \tag{2}$$

By Theorem VI of Section 3.8, since $\partial\varphi/\partial n$ on the sphere is correctly determined and since (2) gives the correct uniform velocity $-U\mathbf{i}$ at infinity, we infer that the form (2) is unique.

The velocity components at $P(r,\theta,\psi)$, for $r\geqslant a$, are given by

$$\begin{cases} q_r=-\dfrac{\partial\varphi}{\partial r}=-U\cos\theta\left(1-\dfrac{a^3}{r^3}\right),\\[2ex] q_\theta=-\dfrac{1}{r}\dfrac{\partial\varphi}{\partial\theta}=U\sin\theta\left(1+\tfrac{1}{2}\dfrac{a^3}{r^3}\right),\\[2ex] q_\psi=-\dfrac{1}{r\sin\theta}\dfrac{\partial\varphi}{\partial\psi}=0.\end{cases}$$

The *stagnation points* of the flow are those points, if any, where the velocity $\mathbf{q}=\mathbf{0}$. These are found from the equations

$$U\cos\theta(1-a^3r^{-3})=0=U\sin\theta(1+\tfrac{1}{2}a^3r^{-3})$$

which, since $r\geqslant a$, are consistently satisfied only by $r=a$, $\sin\theta=0$, i.e. by $r=a$, $\theta=0$ or π. These correspond to the points A, B in Fig. 3.9. We can find the *stagnation pressure* p_0 at A by applying Bernoulli's equation along the streamline XA. Taking p_∞ to be the pressure at infinity where the velocity is $-U\mathbf{i}$, we have

$$\frac{p_\infty}{\rho}+\tfrac{1}{2}U^2=\frac{p_0}{\rho}+0,$$

so that

$$p_0=p_\infty+\tfrac{1}{2}\rho U^2. \tag{3}$$

The pressure at any point $C\,(a,\theta,\psi)$ on the surface of the sphere may be found. At such a point we have

$$q_r=0,\quad q_\theta=\tfrac{3}{2}U\sin\theta,\quad q_\psi=0.$$

Applying Bernoulli's equation along the streamline XAC if p denote the pressure at C then

$$\frac{p}{\rho}+\tfrac{1}{2}(\tfrac{3}{2}U\sin\theta)^2=\frac{p_\infty}{\rho}+\tfrac{1}{2}U^2,$$

so that

$$p=p_\infty+\tfrac{1}{2}\rho U^2(1-\tfrac{9}{4}\sin^2\theta). \tag{4}$$

Now max. $\sin^2 \theta = 1$, so that (4) shows that

$$\text{min. } p = p_\infty - \tfrac{5}{8}\rho U^2.$$

Thus the minimum pressure occurs on the equatorial line $r = a$, $\theta = \frac{1}{2}\pi$. When min. $p = 0$, we have

$$U = \sqrt{(8p_\infty / 5\rho)}. \qquad (5)$$

This indicates that when the speed U of the undisturbed stream is increased, a point will be reached at which the pressure everywhere along the equatorial line $r = a$, $\theta = \frac{1}{2}\pi$ falls to zero. At this stage the fluid will tend to break away from the surface of the sphere along the equatorial line: *cavitation* is said to ensue. (*N.B.* If we take account of the finite aqueous vapour pressure of the fluid at the given temperature of operation, cavitation will actually take place along the equatorial line when this aqueous vapour pressure is reached, before the pressure falls to zero.)

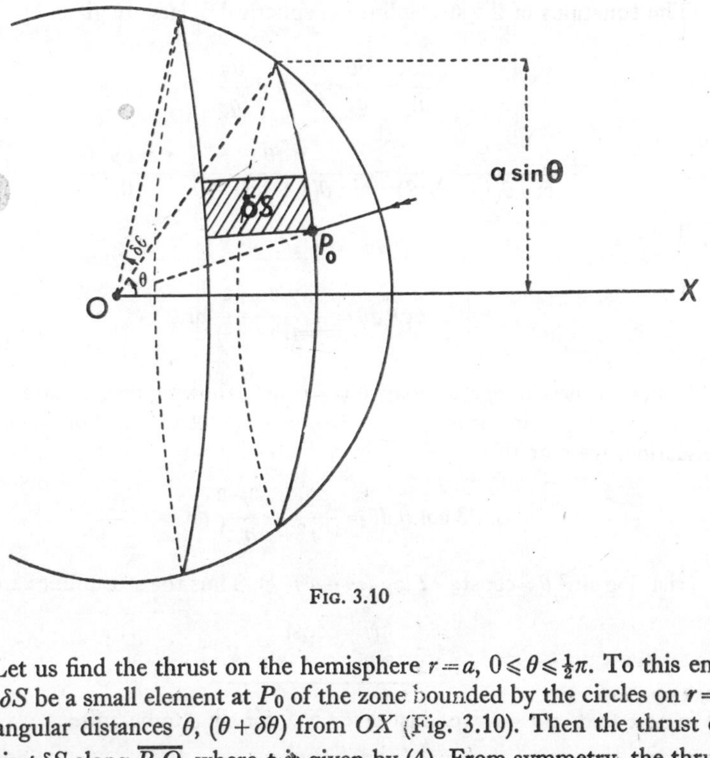

FIG. 3.10

Let us find the thrust on the hemisphere $r = a$, $0 \leqslant \theta \leqslant \frac{1}{2}\pi$. To this end, let δS be a small element at P_0 of the zone bounded by the circles on $r = a$ at angular distances θ, $(\theta + \delta\theta)$ from OX (Fig. 3.10). Then the thrust on δS is $p\delta S$ along $\overline{P_0 O}$, where p is given by (4). From symmetry, the thrust on the zone bounded by the circles is along $\overline{XO}$. The component of thrust on δS along $\overline{XO}$ is $p \cos \theta \delta S$. Hence the total thrust on the zone is along $\overline{XO}$ and of amount

$$\int_{\text{zone}} p \cos \theta \, dS = p \cos \theta \int_{\text{zone}} dS = p \cos \theta (2\pi a^2 \sin \theta \, \delta\theta),$$

since $p \cos \theta$ is, to the first order, constant over the zone. Hence the total thrust on the hemisphere $r=a$, $0 \leqslant \theta \leqslant \frac{1}{2}\pi$ is

$$2\pi a^2 \int_0^{\pi/2} \sin \theta \cos \theta (p_\infty + \tfrac{1}{2}\rho U^2 - \tfrac{9}{8}\rho U^2 \sin^2 \theta) \, d\theta$$
$$= \underline{\pi a^2 (p_\infty - \tfrac{1}{16}\rho U^2)}.$$

(*N.B.* The thrust on the entire sphere, obtained by integrating the same function from $\theta=0$ to π, is easily found to be zero. This result can be generalized. The total thrust on a rigid body of any shape in a uniform stream is zero. This is called *d'Alembert's Paradox.*)

The equations of the streamlines in spherical polars are given by

$$\frac{dr}{q_r} = r\frac{d\theta}{q_\theta} = r \sin \theta \frac{d\psi}{q_\psi}$$

or

$$\frac{dr}{-\cos \theta (1 - a^3 r^{-3})} = \frac{r \, d\theta}{\sin \theta (1 + \frac{1}{2}a^3 r^{-3})} = \frac{r \sin \theta \, d\psi}{0}.$$

Thus

$$\begin{cases} d\psi = 0, \\ 2 \cot \theta \, d\theta = -\dfrac{2r^3 + a^3}{r(r^3 - a^3)} dr. \end{cases}$$

The first of these integrates to give $\psi = \text{const.}$, showing that the streamlines lie in planes passing through the axis of symmetry OX. For the second equation, we note that

$$2 \cot \theta \, d\theta = -\frac{2r + a^3 r^{-2}}{r^2 - a^3 r^{-1}} dr,$$

so that $\log \sin^2 \theta = \text{const.} - 2 \log (r^2 - a^3 r^{-1})$. Thus the streamlines are

$$\underline{\begin{cases} \psi = \text{const.} \\ \sin^2 \theta = Cr/(r^3 - a^3). \end{cases}}$$

We note that on $r=a$, we must take $C=0$ for $\sin \theta$ to be finite.

EXAMPLE 2

Sphere moving with constant velocity in liquid which is otherwise at rest.

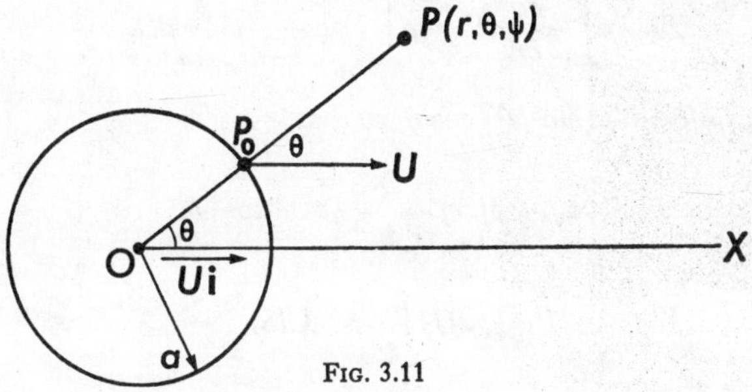

FIG. 3.11

Figure 3.11 shows a solid sphere, centre O, radius a, moving with uniform velocity $U\mathbf{i}$ in incompressible fluid of infinite extent which is at rest at infinity. OX is the axis of symmetry and the direction of the unit vector $\mathbf{i}$. As we require φ to be finite at infinity we take the velocity potential at $P(r, \theta, \psi)$ $(r \geqslant a)$ in the form

$$\varphi(r, \theta) = Ar^{-2} \cos \theta \tag{1}$$

which satisfies the axially symmetric form of Laplace's equation in spherical polar coordinates. From (1),

$$\left\{\begin{aligned} q_r &= -\frac{\partial \varphi}{\partial r} = \frac{2A \cos \theta}{r^3} \\ q_\theta &= -\frac{1}{r}\frac{\partial \varphi}{\partial \theta} = \frac{A \sin \theta}{r^3} \\ q_\psi &= -\frac{1}{r \sin \theta}\frac{\partial \varphi}{\partial \psi} = 0. \end{aligned}\right.$$

Hence at $P_0(a, \theta, \psi)$ on the surface,

$$\left\{\begin{aligned} q_r &= 2Aa^{-3} \cos \theta, \\ q_\theta &= Aa^{-3} \sin \theta, \\ q_\psi &= 0. \end{aligned}\right.$$

Now the velocity of P_0 is $U\mathbf{i}$. Hence $q_r = U \cos \theta$.

$$\therefore\ U \cos \theta = 2Aa^{-3} \cos \theta$$

or

$$A = \tfrac{1}{2}Ua^3.$$

Hence, for $r \geqslant a$,

$$\underline{\varphi(r, \theta) = \tfrac{1}{2}Ua^3 r^{-2} \cos \theta.} \tag{2}$$

To find the K.E. of the fluid, we may use the result of Section 3.8, Theorem IV. Then, denoting by S the surface of the sphere, the K.E. is given by

$$T=\tfrac{1}{2}\rho\int_S \varphi\frac{\partial\varphi}{\partial n}dS=\tfrac{1}{2}\rho\int_S\left\{\varphi\left(-\frac{\partial\varphi}{\partial r}\right)\right\}_{r=a}dS.$$

From (2), $\partial\varphi/\partial r=-(Ua^3/r^3)\cos\theta$, so that

$$\left\{\varphi(-\partial\varphi/\partial r)\right\}_{r=a}=\tfrac{1}{2}aU^2\cos^2\theta.$$

$$\therefore\ T=\tfrac{1}{4}\rho aU^2\int_S\cos^2\theta\,dS.$$

$$\therefore\ T=\tfrac{1}{4}\rho aU^2\int_0^\pi\cos^2\theta\cdot 2\pi a^2\sin\theta\,d\theta$$

$$=\tfrac{1}{6}\rho\pi U^2a^3[-\cos^3\theta]_0^\pi$$

$$=\tfrac{1}{3}\rho\pi a^3U^2=\tfrac{1}{4}M'U^2,$$

where M' is the mass of fluid displaced. The total kinetic energy of the sphere and fluid is thus $\tfrac{1}{2}(M+\tfrac{1}{2}M')U^2$. The quantity $M+\tfrac{1}{2}M'$ is called the *virtual mass* of the sphere.

Example 3

Accelerating sphere moving in fluid at rest at infinity.

A sphere of centre O and radius a moves through an infinite liquid of constant density ρ at rest at infinity, O describing a straight line with velocity $V(t)$. If there are no body forces, show that the pressure p at points on the surface of the sphere in a plane perpendicular to the straight line at a distance x from O measured positively in the direction of v is given by

$$p=p_0-\tfrac{5}{8}\rho V^2+\tfrac{9}{8}\rho V^2\frac{x^2}{a^2}+\tfrac{1}{2}\rho x\frac{dV}{dt},$$

where p_0 is the pressure at infinity.

Deduce that the thrust on the sphere is $\tfrac{1}{2}M'(dV/dt)$, where M' is the mass of the liquid having the volume of the sphere.

(London Univ., B.Sc. (Spec.), 1957)

At $P(r, \theta, \psi)$ in the liquid, assume

$$\varphi(r,\theta,t)=A(t)r^{-2}\cos\theta,$$

which satisfies the spherical polar form of Laplace's equation and ensures conditions of zero velocity at infinity. Using the boundary condition

$$(-\partial\varphi/\partial r)_{r=a}=V(t)\cos\theta,$$

we find

$$A(t)=\tfrac{1}{2}V(t)a^3.$$

Hence

$$\varphi(r,\,\theta,\,t)=\tfrac{1}{2}V(t)a^3r^{-2}\cos\theta. \tag{1}$$

Bernoulli's equation for time-varying flow with no body forces is

$$\frac{p}{\rho}+\tfrac{1}{2}q^2-\frac{\partial\varphi}{\partial t}=f(t), \tag{2}$$

where $\partial\varphi/\partial t$ is to be evaluated subject to the coordinates of P being fixed. At $(r,\,\theta,\,\psi)$, for $r\geqslant a$,

$$\begin{cases} q_r=-\dfrac{\partial\varphi}{\partial r}=Var^{-3}\cos\theta, \\[2ex] q_\theta=-\dfrac{1}{r}\,\dfrac{\partial\varphi}{\partial\theta}=\dfrac{Va^3}{2r^3}\sin\theta, \\[2ex] q_\psi=-\dfrac{1}{r\sin\theta}\,\dfrac{\partial\varphi}{\partial\psi}=0. \end{cases}$$

Hence

$$q^2=\tfrac{1}{4}Va^6r^{-6}(4\cos^2\theta+\sin^2\theta),\quad V=V(t).$$

Now

$$\left(\frac{\partial\varphi}{\partial t}\right)_{P\text{ fixed}}=\tfrac{1}{2}a^3\frac{\partial}{\partial t}\left(\frac{V}{r^2}\cos\theta\right)_{P\text{ fixed}}$$

$$=\tfrac{1}{2}a^3\left\{\frac{\cos\theta}{r^2}\,\frac{dV}{dt}+V\frac{\partial}{\partial t}\left(\frac{\cos\theta}{r^2}\right)\right\}_{P\text{ fixed}}.$$

Since $\cos\theta/r^2=(\mathbf{r}\cdot\mathbf{i})/r^3$,

$$\frac{\partial}{\partial t}\left(\frac{\cos\theta}{r^2}\right)=\mathbf{i}\cdot\frac{\partial\mathbf{r}}{\partial t}\,r^{-3}+(\mathbf{r}\cdot\mathbf{i})\,(-3r^{-4})\,\frac{\partial r}{\partial t}.$$

With P considered fixed, since $\overline{PO}\equiv-\mathbf{r}$,

$$(\partial/\partial t)\,(-\mathbf{r})=\text{vel. of } O=V\mathbf{i},$$

and so

$$\partial\mathbf{r}/\partial t=-V\mathbf{i}.$$

Now from

$$r^2=\mathbf{r}^2,$$

$$2r\frac{\partial r}{\partial t}=2\mathbf{r}\cdot\frac{\partial\mathbf{r}}{\partial t},$$

or

$$\left(\frac{\partial r}{\partial t}\right)_{P\text{ fixed}}=\frac{1}{r}(-V\mathbf{i}\cdot\mathbf{r}).$$

Thus

$$\frac{\partial}{\partial t}\left(\frac{\cos\theta}{r^2}\right)=\frac{V}{r^3}(-1+3\cos^2\theta),$$

and so

$$\left(\frac{\partial\varphi}{\partial t}\right)_{P\text{ fixed}}=\frac{a^3}{2r^3}\left(r\cos\theta\frac{dV}{dt}-V^2+3V^2\cos^2\theta\right). \tag{3}$$

Thus (2) becomes

$$\frac{p}{\rho}+\frac{V^2a^6}{4r^6}(4\cos^2\theta+\sin\theta)-\frac{a^3}{2r^3}\left(r\cos\theta\frac{dV}{dt}-V^2+3V^2\cos^2\theta\right)$$

$=f(t)=p_0/\rho$, using conditions at infinity. Putting $r=a$, $x=a\cos\theta$ in this last result gives, on simplification,

$$\underline{p=p_0-\tfrac{5}{8}\rho V^2+\tfrac{9}{8}\rho V^2\frac{x^2}{a^2}+\tfrac{1}{2}\rho x\frac{dV}{dt}}. \tag{4}$$

The thrust on a circular band on the sphere whose edges are at angular distances θ, $(\theta+\delta\theta)$ from OX is along $\overline{XO}$ and of magnitude $p\cos\theta\,(2\pi a\sin\theta\cdot a\delta\theta)=-2\pi px\delta x$. Thus the total thrust on the sphere is along $\overline{XO}$ and

$$=-2\pi\int_a^{-a}\left(p_0-\tfrac{5}{8}\rho V^2+\tfrac{9}{8}\rho V^2\frac{x^2}{a^2}+\tfrac{1}{2}\rho x\frac{dV}{dt}\right)x\,dx$$

$$=\tfrac{2}{3}\pi\rho a^3\frac{dV}{dt}=\underline{\tfrac{1}{2}M'\frac{dV}{dt}}.$$

EXAMPLE 4

Underwater explosion giving spherical gas bubble.

An explosion takes place under water initially at rest. As a result, a spherical gas bubble of centre O is formed (Fig. 3.12). During expansion, the gas obeys the adiabatic law $pv^\gamma=\text{const.}$, where p is the pressure, v the volume of the gas and γ is a constant. It is required to find the radius R of the gas bubble at any time t after the explosion, taking R_0 as the radius when $t=0$.

At time t, when the gas bubble has radius R, let $r(\geqslant R)$ be the radius of a concentric spherical surface in the water. Since the water is incom-

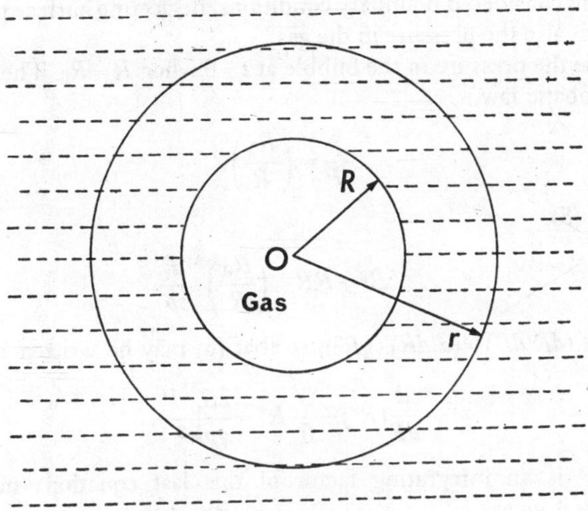

FIG. 3.12

pressible, the equation of continuity applied to the two concentric surfaces is

$$4\pi R^2\dot{R}=4\pi r^2\dot{r}$$

or

$$\dot{r}=R^2\dot{R}/r^2. \tag{1}$$

At the point whose spherical polar coordinates are (r, θ, ψ) on the surface of radius r, we have $\dot{r}=-\partial\varphi/\partial r$, where $\varphi(r, t)$ is the velocity potential. Thus we may take

$$\varphi(r, t)=R^2\dot{R}/r. \tag{2}$$

Bernoulli's equation at (r, θ, ψ) for unsteady incompressible flow under no body forces is

$$\frac{p}{\rho}+\tfrac{1}{2}\dot{r}^2-\left(\frac{\partial\varphi}{\partial t}\right)_r=f(t). \tag{3}$$

Assuming zero pressure at infinity, (3) gives

$$\frac{p}{\rho}+\frac{R^4\dot{R}^2}{2r^4}-\frac{1}{r}\frac{\partial}{\partial t}(R^2\dot{R})=0.$$

Putting $r=R$,

$$\frac{p}{\rho}+\tfrac{1}{2}\dot{R}^2-\frac{1}{R}(R^2\ddot{R}+2R\dot{R}^2)=0,$$

or

$$p=\rho(\tfrac{3}{2}\dot{R}^2+R\ddot{R}). \tag{4}$$

In (4), from considered boundary conditions, neglecting surface tension, we see that p is also the pressure in the gas.

Let P be the pressure in the bubble at $t=0$ when $R=R_0$. Then in virtue of the adiabatic law,

$$\frac{p}{P}=\left(\frac{R_0}{R}\right)^{3\gamma}. \tag{5}$$

From (4), (5),

$$\tfrac{3}{2}\dot{R}^2+R\ddot{R}=\left(\frac{R_0}{R}\right)^{3\gamma}\frac{P}{\rho}. \tag{6}$$

Now $\ddot{R}=\dot{R}(d\dot{R}/dR)=(d/dR)\,(\tfrac{1}{2}\dot{R}^2)$, so that (6) may be written

$$\frac{d}{dR}(\dot{R}^2)+\frac{3}{R}\dot{R}^2=\frac{2R_0^{3\gamma}P}{\gamma\rho R^3}.$$

Clearly R^3 is an integrating factor of this last equation: multiplying through by it gives

$$\frac{d}{dR}(R^3\dot{R}^2)=\frac{2\gamma R_0^3P}{\rho R^{3\gamma-2}}$$

whence

$$R^3\dot{R}^2=\frac{2R_0^{3\gamma}}{3(\gamma-1)}\,\frac{P}{\rho}\{R_0^{-3(\gamma-1)}-R^{-3(\gamma-1)}\}$$

or

$$\dot{R}^2=\frac{2P}{3(\gamma-1)\rho}\left\{\left(\frac{R_0}{R}\right)^3-\left(\frac{R_0}{R}\right)^{3\gamma}\right\}. \tag{7}$$

In the special case when $\gamma=4/3$, a typical value for most gases, the solution may be completed. For

$$\frac{dR}{dt}=\left[\frac{2P}{\rho}\left\{\left(\frac{R_0}{R}\right)^3-\left(\frac{R_0}{R}\right)^4\right\}\right]^{1/2}$$

or

$$dt=\frac{R^2dR}{\{(2PR_0^3/\rho)\,(R-R_0)\}^{1/2}},$$

and substituting $R=R_0+x$, we find on integration

$$t=(2\rho x/PR_0^3)^{1/2}(\tfrac{1}{5}x^2+\tfrac{2}{3}R_0+R_0^2).$$

3.10 Some Special Two-dimensional Flows

Suppose a fluid moves in such a way that its motions in all planes parallel to a given plane in the fluid have the same pattern at any considered instant of time. Such motion is said to be *two-dimensional*, since knowledge of the motion in any one such plane determines that throughout the body of fluid. Also there can be no motion perpendicular to such a plane.

For such a two-dimensional motion, let the z-axis be taken normal to each of the parallel planes in which the patterns of flow are congruent at any instant. Then for irrotational incompressible flow, the equation of continuity in Cartesians (x, y, z) is

$$\frac{\partial^2 \varphi}{\partial x^2}+\frac{\partial^2 \varphi}{\partial y^2}=0 \tag{1}$$

and in cylindrical polars (R, θ, z) it is

$$\frac{\partial^2 \varphi}{\partial R^2}+\frac{1}{R}\frac{\partial \varphi}{\partial R}+\frac{1}{R^2}\frac{\partial \varphi^2}{\partial \theta^2}=0. \tag{2}$$

Both forms and their solutions will be discussed in greater detail in Chapter 5. Here we discuss some of the simpler types of solution arising from equation (2).

Special solutions of (2) may be found by the *method of separation of variables*. This is treated in detail in Chapter 5. Here we content ourselves with the easily verifiable facts that $\varphi=R \cos \theta$, $\varphi=(1/R) \cos \theta$ are special solutions. Hence, since the equation is linear, a more general type of solution is

$$\varphi=(AR+BR^{-1}) \cos \theta. \tag{3}$$

EXAMPLE

A stationary infinite right circular solid cylinder of radius a is placed in a uniform stream, its axis being perpendicular to the direction of flow.

As this problem bears some resemblance to the problem of the sphere in uniform flow (Example 1, Section 3.9), we refer to Fig. 3.9 of Section 3.9 in which the diagram now represents a section of the cylinder. Let P have cylindrical polar coordinates (R, θ, z). The undisturbed velocity potential at P will be $Ux=UR \cos \theta$, so that when the cylinder is inserted we take, for $R \geqslant a$,

$$\varphi(R, \theta)=UR \cos \theta+AR^{-1} \cos \theta, \tag{4}$$

the last term being a suitable perturbation potential which satisfies (2). On $R=a$, we require $\partial\varphi/\partial R=0$, so that $A=Ua^2$.
Thus

$$\varphi(R, \theta)=U \cos \theta(R+a^2R^{-1}). \tag{5}$$

Using (5), analysis may now proceed in like manner to that of flow past a stationary sphere.

3.11 Impulsive Motion

Suppose sudden velocity changes are produced at the boundaries of an incompressible fluid or that impulsive forces are applied to its interior.

Then the disturbances produced in both cases are propagated instantaneously throughout the fluid. (We shall see in Chapter 7 that in the case of a compressible fluid the changes are effected in a finite time, depending on the speed of sound in the medium.)

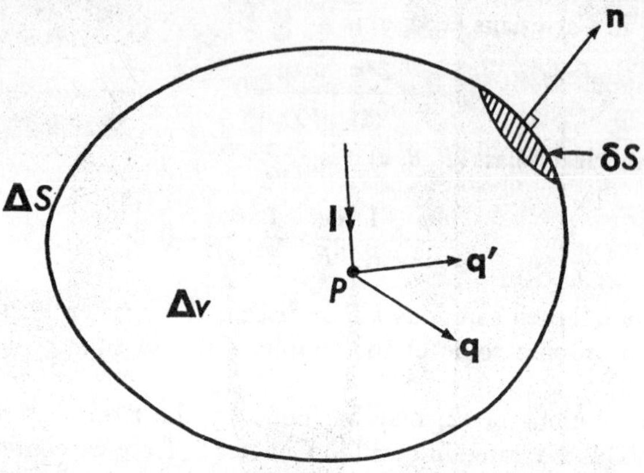

FIG. 3.13

To put the matter on a quantitative basis, consider the incompressible fluid particle of volume Δv bounded by a closed surface ΔS (Fig. 3.13). Suppose the particle is moving initially with velocity $\mathbf{q}$ and that it is subjected to a sudden external impulsive body force $\mathbf{I}$ per unit mass. Let this impulse change the velocity of the particle instantaneously from $\mathbf{q}$ to $\mathbf{q}'$ and let it produce impulsive pressure on the boundary ΔS. If we denote the impulsive pressure at the element δS of ΔS (where $\delta S \ll \Delta S$), by P then since the impulse on a volume element δv of Δv is $\mathbf{I}\rho\delta v$ and that on δS (having unit *outward* normal $\mathbf{n}$) is $-\mathbf{n}P\delta S$, the total impulse applied to the fluid particle is

$$\int_{\Delta v} \rho \mathbf{I}\, dv - \int_{\Delta S} \mathbf{n}P\, dS = \int_{\Delta v} (\rho\mathbf{I} - \nabla P)\, dv.$$

The momentum change is

$$\int_{\Delta v} \rho(\mathbf{q}' - \mathbf{q})\, dv.$$

Hence

$$\int_{\Delta v} (\rho\mathbf{I} - \nabla P)\, dv = \int_{\Delta v} \rho(\mathbf{q}' - \mathbf{q})\, dv.$$

Since Δv is an arbitrarily small volume, our usual assumption of fluid continuity yields the result

$$\rho\mathbf{I} - \nabla P = \rho(\mathbf{q}' - \mathbf{q}) \tag{1}$$

at each point P of the fluid. When there are no externally applied impulses, (1) gives

$$-\nabla P = \rho(\mathbf{q}' - \mathbf{q}).$$

If the motion is irrotational and $\mathbf{q} = -\nabla\varphi$, $\mathbf{q}' = -\nabla\varphi'$, then this last result is satisfied by

$$P = \rho(\varphi' - \varphi), \tag{2}$$

ignoring immaterial integration constants.

Example

An incompressible fluid is contained within the region bounded by two concentric rigid spherical surfaces of radii R_1, $R_2 (R_2 > R_1)$. The fluid is initially at rest. If the inner surface is now given a sudden velocity $U\mathbf{i}$, where $\mathbf{i}$ is a constant unit vector, show that the impulsive thrust on the outer surface is

$$2\pi\rho R_1^3 R_2^3 (R_2^3 - R_1^3)^{-1} U\mathbf{i},$$

where ρ is the fluid density.

For the motion set up assume that at the point having spherical polar coordinates (r, θ, ψ) in the fluid (referred to the direction of $\mathbf{i}$ through the centre as initial line) the velocity potential is

$$\varphi(r, \theta) = (Ar + Br^{-2}) \cos\theta \quad (R_1 \leqslant r \leqslant R_2). \tag{1}$$

On $r = R_2$, $\partial\varphi/\partial r = 0$, so that

$$A - 2BR_2^{-3} = 0. \tag{2}$$

On $r = R_1$, $\partial\varphi/\partial r = -U\cos\theta$, so that

$$A - 2BR_1^{-3} = -U. \tag{3}$$

From (2), (3),

$$\begin{cases} A = \tfrac{1}{2} U R_1^3 R_2^3 (R_2^3 - R_1^3)^{-1} \\ B = U R_1^3 (R_2^3 - R_1^3)^{-1} \end{cases}$$

Hence

$$\varphi(r, \theta) = U R_1^3 (R_2^3 - R_1^3)^{-1} [r + (R_2^3/2r^2)] \cos\theta. \tag{4}$$

Since the impulsive pressure at any point is $P(r, \theta) = \rho\varphi$, that on the surface $r = R_2$ is

$$P(R_2, \theta) = \tfrac{3}{2}\rho U R_1^3 R_2 (R_2^3 - R_1^3)^{-1} \cos\theta. \tag{5}$$

In Fig. 3.14, δS in an element at Q on the surface $r = R_2$ of the circular zone whose boundaries are at angular distances θ, $\theta + \delta\theta$ from $\overline{OX}$, drawn in the direction $\mathbf{i}$. The impulsive thrust on δS is $P(R_2, \theta)\delta S$ along $\overline{OQ}$.

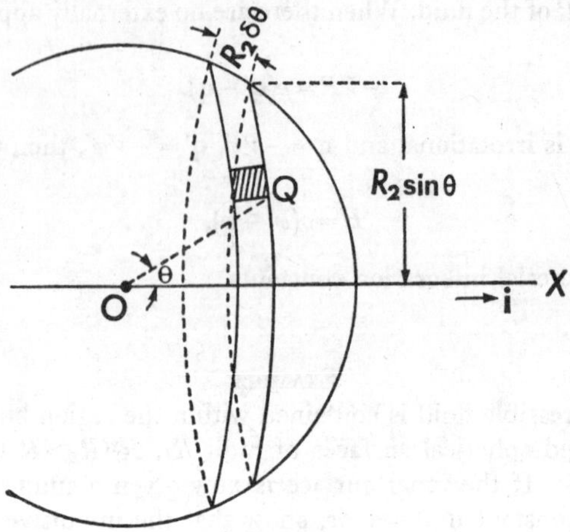

FIG. 3.14

This has a component along **i** of $P(R_2, \theta)\cos\theta\delta S$. The total thrust on the circular zone is along $\overline{OX}$ and its magnitude is

$$\int_{\text{zone}} P(R_2, \theta)\cos\theta\, dS = P(R_2, \theta)\cos\theta \times 2\pi R_2^2 \sin\theta\,\delta\theta.$$

$$= 3\pi\rho U R_1^3 R_2^3 (R_2^3 - R_1^3)^{-1}\cos^2\theta\sin\theta\,\delta\theta.$$

Hence the total thrust on the outer shell is along **i** and of magnitude

$$3\pi\rho U R_1^3 R_2^3 (R_2^3 - R_1^3)^{-1}\int_0^\pi \cos^2\theta\sin\theta\, d\theta$$

$$= \underline{2\pi\rho U R_1^3 R_2^3 (R_2^3 - R_1^3)^{-1}}.$$

3.12 Some Further Aspects of Vortex Motion

We first establish *Kelvin's theorem* expressing the conditions under which the circulation in a closed circuit of fluid particles moving along with the fluid is constant.

For an inviscid fluid the circulation around any closed circuit of fluid particles, moving along with the fluid, remains constant provided that the body forces are conservative and the pressure is a single-valued function of density only.

Proof: In the notation previously used, Euler's equation of motion is

$$\frac{d\mathbf{q}}{dt} = \mathbf{F} - \frac{1}{\rho}\nabla p$$

$$= -\nabla\Omega - \frac{1}{\rho}\nabla p,$$

if Ω is the potential of the conservative body forces. Now let $\mathscr{C}$ be a closed circuit of fluid particles moving along with the fluid and suppose that $\mathbf{r}$, $(\mathbf{r}+d\mathbf{r})$ are the position vectors of neighbouring particles P, P' of $\mathscr{C}$, referred to a fixed origin O. Then the velocity of P is $\mathbf{q}=d\mathbf{r}/dt$. Also the circulation round $\mathscr{C}$ is

$$\Gamma=\oint_{\mathscr{C}} \mathbf{q}\cdot d\mathbf{r}.$$

The time-derivative of Γ following the motion of the fluid is

$$\frac{d\Gamma}{dt}=\oint_{\mathscr{C}} \frac{d}{dt}(\mathbf{q}\cdot d\mathbf{r})=\oint_{\mathscr{C}}\left\{\frac{d\mathbf{q}}{dt}\cdot d\mathbf{r}+\mathbf{q}\cdot\frac{d}{dt}(d\mathbf{r})\right\}.$$

$$=\oint_{\mathscr{C}}\left\{\mathbf{q}\cdot\frac{d}{dt}(d\mathbf{r})-\nabla\Omega\cdot d\mathbf{r}-\frac{1}{\rho}\nabla p\cdot d\mathbf{r}\right\}.$$

We now show that $(d/dt)\,(d\mathbf{r})=d(d\mathbf{r}/dt)=d\mathbf{q}$. This result, though obvious intuitively, needs careful establishment. At time t, P, P' have the position vectors $\mathbf{r}$, $(\mathbf{r}+d\mathbf{r})$, referred to O. Then at time $(t+dt)$, their position vectors are

$$\mathbf{r}+\frac{d\mathbf{r}}{dt}dt,\quad (\mathbf{r}+d\mathbf{r})+\left\{\frac{d}{dt}(\mathbf{r}+d\mathbf{r})\right\}dt.$$

Since $d\mathbf{r}$ denotes a variation from the fluid element at P to that at P', the position vector of P' at time $(t+dt)$ may be written

$$r+\frac{d\mathbf{r}}{dt}dt+d\left(\mathbf{r}+\frac{d\mathbf{r}}{dt}dt\right).$$

Comparison of the two forms establishes the desired result.

We thus have

$$\frac{d\Gamma}{dt}=\oint_{\mathscr{C}}\left(\mathbf{q}\cdot d\mathbf{q}-\nabla\Omega\cdot d\mathbf{r}-\frac{1}{\rho}\nabla p\cdot d\mathbf{r}\right),$$

the differentials denoting variations along the arc of $\mathscr{C}$ from P to P'. Otherwise expressed,

$$\frac{d\Gamma}{dt}=\oint_{\mathscr{C}}\left(d(\tfrac{1}{2}\mathbf{q}^2)-d\Omega-\frac{1}{\rho}dp\right)=-\oint\frac{dp}{\rho}.$$

If now p is a single-valued function of ρ, the last integral is zero and we have

$$\Gamma=\text{const. in the moving circuit.}$$

The following important corollaries stem from Kelvin's theorem:

1. *In a closed circuit $\mathscr{C}$ of fluid particles moving under the same conditions as before,* $\int_S \mathbf{n}\cdot\boldsymbol{\zeta}\,dS$ *is constant, where S is an open surface with $\mathscr{C}$ as rim.*

Proof: For $\Gamma=\oint_{\mathscr{C}} \mathbf{q}\cdot d\mathbf{s}=\int_S \mathbf{n}\cdot\boldsymbol{\zeta}\,dS$,

which establishes the result since Γ is constant.

2. *Under the above conditions, if any portion of the moving fluid once becomes irrotational, then it remains so for all subsequent times.*

Proof: Suppose that at some instant the fluid on S becomes irrotational. Then $\boldsymbol{\zeta}=\mathbf{0}$ at all points of S and the last result shows $\Gamma=0$ at the given instant and hence at all subsequent times. Thus at any later stage,

$$\int_S \mathbf{n}\cdot\boldsymbol{\zeta}\,dS=0.$$

If we now take S to be a non-zero infinitesimal element, say ΔS, then to the first order

$$\Delta S\mathbf{n}\cdot\boldsymbol{\zeta}=0$$

showing that in general $\boldsymbol{\zeta}=\mathbf{0}$ at any point of ΔS, i.e. that the motion stays irrotational.

3. *With the above conditions, the vortex lines move along with the fluid.*

Proof: For a surface S moving with the fluid we have seen that $\int_S \mathbf{n}\cdot\boldsymbol{\zeta}\,dS=$ const. The integral represents the total strength of vortex tubes passing through S. This shows that the vortex tubes move with the fluid. By taking S vanishingly small, we infer that the vortex lines move with the fluid.

Alternative Proof: Let $\mathscr{C}$ be any closed curve drawn on the surface of a vortex tube containing an area S of the tube and not embracing the tube. Then $\mathbf{n}\cdot\boldsymbol{\zeta}=0$ at each point of S. Thus $\oint_{\mathscr{C}} \mathbf{q}\cdot d\mathbf{s}=0$. Now let the circuit $\mathscr{C}$ and the surface S move with the fluid. Then for irrotational flow, the circulation round $\mathscr{C}$ remains zero. This implies that $\int_S \mathbf{n}\cdot\boldsymbol{\zeta}\,dS=0$, i.e. that $\mathbf{n}\cdot\boldsymbol{\zeta}$ remains zero, or that S remains a surface composed of vortex lines, Thus vortex lines and tubes move with the fluid.

Example

Define the vorticity $\boldsymbol{\omega}$ in the motion of a continuous medium with velocity $\mathbf{v}(x, y, z, t)$. Show that for a motion of an inviscid incompressible fluid of uniform density, under gravity, the vorticity satisfies the equation

$$\frac{\partial\boldsymbol{\omega}}{\partial t}+(\mathbf{v}\cdot\nabla)\boldsymbol{\omega}=(\boldsymbol{\omega}\cdot\nabla)\mathbf{v},$$

and explain the significance of each term in this equation.

A motion, symmetric about the axis $z=0$, is described in terms of cylindrical polar coordinates (r, θ, z), the velocity having components $v_r(r, z)$, $v_\theta=0$, $v_z(r, z)$. By evaluating the term $(\boldsymbol{\omega} \cdot \nabla)\mathbf{v}$, or otherwise, show that if the fluid element has vorticity ω_0 when at radius r_0, its vorticity ω at radius r is given by $r_0\omega=r\omega_0$.

(London Univ., B.Sc. Spec. I, 1963)

The notation of the question is slightly different from that of the text. For clarity we define $\boldsymbol{\omega}=\nabla \wedge \mathbf{v}$.

Euler's equation of motion under the gravitational body force $\mathbf{g}$ per unit mass is

$$\frac{d\mathbf{v}}{dt}=\mathbf{g}-\frac{1}{\rho}\nabla p,$$

the time-differentiation following the fluid particle's motion. Since the gravitational force is conservative, $\mathbf{g}=-\nabla V$, where V is a suitable potential function. Also,

$$\frac{d\mathbf{v}}{dt}=\frac{\partial \mathbf{v}}{\partial t}+(\mathbf{v}\cdot\nabla)\mathbf{v}=\frac{\partial \mathbf{v}}{\partial t}+\nabla(\tfrac{1}{2}\mathbf{v}^2)-\mathbf{v}\wedge\boldsymbol{\omega}.$$

Hence $(\partial\mathbf{v}/\partial t)+\nabla(\frac{1}{2}\mathbf{v}^2)-\mathbf{v}\wedge\boldsymbol{\omega}=-\nabla\{V+(p/\rho)\}$. Taking the curl of both sides of this equation,

$$\nabla\wedge\frac{\partial \mathbf{v}}{\partial t}-\nabla\wedge(\mathbf{v}\wedge\boldsymbol{\omega})=\mathbf{0}. \tag{1}$$

Now

$$\nabla\wedge\frac{\partial \mathbf{v}}{\partial t}=\frac{\partial}{\partial t}(\nabla\wedge\mathbf{v})=\frac{\partial \boldsymbol{\omega}}{\partial t}.$$

Also, from Section 1.15,

$$\nabla\wedge(\mathbf{v}\wedge\boldsymbol{\omega})=(\nabla\cdot\boldsymbol{\omega})\mathbf{v}-(\mathbf{v}\cdot\nabla)\boldsymbol{\omega}+(\boldsymbol{\omega}\cdot\nabla)\mathbf{v}-(\nabla\cdot\mathbf{v})\boldsymbol{\omega}.$$

But $\nabla\cdot\boldsymbol{\omega}=\nabla\cdot(\nabla\wedge\mathbf{v})\equiv 0$, and $\nabla\cdot\mathbf{v}=0$—the equation of continuity of an incompressible fluid. Thus (1) becomes

$$\frac{\partial \boldsymbol{\omega}}{\partial t}+(\mathbf{v}\cdot\nabla)\boldsymbol{\omega}=(\boldsymbol{\omega}\cdot\nabla)\mathbf{v}. \tag{2}$$

The L.H.S. of (2) may be written as $d\boldsymbol{\omega}/dt$, denoting a total rate of change of $\boldsymbol{\omega}$ following the fluid. $\partial\boldsymbol{\omega}/\partial t$ denotes the local rate of change of $\boldsymbol{\omega}$. Hence $(\mathbf{v}\cdot\nabla)\boldsymbol{\omega}$ denotes the rate of change of $\boldsymbol{\omega}$ due to the motion of the fluid. In steady motion we should have $\partial\boldsymbol{\omega}/\partial t=0$ so that the rate of change of $\boldsymbol{\omega}$ would be just $(\mathbf{v}\cdot\nabla)\boldsymbol{\omega}$.

Write (2) in the form

$$d\boldsymbol{\omega}/dt=(\boldsymbol{\omega}\cdot\nabla)\mathbf{v}. \tag{3}$$

Now

$$\boldsymbol{\omega}=\nabla\wedge\mathbf{v}=\frac{1}{r}\begin{vmatrix}\hat{\mathbf{r}} & r\hat{\boldsymbol{\theta}} & \mathbf{k}\\ \dfrac{\partial}{\partial r} & \dfrac{\partial}{\partial\theta} & \dfrac{\partial}{\partial z}\\ v_r(r,z) & 0 & v_z(r,z)\end{vmatrix}$$

$$=\hat{\boldsymbol{\theta}}\left(\frac{\partial v_r}{\partial z}-\frac{\partial v_z}{\partial r}\right),$$

and so

$$\boldsymbol{\omega}\cdot\nabla\equiv\boldsymbol{\omega}\cdot\left(\hat{\mathbf{r}}\frac{\partial}{\partial r}+\hat{\boldsymbol{\theta}}\frac{1}{r}\frac{\partial}{\partial\theta}+\mathbf{k}\frac{\partial}{\partial z}\right)$$

$$\equiv\frac{1}{r}\left(\frac{\partial v_r}{\partial z}-\frac{\partial v_z}{\partial r}\right)\frac{\partial}{\partial\theta}.$$

Thus

$$\boldsymbol{\omega}\cdot\nabla\mathbf{v}=\frac{1}{r}\left(\frac{\partial v_r}{\partial z}-\frac{\partial v_z}{\partial r}\right)\frac{\partial}{\partial\theta}(v_r\hat{\mathbf{r}}+v_z\mathbf{k})$$

$$=\frac{v_r}{r}\left(\frac{\partial v_r}{\partial z}-\frac{\partial v_z}{\partial r}\right)\frac{\partial\hat{\mathbf{r}}}{\partial\theta}.$$

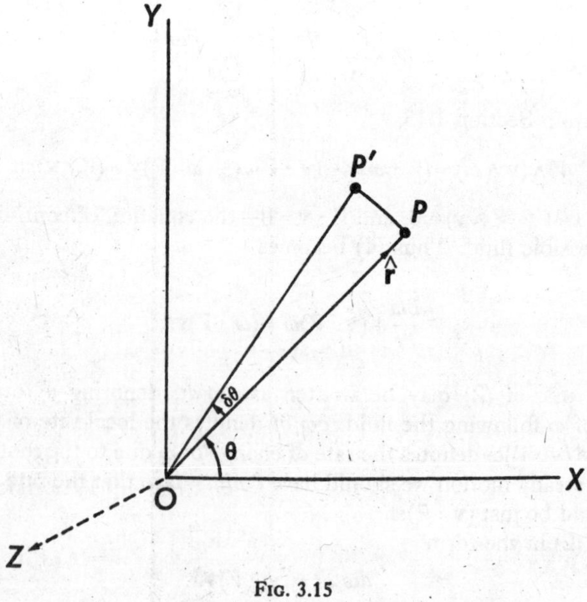

FIG. 3.15

In Fig. 3.15, $\overline{OP}$, $\overline{OP'}$ are unit vectors in the plane $z=0$, so that $\overline{OP}\equiv\hat{\mathbf{r}}$ at angular distance θ, and $\overline{OP'}$ is its value at angular distance $(\theta+\delta\theta)$. Since $\overline{PP'}\equiv\delta\theta\hat{\boldsymbol{\theta}}$, $\partial\hat{\mathbf{r}}/\partial\theta=\hat{\boldsymbol{\theta}}$. Hence

$$\boldsymbol{\omega}\cdot\nabla\mathbf{v}=\frac{v_r}{r}\left(\frac{\partial v_r}{\partial z}-\frac{\partial v_z}{\partial r}\right)\hat{\boldsymbol{\theta}}=\frac{v_r}{r}\boldsymbol{\omega}. \tag{4}$$

From (3), (4),

$$\frac{d\boldsymbol{\omega}}{dt}=\frac{v_r}{r}\boldsymbol{\omega}. \tag{5}$$

Now

$$\mathbf{v}=d\mathbf{r}/dt=v_r\hat{\mathbf{r}}+v_z\mathbf{k},$$

so that $v_r=\mathbf{v}\cdot\hat{\mathbf{r}}$ and (5) becomes

$$r\frac{d\boldsymbol{\omega}}{dt}=(\mathbf{v}\cdot\hat{\mathbf{r}})\,\boldsymbol{\omega}.$$

Since $\mathbf{r}^2=r^2$,

$$\mathbf{r}\cdot\frac{d\mathbf{r}}{dt}=r\frac{dr}{dt}$$

or

$$\hat{\mathbf{r}}\cdot\mathbf{v}=dr/dt.$$

Hence

$$r\frac{d\boldsymbol{\omega}}{dt}=\frac{dr}{dt}\boldsymbol{\omega}. \tag{6}$$

From (6),

$$\left(r\frac{d\boldsymbol{\omega}}{dt}-\boldsymbol{\omega}\frac{dr}{dt}\right)r^{-2}=0$$

or

$$\frac{d}{dt}\left(\frac{\boldsymbol{\omega}}{r}\right)=0.$$

Thus

$$\underline{\boldsymbol{\omega}/r=\text{const.}=\boldsymbol{\omega}_0/r_0.}$$

We make the following observations:

1. The result is obviously true for any conservative body forces $\mathbf{F}$, gravitational or otherwise.

2. The result would hold for a compressible fluid for which $p=p(\rho)$, for the term $(1/\rho)\nabla p$ could then be written as $\nabla\int(dp/\rho)$, the curl of which is zero.

3. If $\boldsymbol{\omega}_0=\mathbf{0}$, then $\boldsymbol{\omega}=\mathbf{0}$, so that the equation $r_0\boldsymbol{\omega}=r\boldsymbol{\omega}_0$ shows that, for the conditions imposed on the fluid, both rotational and irrotational motions are permanent, once set up.

Exercise 3

1. Derive the equation of hydrostatic equilibrium of a fluid

$$\rho \mathbf{F} = \nabla p$$

where ρ is the density (not necessarily uniform), p the pressure, and $\mathbf{F}$ the body force per unit mass.

A mass of incompressible fluid extends from infinity to the surface of a rigid sphere of radius R and centre O. The equation of state is $p = K\rho$ (K constant). Each element is attracted towards O by a force μ/r^2 per unit mass, where μ is a constant, and r is the radial distance from O. Show that the pressure exerted on the surface of the sphere is

$$K\rho_0 \exp(\mu/KR),$$

where ρ_0 is the density at infinity.

(Cambridge Univ., M.T. I., 1960)

2. A perfect incompressible fluid is moving steadily around the outside of a fixed cylinder of radius a and vertical axis Oz. The fluid particles are traversing horizontal circles with centres on Oz, the speed at distance r from Oz being a/r. Show that the motion is irrotational. If the surface of the fluid is open to the atmosphere, and the origin O is chosen so that on the free surface $z = 0$ when $r = a$, prove, by means of Bernoulli's equation or otherwise, that the equation of the free surface is

$$2gz = 1 - a^2/r^2.$$

London Univ., Gen. B.Sc. II, 1957)

3. Explain local rate of change $\partial/\partial t$ and particle rate of change D/Dt and show that for a fluid motion with particle velocity $\mathbf{v}$

$$\frac{D\mathbf{v}}{Dt} = \frac{\partial \mathbf{v}}{\partial t} + (\mathbf{v}\cdot\nabla)\mathbf{v} = \frac{\partial \mathbf{v}}{\partial t} + \nabla(\tfrac{1}{2}v^2) - \mathbf{v}\wedge\boldsymbol{\zeta},$$

where $\boldsymbol{\zeta}$ is the vorticity vector.

The particle velocity $\mathbf{v}$ for a fluid motion referred to rectangular Cartesian axes $O(x, y, z)$ with unit vectors $\mathbf{i}, \mathbf{j}, \mathbf{k}$ is given by

$$\mathbf{v} = v_0\left(\mathbf{i}\cos\frac{\pi x}{2a}\cos\frac{\pi z}{2a} + \mathbf{k}\sin\frac{\pi x}{2a}\sin\frac{\pi z}{2a}\right),$$

where v_0 is a constant. Show that this is a possible motion of an incompressible fluid motion under no body forces in an infinite fixed rigid tube $-a \leqslant x \leqslant a$, $0 \leqslant z \leqslant 2a$ and that the pressure p is given by

$$p = \tfrac{1}{4}\rho v_0^2\{\cos(\pi z/a) - \cos(\pi x/a)\} + \text{const.}$$

(London Univ., B.Sc. Spec. Hons., 1965)

4. A pipe of variable cross-section is given by $r=a(\cosh \alpha z)^{1/4}$, where r, θ, z are cylindrical polar coordinates with the z-axis vertical. Incompressible inviscid fluid is in steady irrotational motion along the pipe, a volume Q of fluid passing every cross-section per unit time. Assuming the vertical component of velocity to depend on z only, and neglecting the effects of the horizontal velocities, show that the pressure will not be a monotonic function of z if $\alpha Q^2 > 4\pi^2 a^4 g$.

(London Univ., B.Sc. Spec. Hons., 1955)

5. Prove that the equation of motion of a homogeneous inviscid liquid moving under forces arising from a potential V may be written in the form

$$\frac{\partial \mathbf{q}}{\partial t} - \mathbf{q} \wedge \boldsymbol{\zeta} = -\operatorname{grad}\left(\frac{p}{\rho} + \tfrac{1}{2}q^2 + V\right),$$

where $\boldsymbol{\zeta} = \operatorname{curl} \mathbf{q}$ is the vorticity.

If the velocity $\mathbf{q}$ referred to cylindrical polar coordinates (r, θ, z) is given by

$$\mathbf{q} = [0, \tfrac{1}{2}\omega r, 0] \quad (0 \leqslant r \leqslant a),$$
$$\mathbf{q} = [0, \tfrac{1}{2}\omega a^2/r, 0], \quad (r \geqslant a),$$

where ω is a constant, prove that the vorticity is given by

$$\boldsymbol{\zeta} = [0, 0, \omega], \quad (0 \leqslant r \leqslant a),$$
$$\boldsymbol{\zeta} = [0, 0, 0], \quad (r \geqslant a).$$

Determine the corresponding pressure distribution if the motion takes place under no external forces, and find the smallest value of the pressure at infinity which will ensure that the pressure is everywhere positive.

(London Univ., B.Sc. Spec. Hons., 1951)

6. A tube AB of uniform fine bore is in the form of the arc of the cycloid $s=4a \sin \psi$, the ends A and B being given by $\psi=0$, $\psi=\tfrac{1}{2}\pi$, respectively. It is fixed in a vertical plane so that A is its lowest point and the tangent there is horizontal. The tube, which is closed at A and open at B, is full of uniform liquid. If the end A is opened, show that the tube empties in time $\pi\sqrt{(2a/g)}$.

(London Univ., B.Sc. Spec. Hons., 1947)

7. A right circular cylinder of radius a stands with its axis vertical and its base attached to an infinite rigid horizontal plane $z=0$. It is surrounded by an ocean of incompressible non-viscous fluid of infinite extent, bounded below by the plane $z=0$ and above by its free surface open to the atmosphere at pressure P_0. The cylinder extends above the free surface of the ocean, whose depth at a large distance from the cylinder is h. If the velocity components of the fluid at the point (x, y, z) are $[-\omega a^2 y/r^2, \omega a^2 x/r^2, 0]$, where $r^2=x^2+y^2$, find (i) the pressure on the cylinder at a height z; (ii) the pressure at a point on the plane $z=0$ distant $r(>a)$ from the axis of the

cylinder. Prove that the depth of the fluid where it touches the cylinder is $h-(a^2\omega^2/2g)$, assuming that $a^2\omega^2<2gh$.

(London Univ., B.Sc. Spec. Hons., 1963)

8. A stream of water at great depth flows with constant velocity V over a plane level bottom. A solid hemisphere of radius a is fixed with its base covering a circular area of the bottom. Show that the downward force exerted by the water on the rest of the bottom is thereby increased by $\frac{11}{32}\pi\rho V^2a^2$.

(London Univ., B.Sc. Spec. Hons., 1955)

9. The velocity potential for the motion due to a sphere of radius a moving with variable velocity U in the direction $\theta=0$ in an infinite uniform inviscid incompressible fluid of density ρ is $\varphi=-(Ua^3/2r^2)\cos\theta$. Assuming a suitable form of Bernoulli's equation, deduce that the pressure at a point on the sphere is given by

$$\frac{p}{\rho}=\frac{p_0}{\rho}+gZ-\tfrac{1}{8}U^2(5-9\cos^2\theta)+\tfrac{1}{2}\dot{U}a\cos\theta,$$

when the only body force is gravity and Z is the vertical coordinate measured *downwards* from a fixed origin.

A uniform sphere of mass M, immersed in such a fluid at rest at infinity, is released from rest. Show that the pressure on its horizontal equator will fall below its initial value if $M>7M'$ where M', which is given to be less than M, is the mass of fluid displaced. (It may be assumed that the virtual mass of the sphere is $\frac{1}{2}M'$.)

(Imperial College Lond., B.Sc. Spec., 1963)

10. The space between two spheres is filled with incompressible fluid. The spheres have radii a, b, $(a<b)$ and move with constant speeds U, V respectively along the line of centres. Show that the instant when the spheres are concentric, the velocity potential is given by

$$\varphi=\frac{\{(a^3U-b^3V)r+\frac{1}{2}(U-V)a^3b^3r^{-2}\}\cos\theta}{b^3-a^3}$$

and that the equation of the streamlines takes the form

$$\{(b^3V-a^3U)r^2-(V-U)a^3b^3r^{-1}\}\sin^2\theta=\text{const.},\quad \psi=\text{const.},$$

where (r,θ,ψ) are spherical polar coordinates with origin at the common centre of the spheres and axis $\theta=0$ along their line of motion.

(Manchester Univ., Faculty of Tech., B.Sc. Tech. II, 1962)

11. Infinite inviscid fluid of constant density is attracted towards a fixed point O by a force $f(r)$ per unit mass, where r is the distance of any point from O. Initially the liquid is at rest, and there is a cavity bounded by a sphere of radius a. If there is no pressure at infinity or in the cavity, prove that the radius R of the cavity at time t is such that

$$\frac{d}{dt}\left[R^3\left(\frac{dR}{dt}\right)^2\right]+2R^2\frac{dR}{dt}\int_R^{\infty}f(r)\,dr=0.$$

If $f(r)=kr^{-3/2}$, where k is a constant, and the cavity is filled after time T, show that $25k^2T^4=4a^5$.

(London Univ., B.Sc. Spec. Hons., 1958)

12. A spherical globule of gas of radius a and at pressure P expands in an infinite mass of liquid of density ρ in which the pressure at infinity is zero. The gas is initially at rest and its pressure and volume are governed by the equation $pv^{4/3}=$constant. Prove that the gas doubles its radius in time

$$(28a/15)\sqrt{(2\rho P)}.$$

(Leeds Univ., B.Sc., 1961)

13. A large mass of incompressible non-viscous fluid contains a spherical air bubble, the air inside the bubble obeying Boyle's law, $pv=$constant. At a great distance from the bubble the pressure is zero. Neglecting the inertia of the system, show that the radius R of the bubble at time t satisfies an equation of the form

$$\frac{d}{dt}\left(R^2\frac{dR}{dt}\right)-\tfrac{1}{2}R\left(\frac{dR}{dt}\right)^2=\frac{k}{R^2},$$

where k is a constant.

(Reading Univ., B.Sc. Spec., 1965)

14. An infinite mass of ideal incompressible fluid is subjected to a force $\mu r^{-7/3}$ per unit mass directed towards the origin. If initially the fluid is at rest and there is a cavity in the form of the sphere $r=a$ in it, show that the cavity will be completely filled after an interval of time $\pi a^{5/3}(10\mu)^{-1/2}$.

(Univ. of Aston in Birm., Gen. B.Sc., 1966)

15. The radial oscillations of a small air bubble in water induced by local pressure changes as it is swept past an obstacle in a stream, may be represented by the radial motion of a spherical bubble in an unbounded inviscid liquid of density ρ that is at rest at a great distance from the bubble with a pressure there given by $p_0\{1+f(t)\}$, where p_0 is a constant and $f(t)$ is a small function of time such that $f(t)\to 0$ as $t\to\pm\infty$. At a great distance from the obstacle the bubble has radius R_0 and is in equilibrium with internal pressure p_0. During the subsequent motion the air is compressed adiabatically in accordance with the law $p\propto\rho^{\gamma}$. Neglecting the inertia of the air, and assuming that departures from equilibrium are small, show that the radius of the bubble at time t is

$$R_0+\left(\frac{p_0}{3\gamma\rho}\right)^{1/2}\int_{-\infty}^{t}f(\tau)\sin\left\{\left(\frac{3p_0\gamma}{\rho R_0^2}\right)(\tau-t)\right\}d\tau.$$

(Cambridge Univ., M.T. II, 1960)

16. Incompressible liquid is in irrotational motion with velocity potential φ. If a portion of the liquid completely occupying the space $r_1 \leqslant r \leqslant r_2$ between concentric spherical surfaces contains no sources or sinks, show that the kinetic energy of the liquid in this region is

$$\tfrac{1}{2}\rho\int_{r=r_2}\varphi\frac{\partial\varphi}{\partial r}dS-\tfrac{1}{2}\rho\int_{r=r_1}\varphi\frac{\partial\varphi}{\partial r}dS.$$

An infinite liquid contains a solid sphere of mass M of the same density as the liquid, and an impulse I is imparted to the sphere. Show that it begins to move with speed $(2I/3M)$. Show that the kinetic energy of the liquid inside the spherical surface of radius $2^{1/3}a$ concentric with the sphere is equal to one half the total kinetic energy of the liquid.

(London Univ., B.Sc. Spec. Hons, 1956)

17. The space between a uniform rigid sphere, of radius a and density σ, and a fixed concentric spherical envelope of radius b is filled with inviscid liquid of constant density ρ. An impulse is applied to the rigid sphere and the system is set in motion from rest. Show that, just after the impulse, the kinetic energy of the liquid and sphere is

$$\tfrac{1}{3}\pi a^3\rho U^2\left(\frac{2\sigma}{\rho}+\frac{b^3+2a^3}{b^3-a^3}\right),$$

where U is the velocity of the sphere. Hence or otherwise, determine the impulse.

(London Univ., B.Sc. Spec. Hons., 1958)

18. The space between two concentric spherical shells of radii a, b $(a<b)$ is filled with liquid of density ρ. If the shells are set in motion, the inner one with velocity U in the x-direction and the outer with velocity V in the y-direction, show that the initial motion of the liquid is given by the velocity potential

$$\varphi=\frac{\{a^3U(1+\frac{1}{2}b^3r^{-3})x-b^3V(1+\frac{1}{2}a^3r^{-3})y\}}{b^3-a^3}$$

where $r^2=x^2+y^2+z^2$, the coordinates being rectangular.

Evaluate the velocity at any point of the liquid and hence, or otherwise, prove that the total momentum communicated has components

$$[\tfrac{4}{3}\pi\rho a^3U,\ \tfrac{4}{3}\pi\rho b^3V,\ 0].$$

(London Univ., B.Sc. Spec. Hons., 1951)

19. The motion of a source-free incompressible fluid of uniform density ρ is irrotational. Prove that the kinetic energy in the singly-connected volume bounded by the surface or surfaces S is $-\tfrac{1}{2}\rho\iint_S\varphi\frac{\partial\varphi}{\partial n}dS$, where φ

is the velocity potential and n is the normal at the surface drawn into the volume considered.

A sphere of radius a moves through a fluid whose boundaries are fixed at a large distance from the sphere. Show that the kinetic energy of the fluid is $\frac{1}{3}\pi a^3 \rho u^2$ at the instant when the sphere has speed u.

Hence or otherwise, show that if the sphere is uniform and of density $\frac{1}{2}\rho$, it will rise with an acceleration $\frac{1}{2}g$ when released from rest in such a fluid, the system being subject to gravity and the acceleration due to gravity being g.

(Bristol Univ., B.Sc. Spec. I, 1961)

20. An inviscid fluid moves under a conservative system of forces. Prove that the circulation in any circuit which moves with the fluid is independent of time.

Obtain an equation for the rate of change of vorticity with respect to time and deduce the permanence of vortex rings.

Prove that in two-dimensional motion the vorticity along a streamline is constant.

(London Univ., B.Sc. Spec. Hons., 1960)

21. An incompressible inviscid fluid moves under an arbitrary body force $\mathbf{F}$ per unit mass. Prove that the generation of vorticity is determined by the equation

$$d\boldsymbol{\omega}/dt = (\boldsymbol{\omega} \cdot \nabla)\mathbf{v} + \text{curl } \mathbf{F},$$

where $\boldsymbol{\omega} = \text{curl } \mathbf{v}$.

(London Univ., B.Sc. Spec. Hons., 1962)

4

SOME THREE-DIMENSIONAL FLOWS

4.1 Introduction

In the previous chapter we discussed a number of problems which were mainly of the three-dimensional type. We now develop special methods for treating other problems in three dimensions. The flows considered in this chapter are entirely incompressible in the sense that the fluids are of constant density throughout.

4.2 Sources, Sinks and Doublets

Suppose at a point O in a fluid the flow is such that it is directed radially outwards from O in all directions and in a symmetrical manner. Then fluid enters the system through O which is termed a *simple source.* If at O the volume entering per unit time is $4\pi m$, where m is a constant, then the *strength of the source* is defined to be m.

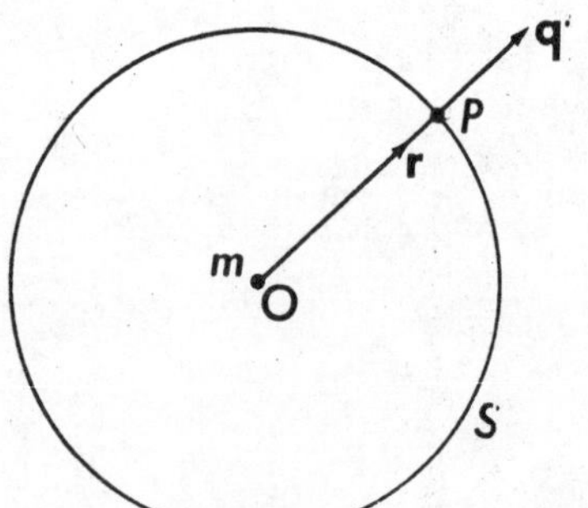

FIG. 4.1

If, however, the flow is such that fluid is directed radially inwards to O from all directions in a symmetrical manner, then fluid leaves the system at O which is termed a *simple sink*. A sink of strength m is a source of strength $-m$.

Figure 4.1 shows a simple source of strength m at O in a fluid which is presumed to contain no other sources or sinks and which would otherwise be at rest. S is the surface of the sphere centre O and radius r and P is a field on S such that $\overline{OP} \equiv \mathbf{r}$. Then the fluid velocity at P is $\mathbf{q}$ along $\overline{OP}$ and the magnitude q is everywhere constant over S. The volume of fluid crossing S per unit time is thus $4\pi r^2 q$: that emitted from O per unit time is $4\pi m$. Since the fluid is incompressible, these two are equal from considerations of continuity. Thus

$$q = \frac{m}{r^2},$$

$$\mathbf{q} = \frac{m}{r^2}\hat{\mathbf{r}}. \tag{1}$$

It is easily shown that curl $\mathbf{q} = \mathbf{0}$ (except at $r=0$) so that the flow is of the potential kind. Let φ be the velocity potential at P. From considerations of symmetry, $\varphi = \varphi(r)$ and from Chapter 1 (Section 1.12),

$$\nabla\varphi = \varphi'(r)\hat{\mathbf{r}}.$$

Thus

$$\varphi'(r) = -m/r^2,$$

and so

$$\varphi(r) = m/r. \tag{2}$$

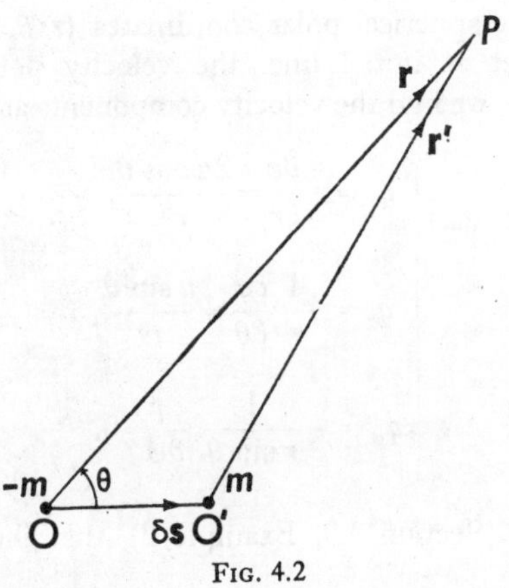

FIG. 4.2

Now suppose we have simple sources of strengths $-m$ at O and m at O' in a fluid (Fig. 4.2), where $\overline{OO'} \equiv \delta\mathbf{s}$. Let P be any other point in the fluid such that $\overline{OP}\equiv\mathbf{r}$, $\overline{O'P}\equiv\mathbf{r}'$. In the following we shall use $r=|\mathbf{r}|$, $r'=|\mathbf{r}'|$, $\delta s=|\delta\mathbf{s}|$. Also let $\mu=m\delta s$. Now suppose that $m\to\infty$, $\delta s\to 0$ in such a way that μ remains finite and constant. Then in the limit the two sources $\pm m$, of infinitely great magnitude and coincident at O, are said to constitute a *three-dimensional doublet* or *dipole* at O. The quantity μ is termed the *moment* or *strength* of the doublet. The quantity $\boldsymbol{\mu}=\mu\hat{\mathbf{a}}$, where $\hat{\mathbf{a}}$ is the unit vector in the direction $\overline{OO'}$ is called the *vector moment* of the doublet. Also the direction of $\hat{\mathbf{a}}$ is that of the *axis* of the doublet.

Let us now find the velocity potential at P due to the doublet at O. For the configuration shown, assuming the flow to be entirely due to $-m$ at O and m at O', the velocity potential at P is

$$\varphi=\frac{m}{r'}-\frac{m}{r}=\frac{m(r-r')}{rr'}$$

$$=\frac{m(r^2-r'^2)}{rr'(r+r')}=\frac{m(\mathbf{r}-\mathbf{r}')\cdot(\mathbf{r}+\mathbf{r}')}{rr'(r+r')}=\frac{m\delta\mathbf{s}\cdot(\mathbf{r}+\mathbf{r}')}{rr'(r+r')}=\frac{\boldsymbol{\mu}\cdot(\mathbf{r}+\mathbf{r}')}{rr'(r+r')}.$$

Now keep O fixed and let $O'\to O$, so that $\mathbf{r}'\to\mathbf{r}$, μ staying constant. Then in the limit we have

$$\varphi=(\boldsymbol{\mu}\cdot\mathbf{r})r^{-3}. \tag{3}$$

With $\angle POO'=\theta$, other equivalent forms for φ are

$$\varphi=(\boldsymbol{\mu}\cdot\hat{\mathbf{r}})r^{-2}=\mu r^{-2}\cos\theta. \tag{3'}$$

The last form shows that $\varphi=\varphi(r,\theta)$, so that the axis of the doublet is an axis of symmetry as is physically obvious.

Taking P to have spherical polar coordinates (r,θ,ψ) referred to the axis of the doublet as initial line, the velocity potential at P being $\varphi(r,\theta)=(\mu\cos\theta)/r^2$, we find the velocity components at P are

$$\begin{cases} q_r=-\dfrac{\partial\varphi}{\partial r}=\dfrac{2\mu\cos\theta}{r^3}, \\[2ex] q_\theta=-\dfrac{1}{r}\dfrac{\partial\varphi}{\partial\theta}=\dfrac{\mu\sin\theta}{r^3}, \\[2ex] q_\psi=-\dfrac{1}{r\sin\theta}\dfrac{\partial\varphi}{\partial\psi}=0. \end{cases}$$

It has been shown (Section 3.9, Example 2), that the equations of the streamlines are

$$\psi=\text{const.},\ r=A\sin^2\theta.$$

Thus the streamlines lie in planes passing through the axis of the doublet.

They are symmetrical not only with respect to OX but also about the plane $\theta=\frac{1}{2}\pi$. Figure 4.3 shows a sketch of them.

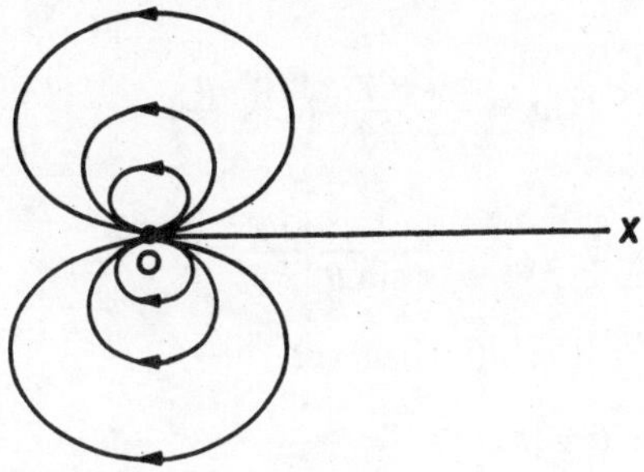

FIG. 4.3

The form $\varphi(r, \theta)=(\mu \cos \theta)/r^2$ shows that φ, at points other than O, must satisfy the axially symmetrical spherical polar form of Laplace's equation.

EXAMPLE 1

Doublet in a uniform stream.

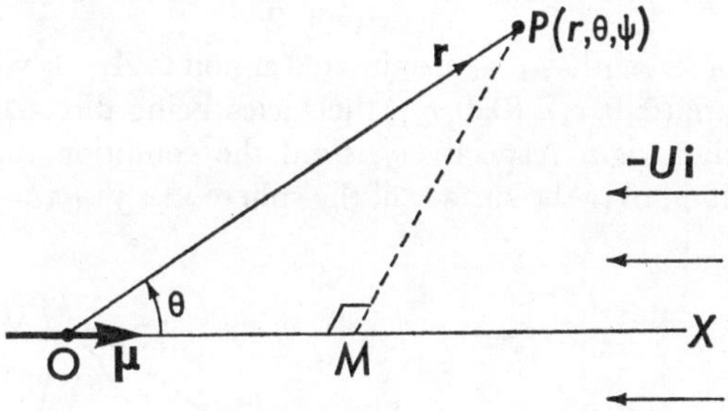

FIG. 4.4

Figure 4.4 shows a doublet of vector moment $\boldsymbol{\mu}=\mu\mathbf{i}$ at O in a uniform stream whose velocity, in the absence of the doublet, would be $-U\mathbf{i}$. We first find the velocity potential $\varphi(r, \theta)$ at the field point P in the fluid having spherical polar coordinates (r, θ, ψ). If PM is the perpendicular from P on OX and if $OM=x$, then the velocity potential at P due to the streamline is $Ux=Ur \cos \theta$. That due to the doublet at O is $(\mu \cos \theta)/r^2$, so that the total velocity potential at P is

$$\varphi(r, \theta)=(Ur+\mu r^{-2}) \cos \theta. \tag{1}$$

Thus the velocity components at P are

$$\left.\begin{aligned} q_r &= -\frac{\partial\varphi}{\partial r} = -\left(U - \frac{2\mu}{r^3}\right), \\ q_\theta &= -\frac{1}{r}\frac{\partial\varphi}{\partial\theta} = \left(U + \frac{\mu}{r^3}\right), \\ q_\psi &= -\frac{1}{r\sin\theta}\frac{\partial\varphi}{\partial\psi} = 0. \end{aligned}\right\} \tag{2}$$

From (2), we see that $q_r = 0$ whenever

(i) $\cos\theta = 0$, i.e. $\theta = \frac{1}{2}\pi$;
(ii) $r = (2\mu/U)^{1/3}$.

Write $a = (2\mu/U)^{1/3}$, so that $\mu = \frac{1}{2}Ua^3$. Then it follows that there is no flow over the surface of the sphere $r = a$. In fact for $r \geqslant a$, with $\mu = \frac{1}{2}Ua^3$ substituted in (2), we obtain precisely the same velocity potential as was obtained for uniform flow past a stationary sphere of radius a (Section 3.9, Example 1). Thus, for the region $r \geqslant a$, the analysis of this sphere problem is the same as that of the dipole of strength $\frac{1}{2}Ua^3$ in the uniform stream of velocity $-U\mathbf{i}$, the axis of the dipole being in the direction of $\mathbf{i}$.

EXAMPLE 2

Doublets of strengths μ_1, μ_2 are situated at points A_1, A_2 whose Cartesian coordinates are $(0, 0, c_1)$, $(0, 0, c_2)$, their axes being directed towards and away from the origin respectively. Find the condition that there is no transport of fluid over the surface of the sphere $x^2 + y^2 + z^2 = c_1c_2$.

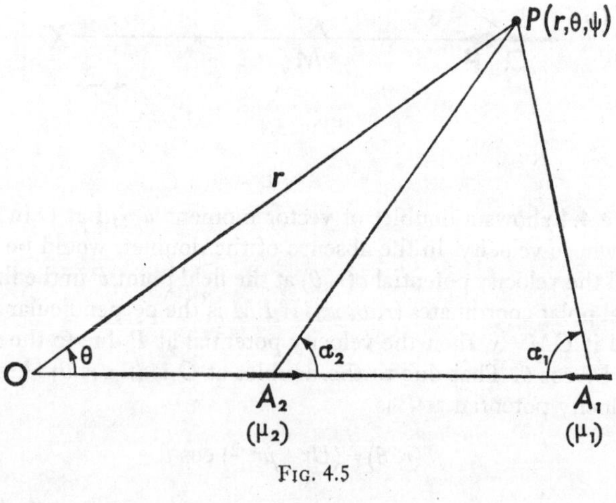

FIG. 4.5

Referred to OA_2A_1 as initial line, P has spherical polar coordinates (r, θ, ψ) (Fig. 4.5). The axes of the doublets at A_1, A_2 make angles α_1, α_2 with A_1P, A_2P. Then the velocity potential at P is

$$\varphi=\frac{\mu_2 \cos \alpha_2}{A_2P^2}+\frac{\mu_1 \cos \alpha_1}{A_1P^2}.$$

Now

$$A_1P^2=r^2-2rc_1 \cos \theta+c_1^2, \quad A_2P^2=r^2-2rc_2 \cos \theta+c_2^2,$$

$$\cos \alpha_1=\frac{c_1-r \cos \theta}{A_1P}, \quad \cos \alpha_2=\frac{r \cos \theta-c_2}{A_2P}.$$

$$\therefore \quad \varphi(r, \theta)=\mu_2(r \cos \theta-c_2)(r^2-2rc_2 \cos \theta+c_2^2)^{-3/2} + \mu_1(c_1-r \cos \theta)(r^2-2rc_1 \cos \theta+c_1^2)^{-3/2}.$$

We require that $\partial\varphi/\partial r=0$ when $r=(c_1c_2)^{1/2}$.

Now

$$\partial\varphi/\partial r=\mu_2\{\cos \theta(r^2-2rc_2 \cos \theta+c_2^2)^{-3/2} - 3(r \cos \theta-c_2)(r-c_2 \cos \theta)(r^2-2rc_2 \cos \theta+c_2^2)^{-5/2}\}$$
$$+\mu_1\{-\cos \theta(r^2-2rc_1 \cos \theta+c_1^2)^{-3/2}-3(c_1-r \cos \theta)(r-c_1 \cos \theta)(r^2-2rc_1 \cos \theta+c_1^2)^{-5/2}\}.$$

Thus we find

$$\mu_2[\cos \theta\{c_1c_2-2c_2(c_1c_2)^{1/2} \cos \theta+c_2^2\}^{-3/2}-3\{(c_1c_2)^{1/2} \cos \theta-c_2\}\{c_1c_2-2c_2(c_1c_2)^{1/2} \cos \theta+c_2^2\}^{-5/2}]$$
$$=\mu_1[\cos \theta\{c_1c_2-2c_1(c_1c_2)^{1/2} \cos \theta+c_1^2\}^{-3/2}+3\{c_1-(c_1c_2)^{1/2} \cos \theta\}\{(c_1c_2)^{1/2}-c_1 \cos \theta\}\{c_1c_2-2c_1(c_1c_2)^{1/2} \cos \theta+c_1^2\}^{-5/2}].$$

Hence

$$\underline{\mu_2/\mu_1=(c_2/c_1)^{3/2}.}$$

EXAMPLE 3

Illustration of line distributions.

Prove that the velocity potential at a point P due to a uniform finite line source AB of strength m per unit length is of the form $\varphi=m \log f$, where

$$f=\frac{r_2+x_2}{r_1+x_1}=\frac{r_1-x_1}{r_2-x_2}=\frac{a+l}{a-l}$$

in which $AB=2l$, $PA=r_1$, $PB=r_2$, $NA=x_1$, $NB=x_2$, N being the foot of the perpendicular from P on the line AB, and $2a$ the length of the major axis of the spheroid through P having A, B as foci.

Show that the velocity at P is $(2ml \cos \alpha)/(a^2-l^2)\mathbf{u}$, where $\mathbf{u}$ is the unit vector along the normal to the spheroid at P and $2\alpha=\angle APB$.

(London Univ., B.Sc. Spec., 1957)

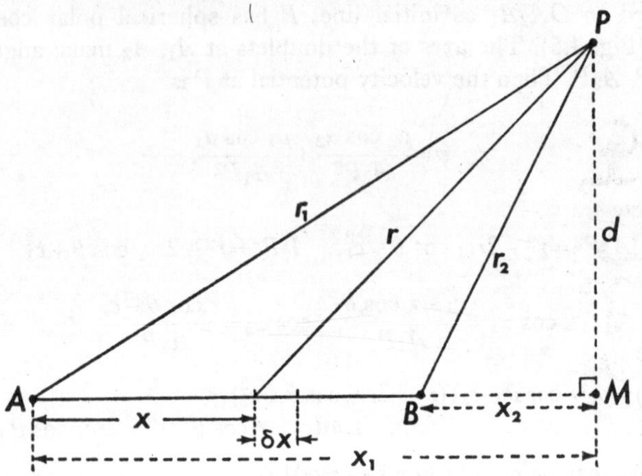

FIG. 4.6

The line-section of length δx in AB at distance x from A is effectively a point source of strength $m\delta x$ giving a velocity potential at P of amount $(m\delta x)/r$, where $AP=r$ (Fig. 4.6). From elementary geometry, $r^2=r_1^2+x^2-2x_1x$, and so the total velocity potential at P due to the entire line distribution is

$$\varphi=m\int_0^{2l}\frac{dx}{r}=m\int_0^{2l}\frac{dx}{\{r_1^2-x_1^2+(x_1-x)^2\}^{1/2}}$$

$$=m\left\{-\operatorname{sh}^{-1}\left(\frac{x_1-x}{d}\right)\right\}_0^{2l}=m\left\{\operatorname{sh}^{-1}\left(\frac{x_1}{d}\right)-\operatorname{sh}^{-1}\left(\frac{x_1-2l}{d}\right)\right\}$$

$$=m\left\{\operatorname{sh}^{-1}\left(\frac{x_1}{d}\right)-\operatorname{sh}^{-1}\left(\frac{x_2}{d}\right)\right\}$$

$$=m\log\left\{\frac{x_1+\sqrt{(x_1^2+d^2)}}{x_2+\sqrt{(x_2^2+d^2)}}\right\}$$

$$=m\log\left(\frac{x_1+r_1}{x_2+r_2}\right).$$

Now $r_1^2-x_1^2=r_2^2-x_2^2=d^2$, so that

$$\frac{r_1+x_1}{r_2+x_2}=\frac{r_2-x_2}{r_1-x_1}=\frac{r_1+x_2+r_1-x_2}{r_1+r_2+x_2-x_1}=\frac{r_1+r_2+2l}{r_1+r_2-2l}.$$

At P on the spheroid through P having A, B as foci, $r_1+r_2=2a$. Hence $(r_1+x_1)/(r_2+x_2)=(r_2-x_2)/(r_1-x_1)=(a+l)/(a-l)$, as required. Thus

$$\varphi=m\log\left(\frac{a+l}{a-l}\right).$$

The equipotentials are given by $\varphi=$const., or

$$\frac{a+l}{a-l}=\text{const.}$$

Thus they are $a=$const., or $r_1+r_2=$const.

These surfaces are confocal ellipsoids of revolution about AB with A, B as foci.

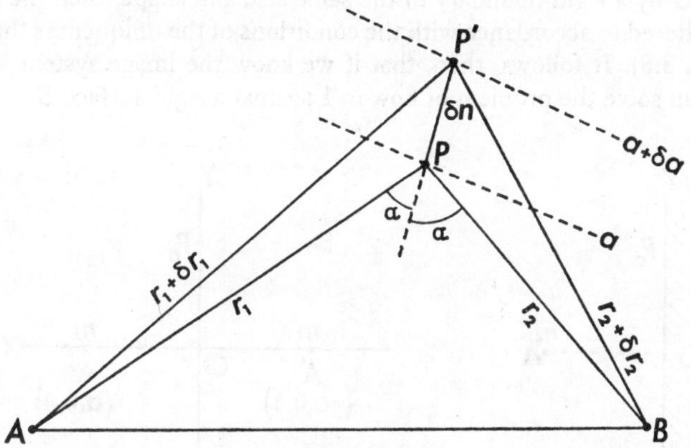

FIG. 4.7

Let P be any point on the ellipsoid specified by parameter a, P' a point on the neighbouring ellipsoid specified by parameter $(a+\delta a)$, where $\overline{PP'}\equiv\delta n\mathbf{u}$ (Fig. 4.7). Then the velocity at P is

$$-\nabla\varphi=-\frac{\partial\varphi}{\partial n}\mathbf{u}$$

$$=-m\mathbf{u}\frac{\partial}{\partial n}\left\{\log(a+l)-\log(a-l)\right\}$$

$$=-m\mathbf{u}\frac{\partial a}{\partial n}\left(\frac{1}{a+l}-\frac{1}{a-l}\right)$$

$$=\frac{2lm}{a^2-l^2}\left(\frac{\partial a}{\partial n}\right)\mathbf{u}.$$

Now the normal at P bisects 2α, the angle between the focal radii AP, PB (Fig. 4.7). The diagram shows

$$r_1+\delta r_1=\{r_1^2+(\delta n)^2+2r_1\delta n\cos\alpha\}^{1/2}\simeq r_1+\delta n\cos\alpha,$$
$$r_2+\delta r_2=\{r_2^2+(\delta n)^2+2r_2\delta n\cos\alpha\}^{1/2}\simeq r_2+\delta n\cos\alpha.$$

Hence $2\delta a=(r_1+\delta r_1)+(r_2+\delta r_2)-(r_1+r_2)=2\delta n\cos\alpha$. Thus $\partial a/\partial n=\cos\alpha$, which yields the stated result.

4.3 Images in a Rigid Infinite Plane

Suppose a surface S can be drawn in a moving fluid in such a way that there is no transport of fluid across S. Let S divide the fluid into two regions labelled 1, 2. Then any system of sources, sinks or doublets in 2 is called an *image system* of the region 1 in S. If we remove the fluid in 2 and replace S by a rigid boundary of the same size and shape, then the flow in 1 is unaltered in accordance with the conditions of the uniqueness theorems (Section 3.8). It follows, then, that if we know the image system for 1 in S, we can solve the problem of flow in 1 against a rigid surface S.

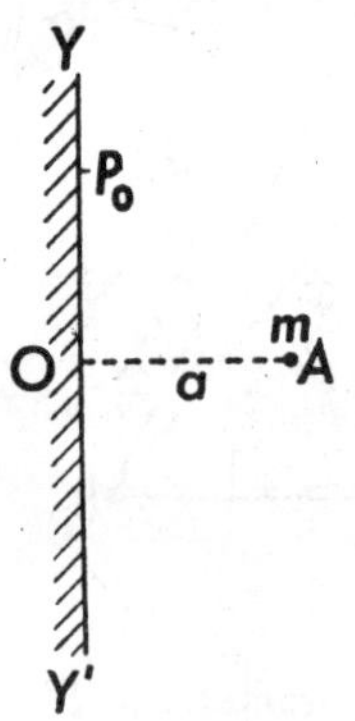

FIG. 4.8(i)

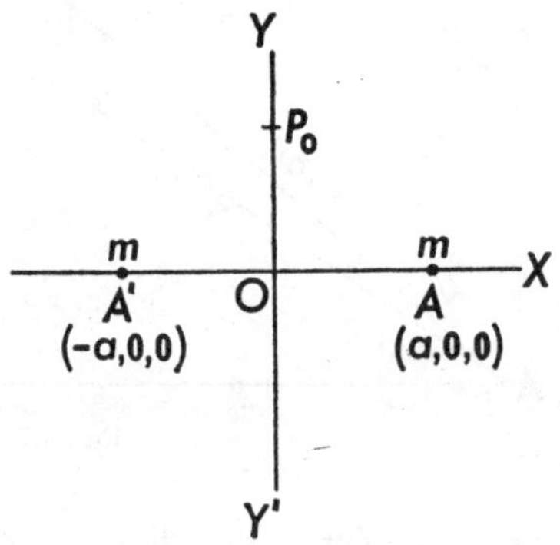

FIG. 4.8(ii)

Figure 4.8(i) shows a simple source of strength m situated at a distance a from an infinite rigid plane YY'. We first show that *the appropriate image system for this is an equal source at A', the optic image of A in the plane.*

To this end, consider Fig. 4.8(ii) in which we have equal sources of strength m at $A(a, 0, 0)$ and at $A'(-a, 0, 0)$. Let P_0 be any point on the plane YY' in Fig. 4.8(ii). Then the fluid velocity at P_0 is

$$=\left(\frac{m}{AP_0^3}\right)\overline{AP_0}+\left(\frac{m}{A'P_0}\right)\overline{A'P_0}$$

$$=\left(\frac{m}{AP_0^3}\right)(\overline{AP_0}+\overline{A'P_0})$$

$$=\left(\frac{2m}{AP_0^3}\right)\overline{OP_0},\quad \text{since } \overline{AP_0}+\overline{A'P_0}=\overline{2OP_0}.$$

This shows that at any point P_0 on YY', the fluid flows tangentially to the plane YY'. Thus there is no transport of fluid across this plane. Thus in

Figs. 4.8(i), (ii), at all corresponding points P_0 on the surfaces YY', $\partial\varphi/\partial n=0$, for the region of flow $x \geqslant 0$. By the uniqueness theorems, we infer, then, that the image of m at A in YY' in Fig. 4.8(i) is m at A', the optic image of A in YY'.

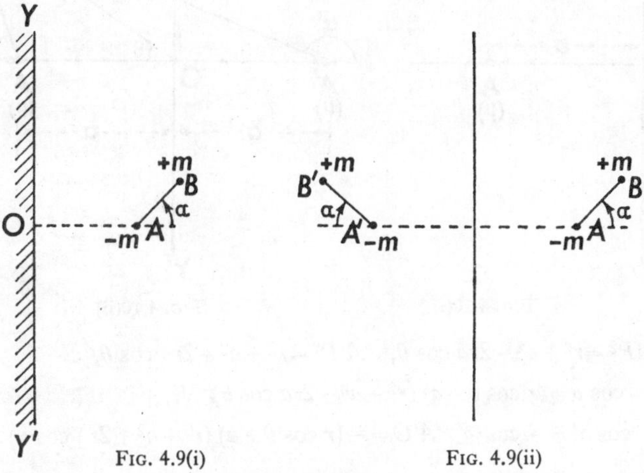

FIG. 4.9(i) FIG. 4.9(ii)

Now consider a pair of sources $-m$ at A, $+m$ at B, close together and on one side of the rigid plane YY' (Fig. 4.9(i)). The image system is $-m$ at A', m at B', where A', B' are the respective optic images of the points A, B in the plane YY' (Fig. 4.9(ii)). In the limit it follows, then, that *the image of a doublet in an infinite rigid plane is an equal doublet symmetrically disposed with respect to the plane.*

EXAMPLE

A three-dimensional doublet of strength μ whose axis is in the direction $\overline{Ox}$ is distant a from the rigid plane $x=0$ which is the sole boundary of liquid of density ρ, infinite in extent. Find the pressure at a point on the boundary distant r from the doublet given that the pressure at infinity is p_∞. Show that the pressure on the plane is least at a distance $a\sqrt{5}/2$ from the doublet.

(University of Aston, Tech. Maths. II, 1966)

Figure 4.10(i) shows the physical model, YY' being a rigid boundary. The appropriate image system is represented in Fig. 4.10(ii) in which P having spherical polar coordinates (r, θ, ψ) is the considered field point in the region $0 \leqslant \theta \leqslant \frac{1}{2}\pi$. Thus the velocity potential at P is

$$\varphi = \frac{\mu \cos \alpha}{AP^2} + \frac{\mu \cos \alpha'}{A'P^2}.$$

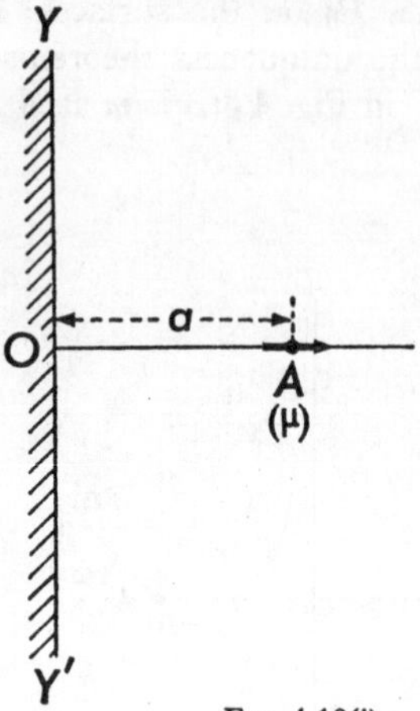

FIG. 4.10(i)

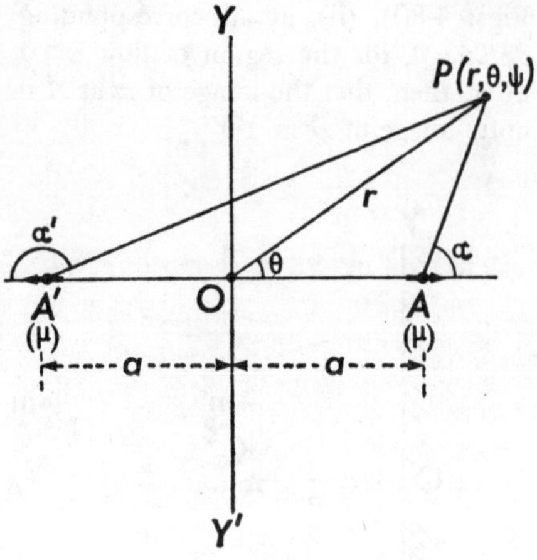

FIG. 4.10(ii)

Now $AP^2 = r^2 + a^2 - 2ra\cos\theta$, $\quad A'P^2 = r^2 + a^2 + 2ra\cos\theta$,

$$\cos\alpha = (r\cos\theta - a)(r^2 + a^2 - 2ra\cos\theta)^{-1/2},$$

$$\cos\alpha' = -\cos\angle PA'O = -(r\cos\theta + a)(r^2 + a^2 + 2ra\cos\alpha)^{-1/2},$$

and so

$$\varphi = \mu\{(r\cos\theta - a)(r^2 + a^2 - 2ra\cos\theta)^{-3/2} - (r\cos\theta + a)(r^2 + a^2 + 2ra\cos\theta)^{-3/2}\}.$$

Thus

$$q_r = -\partial\varphi/\partial r = -\mu\{\cos\theta(r^2 + a^2 - 2ra\cos\theta)^{-3/2} - 3(r - a\cos\theta)(r\cos\theta - a)(r^2 + a^2 - 2ra\cos\theta)^{-5/2} - \cos\theta(r^2 + a^2 - 2ra\cos\theta)^{-3/2} + 3(r + a\cos\theta)(r\cos\theta + a)(r^2 + a^2 + 2ra\cos\theta)^{-5/2}\},$$

$$q_\theta = -\frac{1}{r}\frac{\partial\varphi}{\partial\theta} = -\frac{\mu}{r}\Big\{-r\sin\theta\,(r^2 + a^2 - 2ra\cos\theta)^{-3/2} - 3ra\sin\theta(r\cos\theta - a)(r^2 + a^2 - 2ra\cos\theta)^{-5/2} + r\sin\theta(r^2 + a^2 + 2ra\cos\theta)^{-3/2} - 3ra\sin\theta(r\cos\theta + a)(r^2 + a^2 + 2ra\cos\theta)^{-5/2}\Big\},$$

$$q_\psi = -\frac{1}{r\sin\theta}\frac{\partial\varphi}{\partial\psi} = 0.$$

Thus when $\theta = \frac{1}{2}\pi$,

$$q_r = -6\mu ar(r^2 + a^2)^{-5/2}, \quad q_\theta = 0, \quad q_\psi = 0.$$

Along the streamline through $(r, \pi/2, \psi)$, Bernoulli's equation is

$$\frac{p}{\rho} + \tfrac{1}{2}q^2 = \frac{p_\infty}{\rho},$$

or

$$\frac{p}{\rho}+18\mu^2a^2r^2(r^2+a^2)^{-5}=\frac{p_\infty}{\rho}.$$

Thus

$$p=p_\infty-18\rho\mu^2a^2r^2(r^2+a^2)^{-5},$$

and so

$$dp/dr=36\rho\mu^2a^2r(4r^2-a^2)(r^2+a^2)^{-6}.$$

When $r=\frac{1}{2}a$, $dp/dr=0$. Also for $r=\frac{1}{2}a+$, $dp/dr>0$ and for $r=\frac{1}{2}a-$, $dp/dr<0$. Hence p is a minimum at $r=\frac{1}{2}a$ on the plane, i.e. at a distance $\frac{1}{2}a\sqrt{5}$ from the doublet.

4.4 Images in Solid Spheres*

In Section 3.9 we studied the effect of placing a solid sphere in a uniform stream of incompressible flow. Here we discuss the perturbations obtained when such a sphere is placed in a more general field of flow. We obtain results which do not presuppose axial symmetry of flow.

We first establish some properties of harmonic functions specified by a system of spherical polar coordinates (r, θ, ψ). Such functions satisfy the spherical polar form of Laplace's equation, viz.

$$\nabla^2\varphi\equiv\frac{1}{r^2\sin\theta}\left\{\frac{\partial}{\partial r}\left(r^2\sin\theta\frac{\partial\varphi}{\partial r}\right)+\frac{\partial}{\partial\theta}\left(\sin\theta\frac{\partial\varphi}{\partial\theta}\right)+\frac{\partial}{\partial\psi}\left(\frac{1}{\sin\theta}\frac{\partial\varphi}{\partial\psi}\right)\right\}=0. \quad (1)$$

THEOREM I: If $r^nS_n(\theta,\psi)$ is a harmonic function, so also is $r^{-(n+1)}S_n(\theta,\psi)$.

Proof: Since $\varphi=r^nS_n(\theta,\psi)$ satisfies (1),

$$\frac{\partial}{\partial r}\left(nr^{n+1}S_n\sin\theta\right)+\frac{\partial}{\partial\theta}\left(r^n\sin\theta\frac{\partial S_n}{\partial\theta}\right)+\operatorname{cosec}\theta\frac{\partial}{\partial\psi}\left(r^n\frac{\partial S_n}{\partial\psi}\right)=0$$

or

$$n(n+1)S_n\sin\theta+\frac{\partial}{\partial\theta}\left(\sin\theta\frac{\partial S_n}{\partial\theta}\right)+\operatorname{cosec}\theta\frac{\partial^2S_n}{\partial\psi^2}=0. \quad (2)$$

Also, taking

$$\varphi=r^{-(n+1)}S_n(\theta,\psi),$$

$$r^2\sin\theta\nabla^2\varphi=\frac{\partial}{\partial r}\left\{-(n+1)r^{-n}\sin\theta S_n\right\}+r^{-(n+1)}\frac{\partial}{\partial\theta}\left(\sin\theta\frac{\partial S_n}{\partial\theta}\right)+\operatorname{cosec}\theta\, r^{-(n+1)}\frac{\partial^2S_n}{\partial\psi^2}$$

$$=r^{-(n+1)}\left\{(n+1)nS_n\sin\theta+\frac{\partial}{\partial\theta}\left(\sin\theta\frac{\partial S_n}{\partial\theta}\right)+\operatorname{cosec}\theta\frac{\partial^2S_n}{\partial\psi^2}\right\}$$

$=0$, in virtue of (2).

This establishes the result.

* See F. Chorlton: 'Some Potential Problems Involving Spheres', *Mathematical Gazette*, May, 1967.

THEOREM II: If $\varphi(r, \theta, \psi)$ has an expansion of the form

$$\varphi(r, \theta, \psi)=\sum_{n=0}^{\infty} \alpha_n r^n S_n(\theta, \psi),$$

the series on the right being uniformly convergent with respect to r, then for constant λ,

$$\frac{1}{r^\lambda}\int_0^r R^{\lambda-1}\varphi(R, \theta, \psi)\,dR=\sum_{n=0}^{\infty}\frac{\alpha_n r^n S_n(\theta, \psi)}{n+\lambda}.$$

Proof:

$$R^{\lambda-1}\varphi(R, \theta, \psi)=\sum_{n=0}^{\infty} \alpha_n R^{n+\lambda-1} S_n(\theta, \psi).$$

Hence

$$\int_0^r R^{\lambda-1}\varphi(R, \theta, \psi)\,dR=\sum_{n=0}^{\infty}\frac{\alpha_n r^{n+\lambda} S_n(\theta, \psi)}{n+\lambda},$$

whence the result follows.

THEOREM III: Weiss's sphere theorem*.

Let $\varphi(r, \theta, \psi)$ be the velocity potential at a point P having spherical polar coordinates (r, θ, ψ) in an incompressible fluid having irrotational motion, there being no rigid boundaries. Also suppose φ has no singularities within the region $r \leqslant a$. Then when a solid of radius a is introduced into the flow with its centre at the origin of coordinates, the new velocity potential at P in the fluid is

$$\varphi(r, \theta, \psi)+\frac{a}{r}\varphi\left(\frac{a^2}{r}, \theta, \psi\right)-\frac{1}{a}\int_0^{a^2/r}\varphi(R, \theta, \psi)\,dR \quad (r>a).$$

Proof: Suppose φ is regular near the origin; we assume it has an expansion of the form

$$\varphi(r, \theta, \psi)=\sum_{n=0}^{\infty} \alpha_n r^n S_n(\theta, \psi),$$

where each $r^n S_n(\theta, \psi)$ is harmonic and each $_n$ is known.

When the sphere is introduced into the stream, it will produce a perturbation potential which will be zero at infinity and which must be harmonic. A suitable form to choose for the perturbation potential in virtue of Theorem I is $\sum_{n=0}^{\infty} \beta_n r^{-(n+1)} S_n(\theta, \psi)$ where the coefficients β_n are undertermined. Thus the new velocity potential P is $\Phi(r, \theta, \psi)$, where

* P. WEISS, *Proc. Camb. Phil. Soc.*, 40 (1945).

$$\Phi(r, \theta, \psi) = \sum_{n=0}^{\infty} \alpha_n r^n S_n(\theta, \psi) + \sum_{n=0}^{\infty} \beta_n r^{-(n+1)} S_n(\theta, \psi).$$

On $r = a$, we require $\partial\Phi/\partial r = 0$. Thus equating to zero the coefficient of each $S_n(\theta, \psi)$,

$$\alpha_n n a^{n-1} - (n+1)\beta_n a^{-(n+2)} = 0 \quad (n = 0, 1, \ldots)$$

or

$$\beta_n = \left(1 - \frac{1}{n+1}\right) a^{2n+1} \alpha_n \quad (n = 0, 1, \ldots).$$

$$\text{Perturbation potential} = \frac{a}{r} \sum_{n=0}^{\infty} \alpha_n \left(\frac{a^2}{r}\right)^n S_n - \frac{1}{a} \sum_{n=0}^{\infty} \frac{\alpha_n (a^2/r)^{n+1} S_n}{n+1}$$

$$= \frac{a}{r}\varphi\left(\frac{a^2}{r}, \theta, \psi\right) - \frac{1}{a}\int_0^{a^2/r} \varphi(R, \theta, \psi)\, dR,$$

the integral following by taking $\lambda = 1$ in Theorem II. This establishes the result.

Example 1

Image of a source in a solid sphere.

The velocity potential at (r, θ, ψ) in a fluid due to a simple source of strength m at $(f, 0, 0)$, is

$$\varphi(r, \theta) = m(r^2 - 2rf \cos\theta + f^2)^{-1/2}$$

Introducing a solid sphere in the region $r \leqslant a$, where $a < f$, we obtain, using Weiss's sphere theorem, a perturbation potential of

$$\frac{ma/f}{\sqrt{[r^2 - 2r(a^2/f)\cos\theta + (a^2/f)^2]}} - \frac{m}{a}\int_0^{a^2/r} \frac{dR}{\sqrt{(R^2 - 2Rf\cos\theta + f^2)}},$$

showing that the image system of a point source of strength m placed at $f(> a)$ from the centre of the solid sphere consists of a source of strength ma/f at the inverse point in the sphere, together with a continuous line distribution of sinks of strength $-m/a$ per unit length extending from the centre to the inverse point.

Example 2

Image of a doublet in a sphere when the axis of the doublet passes through the centre of the sphere.

Figure 4.11 shows a doublet AB with its axis $\overline{BA}$ pointing towards the centre O of a sphere of radius a. $OA = f$, $OB = f + \delta f$, and A', B' are the points in the sphere inverse to A and B so that $OA' = a^2/f$, $OB' = a^2/(f + \delta f)$. At A and B we associate simple sources of strengths m, $-m$ so that the

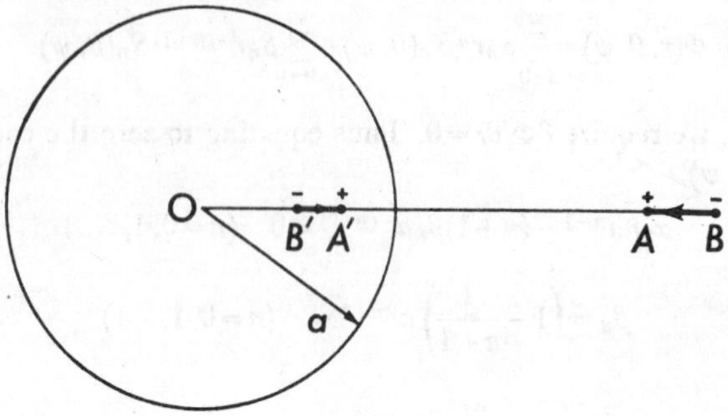

FIG. 4.11

strength of the doublet is $\mu = m\delta f$, where $m \to \infty$, $\delta f \to 0$ in such a way that μ remains finite. From the previous case, the image of m at A consists of ma/f at A' together with a continuous line distribution from O to A' of sinks of density $-m/a$ per unit length, and the image of $-m$ at B consists of $-ma/(f+\delta f)$ at B' together with a continuous line distribution from O to B' of sources of density m/a per unit length. Clearly the line distributions cancel out leaving the source at A' of ma/f and that at B' of $-ma/(f+\delta f)$. Now

$$B'A' = \frac{a^2}{f} - \frac{a^2}{f+\delta f} = \frac{a^2}{f}\left\{1 - \left(1 + \frac{\delta f}{f}\right)^{-1}\right\} \simeq \frac{a^2 \delta f}{f^2}.$$

Thus to the first order we have a doublet at A' of strength

$$\frac{a^2 \delta f}{f^2}\left(\frac{ma}{f}\right) = \frac{ma^3}{f^3}\,\delta f = \frac{\mu a^3}{f^3}.$$

Hence in the limit as $\delta f \to 0$, we obtain a doublet at A of strength μ with its axis towards O and a doublet at A' of strength $\mu a^3/f^3$ with its axis directed away from O.

4.5 Axi-Symmetric Flows; Stokes's Stream Function

We have encountered this type of flow in Section 3.9. Thus uniform flow past a stationary sphere or a sphere moving with uniform velocity in a fluid at rest furnish examples of axi-symmetric flows.

Suppose the z-axis be taken as axis of symmetry in an axi-symmetric flow and suppose that a point P in the fluid may be specified by cylindrical polar coordinates (R, θ, z). Then at P all scalar and vector functions associated with the flow are independent of θ. The equation of continuity $\nabla \cdot \mathbf{q} = 0$ for incompressible flow assumes the form

$$\frac{\partial}{\partial R}(Rq_R) + \frac{\partial}{\partial z}(Rq_z) = 0, \tag{1}$$

taking $\mathbf{q}=q_R\hat{\mathbf{R}}+q_z\mathbf{k}$. If we can find a scalar function $\psi(R, z)$ such that

$$\left.\begin{aligned} q_R &= \frac{1}{R}\frac{\partial\psi}{\partial z} \\ q_z &= -\frac{1}{R}\frac{\partial\psi}{\partial R} \end{aligned}\right\} \tag{2}$$

then on substituting equations (2) into (1), we find the latter is identically satisfied. Such a function ψ is called *Stokes's stream function*. We now obtain a physical meaning for ψ.

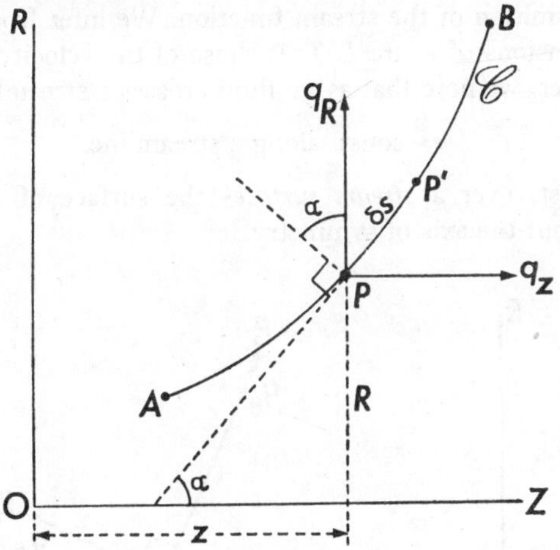

FIG. 4.12

Figure 4.12 shows a meridian section of the type of flow under discussion. There is no velocity component perpendicular to the meridian plane. In the section, AB is an arc of a plane curve $\mathscr{C}$, P being the point on it distant R from OZ and z from OR. PP' is of length δs. We seek to find the volume of fluid crossing the surface of revolution of AB about OZ per unit time from right to left. Denoting by α the angle which the tangent at P to $\mathscr{C}$ makes with OZ, the normal component of velocity at P from right to left is

$$\begin{aligned} q_R\cos\alpha - q_z\sin\alpha &= \frac{1}{R}\left(\cos\alpha\frac{\partial\psi}{\partial z}+\sin\alpha\frac{\partial\psi}{\partial R}\right) \\ &= \frac{1}{R}\left(\frac{\partial z}{\partial s}\frac{\partial\psi}{\partial z}+\frac{\partial R}{\partial s}\frac{\partial\psi}{\partial R}\right)=\frac{1}{R}\frac{\partial\psi}{\partial s}. \end{aligned}$$

Thus the volume of fluid which crosses the surface of revolution at PP' about OZ per unit time is

$$\frac{1}{R}\frac{\partial\psi}{\partial s}\times 2\pi R\delta s = 2\pi\delta s\frac{\partial\psi}{\partial s} = 2\pi\delta\psi,$$

where $\delta\psi = \psi_{P'} - \psi_P$. Hence the total volume of fluid crossing the surface of revolution of the arc AB about OZ per unit time from right to left is

$$\int_A^B 2\pi\, d\psi = 2\pi(\psi_B - \psi_A).$$

This quantity depends only on the positions of A and B in the meridian section and not at all on the shape of $\mathscr{C}$. If $\mathscr{C}$ falls on the axis OZ, it is convenient to take $\psi_A = 0$, so that the volume crossing the surface of revolution of AP per unit time from right to left is $2\pi\psi$. Many writers take this as the definition of the stream function. We infer from equations (2) that the dimensions of ψ are L^3T^{-1}: those of the velocity potential φ are L^2T^{-1}. Further, we note that as no fluid crosses a streamline,

$$\psi = \text{const. along a streamline.}$$

Also, ψ=const. over a *stream surface*—the surface of revolution of a streamline about the axis of symmetry.

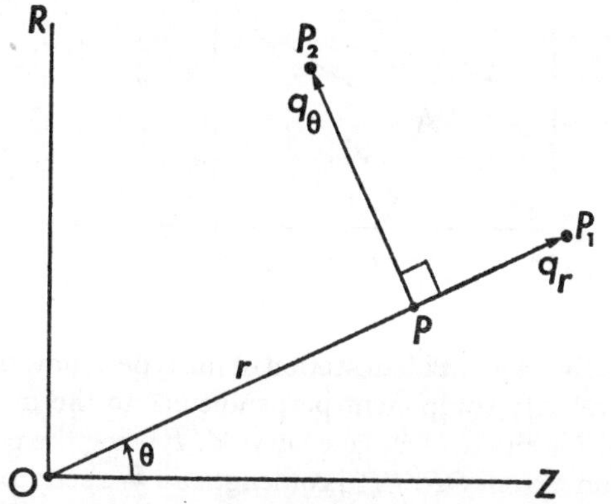

FIG. 4.13

It is often useful to obtain velocity components in terms of spherical polar coordinates and derivatives of the stream function. Figure 4.13 shows a meridian section through the axis of symmetry OZ of the flow, P being at distance r from O and at angular distance θ from $\overline{OZ}$. The azimuthal coordinate is here redundant. The fluid velocity components at P are q_r along $\overline{OP}$ and q_θ at right-angles to it in the sense of θ increasing. The lengths PP_1 and PP_2 are respectively δr and $r\delta\theta$. (θ must not be confused with the angular coordinate used previously for cylindrical polar co-ordinates.)

If we first take $\delta s = \delta r$, the area of the surface of revolution of δs about OZ is $2\pi r \sin\theta\delta r$ and the volume of fluid crossing this per unit time from right to left is $2\pi r \sin\theta\delta r q_\theta$. Hence if ψ is the stream function at P, then

$$2\pi\delta\psi = 2\pi r \sin\theta\delta r q_\theta,$$

so that

$$q_\theta = \frac{1}{r\sin\theta}\frac{\partial\psi}{\partial r}. \tag{3}$$

If we now take $\delta s = r\delta\theta$, the area of the surface of revolution about OZ is $2\pi r\sin\theta\cdot r\delta\theta = 2\pi r^2\sin\theta\delta\theta$. The volume of fluid crossing this per unit time from right to left is $(2\pi r^2\sin\theta\delta\theta)(-q_r)$ and so

$$2\pi\delta\psi = (2\pi r^2\sin\theta\delta\theta)(-q_r)$$

or

$$q_r = -\frac{1}{r^2\sin\theta}\frac{\partial\psi}{\partial\theta}. \tag{4}$$

4.5.1 Some Special Forms of the Stream Function for Axi-Symmetric Irrotational Motions

The function ψ may be evaluated for a number of simple cases. The value of ψ for problems involving components of such standard distributions can then be obtained by superposition of the values of ψ for the separate components. The justification of superposing values of ψ depends on the fact that ψ satisfies a linear second-order partial differential equation, viz.

$$\frac{\partial^2\psi}{\partial R^2} - \frac{1}{R}\frac{\partial\psi}{\partial R} + \frac{\partial^2\psi}{\partial z^2} = 0 \tag{5}$$

in cylindrical polar coordinates (R, θ, z) when the flow is irrotational. This equation is derived from equations (2) using

$$q_R = -\frac{\partial\varphi}{\partial R},\quad q_z = -\frac{\partial\varphi}{\partial z}$$

so that $\partial q_R/\partial z = \partial q_z/\partial R$. The velocity potential φ satisfies Laplace's equation in this coordinate system in the form

$$\frac{\partial^2\varphi}{\partial R^2} + \frac{1}{R}\frac{\partial\varphi}{\partial R} + \frac{\partial^2\varphi}{\partial z^2} = 0. \tag{6}$$

Comparison of (5), (6) shows that ψ is not a harmonic function.

The usefulness of evaluating ψ for a given system compounded of standard components, for each of which ψ is separately known, lies in the fact that it enables streamlines or stream surfaces to be found quickly from ψ = const. In all cases a stream surface ψ = const. may be replaced by a rigid surface. Also the equations (2), (3), (4) permit of the calculation of velocity components in cylindrical and spherical polar coordinates.

(i) *Uniform Stream*

Suppose we have a flow whose velocity is $-U\mathbf{k}$, where U is a constant. Let P be the point in the stream having cylindrical polar coordinates (R, θ, z) and let P_0 be its projection on the z-axis. The dispositions of the coordinate axes are as shown in Fig. 4.12. Taking ψ to be the stream function associated with P and zero that for P_0, since the volume flowing from right to left through the circular disc obtained by revolving PP_0 about the z-axis is $\pi R^2 U$, it follows that

$$2\pi\psi = \pi R^2 U$$

or

$$\psi = \tfrac{1}{2}R^2 U.$$

This is immediately expressible in spherical polar coordinates, taking $R = r \sin \angle POZ$ where $r = OP$.

(ii) *Simple Source*

For a simple source of strength m at the origin, the value of ψ at a point P may be found by using the equations (3), (4). In the spherical polar coordinate notations used, $q_r = m/r^2$, $q_\theta = 0$ so that these equations give

$$\frac{\partial\psi}{\partial r} = 0, \quad \frac{\partial\psi}{\partial\theta} = -m \sin\theta.$$

Hence

$$d\psi(r, \theta) = (\partial\psi/\partial r)\, dr + (\partial\psi/\partial\theta)\, d\theta = -m \sin\theta\, d\theta.$$

Thus $\underline{\psi(r, \theta) = m \cos\theta}$, in simplest form.

N.B. An alternative treatment is given in the worked example at the end of this chapter.

(iii) *Doublet at O, Axis along $\overline{OZ}$*

In spherical polar coordinates the velocity potential at P is

$$\varphi(r, \theta) = \frac{M \cos\theta}{r^2},$$

where M is the strength of the doublet. Then

$$\left.\begin{aligned} q_r &= -\frac{\partial\varphi}{\partial r} = \frac{2M\cos\theta}{r^3} \\ q_\theta &= -\frac{1}{r}\frac{\partial\varphi}{\partial\theta} = \frac{M\sin\theta}{r^3}. \end{aligned}\right\}$$

Thus equations (3), (4) give

$$\frac{\partial\psi}{\partial r}=\frac{M\sin^2\theta}{r^2},\quad \frac{\partial\psi}{\partial\theta}=-\frac{2M\sin\theta\cos\theta}{r}$$

and so

$$d\psi(r,\theta)=M\{(r^{-2}\sin^2\theta)\,dr-2r^{-1}\sin\theta\cos\theta\,d\theta\},$$

giving

$$\underline{\psi(r,\theta)=-(M/r)\sin^2\theta.}$$

(iv) *Uniform Line Source along* $\overline{OZ}$

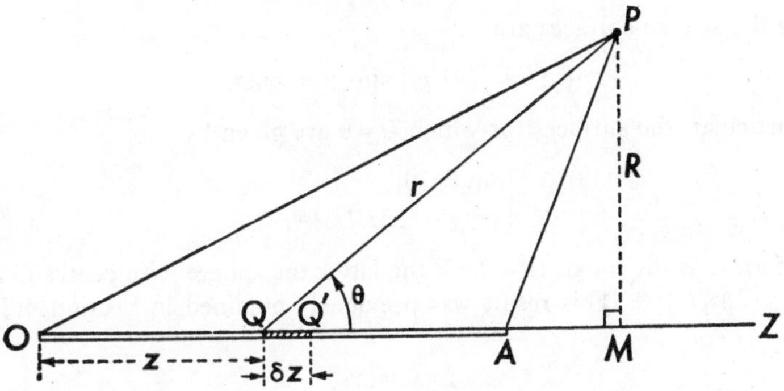

FIG. 4.14

Figure 4.14 depicts a uniform line source of fluid extending along OZ from O to A. Let Q, Q' be neighbouring points on OA, where $OQ=z$, $OQ'=z+\delta z$. Also suppose $QP=r$, $\angle PQA=\theta$. If m denotes the strength per unit length of the line source, then QQ' is effectively a simple source at Q of strength $m\delta z$. Hence the contribution to ψ at P from this source is $\delta\psi$ where

$$\delta\psi=m\delta z\cos\theta,$$

using (ii) above. Denoting $OA=l$, $AM=b$, $PM=R$, all of which are constant elements in the following integration procedure, we have

$$\delta\psi=m\delta z(l+b-z)\{(l+b-z)^2+R^2\}^{-1/2},$$

so that

$$\psi=m\int_0^l\frac{(l+b-z)}{\{(l+b-z)^2+R^2\}^{1/2}}\,dz$$

$$=m[\{(l+b)^2+R^2\}^{1/2}-\{b^2+R^2\}^{1/2}]$$

$$=\underline{m(OP-AP).}$$

This result shows that the stream surfaces, $\psi = \text{const.}$, are given by

$$OP - AP = \text{const.}$$

These are confocal hyperboloids of revolution about OZ, having O and A as foci.

(v) *Doublet in Uniform Stream*

Suppose the doublet specified in (iii) is in a uniform stream whose undisturbed velocity is $-U\mathbf{k}$. Then, combining the results of (i), (iii), the stream function at P in spherical polars is

$$\psi(r, \theta) = \tfrac{1}{2}Ur^2 \sin^2 \theta - (M/r) \sin^2 \theta.$$

Thus the stream surfaces are

$$\{\tfrac{1}{2}Ur^2 - (M/r)\} \sin^2 \theta = \text{const.}$$

In particular, the surfaces for which $\psi = 0$ are given by

$$\begin{cases} \sin \theta = 0, \\ \qquad r = (2M/U)^{1/3}. \end{cases}$$

The former is the z-axis ($\theta = 0, \pi$), the latter the sphere with centre O and radius $(2M/U)^{1/3}$. This result was previously obtained in Section 4.2.

Example

Define the Stokes stream function $\psi(r, \theta)$ (in spherical polar coordinates (r, θ, φ) for the axi-symmetric flow of an incompressible fluid. Determine the stream functions corresponding (i) to a uniform stream U parallel to the axis $\theta = 0$, and (ii) to the spherically symmetric radial velocity field from a point source at the origin, the total outward flux being $4\pi m$.

The equation

$$r \sin \theta = 2a \cos \tfrac{1}{2}\theta$$

represents the surface of a rigid blunt-nosed cylinder, symmetric about the axis $\theta = 0$. Inviscid fluid flows irrotationally past this cylinder, the velocity far from the cylinder being U parallel to the cylinder axis. Show that the sum of the two stream functions (i) and (ii) above, with $m = a^2U$, may be used to represent this flow, and find the pressure distribution, as a function of θ, on the cylinder surface.

(Cambridge Univ., Prelim. Exam. to Maths. Tripos II, 1964)

By the method used in the text, the stream function due to the uniform stream is, in this case, $-\tfrac{1}{2}r^2U \sin^2 \theta$.

The stream function due to the simple source of strength $4\pi m$ at the origin can also be obtained from the standard result, but it is instructive to find it by an alternative process. Let P be the point (r, θ, φ) and P_0 the point on $\theta = 0$ distant r from O. When the circular arc centre O and radius $OP = OP_0 = r$ is revolved once round the line $\theta = 0$, it generates a cap of a sphere whose area is equal to its projection on the corresponding enveloping cylinder with axis along $\theta=0$. Since the length of the projection is $r(1-\cos\theta)$, the area of the cap is $2\pi r^2(1-\cos\theta)$. In unit time $4\pi m$ units of volume of fluid emerge from O. Hence the volume flux per unit time across the cap is

$$\frac{2\pi r^2(1-\cos\theta)}{4\pi r^2}\times 4\pi m = 2m(1-\cos\theta)$$

from left to right. Thus $-2m(1-\cos\theta)$ is the flux from right to left and as this is $2\pi\psi$, we have

$$\psi=-m(1-\cos\theta), \text{ for the source.}$$

Thus the total stream function is

$$\psi=-m(1-\cos\theta)-\tfrac{1}{2}Ur^2\sin^2\theta.$$

Taking $m=a^2U$,

$$\psi=-U\{a^2(1-\cos\theta)+\tfrac{1}{2}r^2\sin^2\theta\}. \tag{1}$$

The stream surfaces are given by $\psi=$const. or

$$2a^2\cos^2\tfrac{1}{2}\theta-\tfrac{1}{2}r^2\sin^2\theta=\text{const.} \tag{2}$$

The surface or surfaces for which the constant on the R.H.S. of (2) is zero are given by

$$r\sin\theta=2a\cos\tfrac{1}{2}\theta. \tag{3}$$

Equation (3) is satisfied by

$$\begin{cases}\cos\tfrac{1}{2}\theta=0, & \text{i.e. } \theta=\pi;\\ r=a\operatorname{cosec}\tfrac{1}{2}\theta.\end{cases}$$

Thus the given cylinder can be fitted to the stream surface (3).

The velocity components q_r, q_θ at any point of the fluid are found from (1). Thus

$$\begin{cases} q_r=-\dfrac{1}{r^2\sin\theta}\dfrac{\partial\psi}{\partial\theta}=U\left\{\dfrac{a^2}{r^2}+\cos\theta\right\}\\[2ex] q_\theta=\dfrac{1}{r\sin\theta}\dfrac{\partial\psi}{\partial r}=-U\sin\theta.\end{cases}$$

As $r\to\infty$, $q_r\to U\cos\theta$, $q_\theta=-U\sin\theta$. Thus at infinite distance from $\theta=0$, the flow is uniform with velocity U parallel to $\theta=0$. Thus (1) is the appropriate stream function for flow past the given cylinder.

The fluid speed q at any point is

$$q=\sqrt{(q_r^2+q_\theta^2)}=U\sqrt{\{1+2(a/r)^2\cos\theta+(a/r)^4\}}.$$

On the cylinder surface, $r=a\operatorname{cosec}\tfrac{1}{2}\theta$ and so we obtain

$$q=\tfrac{1}{2}U\sqrt{(5+2\cos\theta-3\cos^2\theta)^{1/2}}.$$

Now apply Bernoulli's equation along the entire surface $\psi=0$. Taking $p=p_\infty$ at $r=\infty$ on this surface,

$$\frac{p}{\rho}+\tfrac{1}{2}q^2=\frac{p_\infty}{\rho}+\tfrac{1}{2}U^2,$$

or

$$\underline{p=p_\infty-\tfrac{1}{8}\rho U^2(1+2\cos\theta-3\cos^2\theta).}$$

EXERCISE 4

1. Define a streamline and say when streamlines are paths of fluid particles.

A point source of strength m is caused to move with uniform speed U through an inviscid incompressible fluid at rest at infinity. If r denotes the distance from the source and θ is the angle that the radius vector from the source makes with the direction of motion, prove that the streamlines are $\theta=$constant, but that the paths of the particles are given by

$$Ur^2\sin^2\theta=-2m\cos\theta+A, \text{ where } A \text{ is constant.}$$

(London Univ., B.Sc. Spec. Hons., 1962)

2. A simple source, strength m, is fixed at the origin O in a uniform stream of incompressible fluid moving with velocity $U\mathbf{i}$. Show that the velocity potential φ at any point P of the stream is $(m/r)-Ur\cos\theta$, where $OP=r$ and θ is the angle $\overline{OP}$ makes with the direction $\mathbf{i}$. Find the differential equation of the streamlines and show that they lie on the surfaces

$$Ur^2\sin^2\theta-2m\cos\theta=\text{const.}$$

3. An infinite mass of liquid of constant density ρ lies to one side of a rigid impermeable infinite plane and a three-dimensional doublet of strength μ is situated in the liquid, at a distance d from the plane and having its axis directed straight towards the wall. Assuming the pressure in the liquid at infinite distance from the doublet to be zero, show that the doublet is attracted towards the wall with a force $3\pi\mu^2\rho/(2d^4)$.

4. Three-dimensional doublets of strengths μ and $(a/c)^3\mu$ are situated in an infinite expanse of liquid of constant density at the points having Cartesian coordinates $(c, 0, 0)$ and $(a^2/c, 0, 0)$ respectively. Their axes are directed towards and away from the origin respectively. Find the velocity potential and hence show that no fluid crosses the sphere centre $(0, 0, 0)$ and radius a.

5. Derive the stream functions for (i) a uniform flow of magnitude U, (ii) a uniform line source of strength m. Show that a combination of these two stream functions can be used to give the two-dimensional flow past a certain bluff semi-infinite body and express the thickness of the body in terms of m, U.

(Durham Univ., Gen. B.Sc., 1963)

6 In an irrotational motion of an inviscid liquid of constant density ρ there is a source with a rate of flow $4\pi Ua^2$ at the origin O together with a uniform stream with velocity U; the direction of U is such that the point with coordinates $(a, 0, 0)$ with respect to polar coordinates (r, θ, φ) is a point of stagnation. Show that the velocity at any point of the surface $r=a \sec(\theta/2)$ is tangential to the surface, and if there is no body force, find an expression for the pressure at a general point on this surface.

(London Univ., B.Sc. Spec., 1958)

7. Show that stream functions for a source and a uniform stream can be expressed in the forms $m \cos \theta$ and $-\frac{1}{2}Ur^2 \sin^2 \theta$ respectively. A sink and a source, having strengths of equal magnitude $4\pi m$, are situated at $x=\pm a$ respectively in an otherwise uniform stream U parallel to the positive direction of the x-axis. Show that particles originating at distance b from the axis cross the plane $x=0$ at distance $a\sqrt{3}$ from the x-axis if $b=\sqrt{(3a^2-2m/U)}$. Sketch the streamlines and explain the absence of a real value of b in the case $2m>3Ua^2$.

(Imperial Coll., B.Sc. Spec., 1963)

8. A three-dimensional doublet points in the direction of flow of a uniform stream in which it is situated. Show that one of the equipotential surfaces is a sphere and show that the stagnation points lie in this sphere. Taking the centre of the doublet as the origin and its axis as the line $\theta=0$, show that the equations of the streamlines in spherical polar coordinates are of the form $(2a^3+r^3) \sin^2 \theta = Cr$, where a is the radius of the sphere.

9. The three-dimensional motion of a liquid is symmetrical with respect to the axis $\theta=0$ of spherical polar coordinates, and the velocity components are $[u, v]$ in directions associated with the coordinates (r, θ). Obtain the equation of continuity and deduce that u and v can be expressed in terms of the stream function, ψ.

Show that, if the motion is irrotational, then ψ satisfies the equation

$$r^2 \frac{\partial^2 \psi}{\partial r^2} = \cot \theta \frac{\partial \psi}{\partial \theta} - \frac{\partial^2 \psi}{\partial \theta^2}.$$

Show that if

$$\psi = \tfrac{1}{2}U(a^4/r^2) \sin^2 \theta \cos \theta - \tfrac{1}{2}Ur^2 \sin^2 \theta,$$

the motion is irrotational.

(Leeds Univ., B.Sc., 1962)

5

SOME TWO-DIMENSIONAL FLOWS

5.1 Meaning of Two-Dimensional Flow

Suppose a fluid moves in such a way that at any given instant the flow pattern in a certain plane is the same as that in all other parallel planes within the fluid. Then the flow is said to be *two-dimensional.* If we take any one of the parallel planes to be the plane $z=0$, then at any point in the fluid having Cartesian coordinates (x, y, z), all physical quantities (velocity, pressure, density, etc.) associated with the fluid are independent of z.

In this chapter we consider special methods for treating this class of problem. We confine attention to inviscid incompressible fluids throughout.

5.2 Use of Cylindrical Polar Coordinates

If the flow is irrotational, then the equation satisfied by the velocity potential φ at any point having cylindrical polar coordinates (R, θ, z) is

$$\frac{1}{R}\frac{\partial}{\partial R}\left(R\frac{\partial \varphi}{\partial R}\right)+\frac{1}{R^2}\frac{\partial^2 \varphi}{\partial \theta^2}+\frac{\partial^2 \varphi}{\partial z^2}=0,$$

which is Laplace's equation in these coordinates. If the motion is two-dimensional and the coordinate axes so chosen that all physical quantities associated with the flow are independent of z, then

$$\frac{1}{R}\frac{\partial}{\partial R}\left(R\frac{\partial \varphi}{\partial R}\right)+\frac{1}{R^2}\frac{\partial^2 \varphi}{\partial \theta^2}=0. \tag{1}$$

In order to obtain special solutions of this equation we try separating the variables by putting $\varphi=f(R)g(\theta)$ to give

$$g(\theta)\frac{1}{R}\frac{d}{dR}\{Rf'(R)\}+\frac{1}{R^2}f(R)g''(\theta)=0$$

or

$$\frac{R(d/dR)\{Rf'(R)\}}{f(R)}=-\frac{g''(\theta)}{g(\theta)}. \tag{2}$$

The L.H.S. of (2) is a function of R only: the R.H.S. is a function of θ only. Thus each is constant. Let n^2 be the value of the constant. Then we obtain the ordinary differential equations

$$R^2f''(R)+Rf'(R)-n^2f(R)=0, \tag{3}$$

$$g''(\theta)+n^2g(\theta)=0. \tag{4}$$

Equation (3) is of the Euler-homogeneous type and is solved by putting $R=e^t$ when the equation reduces to

$$\frac{d^2f}{dt^2}-n^2f=0,$$

giving the solutions

$$f=e^{\pm nt}$$

or

$$f(R)=R^{\pm n} \quad (n\neq 0).$$

Equation (4) has the special solutions

$$g(\theta)=\begin{matrix}\cos\\ \sin\end{matrix}\, n\theta \quad (n\neq 0).$$

Thus special solutions to equation (1) are

$$\varphi(R,\theta)=(A_nR^n+B_nR^{-n})\,(C_n\cos n\theta+E_n\sin n\theta).$$

A more general type of solution is obtained by superposition:

$$\varphi(R,\theta)=\sum_n(A_nR^n+B_nR^{-n})\,(C_n\cos n\theta+E_n\sin n\theta).$$

This summation, over n, may be finite or infinite and need not be restricted to integral values of n.

In the special case of $n=0$, rejected in the above development,

$$f=k_1+k_2t=k_1+k_2\log R,$$
$$g=k_3+k_4\theta,$$

so that

$$\varphi=(k_1+k_2\log R)\,(k_3+k_4\theta).$$

The particular case of $n=1$ yields the important special solutions

$$\varphi = R\cos\theta,\quad \varphi = R\sin\theta,\quad \varphi = \frac{1}{R}\cos\theta,\quad \varphi = \frac{1}{R}\sin\theta.$$

We now illustrate the use of these results.

Example 1

Uniform flow past a fixed infinite circular cylinder.

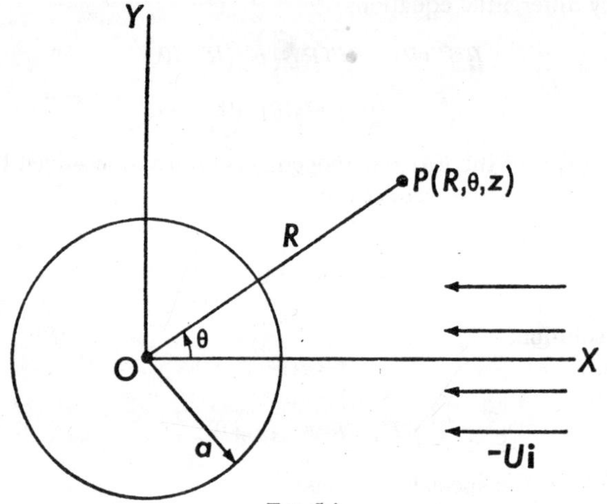

Fig. 5.1

Figure 5.1 shows a uniform stream of undisturbed velocity $-U\mathbf{i}$ flowing past the fixed cylinder $R=a$. P is a point in the fluid having cylindrical polar coordinates (R, θ, z) and Cartesian coordinates (x, y, z), the common coordinate z being redundant as the flow is two-dimensional. The velocity potential due to the uniform stream is

$$Ux = UR\cos\theta.$$

When the cylinder is introduced, it will produce a perturbation of the flow. This perturbation must be such as to satisfy Laplace's equation and to become vanishingly small for large R. The simplest harmonic function satisfying these requisites is clearly $R^{-1}\cos\theta$. Let us, then, try to find A so that

$$\varphi(R, \theta) = UR\cos\theta + AR^{-1}\cos\theta$$

is the required velocity potential. On the cylinder we require that $\partial\varphi/\partial R = 0$ for all θ which gives at once $A = Ua^2$.
Thus

$$\varphi(R, \theta) = U\cos\theta(R + a^2R^{-1}).$$

Hence the velocity components at P are

$$\begin{cases} q_R = -\dfrac{\partial\varphi}{\partial R} = -U\cos\theta\left\{1-\left(\dfrac{a}{R}\right)^2\right\}, \\ q_\theta = -\dfrac{1}{R}\dfrac{\partial\varphi}{\partial\theta} = U\sin\theta\left\{1+\left(\dfrac{a}{R}\right)^2\right\}, \\ q_z = -\dfrac{\partial\varphi}{\partial z} = 0. \end{cases}$$

As $R\to\infty$, $q_R\to -U\cos\theta$, $q_\theta\to U\sin\theta$, appropriately.

EXAMPLE 2

A cylinder of infinite length and nearly circular section moves through an infinite volume of liquid with a velocity U at right-angles to its axis and in the direction of the x-axis. If its section is specified by the equation

$$R = a(1+\varepsilon\cos n\theta),$$

where ε is small, show that the approximate value of the velocity potential is

$$Ua\left\{\frac{a}{R}\cos\theta + \varepsilon\left(\frac{a}{R}\right)^{n+1}\cos(n+1)\theta - \varepsilon\left(\frac{a}{R}\right)^{n-1}\cos(n-1)\theta\right\}.$$

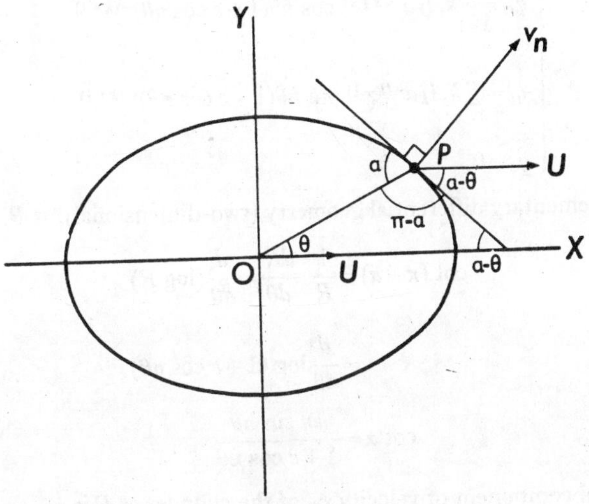

FIG. 5.2

Figure 5.2 shows a section of the cylinder, the tangent at a point on it making angles α, $(\pi-\alpha)$ with the radius vector drawn from O. At large radial distances R from OZ, the fluid velocity becomes vanishingly small.

Thus suitable harmonic functions are of the forms $R^{-k}\cos k\theta$, $R^{-k}\sin k\theta$ $(k=1, 2, 3\ldots)$. Let us try the solution

$$\varphi(R, \theta)=\sum_{k=1}^{\infty} R^{-k}(A_k \cos k\theta+B_k \sin k\theta).$$

At $\theta=0$ on the boundary, $q_\theta=0$, which is satisfied by taking $B_k=0$ $(k=1, 2, \ldots)$. Further the function

$$\varphi(R, \theta)=\sum_{k=1}^{\infty} A_k R^{-k}\cos k\theta \tag{1}$$

is appropriately an even function of θ.

At any point (R, θ, z) in the fluid,

$$\begin{cases} q_R=-\dfrac{\partial\varphi}{\partial R}=\displaystyle\sum_{k=1}^{\infty} kA_k R^{-(k+1)}\cos k\theta, \\ q_\theta=-\dfrac{1}{R}\dfrac{\partial\varphi}{\partial\theta}=\displaystyle\sum_{k=1}^{\infty} kA_k R^{-(k+1)}\sin k\theta, \\ q_z=-\dfrac{\partial\varphi}{\partial z}=0. \end{cases}$$

Hence at the point P on the surface where $R=a(1+\varepsilon\cos n\theta)$,

$$\begin{cases} q_R=\displaystyle\sum_{k=1}^{\infty} kA_k a^{-(k+1)}\cos k\theta(1+\varepsilon\cos n\theta)^{-(k+1)}, \\ q_\theta=\displaystyle\sum_{k=1}^{\infty} kA_k a^{-(k+1)}\sin k\theta(1+\varepsilon\cos n\theta)^{-(k+1)}, \\ q_z=0. \end{cases}$$

From elementary differential geometry (two-dimensional), at P

$$\cot(\pi-\alpha)=\frac{1}{R}\frac{dR}{d\theta}=\frac{d}{d\theta}(\log R)$$

$$=\frac{d}{d\theta}\log(1+\varepsilon\cos n\theta)$$

so that

$$\cot\alpha=\frac{\varepsilon n\sin n\theta}{1+\varepsilon\cos n\theta}.$$

The normal component of velocity v_n of the cylinder at P is

$$v_n=U\sin(\alpha-\theta)=U(\sin\alpha\cos\theta-\cos\alpha\sin\theta)$$

$$=\frac{U\{\cos\theta(1+\varepsilon\cos n\theta)-\sin\theta\cdot\varepsilon n\sin n\theta\}}{\{(1+\varepsilon\cos n\theta)^2+\varepsilon^2n^2\sin^2 n\theta\}^{1/2}}.$$

As there is no transport of fluid across the surface, the component of fluid velocity in the same direction at P must also be v_n, so that

$$v_n = q_R \sin\alpha + q_\theta \cos\alpha$$

$$= \frac{\sum_{k=1}^{\infty} kA_k a^{-(k+1)}(1+\varepsilon\cos n\theta)^{-(k+1)}\{\cos k\theta(1+\varepsilon\cos n\theta) + \sin k\theta\cdot\varepsilon n\sin n\theta\}}{\{(1+\varepsilon\cos n\theta)^2+\varepsilon^2 n^2\sin^2 n\theta\}^{1/2}}.$$

Equating the two forms for v_n gives, on simplification,

$$\sum_{k=1}^{\infty} kA_k a^{-(k+1)}(1+\varepsilon\cos n\theta)^{-(k+1)}\{\cos k\theta(1+\varepsilon\cos n\theta)+\sin k\theta\cdot\varepsilon n\sin n\theta\}$$
$$= U\{\cos\theta(1+\varepsilon\cos n\theta)-\sin\theta\cdot\varepsilon n\sin n\theta\}.$$

Correct to the first order in ε this gives the approximation

$$\sum_{k=1}^{\infty} kA_k[\cos k\theta+\tfrac{1}{2}\varepsilon\{(n-k)\cos(n-k)\theta-(n+k)\cos(n+k)\theta\}]$$
$$\simeq U[\cos\theta+\tfrac{1}{2}\varepsilon\{\cos(n+1)\theta+\cos(n-1)\theta\}-\tfrac{1}{2}\varepsilon n\{\cos(n-1)\theta-\cos(n+1)\theta\}].$$

Equating coefficients of $\cos\theta$,

$$U = b_1/a^2.$$

Equating coefficients of $\cos(n-1)\theta$,

$$\tfrac{1}{2}U\varepsilon+\tfrac{1}{2}U\varepsilon n=(n+1)\frac{b_{n-1}}{a^{n+2}}-\tfrac{1}{2}(n+1)\varepsilon\left(\frac{b_1}{a^2}\right),$$

so that $b_{n+1} = U\varepsilon a^{n+2}$.

All other b's are zero. Substitution for b_1, b_{n-1}, b_{n+1} yields the required result.

5.3 The Stream Function

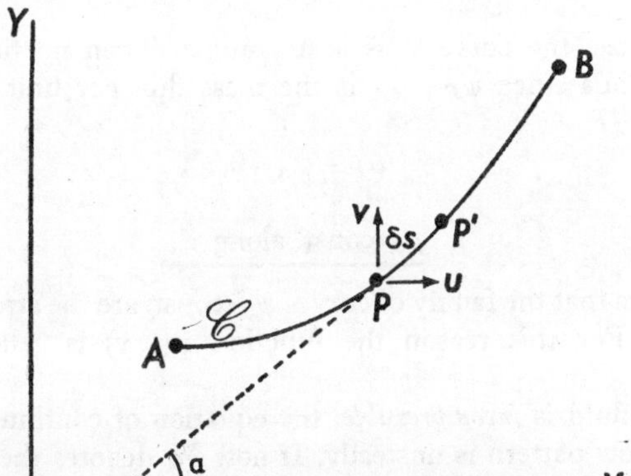

Fig. 5.3

$\mathscr{C}$ is an arc of a curve in the (x,y)-plane joining two points A, B (Fig. 5.3). Fluid (compressible or incompressible) is constrained to flow in two-dimensional steady motion parallel to this plane. At $P(x,y)$ on the curve the velocity components are $[u,v]$. P' is the neighbouring point $(x+\delta x, y+\delta y)$ on $\mathscr{C}$ so that $PP'=\delta s$. If the tangent to $\mathscr{C}$ at P makes angle α with OX, then the velocity component at P along the normal from right to left as we travel from A to B is $v\cos\alpha - u\sin\alpha$. Hence the mass of fluid which crosses unit thickness of the surface element through PP' normal to the plane of flow per unit time is

$$\rho(v\cos\alpha - u\sin\alpha)\delta s = \rho(v\delta x - u\delta y).$$

This mass flux is constant: denote it by $\delta\psi$. Then

$$\delta\psi = \rho(v\delta x - u\delta y)$$

or

$$d\psi = \rho(v\,dx - u\,dy). \tag{1}$$

The total mass flux per unit thickness per unit time from right to left across the arc AB is

$$\psi_B - \psi_A = \int_A^B \rho(v\,dx - u\,dy). \tag{2}$$

Clearly when A and B are fixed, the integral (2) is independent of the shape of the curve $\mathscr{C}$ joining them. Thus $d\psi$ is a perfect differential. This may also be seen from the fact that for steady flow the equation of continuity $\nabla\cdot(\rho\mathbf{q})=0$ becomes

$$\frac{\partial}{\partial x}(\rho u) + \frac{\partial}{\partial y}(\rho v) = 0.$$

Thus

$$\frac{\partial\psi}{\partial x} = \rho v, \quad \frac{\partial\psi}{\partial y} = -\rho u. \tag{3}$$

Now suppose the curve $\mathscr{C}$ is a streamline. Then no fluid crosses its boundary. Thus since $\psi_P - \psi_A$ is the mass flux per unit time per unit thickness of AP,

$$\psi_P - \psi_A = 0,$$

or

$$\underline{\psi = \text{const. along } \mathscr{C}.}$$

It thus follows that the family of curves $\psi=$const. are the streamlines in the plane $z=0$. For this reason the function $\psi(x,y)$ is called the *stream function*.

When the fluid is *incompressible*, the equation of continuity is $\nabla\cdot\mathbf{q}=0$ even if the flow pattern is unsteady. If now $\delta\psi$ denotes the *volume* of the fluid crossing unit thickness of PP' from right to left per unit time, the equations (1), (2), (3) become

$$d\psi = v\,dx - u\,dy, \tag{1'}$$

$$\psi_B - \psi_A = \int_A^B (v\,dx - u\,dy), \tag{2'}$$

$$\partial\psi/\partial x = v, \quad \partial\psi/\partial y = -u. \tag{3'}$$

If we further suppose the flow to be irrotational, then there exists a velocity potential φ such that

$$u = -\partial\varphi/\partial x, \quad v = -\partial\varphi/\partial y. \tag{4}$$

The equation of continuity is

$$\frac{\partial u}{\partial x} + \frac{\partial v}{\partial y} = 0. \tag{5}$$

The equations (3′), (4), (5) now yield the following important results:

$$\frac{\partial\varphi}{\partial x} = \frac{\partial\psi}{\partial y}, \quad \frac{\partial\varphi}{\partial y} = -\frac{\partial\psi}{\partial x}, \tag{6}$$

$$\frac{\partial^2\varphi}{\partial x^2} + \frac{\partial^2\varphi}{\partial y^2} = 0 \tag{7}$$

$$\frac{\partial^2\psi}{\partial x^2} + \frac{\partial^2\psi}{\partial y^2} = 0 \tag{8}$$

$$\frac{\partial\varphi}{\partial x}\frac{\partial\psi}{\partial x} + \frac{\partial\varphi}{\partial y}\frac{\partial\psi}{\partial y} = 0. \tag{9}$$

Thus for irrotational incompressible two-dimensional flow (steady or unsteady), $\varphi(x, y)$, $\psi(x, y)$ are harmonic functions and the families of curves φ = constant (equipotentials) and ψ = constant (streamlines) intersect orthogonally. This last result follows from the fact that (9) expresses the perpendicularity of the vectors $\nabla\varphi, \nabla\psi$.

5.4 The Complex Potential for Two-Dimensional, Irrotational, Incompressible Flow

At this stage we introduce the notion of a function of a complex variable. We shall see that the complex function $w = \varphi + i\psi$ has far-reaching properties which facilitate solution of many types of two-dimensional flow problems.

We first reproduce some elementary theory of functions of a complex variable. The reader will find a fuller treatment in *Functions of a Complex Variable*, by E. G. Phillips (Oliver and Boyd).

Suppose that $z = x + iy$ and that

$$w = f(z) = \varphi(x, y) + i\psi(x, y)$$

where x, y, φ, ψ are all real and $i=\sqrt{(-1)}$. We write $\varphi=\text{Re}(w)$, $\psi=\text{Im}(w)$. Also we suppose that φ and ψ and their first derivatives are everywhere continuous within a given region. If at any point of the region specified by z the derivative $dw/dz=f'(z)$ is unique, then w is said to be *analytic* or *regular* at that point. If the derivative is unique throughout the region, then w is said to be analytic or regular throughout the region. By definition, the derivative of w at z is

$$\frac{dw}{dz}=f'(z)=\lim_{\delta x,\delta y\to 0}\left\{\frac{\delta\varphi+i\delta\psi}{\delta x+i\delta y}\right\},$$

where $\delta\varphi=\varphi(x+\delta x, y+\delta y)-\varphi(x,y)$, $\delta\psi=\psi(x+\delta x, y+\delta y)-\psi(x,y)$. This limit may be evaluated in infinitely many ways according to the nature of the differentials δx, δy. If we first keep y constant so that $\delta y=0$, then

$$\frac{dw}{dz}=\lim_{\delta x\to 0}\left\{\frac{\delta\varphi+i\delta\psi}{\delta x}\right\}=\frac{\partial\varphi}{\partial x}+i\frac{\partial\psi}{\partial x}. \tag{1}$$

If, however, we keep x constant so that $\delta x=0$, then

$$\frac{dw}{dz}=\lim_{\delta y\to 0}\left\{\frac{\delta\varphi+i\delta\psi}{i\delta y}\right\}=-i\frac{\partial\varphi}{\partial y}+\frac{\partial\psi}{\partial y}. \tag{2}$$

If w is an analytic function, the two values of dw/dz expressed by (1) and (2) must be equal. Then, equating real and imaginary parts, we obtain at once

$$\frac{\partial\varphi}{\partial x}=\frac{\partial\psi}{\partial y},\quad \frac{\partial\varphi}{\partial y}=-\frac{\partial\psi}{\partial x}. \tag{3}$$

The equations (3) are called the *Cauchy-Riemann equations*. It can be shown that (provided all first order derivatives are continuous) they express the necessary and sufficient conditions for w to be analytic at a point. If they hold at all points of a region they express the necessary and sufficient conditions for w to be analytic throughout the region. The functions φ, ψ are termed *conjugate functions*.

If we now assign to φ and ψ the meanings of velocity potential and stream function for two-dimensional, irrotational, incompressible, inviscid flow, then $w=\varphi+i\psi$ is called the *complex velocity potential*. Thus the real and imaginary parts of w are the velocity potential and stream function, respectively. Comparison of the Cauchy-Riemann equations (3) above with equations (6) of Section 4.3 shows that the complex velocity potential is analytic. Also equation (1) above shows that

$$dw/dz=-u+iv,$$

since $u=-\partial\varphi/\partial x=-\partial\psi/\partial y$, $v=-\partial\varphi/\partial y=\partial\psi/\partial x$. Also we have

$$|dw/dz|=\sqrt{(u^2+v^2)}=q,$$

giving the speed of the fluid at any point.

EXAMPLE

Discuss the flow for which $w=z^2$.

We have $\varphi+i\psi=(x+iy)^2=x^2-y^2+2ixy$.

Hence

$$\begin{cases}\varphi=x^2-y^2,\\ \psi=2xy.\end{cases}$$

The equipotentials are the rectangular hyperbolae

$$x^2-y^2=\text{const.}$$

having asymptotes $y=\pm x$. The streamlines are the rectangular hyperbolae

$$xy=\text{const.}$$

having asymptotes $x=0$, $y=0$. The two families of curves cut each other orthogonally. Figure 5.4 shows both families.

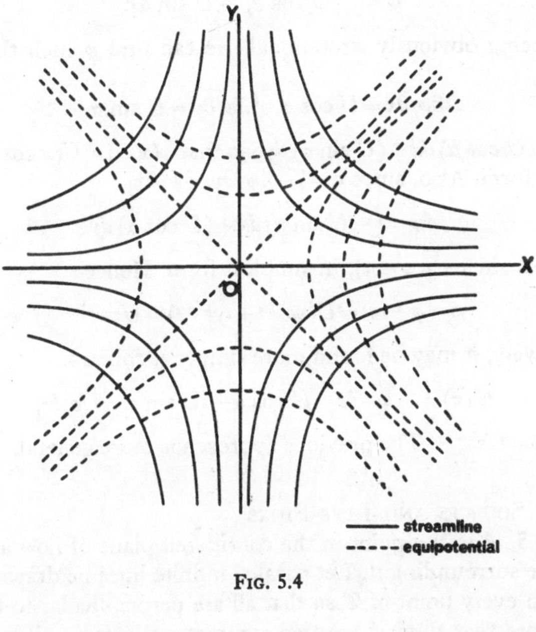

FIG. 5.4

Since there is no flow over a streamline, any one may be taken as a rigid barrier. In particular, the axes $x=0$, $y=0$ may be taken as rigid boundaries. Since $dw/dz=2z$ is zero only at the origin, this is the only stagnation point.

It is easily confirmed that both φ and ψ are harmonic and so from the first fact we infer that the flow is irrotational.

5.5 Complex Velocity Potentials for Standard Two-Dimensional Flows

For further developments, it is convenient to evaluate the complex potential w for some of the more common types of two-dimensional flows. We consider in turn the flow patterns due to a uniform stream, a line source and sink, and a line doublet. Throughout this section the flow is assumed to be inviscid, irrotational, incompressible and two-dimensional.

5.5.1 Uniform Stream

(i) We first consider the uniform stream having velocity $-U\mathbf{i}$. This gives rise to a velocity potential $\varphi = Ux$. Clearly, since $x = \text{Re}(z)$,

$$w = Uz = U(x+iy).$$

The stream function $\psi = \text{Im}(w) = Uy$, so that the lines $y = \text{const.}$ are the streamlines.

(ii) Secondly, suppose the uniform stream is incident to the positive x-axis at angle α, so that

$$\mathbf{q} = [-U\cos\alpha,\ -U\sin\alpha].$$

The flow being obviously irrotational, we can find φ such that $\mathbf{q} = -\nabla\varphi$ and hence

$$\partial\varphi/\partial x = U\cos\alpha, \quad \partial\varphi/\partial y = U\sin\alpha.$$

Thus $d\varphi = (U\cos\alpha)\,dx + (U\sin\alpha)\,dy$, so that $\varphi(x, y) = U(x\cos\alpha + y\sin\alpha)$, in simplest form. Also, since $\mathbf{q} = [-\partial\psi/\partial y,\ \partial\psi/\partial x]$,

$$d\psi = (-U\sin\alpha)\,dx + (U\cos\alpha)\,dy$$

or $\psi = U(-x\sin\alpha + y\cos\alpha)$, in simplest form. Hence

$$w = \varphi + i\psi = U(xe^{-i\alpha} + iye^{-i\alpha}) = Uze^{-i\alpha}.$$

Alternatively, w may be found more simply as follows:

$$w'(z) = -u + iv = U\cos\alpha - iU\sin\alpha = Ue^{-i\alpha},$$

so that $w(z) = Uze^{-i\alpha}$. The previous approach is more general.

5.5.2 Line Sources and Line Sinks

In Fig. 5.5, A is any point in the considered plane of flow and $\mathscr{C}$ is any closed curve surrounding it. Let parallel infinite lines be drawn through A and through every point of $\mathscr{C}$ so that all are perpendicular to the plane of flow. Suppose that fluid is emitted symmetrically from all points of the infinite line through A, the rate of emission from all such points A on the line being the same everywhere and parallel to the plane of flow. Then the line through A is called a *line source*. If fluid drains away through such a line and under the same conditions of symmetry, then the line is called a *line sink*.

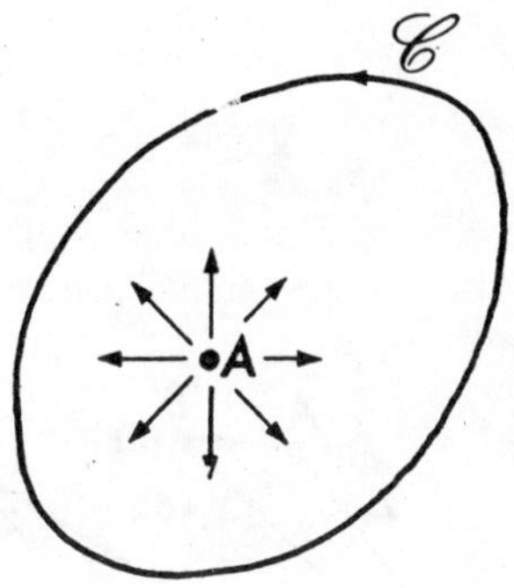

FIG. 5.5

Suppose the infinite line through A is a source and that it emits fluid at the rate of $2\pi m\rho$ units of mass per unit length of the source per unit time. Then this is the mass flux per unit time per unit length of the infinite cylinder whose trace in the plane of flow is the closed curve $\mathscr{C}$ and whose generators are normal to this plane. We define the *strength of the line source* to be m. If we now take for $\mathscr{C}$ the circle centre A and radius r, then the speed of flow q is everywhere the same on $\mathscr{C}$ and the mass flux is now $2\pi r\rho q$ per unit length. Thus $2\pi m\rho = 2\pi r\rho q$ or

$$\underline{q = m/r.}$$

On such a circle, the velocity potential is clearly $\varphi(r)$, so that

$$-\frac{\partial\varphi}{\partial r} = \frac{m}{r}$$

or

$$\underline{\varphi = -m\log r,} \quad \text{in simplest form.}$$

If the considered field-point P on the circle has polar coordinates (r, θ) w.r.t. A (origin), and if $\overline{OP} \equiv z = re^{i\theta}$, then since $\log r = \log|z|$, we see at once that the complex velocity potential of the line source through A of uniform strength m is

$$\underline{w = -m\log z = -m\{\log r + i\theta\}.}$$

Thus $\psi = -m\theta$, showing that the streamlines, $\psi = \text{const.}$, in any plane of flow are the straight lines $\theta = \text{const.}$, the equipotentials, $\varphi = \text{const.}$, being the concentric circles $r = \text{const.}$ centred on A.

If the line source of strength m were situated at $z = z_1$ instead of at $z = 0$, since in the above $r = |z - z_1|$, we should have for the complex velocity potential

$$\underline{w = -m\log(z - z_1).}$$

5.5.3 Line Doublets

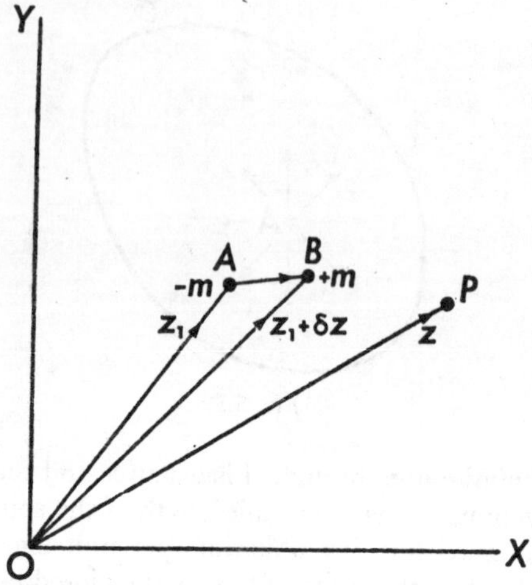

Fig. 5.6

Figure 5.6 shows a uniform line source of strength $-m$ at A, one of strength $+m$ at B and the considered field point P, all in the same plane of flow, where

$$\overline{OA} \equiv z_1, \quad \overline{OB} \equiv z_1 + \delta z_1, \quad \overline{OP} \equiv z.$$

Then $\overline{AP} \equiv z - z_1$, $\overline{BP} \equiv z - z_1 - \delta z_1$. Hence the complex potential at P due to the two line sources at A, B is

$$w = m \log (z - z_1) - m \log (z - z_1 - \delta z_1).$$

Now

$$\begin{aligned} \log (z - z_1 - \delta z_1) &= \log (z - z_1) - \delta z_1(\partial/\partial z) [\log (z - z_1)] + 0\{(\delta z)^2\} \\ &= \log (z - z_1) - \delta z_1 (z - z_1)^{-1} + 0\{(\delta z)^2\}. \end{aligned}$$

Thus if we assume that $|\delta z_1|$ is so small that $|\delta z_1|^2$, $|\delta z_1|^3$, ... are negligible, then to the first order,

$$w = -m\delta z_1 (z - z_1)^{-1} = m |\delta z_1| e^{i\alpha} (z - z_1)^{-1},$$

if AB makes angle α with OX. Now suppose that m becomes very large and $|\delta z_1|$ very small, but in such a way that the quantity $\mu = m |\delta z_1|$ remains finite and constant. Then the juxtaposed equal and opposite line sources at A are said to comprise a *line doublet of strength μ per unit length.* $\overline{AB}$ gives the direction of the axis of the doublet. Thus, in the limit, the complex potential at P is

$$\underline{w = \mu e^{i\alpha} (z - z_1)^{-1}.}$$

EXAMPLE

Discuss the flow due to a uniform line doublet at O of strength μ per unit length, its axis being along $\overline{OX}$.

The complex potential at $P(x, y)$ is

$$w = \frac{\mu}{z} = \frac{\mu(x - iy)}{x^2 + y^2}$$

so that

$$\begin{cases} \varphi = \dfrac{\mu x}{x^2 + y^2}, \\ \psi = -\dfrac{\mu y}{x^2 + y^2}. \end{cases}$$

Thus the equipotentials, $\varphi = \text{const.}$, are the coaxial circles

$$x^2 + y^2 = 2k_1 x$$

and the streamlines, $\psi = \text{const.}$, are the coaxial circles

$$x^2 + y^2 = 2k_2 y.$$

The first family have centres $(k_1, 0)$ and radii k_1: the second centres $(0, k_2)$ and radii k_2. Figure 5.7 shows some of each in sketch. The two families are mutually orthogonal.

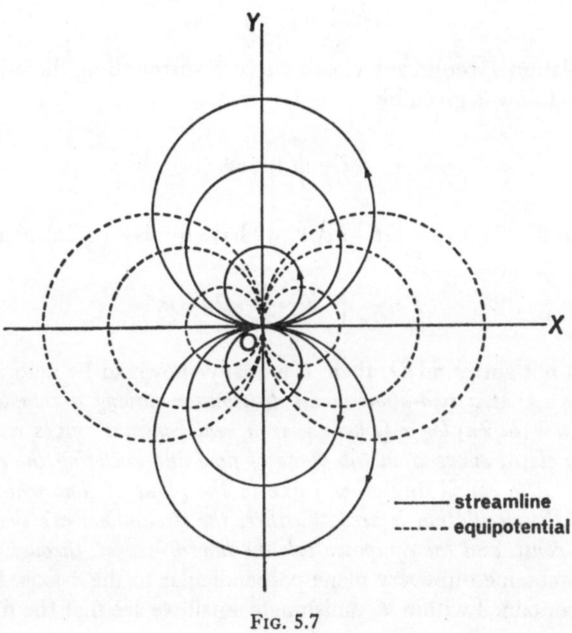

FIG. 5.7

5.5.4 Line Vortices

If for a two-dimensional flow $\mathbf{q}=u\mathbf{i}+v\mathbf{j}$, where $u=u(x,y)$, $v=v(x,y)$, then the vorticity vector $\boldsymbol{\zeta}$ is given by

$$\boldsymbol{\zeta}=\nabla\wedge\mathbf{q}=\left(\frac{\partial v}{\partial x}-\frac{\partial u}{\partial y}\right)\mathbf{k},$$

showing that *in two-dimensional flow the vorticity vector is perpendicular to the plane of flow.*

We now discuss the two-dimensional flow for which

$$w=(i\kappa/2\pi)\log z,$$

where κ is a real constant. Putting $z=re^{i\theta}$, we have

$$\begin{cases}\varphi=-\kappa\theta/2\pi,\\ \psi=(\kappa/2\pi)\log r.\end{cases}$$

Thus the streamlines in the plane of flow are the concentric circles $r=$const. and the equipotentials are the radii vectors $\theta=$const. through the origin. The two families are mutually orthogonal and φ,ψ are both harmonic functions. The radial and transverse velocity components are

$$\begin{cases}q_r=-\dfrac{\partial\varphi}{\partial r}=0,\\[2mm] q_\theta=-\dfrac{1}{r}\dfrac{\partial\varphi}{\partial\theta}=\dfrac{\kappa}{2\pi r}.\end{cases}$$

The circulation Γ round any closed curve $\mathscr{C}$ surrounding the origin and in the plane of flow is given by

$$\Gamma=\oint_{\mathscr{C}}\mathbf{q}\cdot d\mathbf{s}.$$

Taking $\mathbf{q}=\hat{\boldsymbol{\theta}}(\kappa/2\pi r)$, $d\mathbf{s}=dr\hat{\mathbf{r}}+r\,d\theta\hat{\boldsymbol{\theta}}$, we have $\mathbf{q}\cdot d\mathbf{s}=(\kappa/2\pi)\,d\theta$ and so

$$\Gamma=\frac{\kappa}{2\pi}\oint_{\mathscr{C}}d\theta=\frac{\kappa}{2\pi}\times 2\pi=\kappa.$$

If $\mathscr{C}$ does not surround O, then Γ is easily shown to be zero. Hence we have shown that *a two-dimensional distribution having a complex velocity potential $w=(i\kappa/2\pi)\log z$ (where κ is a real constant) gives a circulation round any closed curve $\mathscr{C}$ in the plane of flow and enclosing the origin O of amount κ. Also round any other curve in the plane of flow which does not enclose O the circulation is zero. Further, the streamlines are the concentric circles $r=$const. and the equipotentials the lines $\theta=$const. through O.* Such a pattern is obtained in every plane perpendicular to the z-axis. By making the area contained within $\mathscr{C}$ vanishingly small we see that the distribution producing such a flow must be uniform along the z-axis. Such a distribution along the z-axis is called a *uniform line vortex of strength* κ.

We arrive, then, at the following definition of a uniform line vortex. It is *a uniform distribution along an infinite line such that the circulation round any curve* $\mathscr{C}$ *in any plane perpendicular to that line is a constant* κ *when* $\mathscr{C}$ *encloses the intersection of the line and plane and is zero when* $\mathscr{C}$ *does not contain the intersection.* The *strength* of such a uniform line vortex is defined to be κ and its complex velocity potential is $(i\kappa/2\pi)\log z$ when the origin is taken at the intersection of the plane with the line.* When the line intersects the plane at $z=z_0$, the complex potential is easily seen to be

$$w=(i\kappa/2\pi)\log(z-z_0).$$

This results by moving the origin to $z=z_0$.

In passing we note that for the distribution $w=(i\kappa/2\pi)\log z$, the circulation Γ round any closed curve $\mathscr{C}$, not necessarily plane and not necessarily in any plane of flow is still κ when $\mathscr{C}$ circuits the z-axis and zero when it does not. For, in this case,

$$\mathbf{q}=\hat{\boldsymbol{\theta}}(\kappa/2\pi r),\quad d\mathbf{s}=dr\hat{\mathbf{r}}+r\,d\theta\hat{\boldsymbol{\theta}}+dz\mathbf{k},$$

so that $\mathbf{q}\cdot d\mathbf{s}=(\kappa/2\pi)\,d\theta$, as before.

For the distribution $w=(i\kappa/2\pi)\log z$, we saw that the streamlines in the plane of flow are the concentric circles $r=$const., centre O. It follows that we may introduce a solid boundary on the circle $r=a$. Thus the complex velocity potential

$$w=(i\kappa/2\pi)\log z,\quad |z|\geqslant a$$

will give irrotational flow outside the cylinder $|z|=a$ of infinite length, the fluid being at rest at infinity and having a circulation of amount κ about the cylinder.

5.6 Some Worked Examples

Example 1

Find the equations of the streamlines due to uniform line sources of strength m through the points $A(-c,0)$, $B(c,0)$ and a uniform line sink of strength $2m$ through the origin.

Let P be the point such that $\overline{OP}\equiv z=x+iy$. Then

$$\overline{AP}\equiv z+c,\quad \overline{BP}\equiv z-c$$

and so the complex potential at P is

* Some writers define the strength of the vortex to be such that the circulation round $\mathscr{C}$ is $2\pi\kappa$ when κ is the strength. This would remove the factor 2π from the form of w.

$$w = -m\log(z+c) - m\log(z-c) + 2m\log z$$

$$= m\log\left(\frac{z^2}{z^2-c^2}\right)$$

$$= m\log\left(\frac{x^2-y^2+2ixy}{x^2-y^2-c^2+2ixy}\right)$$

$$= m\log\left\{\frac{(x^2-y^2+2ixy)(x^2-y^2-c^2-2ixy)}{(x^2-y^2-c^2)^2+4x^2y^2}\right\}$$

$$= m\log\left\{\frac{(x^2+y^2)^2-c^2(x^2-y^2)-2ixyc^2}{(x^2-y^2-c^2)^2+4x^2y^2}\right\}.$$

$$\therefore\quad \psi = \mathrm{Im}(w) = m\tan^{-1}\left\{\frac{-2xyc^2}{(x^2+y^2)^2-c^2(x^2-y^2)}\right\}.$$

The streamlines are given by $\psi=$const., i.e. by

$$-\frac{2xyc^2}{(x^2+y^2)^2-c^2(x^2-y^2)} = \text{const.}$$

or

$$\underline{(x^2+y^2)^2 = c^2(x^2-y^2+kxy).}\quad (k=\text{const.})$$

EXAMPLE 2

Describe the irrotational motion of an incompressible liquid for which the complex potential is $w=i\kappa\log z$.

Two parallel line vortices of strengths $\kappa_1, \kappa_2 (\kappa_1+\kappa_2\neq 0)$ in unlimited liquid cross the z-plane at points A, B respectively. The centre of mass of masses κ_1 at A and κ_2 at B is G. Show that if the motion of the liquid is due solely to these vortices, G is a fixed point about which A, B move in circles with angular velocity $(\kappa_1+\kappa_2)/(AB)^2$.

Show also that the fluid speed at any point P in the z-plane is $(\kappa_1+\kappa_2)CP/\{AP\cdot BP\}$, where C is the centre of mass of masses κ_2 at A, κ_1 at B.

(London Univ., B.Sc. Spec., 1957)

$$w = i\kappa\log z = i\kappa\log(re^{i\theta}) = -\kappa\theta + i\kappa\log r.$$

Hence

$$\varphi = -\kappa\theta,\quad \psi = \kappa\log r.$$

This shows that the streamlines are concentric circles centred on O, the equipotentials being the radii through O cutting the circles orthogonally.

Also, at (r, θ) in the plane of flow the velocity components are

$$q_r = 0, \quad q_\theta = -\frac{1}{r}\frac{\partial \varphi}{\partial \theta} = \frac{\kappa}{r}.$$

The flow is due to a line vortex through O of strength κ.

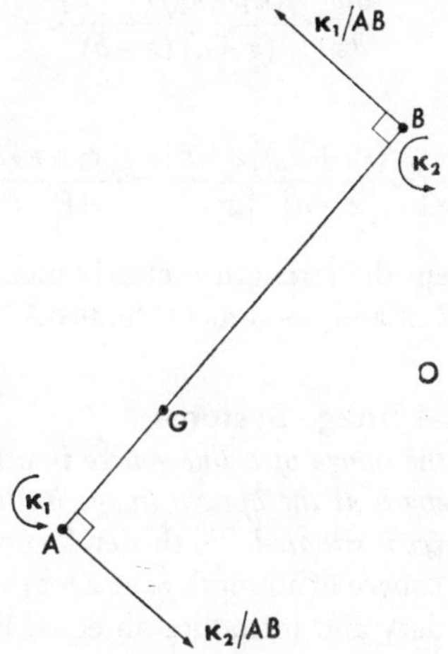

FIG. 5.8

Figure 5.8 shows the line vortices at A, B. κ_1 at A gives B a velocity κ_1/AB in the sense shown. κ_2 at B gives A a velocity κ_2/AB in the sense shown. Associating masses κ_1 at A, κ_2 at B, we see that the total linear momentum of the system in the direction of the velocity of B is

$$\kappa_1\left(-\frac{\kappa_2}{AB}\right) + \kappa_2\left(\frac{\kappa_1}{AB}\right) = 0.$$

Thus the centroid G of the two masses does not move. The diagram shows that the velocity of B relative to A is $(\kappa_1 + \kappa_2)/AB$ so that the angular velocity of AB is $(\kappa_1 + \kappa_2)/(AB)^2$. Thus the line rotates with this angular velocity about G.

Let O be the fixed origin of coordinates and suppose $\overline{OA} \equiv a$, $\overline{OB} \equiv b$, a and b being complex numbers. If P be the point for which $\overline{OP} \equiv z$, then $\overline{AP} \equiv z - a$, $\overline{BP} \equiv -b$ and the complex potential at P is

$$w = i\kappa_1 \log(z-a) + i\kappa_2 \log(z-b).$$

Thus

$$\frac{dw}{dz} = \frac{i\kappa_1}{z-a} + \frac{i\kappa_2}{z-b} = \frac{i\{\kappa_1(z-b) + \kappa_2(z-a)\}}{(z-a)(z-b)}.$$

If C denotes the centroid of κ_2 at A, κ_1 at B and if $\overline{OC} \equiv \bar{z}$, then

$$(\kappa_1+\kappa_2)\overline{CP}=\kappa_2\overline{AP}+\kappa_1\overline{BP}$$

or

$$(\kappa_1+\kappa_2)(z-\bar{z})=\kappa_2(z-a)+\kappa_1(z-b).$$

Hence

$$\frac{dw}{dz}=\frac{i(\kappa_1+\kappa_2)(z-\bar{z})}{(z-a)(z-b)}.$$

Thus the speed of P is

$$\left|\frac{dw}{dz}\right|=\frac{(\kappa_1+\kappa_2)|z-\bar{z}|}{|z-a|\ |z-b|}=\frac{(\kappa_1+\kappa_2)PC}{AP\cdot BP}.$$

(*N.B.* In this problem the strength κ clearly means a circulation round the vortex of amount $2\pi\kappa$ and not κ as in the text.)

5.7 Two-Dimensional Image Systems

We first show that *the image of a line source in a rigid infinite plane is a line source of equal strength at the optical image in the plane of the point at which the given line source is situated.* To this end, suppose the plane is $x=0$ and that there is a line source of strength m at $z=z_1$ where $\mathrm{Re}(z_1)\geqslant 0$. If we now remove the boundary and introduce an equal line source at $z=-\bar{z}_1$, the mirror image of z_1 in $x=0$, then the complex velocity potential at P, where $OP\equiv z$, is

$$w=-m\log(z-z_1)-m\log(z+\bar{z}_1).$$

Then

$$dw/dz=-m(z-z_1)^{-1}-m(z+\bar{z}_1)^{-1}.$$

At $z=iy$ on the plane $x=0$,

$$\begin{aligned}dw/dz&=-m\{(iy-z_1)^{-1}+(iy+\bar{z}_1)^{-1}\}\\&=m\{(z_1-iy)^{-1}-(\bar{z}_1+iy)^{-1}\}.\end{aligned}$$

Since $(\bar{z}_1+iy)^{-1}$ is the complex conjugate of $(z_1-iy)^{-1}$,

$$\mathrm{Re}(dw/dz)=0,$$

showing that on the plane $x=0$ the velocity component $u=0$. Thus no fluid crosses this plane and so it can be replaced by a rigid barrier.

To find the image of a two-dimensional doublet in a rigid plane OY, consider Fig. 5.9 in which there are line sources at the points A and B, taken very close together, of strengths $-m$ and $+m$ per unit length. Their respective images in OY are $-m$ at A', $+m$ at B' where A', B' are the reflections of A, B in OY. The line $\overline{AB}$ makes angle α with OX. Thus $\overline{A'B'}$ makes angle $(\pi-\alpha)$ with $\overline{OX}$. In the limit as $m\to\infty$, $AB\to 0$, we have equal line doublets at A, B with their axes inclined at α, $(\pi-\alpha)$ to $\overline{OX}$.

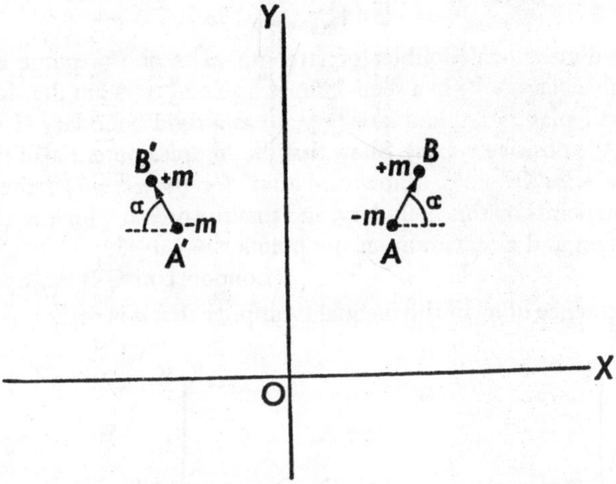

FIG. 5.9

To determine the image of a line vortex in an infinite plane, consider such a vortex of strength κ at $z=z_1$, where $\mathrm{Re}(z_1)>0$. Introduce a line vortex of strength $-\kappa$ at $z=-\bar{z}_1$, the reflection of z_1 in the plane $x=0$. Then the complex potential at z is

$$w=(i\kappa/2\pi)\,\{\log(z-z_1)-\log(z+\bar{z}_1)\},$$

so that

$$dw/dz=(i\kappa/2\pi)\,\{(z-z_1)^{-1}-(z+\bar{z}_1)^{-1}\}.$$

Thus at $z=iy$ on the plane $x=0$,

$$dw/dz=(\kappa/2\pi)\,[-i(\bar{z}_1+iy)^{-1}-i(\bar{z}_1-iy)^{-1}].$$

Since $-i(\bar{z}_1+iy)^{-1}, i(\bar{z}_1-iy)^{-1}$ are complex conjugates,

$$\mathrm{Re}(dw/dz)=0 \text{ on } x=0,$$

showing that no fluid crosses the plane. Hence a rigid plane infinite surface may be introduced into the position $x=0$ and the *appropriate image of a line vortex in such a plane is an equal and opposite line vortex at the mirror image in the plane.*

(*N.B.* The following is an alternative development. We have

$$\psi=\mathrm{Im}(w)=(\kappa/2\pi)\,\{\log|z-z_1|-\log|z+\bar{z}_1|\}.$$

When z lies on the plane $x=0$,

$$\log|z-z_1|=\log|z+\bar{z}_1|, \quad \text{so that } \psi=0 \text{ accordingly.)}$$

EXAMPLE

A two-dimensional doublet of strength $\mu\mathbf{i}$ is at the point $z=ia$ in a stream of velocity $-V\mathbf{i}$ in a semi-infinite liquid of constant density occupying the half-plane $y>0$ and having $y=0$ as a rigid boundary ($\mathbf{i}$ is the unit vector in the positive x-axis). Show that the complex potential of the motion is $w=Vz+2\mu z/(z^2+a^2)$. Show also that, for $0<\mu<4a^2V$, there are no stagnation points on this boundary, and that the pressure on it is a minimum at the origin and a maximum at the points $z=\pm a\sqrt{3}$.

(London Univ., B.Sc. Spec., 1957)

The presence of a^2 in the inequality implies that a is real.

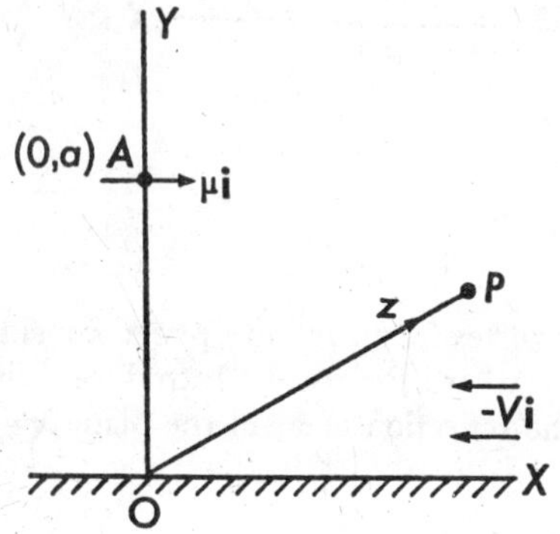

FIG. 5.10(i)

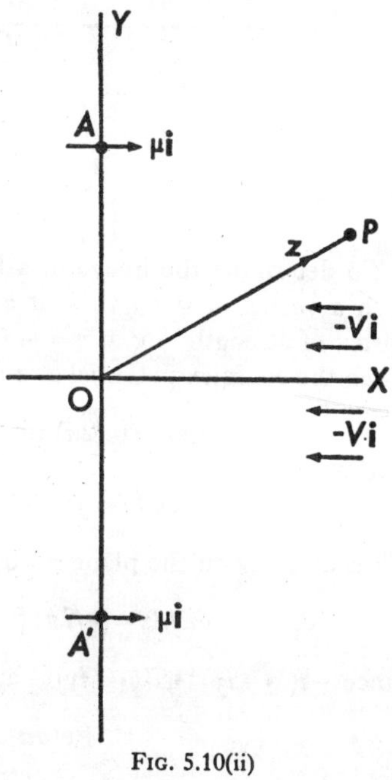

FIG. 5.10(ii)

Figure 5.10(i) displays the physical model, Fig. 5.10(ii) the image system consisting of equal line doublets of strength $\mu\mathbf{i}$ at $A(0, a)$, $A'(0, -a)$. The velocity potential due to the uniform stream is Vx and so its complex velocity potential is Vz. In Fig. 5.10(ii), if $\overline{OP}\equiv z$, then $\overline{AP}\equiv z-ia$, $\overline{A'P}\equiv z+ia$. Thus the complex potentials due to the line doublets at A, A' are respectively

$$\mu(z-ia)^{-1}, \quad \mu(z+ia)^{-1},$$

and so the total velocity potential at P is

$$w=Vz+\mu(z-ia)^{-1}+\mu(z+ia)^{-1}=\underline{Vz+2\mu z(z^2+a^2)^{-1}}.$$

We have

$$dw/dz = V + 2\mu(a^2 - z^2)(a^2 + z^2)^{-2},$$

so that when $z = x$,

$$dw/dz = V + 2\mu(a^2 - x^2)(a^2 + x^2)^{-2}.$$

This gives the speed of the fluid at any point $(x, 0)$ on the boundary $y = 0$. Equating this expression to zero gives the equation

$$Vx^4 + 2(Va^2 - \mu)x^2 + (Va^4 + 2\mu a^2) = 0,$$

a quadratic in x^2 whose discriminant is

$$\Delta = 4\{(Va^2 - \mu)^2 - V(Va^4 + 2\mu a^2)\} = 4\mu(\mu - 4a^2V).$$

Clearly $\Delta < 0$ when $0 < \mu < 4a^2V$, showing that the quadratic equation has no real roots. This means there are no stagnation points on the boundary $y = 0$.

Applying Bernoulli's equation along the streamline $y = 0$ at $(x, 0)$ on the boundary, we have, in the usual notation,

$$\frac{p}{\rho} + \frac{1}{2}\left\{V + \frac{2\mu(a^2 - x^2)}{(a^2 + x^2)^2}\right\}^2 = \text{const.}$$

Thus p is a maximum when $V + 2\mu(a^2 - x^2)(a^2 + x^2)^{-2}$ is a minimum and conversely. Put

$$y = \{V + 2\mu(a^2 - x^2)(a^2 + x^2)^{-2}\}^2$$

or

$$y^{1/2} = V + 2\mu(a^2 - x^2)(a^2 + x^2)^{-2}.$$

Then

$$\tfrac{1}{2}y^{-1/2}y' = -4\mu x(3a^2 - x^2)(a^2 + x^2)^{-3}.$$

Thus $y' = 0$ when $x = 0$ or $x = \pm a\sqrt{3}$. The expression for $\frac{1}{2}y^{-1/2}y'$ shows that $y' < 0$ when $x = 0+$ and $y' > 0$ when $x = 0-$. This shows that y is a maximum and hence p a minimum at the origin. The same expression shows that when $x = a\sqrt{3}+$, $y' > 0$ and when $x = a\sqrt{3}-$, $y' < 0$ so that at $x = a\sqrt{3}$, y is a minimum and so p is a maximum. At $x = -a\sqrt{3}+$, $y' > 0$ and when $x = -a\sqrt{3}-$, $y' < 0$ showing p is also a maximum at $x = -a\sqrt{3}$.

5.8 The Milne-Thomson Circle Theorem*

If a right circular cylinder of infinite length is introduced into a given two-dimensional flow, its axis being aligned with the coordinate axis perpendicular to the plane of flow, it is possible to derive a general expression for the new velocity potential outside the cylinder. To this end we establish a powerful theorem due to Professor L. M. Milne-Thomson. We first, however, explain some terminology involving conjugates of complex functions.

* MILNE-THOMSON, *Proc. Camb. Phil. Soc.*, **36** (1940).

Suppose t is a real variable and that

$$f(t)=U(t)+iV(t),$$

where $U(t), V(t)$ are real functions when t is real. Then $f(t)$ is here a complex function of the real variable t. Its conjugate $\bar{f}(t)$ is defined to be

$$\bar{f}(t)=U(t)-iV(t).$$

If we now replace the real variable t by the complex variable $z(=x+iy)$, then $f(z), \bar{f}(z)$ are defined to be

$$f(z)=U(z)+iV(z),$$
$$\bar{f}(z)=U(z)-iV(z).$$

Further, $f(\bar{z}), \bar{f}(\bar{z})$ are given by

$$f(\bar{z})=U(\bar{z})+iV(\bar{z}),$$
$$\bar{f}(\bar{z})=U(\bar{z})-iV(\bar{z}).$$

Comparing the forms above for $f(z)$, $\bar{f}(\bar{z})$, we see that, since $z=x+iy$, $\bar{z}=x-iy$, the value of $\bar{f}(\bar{z})$ is obtained from $f(z)$ by replacing i throughout by $-i$. In other words $\bar{f}(\bar{z})$ is the complex conjugate of $f(z)$ and we may write $\bar{f}(\bar{z})=\overline{f(z)}$.

We next introduce the idea of singularities in two-dimensional hydrodynamical flows. It will be recalled that the complex potentials due to a line source, a line doublet and a line vortex each have the respective forms

$$-m\log(z-z_1),\quad \frac{\mu e^{i\alpha}}{z-z_1},\quad \frac{i\kappa}{2\pi}\log(z-z_1),$$

in the usual notation. Each of these functions has a singularity at $z=z_1$: elsewhere each is analytic. In the above case the singularities of $f(z)$ are those of $U(z)$ and $V(z)$ and hence they are also those of $\bar{f}(z)$.

The statement of Milne-Thomson's circle theorem is:

Let $f(z)$ be the complex velocity potential for a flow having no rigid boundaries and such that there are no singularities of flow within the circle $|z|=a$. Then, on introducing the solid circular cylinder $|z|=a$ into the flow, the new complex velocity potential is given by $w=f(z)+\bar{f}(a^2/z)$ for $|z|\geqslant a$.

Proof: All the singularities of $f(z)$ (at which line sources, sinks, doublets or vortices may be present) occur in the region $|z|>a$, and so the singularities of $f(a^2/z)$ lie in $|z|<a$. Hence the singularities of $\bar{f}(a^2/z)$ also lie in $|z|<a$. Thus $f(z)$ and $f(z)+\bar{f}(a^2/z)$ both have the same singularities in the region $|z|>a$ and so both functions, considered as complex velocity potentials, may be ascribed to the same hydrodynamical distributions in the region $|z|>a$.

On the circle $|z|=a$, we take $z=ae^{i\theta}$. Then $a^2/z=ae^{-i\theta}$ and so

$$w=f(z)+f(a^2/z)=f(ae^{i\theta})+\bar{f}(ae^{-i\theta})=f(ae^{i\theta})+\overline{f(ae^{i\theta})}.$$

Thus on $|z|=a, w$ is the sum of a complex quantity and its complex conjugate and is therefore a real number. Hence $\psi=\text{Im}(w)=0$ on $|z|=a$. This shows that the circular boundary is a streamline across which no fluid flows. Hence $|z|=a$ is a possible boundary for the new flow and $w=f(z)+\bar{f}(a^2/z)$ is the appropriate complex velocity potential for the new flow.

5.8.1 Some Applications of the Circle Theorem

The following examples show how the circle theorem is applied to determine modified flows when a long circular cylinder is introduced into a given two-dimensional flow.

Example 1

Uniform flow past a stationary cylinder.

We have seen that a uniform stream having velocity $-U\mathbf{i}$ gives rise to a complex potential Uz. Thus we take

$$f(z)=Uz.$$

Then

$$\bar{f}(z)=Uz,$$

and so

$$\bar{f}\left(\frac{a^2}{z}\right)=\frac{Ua^2}{z}.$$

Thus on introducing the cylinder of circular section $|z|=a$ into the stream, the complex potential for the region $|z|\geqslant a$ becomes

$$w=f(z)+\bar{f}(a^2/z)=U(z+a^2z^{-1}).$$

Taking $z=re^{i\theta}$ and equating real and imaginary parts,

$$\begin{cases}\varphi=U\cos\theta(r+a^2r^{-1}),\\ \psi=U\sin\theta(r-a^2r^{-1}).\end{cases}$$

The reader should compare these results with those obtained in Section 5.1, Example 1.

Example 2

Uniform stream at incidence α to $\overline{OX}$.

We have seen that the complex potential for such a stream of velocity U is $Uze^{-i\alpha}$ (Section 5.5.1). Taking

$$f(z)=Uze^{-i\alpha},$$
$$\bar{f}(z)=Uze^{i\alpha},$$

and

$$\bar{f}(a^2/z)=Ue^{i\alpha}(a^2/z).$$

Hence when the cylinder of section $|z|=a$ is introduced, the complex potential in $|z|\geqslant a$ becomes

$$\underline{w=U\{ze^{-i\alpha}+(a^2/z)e^{i\alpha}\}.}$$

Example 3

Image of a line source in a circular cylinder.

Suppose there is a uniform line source of strength m per unit length through the point $z=d$, where $d>a$. Then the complex potential at a point z is

$$f(z)=-m\log(z-d).$$

Then

$$\bar{f}(z)=-m\log(z-d),$$
$$\bar{f}(a^2/z)=-m\log\{(a^2/z)-d\}.$$

On introducing the circular cylinder of section $|z|=a$, the complex velocity potential in the region $|z|\geqslant a$ becomes

$$w=-m\log(z-d)-m\log\{(a^2/z)-d\}.$$

This may be written in the alternative form

$$w=-m\log(z-d)-m\log\{z-(a^2/d)\}+m\log z+\text{const.},$$

the constant (real or complex) being immaterial from the viewpoint of analysing the flow. The term $-m\log\{z-(a^2/d)\}$ is recognized as the complex velocity potential due to a uniform line source of strength m per unit length through the point $z=a^2/d$. This point $z=a^2/d$ is the inverse of the point $z=d$ in the circle $|z|=a$. The term $m\log z$ is the complex potential due to a line sink of strength m per unit length through the centre. We have thus confirmed that *the image of a line source in a right circular cylinder is an equal line source through the inverse point in the circular section in the plane of flow together with an equal line sink through the centre of the section.*

Example 4

Line doublet parallel to the axis of a right circular cylinder.

Suppose there is a uniform line doublet of strength μ per unit length through the point $z=d>a$. Suppose further that the axis of the line doublet is set at angle α to $\overline{OX}$. Then from Section 5.5.3, the complex velocity potential due to this distribution is

$$f(z)=\frac{\mu e^{-i\alpha}}{z-d}.$$

Thus

$$\bar{f}(z)=\frac{\mu e^{-i\alpha}}{z-d}$$

and

$$\bar{f}\left(\frac{a^2}{z}\right)=\frac{\tilde{\mu}e^{i\alpha}}{(a^2/z)-d}.$$

Thus the total complex velocity potential obtained when the circular cylinder of section $|z| = a$ is introduced is

$$w = \frac{\mu e^{i\alpha}}{z - d} + \frac{\mu e^{-i\alpha}}{(a^2/z) - d}.$$

Example 5

A vortex of circulation $2\pi\kappa$ is at rest at the point $z = na$ ($n > 1$), in the presence of a plane circular boundary $|z| = a$, around which there is a circulation $2\pi\lambda\kappa$. Show that $\lambda = 1/(n^2 - 1)$.

Show that there are two stagnation points on the circular boundary $z = ae^{i\theta}$, symmetrically placed about the real axis in the quadrants nearest to the vortex, given by $\cos\theta = (3n^2 - 1)/(2n^3)$, and prove that θ is real.

(London Univ., B.Sc. Spec., 1956)

The complex velocity potential due to the vortex alone is

$$f(z) = i\kappa \log(z - na).$$

Thus

$$\bar{f}(z) = -i\kappa \log(z - na)$$

and so

$$\bar{f}(a^2/z) = -i\kappa \log\{(a^2/z) - na\}.$$

When the cylinder of section $|z| = a$ is introduced into the field of flow due to the vortex, the complex potential is $f(z) + \bar{f}(a^2/z)$. In addition there is a circulation $2\pi\lambda\kappa$ round the cylinder, which can be ascribed to a line vortex through $z = 0$ of strength $2\pi\lambda\kappa$, giving a contribution to the velocity potential in the region $|z| \geqslant a$ of $i\lambda\kappa \log z$. Thus the total velocity potential at z, for $|z| \geqslant a$, is

$$\begin{aligned} w(z) &= i\kappa \log(z - na) - i\kappa \log\{(a^2/z) - na\} + i\lambda\kappa \log z \\ &= i\kappa \log(z - na) - i\kappa \log\{z - (a/n)\} + i(\lambda + 1)\kappa \log z + \text{const.} \end{aligned}$$

Thus the motion is equivalent to that produced by vortices of strengths

$$\kappa \text{ at } z = na; \quad -\kappa \text{ at } z = a/n; \quad (\lambda + 1)\kappa \text{ at } z = 0.$$

Figure 5.11 shows this distribution. A is the point where $z = na$; A' the point where $z = a/n$.

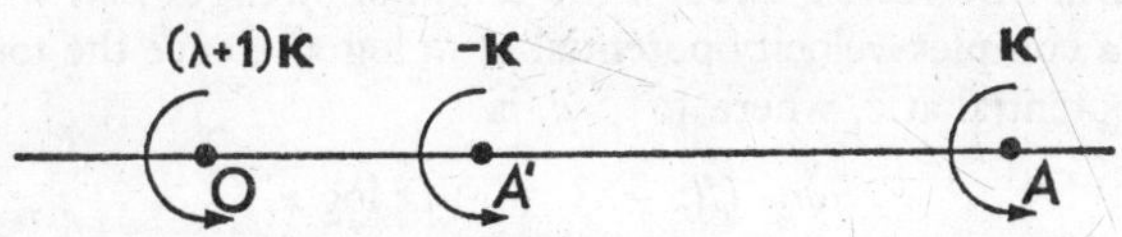

FIG. 5.11

From the diagram we see that the velocity of A (measured in the positive y-direction) is

$$\frac{(\lambda+1)\kappa}{na}+\frac{(-\kappa)}{na-(a/n)}.$$

This is zero if

$$\underline{\lambda=1/(n^2-1).} \tag{1}$$

We find

$$\frac{dw}{dz}=i\kappa\left\{\frac{1}{z-na}-\frac{1}{z-(a/n)}+\frac{\lambda+1}{z}\right\}.$$

Putting $z=ae^{i\theta}$ gives, on simplification,

$$\frac{dw}{dz}=i\kappa e^{-i\theta}\frac{-2n^3\cos\theta+3n^2-1}{a(n^2-1)(n^2-2n\cos\theta+1)}.$$

Hence on $z=ae^{i\theta}$, $|dw/dz|=0$ if

$$\underline{\cos\theta=\frac{3n^2-1}{2n^3}.} \tag{2}$$

Taking

$$f(n)=\frac{3n^2-1}{2n^3}=\frac{3}{2n}-\frac{1}{2n^3},$$

$$f'(n)=-\frac{3}{2n^2}+\frac{3}{2n^4}=\frac{3(1-n^2)}{2n^4}<0 \text{ for } n>1.$$

Thus $f(n)$ is a monotonically decreasing function. Now $f(1)=1$, so that $f(n)<1$ for $n>1$. But $f(n)\to 0$ as $n\to\infty$. Thus for $n>1$, $0\leqslant f(n)<1$. This shows that the values of θ are real, since $|\cos\theta|\leqslant 1$. One root of (2) will be $\theta=\alpha$ where α is a number satisfying $0<\alpha\leqslant\frac{1}{2}\pi$. The other root will be $\theta=-\alpha$.

Example 6

The Magnus effect.

Let a long infinite cylinder of radius a be placed in a uniform stream having velocity $-U\mathbf{i}$, as in Example 1 of this section. Then $U(z+a^2/z)$ is the complex velocity potential. Now suppose that in addition a circulation round the cylinder of $2\pi\kappa$ is produced. This circulation is ascribable to a uniform line vortex through the origin of strength $2\pi\kappa$ which corresponds to a complex velocity potential of $i\kappa\log z$. Hence the total complex velocity potential at z, where $|z|>a$, is

$$w=U(z+a^2z^{-1})+i\kappa\log z.$$

The stagnation points are found from $dw/dz=0$ or

$$U\left\{1-\left(\frac{a}{z}\right)^2\right\}+\frac{i\kappa}{z}=0.$$

They are given by the points

$$z=a\left[-i\left(\frac{\kappa}{2Ua}\right)\pm\left\{1-\left(\frac{\kappa}{2Ua}\right)^2\right\}^{1/2}\right].$$

Let us suppose that the magnitudes of κ, U are so adjusted that $|\kappa/2Ua|<1$. Then we can find a real angle β such that $\sin\beta=(\kappa/2Ua)$ and the two stagnation points are then given by

$$z=a(i\sin\beta\pm\cos\beta),$$

i.e.

$$z_1=ae^{-i\beta},\quad z_2=-ae^{i\beta}.$$

Thus, provided that the above inequality holds, there are two distinct stagnation points at $z=z_1$, $z=z_2$ on the surface of the cylinder. Figure 5.12 shows the pattern of streamlines formed in such a case, A_1, A_2 being the stagnation points.

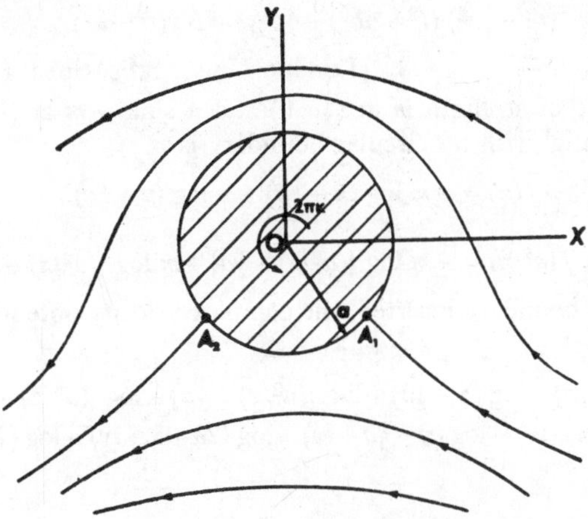

FIG. 5.12

At A_1, A_2, the pressure is a maximum and so the effect produced on the cylinder is a tendency to lifting in the direction $\overline{OY}$. In Example 1 of Section 5.9 we calculate this force. In the case when $\kappa=0$, the streamline flow is symmetrical about the plane $y=0$ and there is no such tendency. Circulation may be produced in practice by rotating the cylinder about its axis. The viscosity of the real fluid would then produce such circulation. This lifting effect produced by the circulation is called the *Magnus effect.* It explains the swerve of a 'sliced' ball in golf.

5.8.2 EXTENSION OF THE CIRCLE THEOREM

The author has developed the following extension of the Milne-Thomson circle theorem. It has potentialities in certain problems in which two-dimensional distributions are prescribed within a right circular cylindrical boundary.

Let $f(z)$ be the complex velocity potential for a flow having no rigid boundaries and such that there are no singularities of flow outside the circle $|z|=a$. Then on introducing the rigid cylindrical surface of section $|z|=a$ into the flow and leaving the distribution of singularities within its interior unchanged, the new complex velocity potential for the fluid motion within the boundary becomes $w=f(z)+\bar{f}(a^2/z)$ for $|z|\leqslant a$.

The proof of this theorem is similar to that for the main circle theorem from which it is derived and will not be given here. Another extension, relevant in electrostatics is found in Example 17 at the end of this chapter.

EXAMPLE

A source and a sink of equal strength are placed at the points $(\pm\frac{1}{2}a, 0)$ within a fixed circular boundary $x^2+y^2=a^2$. Show that the streamlines are given by

$$(r^2-\tfrac{1}{4}a^2)(r^2-4a^2)-4a^2y^2=Ky(r^2-a^2).$$

(London Univ., B.Sc. Spec. Hons., 1921)

For a source of strength m at $(\frac{1}{2}a, 0)$ and a sink $-m$ at $(-\frac{1}{2}a, 0)$, the complex potential with no circular boundary is

$$f(z)=-m\log(z-\tfrac{1}{2}a)+m\log(z+\tfrac{1}{2}a).$$

Hence

$$\bar{f}(a^2/z)=-m\log\{(a^2/z)-\tfrac{1}{2}a\}+m\log\{(a^2/z)+\tfrac{1}{2}a\}.$$

Thus with the boundary inserted, the complex velocity potential at points in its interior is

$$w=m[\log(z+\tfrac{1}{2}a)-\log(z-\tfrac{1}{2}a)+\log\{(a^2/z)+\tfrac{1}{2}a\}-\log\{(a^2/z)-\tfrac{1}{2}a\}]$$
$$=m[\log(x+\tfrac{1}{2}a+iy)-\log(x-\tfrac{1}{2}a+iy)+\log(2a+x+iy)-\log(2a-x-iy)].$$

Thus

$$\psi=m\left\{\tan^{-1}\left(\frac{y}{x+\frac{1}{2}a}\right)-\tan^{-1}\left(\frac{y}{x-\frac{1}{2}a}\right)+\tan^{-1}\left(\frac{y}{2a+x}\right)+\tan^{-1}\left(\frac{y}{2a-x}\right)\right\}$$

$$=m\left\{-\tan^{-1}\left(\frac{\dfrac{y}{x-\frac{1}{2}a}-\dfrac{y}{x+\frac{1}{2}a}}{1+\dfrac{y^2}{x^2-\frac{1}{4}a^2}}\right)+\tan^{-1}\left(\frac{\dfrac{y}{2a-x}+\dfrac{y}{2a+x}}{1-\dfrac{y^2}{4a^2-x^2}}\right)\right\}$$

$$=m\left\{\tan^{-1}\left(\frac{4ay}{4a^2-r^2}\right)-\tan^{-1}\left(\frac{ay}{r^2-\frac{1}{4}a^2}\right)\right\}$$

$$=m\tan^{-1}\left\{\frac{5ay(r^2-a^2)}{(4a^2-r^2)(r^2-\frac{1}{4}a^2)+4a^2y^2}\right\}.$$

Equating ψ to constant value and simplifying gives the required result.

5.9 The Theorem of Blasius

When a long cylinder is placed with its generators perpendicular to the incident stream of a moving fluid containing hydrodynamical singularities, it experiences forces tending to produce translation and rotation of the cylinder. These effects are calculated using the following theorem due to Blasius. The cylinder can be of any general section. In practice it can be an aerofoil. We confine attention to the case when the fluid is incompressible.

An incompressible fluid moves steadily and irrotationally under no external forces parallel to the z-plane past a fixed cylinder whose section in that plane is bounded by a closed curve $\mathscr{C}$. The complex potential for the flow is w. Then the action of the fluid pressure on the cylinder is equivalent to a force per unit length having components [X, Y] and a couple per unit length of moment M, where

$$Y+iX=-\frac{\rho}{2}\oint_{\mathscr{C}}\left(\frac{dw}{dz}\right)^2 dz; \quad M=\text{Re}\left\{-\frac{\rho}{2}\oint_{\mathscr{C}} z\left(\frac{dw}{dz}\right)^2 dz\right\}.$$

Proof:

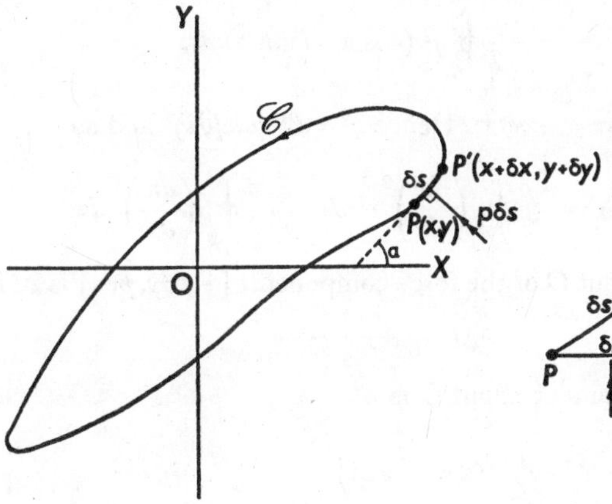

FIG. 5.13(i)

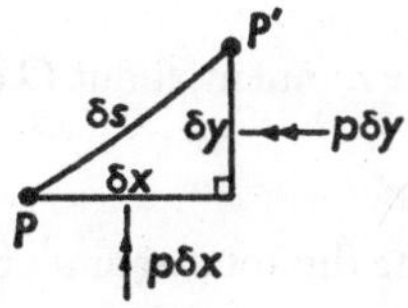

FIG. 5.13(ii)

Figure 5.13(i) shows the section $\mathscr{C}$ of the cylinder in the plane XOY. PP' is an arc element of $\mathscr{C}$ of length δs. Then if p denotes the pressure at P, the force on unit length of the section δs is $p\delta s$ normal to $\mathscr{C}$. If α is the inclination to $\overline{OX}$ of the tangent at P to $\mathscr{C}$, then the x- and y-components of this force are

$$-p\delta s \sin\alpha=-p\delta y; \quad p\delta s\cos\alpha=p\delta x.$$

Hence

$$X=-\oint_{\mathscr{C}} p\,dy; \quad Y=\oint_{\mathscr{C}} p\,dx$$

so that

$$Y+iX=\oint_{\mathscr{C}} p\,(dx-i\,dy).$$

Now Bernoulli's equation in this context is

$$\frac{p}{\rho}+\tfrac{1}{2}q^2=C,$$

where q is the fluid velocity, p the pressure, ρ the density. Hence

$$p=\rho C-\tfrac{1}{2}\rho q^2,$$

and so

$$Y+iX=\rho C\oint_{\mathscr{C}}(dx-i\,dy)-\frac{\rho}{2}\oint_{\mathscr{C}} q^2(dx-i\,dy)$$

$$=-\frac{\rho}{2}\oint_{\mathscr{C}} q^2\left(\frac{dx}{ds}-i\frac{dy}{ds}\right)ds$$

$$=-\frac{\rho}{2}\oint_{\mathscr{C}} q^2\,(\cos\alpha-i\sin\alpha)\,ds.$$

But $dw/dz=-u+iv=-qe^{-i\alpha}$. Hence $q^2=e^{2i\alpha}(dw/dz)^2$ and so

$$Y+iX=-\frac{\rho}{2}\oint_{\mathscr{C}}\left(\frac{dw}{dz}\right)^2 e^{i\alpha}ds=-\frac{\rho}{2}\oint_{\mathscr{C}}\left(\frac{dw}{dz}\right)^2 dz.$$

The moment about O of the force components $[-p\delta y, p\delta x]$ is δM where

$$\delta M=px\delta x+py\delta y,$$

so that the total moment about O is

$$M=\oint_{\mathscr{C}} p(x\,dx+y\,dy)$$

$$=\oint_{\mathscr{C}}(\rho C-\tfrac{1}{2}\rho q^2)\,(x\,dx+y\,dy)$$

$$=-\frac{\rho}{2}\oint_{\mathscr{C}} q^2(x\,dx+y\,dy).$$

Now

$$\begin{aligned}z(dw/dz)^2\,dz&=(x+iy)q^2e^{-2i\alpha}(dx+i\,dy)\\&=(x+iy)q^2e^{-2i\alpha}e^{i\alpha}\,ds\\&=q^2(x+iy)\,(\cos\alpha-i\sin\alpha)\,ds\\&=q^2(x+iy)\,(dx-i\,dy)\\&=q^2(x\,dx+y\,dy)+iq^2(y\,dx-x\,dy).\end{aligned}$$

Hence

$$M = \text{Re}\left\{-\frac{\rho}{2}\oint_{\mathscr{C}} z\left(\frac{dw}{dz}\right)^2 dz\right\}.$$

Evaluation of the integrals for $Y+iX$ and for M is effected using the residue calculus.

EXAMPLE 1

Infinite circular cylinder in uniform stream, with circulation.

Consider again Example 6 of Section 5.8.1. We found the complex potential in the form

$$w = U(z + a^2 z^{-1}) + i\kappa \log z.$$

Thus

$$\frac{dw}{dz} = U\left(1 - \frac{a^2}{z^2}\right) + \frac{i\kappa}{z},$$

and so

$$Y + iX = -\frac{\rho}{2}\oint_{\mathscr{C}}\left(\frac{dw}{dz}\right)^2 dz$$

$$= -\frac{\rho}{2}\oint_{\mathscr{C}}\left[U + i\frac{\kappa}{z} - U\frac{a^2}{z^2}\right]^2 dz$$

$$= -\tfrac{1}{2}\rho \times 2\pi i \times \text{residue of integrand at } z=0$$

$$= -\tfrac{1}{2}\rho \times 2\pi i \times 2iU\kappa = \underline{2\pi\rho\kappa U}.$$

The moment about O is

$$M = -\frac{\rho}{2}\,\text{Re}\oint_{\mathscr{C}} z\left(\frac{dw}{dz}\right)^2 dz$$

$$= -\frac{\rho}{2}\,\text{Re}\oint_{\mathscr{C}} z\left[U + \frac{i\kappa}{z} - U\frac{a^2}{z^2}\right]^2 dz.$$

Now coeff. of $(1/z)$ in integrand $=$ coeff. of $(1/z^2)$ in $[U + (i\kappa/z) - (Ua^2/z^2)]^2$

$$= -\kappa^2 - 2U^2a^2.$$

$$\therefore\ M = -\tfrac{1}{2}\rho \times \text{Re}\{2\pi i(-\kappa^2 - 2U^2a^2)\} = 0.$$

Hence

$$X = 0, \quad Y = 2\pi\rho\kappa U, \quad M = 0.$$

This shows that the cylinder experiences an uplifting force.

EXAMPLE 2

Verify that $w = ik \log\{(z - ia)/(z + ia)\}$ is the complex potential of a steady flow of liquid about a circular cylinder, the plane $y = 0$ being a rigid boundary. Find the force exerted by the liquid on unit length of the cylinder.

(London Univ., B.Sc. Spec., 1955)

Putting $w = \varphi + i\psi$, we see that

$$\psi = k \log \left| \frac{z - ia}{z + ia} \right|.$$

The streamlines $\psi = \text{const.}$ are thus given by

$$\left| \frac{z - ia}{z + ia} \right| = \text{const.}$$

These are non-intersecting coaxial circles having $z = \pm ia$ as the limiting points. In particular the straight line $|z - ia| = |z + ia|$, or $y = 0$, is one and may be taken as a rigid boundary. Any one circle $\mathscr{C}$ of the system is

$$|z - ia| = \lambda |z + ia| \quad (\lambda \neq 1).$$

This establishes the result of the first part of the question. Figure 5.14 shows the circular section $\mathscr{C}$ of the cylinder and the rigid plane.

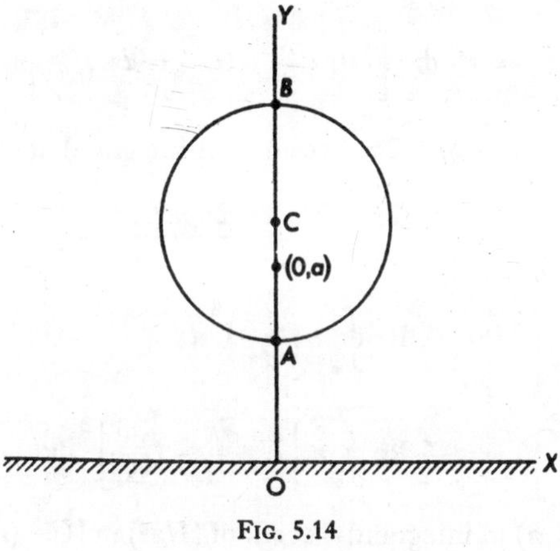

FIG. 5.14

Since

$$w = ik\{\log (z - ia) - \log (z + ia)\},$$

$$\frac{dw}{dz} = ik\left(\frac{1}{z - ia} - \frac{1}{z + ia}\right) = -\frac{2ka}{z^2 + a^2}.$$

In the notation of the text,

$$Y + iX = -\tfrac{1}{2}\rho \oint_{\mathscr{C}} \frac{4k^2a^2}{(z^2 + a^2)^2}\, dz = -2k^2a^2\rho \oint_{\mathscr{C}} \frac{dz}{(z^2 + a^2)^2}.$$

The only poles of the integrand are at $z=\pm ia$: of these only $z=ia$ lies within $\mathscr{C}$. We thus find the residue at $z=ia$.

$$\frac{1}{z^2+a^2}=\frac{1}{(z-ia)(z+ia)}=\frac{1}{2ia}\left(\frac{1}{z-ia}-\frac{1}{z+ia}\right).$$

Hence

$$\frac{1}{(z^2+a^2)^2}=-\frac{1}{4a^2}\left\{\frac{1}{(z-ia)^2}+\frac{1}{(z+ia)^2}-\frac{2}{(z-ia)(z+ia)}\right\}$$

$$=-\frac{1}{4a^2}\left\{\frac{1}{(z-ia)^2}+\frac{1}{(z+ia)^2}-\frac{1}{2ia}\left(\frac{1}{z-ia}-\frac{1}{z+ia}\right)\right\}.$$

Thus the residue of $(z^2+a^2)^{-2}$ at $z=ia$ is $1/(8ia^3)$. Hence

$$Y+iX=-2k^2a^2\rho\times 2\pi i\times 1/(8ia^3)=-(\pi k^2\rho/2a).$$

Hence $\underline{X=0, \quad Y=-(\pi\rho k^2/2a)}$.

This shows that the liquid exerts a downward force on the cylinder of amount $(\pi\rho k^2/2a)$ per unit length.

5.10 The Use of Conformal Transformation

Suppose z and t are two complex variables defined by

$$z=x+iy, \quad t=\xi+i\eta,$$

where x, y, ξ, η are real variables. Then we can form two Argand diagrams for the loci of z and t. In the first x is plotted horizontally and y vertically. In the second, ξ is plotted horizontally and η vertically.

Now suppose that z describes a certain curve $\mathscr{C}$ in the (x, y)-plane and suppose that t is related to z by means of the transformation

$$t=g(z).$$

If $g(z)$ is a single-valued function of z, then to each point in the z-plane we can obtain a corresponding point in the t-plane. In this way the curve $\mathscr{C}$ in the z-plane may be mapped into a curve $\mathscr{C}'$ in the t-plane.

Let us now make the supposition that the function $g(z)$ is analytic, and let P, P_1, P_2 be neighbouring points in the z-plane such that

$$\overline{OP}\equiv z, \quad \overline{OP_1}\equiv z+\delta z_1, \quad \overline{OP_2}\equiv z+\delta z_2.$$

Under the given transformation $t=f(z)$, suppose that P, P_1, P_2 map into the points Q, Q_1, Q_2 in the t-plane, where

$$\overline{OQ}\equiv t, \quad \overline{OQ_1}\equiv t+\delta t_1, \quad \overline{OQ_2}\equiv t+\delta t_2.$$

It is assumed that $|\delta z_1|$, $|\delta z_2|$, $|\delta t_1|$, $|\delta t_2|$ are *small*. The points are shown in Fig. 5.15(i), (ii).

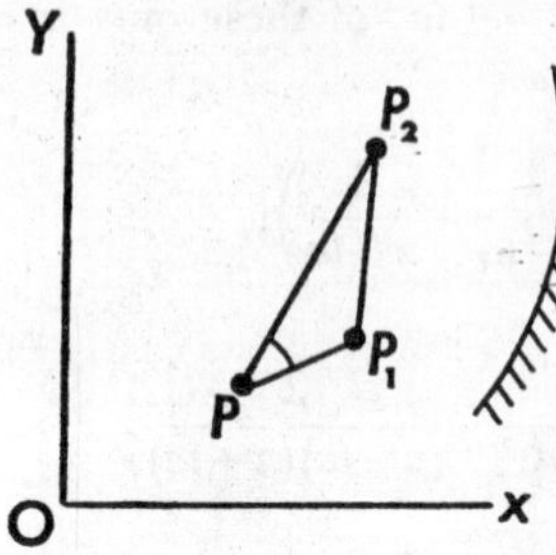

FIG. 5.15(i)

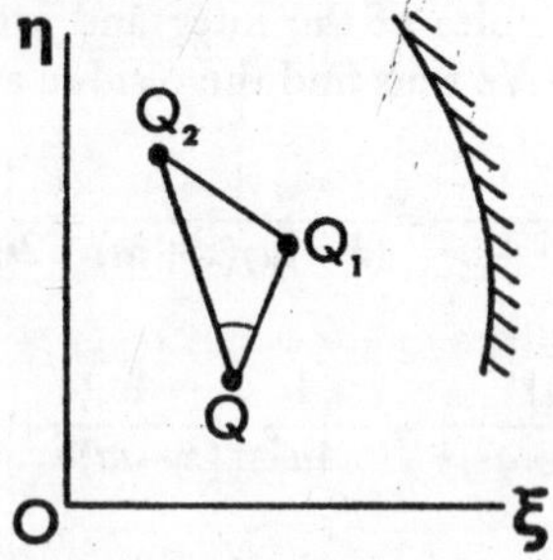

FIG 5.15(ii)

Since $g(z)$ is analytic, dt/dz is unique at P. Thus, to the first order of smallness,

$$\frac{\delta t_1}{\delta z_1}=\frac{\delta t_2}{\delta z_2},$$

or

$$\frac{\delta t_1}{\delta t_2}=\frac{\delta z_1}{\delta z_2}.$$

Hence

$$\left|\frac{\delta t_1}{\delta t_2}\right|=\left|\frac{\delta z_1}{\delta z_2}\right| \tag{1}$$

and

$$\arg \delta t_1-\arg \delta t_2=\arg \delta z_1-\arg \delta z_2. \tag{2}$$

From (1), we see that

$$\frac{QQ_1}{QQ_2}=\frac{PP_1}{PP_2} \tag{1$'$}$$

and from (2), on reversing the signs of each side,

$$\angle Q_2QQ_1=\angle P_2PP_1. \tag{2$'$}$$

The equations (1′), (2′) show that the triangles Q_2QQ_1, P_2PP_1 are similar. Otherwise expressed: *within the neighbourhood of any point P in the z-plane and its corresponding mapping in the t-plane, angles remain unaltered (see equation (2′)) and so do the ratios of corresponding linear dimensions, when the mapping takes place under a transformation of the form $t=f(z)$, where $f(z)$ is analytic at P and within its neighbourhood.* Such a transformation is said to be *conformal.*

5.10.1 Some Hydrodynamical Aspects of Conformal Transformation

Now suppose that we have a two-dimensional incompressible flow, the plane of flow being the z-plane of Fig. 5.15(i). On applying the conformal transformation $t=g(z)$, the new plane of flow becomes the t-plane of

Fig. 5.15(ii). Let ρ be the density of the fluid in both cases. Suppose, further, that $\mathscr{C}$ is a rigid boundary in the z-plane which maps into the curve $\mathscr{C}'$ in the t-plane. Let the complex velocity potential for the z-plane be

$$w=f(z)=\varphi+i\psi,$$

where the real functions $\varphi(x, y)$, $\psi(x, y)$ are the usual velocity potential and stream function respectively. By means of the transformation $t=g(z)$ we can express w as a function $\bar{f}(t)$ of the form

$$w=\bar{f}(t)=\bar{\phi}+i\bar{\psi}$$

where $\bar{\phi}=\bar{\phi}(\xi, \eta)$, $\bar{\psi}=\bar{\psi}(\xi, \eta)$. At corresponding points t, z, w takes the same values so that

$$\varphi=\bar{\phi}, \quad \psi=\bar{\psi}.$$

Now $\mathscr{C}$ is a rigid boundary and so also a streamline for which $\psi=$const. Thus along $\mathscr{C}'$, $\bar{\psi}=$const. showing that $\mathscr{C}'$ is a streamline and hence also a rigid boundary. (Conjugate notation is *not* used here.)

In theory w can be expressed as a function of t by eliminating z from $w=f(z)$ and $t=g(z)$ but it is often more convenient to use the result

$$\frac{dw}{dt}=\frac{dw}{dz}\frac{dz}{dt}=\frac{1}{g'(z)}\frac{dw}{dz}.$$

Hence

$$\left|\frac{dw}{dz}\right|^2=\left|\frac{dw}{dt}\right|^2\left|\frac{dt}{dz}\right|^2.$$

Referring to Fig. 5.14(i), the kinetic energy of unit depth of the liquid element within the triangle PP_1P_2 is

$$\begin{aligned}\delta T &= \tfrac{1}{2}\rho\times\text{Area of triangle } PP_1P_2\times(\text{Speed of } P)^2\\ &=\tfrac{1}{2}\rho\cdot(\triangle PP_1P_2)\cdot\left|\frac{dw}{dz}\right|^2\\ &=\tfrac{1}{2}\rho\cdot(\triangle PP_1P_2)\cdot\left|\frac{dw}{dt}\right|^2\left|\frac{dt}{dz}\right|^2\\ &=\tfrac{1}{2}\rho\cdot(\triangle QQ_1Q_2)\cdot\left|\frac{dw}{dt}\right|^2.\end{aligned}$$

The last expression is the kinetic energy of unit depth of the liquid element within the triangle QQ_1Q_2 in Fig. 5.14(ii). By summation over the entire fluid domains within the z- and t-planes we clearly have:

Total K.E. of liquid in z-plane (per unit depth)
$\qquad$ = Total K.E. of liquid in t-plane (per unit depth).

In general, $|dw/dz| \neq |dw/dt|$, so that velocities at corresponding points in the z- and t-planes are not equal. The total kinetic energies of corresponding areas are the same, but they are differently distributed over the corresponding areas.

We now establish a number of theorems concerning the conformal transformation of line distributions.

THEOREM I: Under conformal transformation a uniform line source maps into another uniform line source of the same strength.

Proof: Let there be a uniform line source of strength m per unit length through the point $z=z_0$ and suppose the conformal transformation $t=g(z)$ is made from the z- to the t-plane so that the point $z=z_0$ maps into the point $t=t_0$. Let $\mathscr{C}$ be a closed curve in the z-plane containing the point $z=z_0$ and suppose $\mathscr{C}$ maps into $\mathscr{C}'$ in the t-plane. Then $t=t_0$ lies within $\mathscr{C}'$. The complex potential w is the same for both systems and has the forms

$$\begin{aligned} w &= \varphi+i\psi \text{ for the } z\text{-plane;} \\ &= \varphi'+i\psi' \text{ for the } t\text{-plane.} \end{aligned}$$

Thus $\varphi=\varphi'$, $\psi=\psi'$. Since ψ is the same at corresponding points of $\mathscr{C}, \mathscr{C}'$,

$$\oint_{\mathscr{C}} d\psi = \oint d\psi'. \tag{1}$$

Now in the z-plane, $w=-m\log(z-z_0)$, $dw=-m\,dz/(z-z_0)$. Hence

$$\oint_{\mathscr{C}} dw = -m\oint_{\mathscr{C}} \frac{dz}{z-z_0} = -m\times 2\pi i,$$

since the integrand has a residue of 1 at $z=z_0$. But $dw=d\varphi+i\,d\psi$, so that equating imaginary parts gives

$$\oint_{\mathscr{C}} d\psi = -2\pi m. \tag{2}$$

The numerical value of this is the volume of fluid crossing unit thickness of $\mathscr{C}$ per unit time. Thus (1), (2) show that the same volume crosses unit thickness of $\mathscr{C}'$ per unit time which implies an equal line source of strength m per unit length through $t=t_0$.

THEOREM II: Under conformal transformation a uniform line vortex maps into another uniform line vortex of the same strength.

Proof: Let there be a uniform line vortex of strength κ per unit length through $z=z_0$ and suppose the conformal transformation $t=g(z)$ is made from the z- to the t-plane so that the point $z=z_0$ maps into $t=t_0$. Let $\mathscr{C}$ be a closed curve containing $z=z_0$ and let $\mathscr{C}'$ be its map in the t-plane. Then

$\mathscr{C}'$ contains $t=t_0$. The complex potential is the same for both systems and has the forms

$$w=\varphi+i\psi \text{ for the } z\text{-plane,}$$
$$=\varphi'+i\psi' \text{ for the } t\text{-plane.}$$

Thus $\varphi=\varphi'$, $\psi=\psi'$. Since φ is the same at corresponding points of $\mathscr{C}, \mathscr{C}'$,

$$\oint_{\mathscr{C}} d\varphi=\oint_{\mathscr{C}'} d\varphi'. \tag{1}$$

Now in the z-plane the complex potential is $w=(i\kappa/2\pi)\log(z-z_0)$, so that

$$\oint_{\mathscr{C}} dw=\frac{i\kappa}{2\pi}\oint\frac{dz}{z-z_0}=\frac{i\kappa}{2\pi}\times 2\pi i=-\kappa.$$

Equating real parts gives

$$-\oint_{\mathscr{C}} d\varphi=+\kappa. \tag{2}$$

The integral on the L.H.S. is the circulation round $\mathscr{C}$. Equations (1), (2) thus show that the circulation round $\mathscr{C}'$ is also κ. Thus, on account of the arbitrariness of $\mathscr{C}, \mathscr{C}'$, the line source through $z=z_0$ of strength κ per unit length maps into an equal line source through $t=t_0$.

THEOREM III: Under conformal transformation a uniform line doublet maps into another uniform line doublet of different strength.

Proof: Let there be a uniform line doublet of strength μ per unit length through P where $z=z_0$ and suppose that under the conformal transformation $t=g(z)$, P maps into Q where $t=t_0$. Let the line doublet be replaced by equivalent line sources of strengths $-m$, $+m$ per unit length through P, P' where $\overline{PP'}\equiv\delta z$, $\mu=m|\delta z|$ and $\overline{PP'}$ is in the direction of the axis of the line doublet. Suppose P' maps into Q'. Then the line sources of strengths $-m$, $+m$ per unit length through P, P' map into ones of strengths $-m$, $+m$ per unit length through Q, Q'. If $\overline{QQ'}\equiv\delta t$, then $\delta t=g'(z)\delta z$, so that

$$|\delta t|=|g'(z)|\cdot|\delta z|, \quad \arg\delta t=\arg g'(z)+\arg\delta z.$$

Hence the two line sources through Q, Q' give a line doublet at Q of strength μ' where

$$\mu'=m|\delta t|=\mu|g'(z)|.$$

The inclination of the axis of the line doublet to the real axis is increased by $\arg g'(z)$.

5.10.2 Some Worked Examples

Example 1

A uniform line doublet of strength μ per unit length is situated at the point $t=i$ in the t-plane, its axis pointing in the direction of the positive real axis. There is a rigid boundary along the entire length of the real axis and fluid occupies the entire region $\text{Im}(t)\geqslant 0$. By applying the transformation $t=e^z$, determine the corresponding motion in the z-plane.

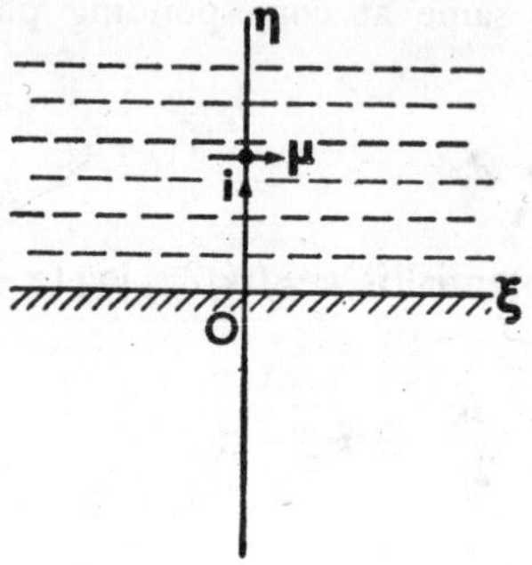

FIG. 5.16(i)

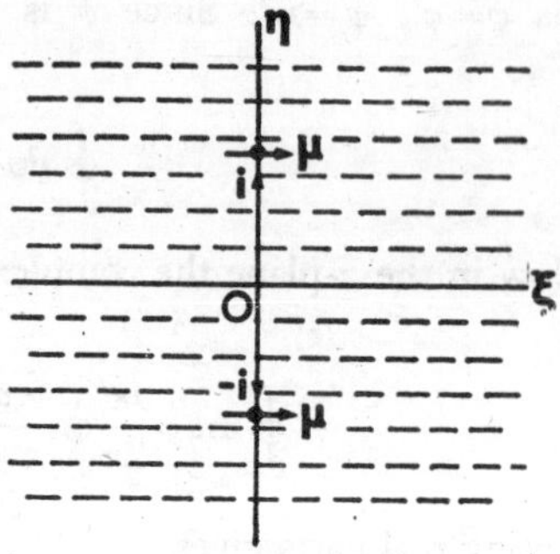

FIG. 5.16(ii)

Figure 5.16(i) shows the physical model and Fig. 5.16(ii) the corresponding image system with equal line doublets of strengths μ per unit length at the points where $t=\pm i$. The complex potential is thus

$$w=\frac{\mu}{t-i}+\frac{\mu}{t+i}=\frac{2\mu t}{t^2+1},\quad \text{Im}(t)\geqslant 0.$$

Putting $t=e^z$, we find

$$w=\frac{2\mu e^z}{e^{2z}+1}=\mu\,\text{sech}\,z.$$

Writing $t=\xi+i\eta$, we have

$$z=\log\,(\xi+i\eta).$$

Thus the line $\eta=0$ becomes

$$z=\log\xi=\log|\xi|+i\arg\xi,$$

so that $y=\arg\xi$,
i.e.

$$\left.\begin{aligned} y&=0 \text{ for } \xi>0,\\ &=\pi \text{ for } \xi<0. \end{aligned}\right\}$$

Further, when $t=i$, $z=\log i=\log|i|+i\arg i=\frac{1}{2}\pi i$. Now $dz/dt=1/t=-i$, when $t=i$. Thus, for $t=i$, $|dz/dt|=1$, $\arg\,(dz/dt)=-\frac{1}{2}\pi$. Thus the uniform line doublet of strength μ per unit length at $t=i$ maps into one of the same strength at $z=\frac{1}{2}\pi i$, its axis making an angle of $-\frac{1}{2}\pi$ with the positive direction of the real axis. Hence we have shown that the complex velocity potential $w=\mu\,\text{sech}\,z$ determines the fluid motion between the rigid planes $y=0$, $y=\pi$ when a uniform line doublet of strength μ per unit length and having its axis pointing in the negative y-direction is introduced at the point $(0, \frac{1}{2}\pi)$ (Fig. 5.17).

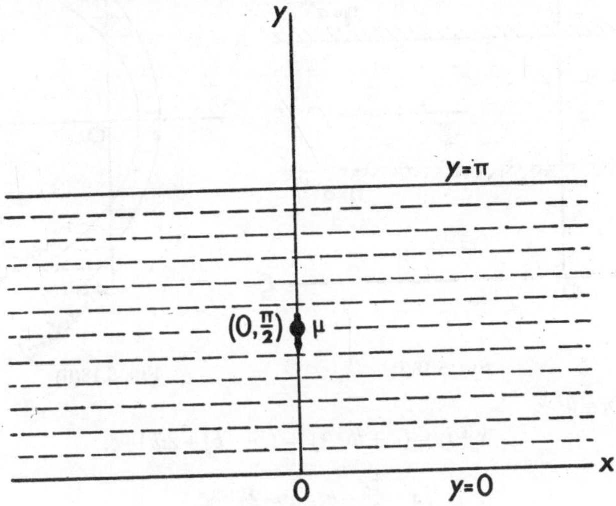

FIG. 5.17

EXAMPLE 2

An inviscid fluid of constant density ρ is in irrotational two-dimensional motion. Show that the fluid velocity $[u, v]$ is given in terms of a function $w(z)$ of a complex variable $z=x+iy$ by the relation $-u+iv=w'(z)$.

The flow in a complex ζ-plane ($\zeta=\xi+i\eta$) is a steady stream of speed V parallel to the positive ξ-axis in a channel given by $b^{1/2} \leqslant \eta \leqslant a^{1/2}$. Find the complex potential of the motion.

Apply the transformation $\zeta=z^{1/2}$ to this flow, describing the boundaries and streamlines of the transformed channel in the z-plane, and show that in the absence of body forces the circles $|z|=$const. are curves of equal pressure.

If the pressure at infinity is just sufficient to make the pressure in the channel everywhere positive, show that the pressure at points in the channel on the circle $|z|=a$ is $\frac{1}{8}\rho V^2\{(1/b)-(1/a)\}$.

(London Univ., B.Sc. Spec. Hons., 1956)

Figure 5.18(i) shows the flow in the ζ-plane. The velocity potential is $-V\xi=-V\text{Re}(\zeta)$, so that the complex velocity potential is

$$w=-V\zeta=-V(\xi+i\eta).$$

Applying the transformation $\zeta=z^{1/2}$, we find

$$w=-Vz^{1/2}=-V(x+iy)^{1/2}.$$

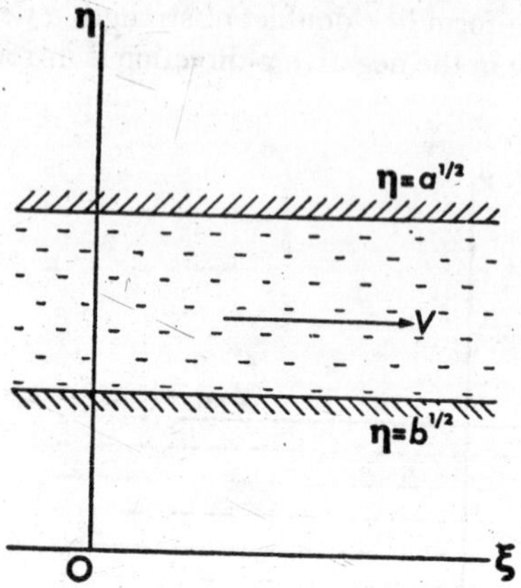

FIG. 5.18(i)

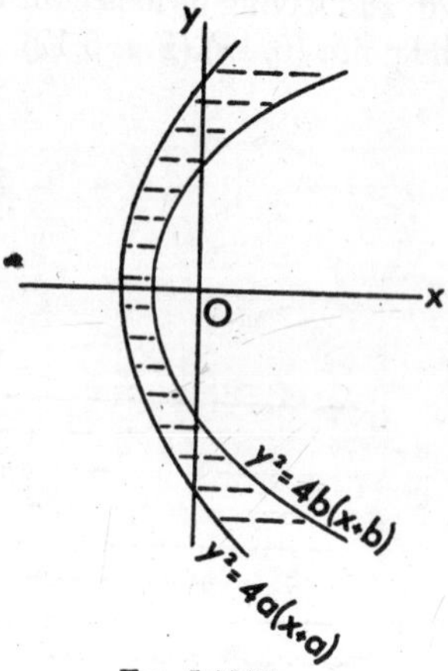

FIG. 5.18(ii)

When $\eta = b^{1/2}$,

$$x + iy = (\xi + ib^{1/2})^2 = (\xi^2 - b) + 2ib^{1/2}\xi,$$

so that

$$x = \xi^2 - b, \quad y = 2b^{1/2}\xi.$$

Eliminating the parameter ξ gives

$$y^2 = 4b(x + b)$$

which is the equation of a parabola, vertex $(-b, 0)$ and focus $(0, 0)$. Since $-\infty \leqslant \xi \leqslant \infty$, the parametric equations of the curve show that the entire line $\eta = b^{1/2}$ maps into the entire parabola $y^2 = 4b(x + b)$.

Similarly, the whole line $\eta = a^{1/2}$ maps into the entire parabola

$$y^2 = 4a(x + a).$$

The region of flow between the lines maps into the region between the two parabolas (Fig. 5.18(ii).)

Taking $z = re^{i\theta}$, $w = \varphi + i\psi$,

$$\varphi + i\psi = -Vz^{1/2} = -Vr^{1/2}e^{i\theta/2}.$$

Thus

$$\varphi = -Vr^{1/2}\cos(\theta/2), \quad \psi = -Vr^{1/2}\sin(\theta/2).$$

The streamlines are the curves $\psi = c$ (=const.) or

$$r = \frac{c^2}{\sin^2(\theta/2)} = \frac{2c^2}{1 - \cos\theta}.$$

$$\therefore\ 2c^2 = \sqrt{(x^2 + y^2)} - x,$$

or

$$y^2 = 4c^2(c^2 + x^2).$$

This last equation shows that the streamlines are confocal parabolas having (0, 0) as focus. Taking $c=b^{1/2}$ and $c=a^{1/2}$ recovers the parabolic boundaries.

From $\varphi=-Vr^{1/2}\cos(\theta/2)$, the speed q is given by

$$q^2=\left(\frac{\partial\varphi}{\partial r}\right)^2+\left(\frac{1}{r}\frac{\partial\varphi}{\partial\theta}\right)^2$$

$$=\{\tfrac{1}{2}Vr^{-1/2}\cos(\theta/2)\}^2+\{\tfrac{1}{2}Vr^{-1/2}\sin(\theta/2)\}^2,$$

and so

$$q=\tfrac{1}{2}Vr^{-1/2}.$$

Thus, on $r=a$, $q=\tfrac{1}{2}Va^{-1/2}$.

The appropriate form of Bernoulli's equation along a streamline is

$$\frac{p}{\rho}+\tfrac{1}{2}q^2=K,$$

or $p=\rho\{K-(V^2/8a)\}$ on the circle $|z|=a$. This confirms that the circles $|z|=\text{const.}$ are iso-pressure curves.

Along any streamline,

$$\frac{p}{\rho}+\tfrac{1}{2}q^2=K.$$

If q denotes the fluid speed at (r,θ) on a streamline, then $q=\tfrac{1}{2}Vr^{-1/2}$ and so $(p/\rho)+(V^2/8r)=K=p_\infty/\rho$.

Thus

$$p=p_\infty-\frac{1}{8}\left(\frac{\rho V^2}{r}\right).$$

Now

$$\min r=b,$$

so that

$$\min p=p_\infty-\frac{1}{8}\left(\frac{\rho V^2}{b}\right).$$

This is zero if

$$p_\infty=\frac{1}{8}\left(\frac{\rho V^2}{b}\right).$$

Hence

$$p=\frac{1}{8}\rho V^2\left(\frac{1}{b}-\frac{1}{r}\right).$$

On the circle $|z|=a$,

$$p=\frac{1}{8}\rho V^2\left(\frac{1}{b}-\frac{1}{a}\right).$$

5.11 The Schwarz–Christoffel Transformation

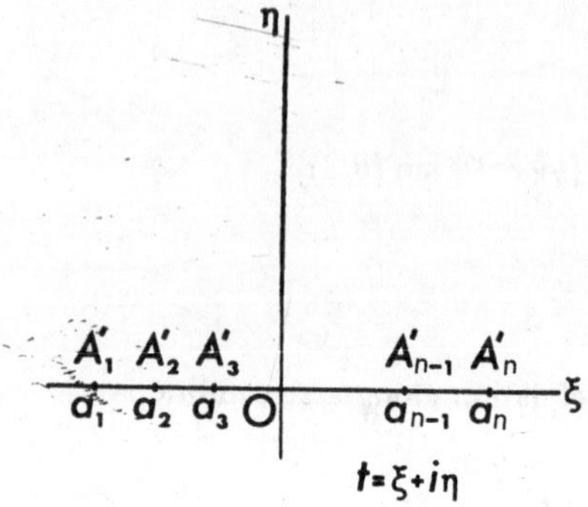

FIG. 5.19(i)

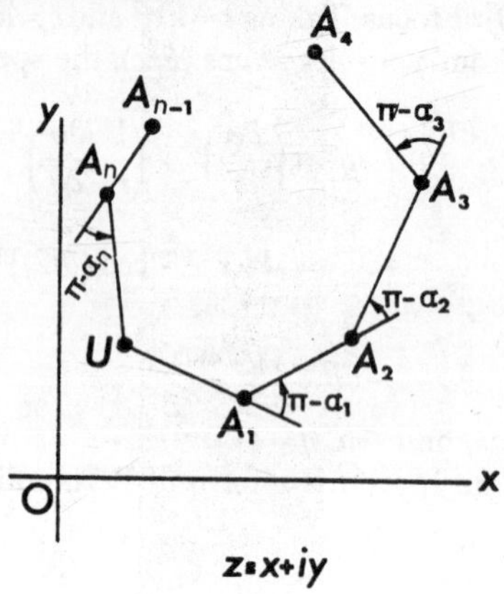

FIG. 5.19(ii)

Figure 5.19(i) shows n points $A_1', A_2', \ldots, A_n'$ on the real axis of the complex t-plane, where $t=\xi+i\eta$, ξ and η being real. The points A_r' are specified by $t=a_r\,(r=1, 2, \ldots, n)$ and the ordering is such that $a_1<a_2<\ldots<a_{n-1}<a_n$. Now suppose that the t-plane is mapped on to the z-plane $(z=x+iy)$ by means of the transformation

$$dz/dt=K(t-a_1)^{(\alpha_1/\pi)-1}(t-a_2)^{(\alpha_2/\pi)-1}\ldots(t-a_n)^{(\alpha_n/\pi)-1},$$

which is conformal everywhere save at $t=a_r\,(r=1, 2, \ldots, n)$. The numbers $\alpha_1, \alpha_2, \ldots, \alpha_n$ are real constants, but K is a constant which may be complex.

Let δt be a change in t and δz the corresponding change in z so that

$$\delta z=\delta t K(t-a_1)^{(\alpha_1/\pi)-1}\ldots(t-a_n)^{(\alpha_n/\pi)-1}.$$

Then $\arg\delta z=\arg\delta t+\arg K+[(\alpha_1/\pi)-1]\arg(t-a_1)$

$$+[\alpha_2/\pi)-1]\arg(t-a_2)+\ldots+[(\alpha_n/\pi)-1]\arg(t-a_n). \quad (1)$$

We first suppose t travels from $-\infty$ to a_1- along $\eta=0$. Then

$$\arg\delta t=0; \quad \arg(t-a_1)=\arg(t-a_2)=\ldots=\arg(t-a_n)=\pi.$$

Hence

$$(\arg\delta z)_1=\arg K+(\alpha_1-\pi)+(\alpha_2-\pi)+\ldots+(\alpha_n-\pi)=\text{const.} \quad (2)$$

Equation (2) shows that z describes a straight-line segment UA_1 as t travels from $-\infty$ to a_1- along $\eta=0$.

Next, let t travel from a_1+ to a_2- along $\eta=0$. Then $\arg\delta t=0$; $\arg(t-a_1)=0$; $\arg(t-a_2)=\arg(t-a_3)=\ldots=\arg(t-a_n)=\pi$. Then

$$(\arg\delta z)_2=\arg K+0+(\alpha_2-\pi)+(\alpha_3-\pi)+\ldots+(\alpha_n-\pi)=\text{const.} \quad (3)$$

Equation (3) shows that z describes a straight-line segment A_1A_2 as t travels from a_1+ to a_2- along $\eta=0$.

From (2), (3),

$$(\arg \delta z)_2 - (\arg \delta z)_1 = \pi - \alpha_1,$$

showing that the angle between the directions $\overline{UA_1}, \overline{A_1A_2}$ is $\pi - \alpha_1$ (Fig. 5.19(ii)). This means that as t travels along the axis $\eta = 0$ from $-\infty$ past A_1, the direction of motion of z changes by $\pi - \alpha_1$ in the positive sense, as the course changes from $\overline{UA_1}$ to $\overline{A_1A_2}$.

Similarly as t moves from a_2+ to a_3- along $\eta = 0$, $\arg \delta t = 0$; $\arg (t - a_1) = \arg (t - a_2) = 0$; $\arg (t - a_3) = \ldots = \arg (t - a_n) = \pi$. Thus

$$(\arg \delta z)_3 = \arg K + 0 + 0 + (\alpha_3 - \pi) + \ldots + (\alpha_n - \pi) = \text{const.} \tag{4}$$

Equation (4) shows that z describes a straight-line segment A_2A_3 as t travels from a_2+ to a_3- along $\eta = 0$. From (3), (4),

$$(\arg \delta z)_3 - (\arg \delta z)_2 = \pi - \alpha_2.$$

This shows that the direction of motion of z changes by $\pi - \alpha_2$ in the positive sense as z passes through A_2.

Continuing in this way, we see that *z describes the sides $UA_1, A_1A_2, A_2A_3, \ldots, A_{n-1}A_n, A_nU$ of a closed polygon in the z-plane as t moves along the real axis from $-\infty$ to a_1, a_1 to a_2, a_2 to $a_3, \ldots, a_{n-1}$ to a_n, a_n to $+\infty$, respectively*. The angles turned through in the positive sense at $A_1, A_2, \ldots, A_n$ are $\pi - \alpha_1, \pi - \alpha_2, \ldots, \pi - \alpha_n$, so that the internal angles of the polygon at these points are $\alpha_1, \alpha_2, \ldots, \alpha_n$.

Now suppose $\alpha_1, \alpha_2, \ldots, \alpha_n$ are so chosen that

$$\alpha_1 + \alpha_2 + \ldots + \alpha_n = (n-2)\pi.$$

Then $A_1A_2 \ldots A_n$ is a closed polygon, the portion A_nUA_1 being a straight line. Further, since the half-plane $\text{Im}(t) > 0$ is on the left of an observer travelling along the positive direction of the axis $\eta = 0$, the corresponding area in the z-plane is the interior of the polygon.

Thus we have established the following result: *If $a_1 < a_2 < a_3 < \ldots < a_n$ are n distinct points on the line* $\text{Im}(t) = 0$ *in the t-plane, and if $\alpha_1, \alpha_2, \ldots, \alpha_n$ are n real constants satisfying*

$$\alpha_1 + \alpha_2 + \ldots + \alpha_n = (n-2)\pi,$$

where n is a positive integer, then the transformation

$$dz/dt = K(t - a_1)^{(\alpha_1/\pi)-1}(t - a_2)^{(\alpha_2/\pi)-1} \ldots (t - a_n)^{(\alpha_n/\pi)-1}$$

or

$$z = L + K\int (t - a_1)^{(\alpha_1/\pi)-1}(t - a_2)^{(\alpha_2/\pi)-1} \ldots (t - a_n)^{(\alpha_n/\pi)-1}\, dt$$

where K, L are constants which may be complex, maps the entire line $\text{Im}(t) = 0$ *on to the boundary $A_1A_2 \ldots A_nA_1$ of a closed polygon of n sides in such a way that the points $t = a_k$ transform into A_k at which the internal angles are $\alpha_k (k = 1, 2, \ldots, n)$ and the half-plane* $\text{Im}(t) > 0$ *transforms into the interior of the polygon.*

This transformation is called the *Schwarz–Christoffel transformation.*

It is important to examine what happens when one of the points on the real axis in the t-plane is at infinity. Suppose that, in Fig. 5.19(i), $a_n = +\infty$. In the transformation

$$dz/dt = K(t-a_1)^{(\alpha_1/\pi)-1}(t-a_2)^{(\alpha_2/\pi)-1}\ldots(t-a_n)^{(\alpha_n/\pi)-1},$$

put $K = A(-a_n)^{1-(\alpha_1/\pi)}$. Then

$$K(t-a_n)^{(\alpha_n/\pi)-1} = A\{(1-(t/a_n)\}^{(\alpha_n/\pi)-1} \to A \text{ as } a_n \to \infty.$$

The Schwarz–Christoffel transformation now becomes

$$dz/dt = A(t-a_1)^{(\alpha_1/\pi)-1}(t-a_2)^{(\alpha_2/\pi)-1}\ldots(t-a_{n-1})^{(\alpha_{n-1}/\pi)-1}.$$

This shows that when $a_n = \infty$, the factor $(t-a_n)^{(\alpha_n/\pi)-1}$ is suppressed. A similar result obtains when $a_n = -\infty$.

Example

An incompressible liquid moves irrotationally in two dimensions within a region that has two boundaries C_1 and C_2 defined as follows:

$$C_1 \ : \ y = a, \ -\infty < x < \infty;$$

$$C_2 \ : \ \begin{cases} y = 0, \ -\infty < x \leqslant 0; \\ y = -x \tan \gamma\pi, \ 0 \leqslant x \leqslant x_1; \\ y = -(b-a), \ x_1 \leqslant x < \infty, \end{cases}$$

where $b - a = x_1 \tan \gamma\pi$, x_1 being >0 and $0 < \gamma < \frac{1}{2}$.

At $y = -\infty$ the liquid is flowing in the positive x-direction with speed U and it always remains in contact with any side walls. Show that, with the usual notation, the flow may be determined from the equations

$$t = \exp\left(-\frac{\pi w}{aU}\right), \quad z = \frac{b}{\pi}\int_1^t \frac{(t-1)^\gamma}{t(t-\alpha)^\gamma}\,dt$$

where $b = a\alpha^\gamma$. Deduce the corresponding results for the case when $x_1 \to \infty$.

(London Univ., B.Sc. Spec., 1957)

Figure 5.20(i) depicts the flow in the z-plane. From considerations of continuity, since U is the entry velocity at the section $B_\infty C_\infty$, that at the exit section $A_\infty F_\infty$ is Ua/b. At the entry section the fluid may be considered to be created by a line source, the rate at which fluid crosses the section being Ua. Similarly at the exit section the fluid may be considered to be drained away by a line sink at the same rate.

The infinite polygon $A_\infty B_\infty C_\infty DEF_\infty A_\infty$ has internal angles $0, 0, (1+\gamma)\pi, (1-\gamma)\pi$ when we regard A_∞, F_∞ as coincident and B_∞, C_∞ as coincident. Let this polygon be mapped on to the full line $A'_\infty B'C'D'E'F'_\infty$ in the t-plane, as in Fig. 5.19(ii), this line being $\text{Im}(t) = 0$. The points on the

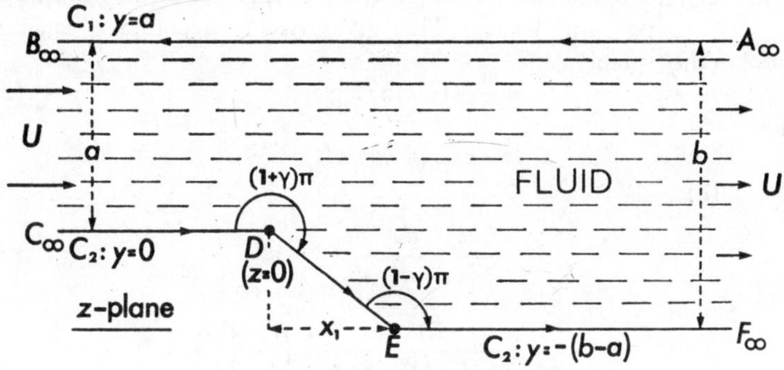

FIG. 5.20(i) z-plane

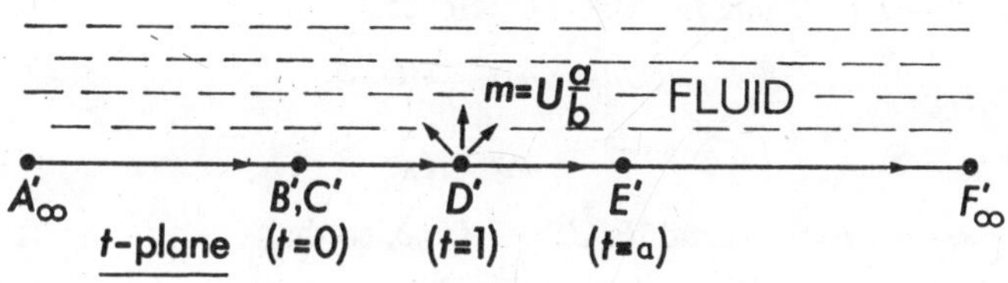

FIG. 5.20(ii) t-plane

line are chosen in the following manner. The required form for z suggests the transformation

$$
\begin{aligned}
dz/dt &= K(t-1)^{\gamma} t^{-1}(t-\alpha)^{-\gamma} \\
&= K(t-1)^{(1+\gamma)-1}(t-0)^{0-1} \cdot (t-\alpha)^{(1-\gamma)-1}.
\end{aligned}
$$

This is secured by taking A'_∞ as the point $t=-\infty$, B' and C' at $t=0$, D' at $t=1$, E' at $t=\alpha$, F'_∞ at $t=+\infty$, the primed letters in the t-plane being the mappings of the unprimed ones in the z-plane. Such a transformation is possible for the following reasons:

(i) The mappings of A_∞, F_∞ stay at $\pm\infty$ and so cause no contributions in the form of dz/dt.

(ii) The internal angle of the polygon at B_∞, C_∞ is 0, that at D is $(1+\gamma)\pi$, that at E is $(1-\gamma)\pi$, that at A_∞, F_∞ is 0. The sum comes to 2π appropriately.

We thus have

$$
z = K\int_1^t \frac{(t-1)^{\gamma}}{t(t-\alpha)^{\gamma}}\,dt + L.
$$

When $t=1$, $z=0$ and so $L=0$. Thus

$$
z = K\int_1^t \frac{(t-1)^{\gamma}}{t(t-\alpha)^{\gamma}}\,dt. \tag{1}
$$

The equivalent line source at $z=-\infty$ maps a line source B', C' which must emit a volume of fluid of Ua per unit time into the region $\text{Im}(t) \geqslant 0$.

Since the region is spread over an angle π, the strength of the line source at B', C' is Ua/π per unit length. This gives rise to a complex velocity potential in the t-plane

$$w = -(Ua/\pi)\log t, \tag{2}$$

whence

$$t = \exp(-\pi w/aU).$$

From (1), (2),

$$\frac{dw}{dz} = \frac{dw}{dt}\,\frac{dt}{dz}$$

$$= -\frac{Ua}{\pi t}\cdot\frac{t(t-\alpha)^\gamma}{K(t-1)^\gamma} = -\frac{Ua}{\pi K}\left(\frac{t-\alpha}{t-1}\right)^\gamma.$$

When $z = -\infty$, $t = 0$ and $dw/dz = -U$. Hence

$$-(Ua/\pi K)\alpha^\gamma = -U$$

or

$$a\alpha^\gamma = K\pi. \tag{3}$$

When $z = +\infty$, $t = +\infty$ and $dw/dz = -Ua/b$, so that

$$-\frac{Ua}{b} = -\frac{aU}{K\pi}$$

or

$$K = \frac{b}{\pi}. \tag{4}$$

From (3), (4),

$$b = a\alpha^\gamma \tag{5}$$

and (1) gives

$$z = \frac{b}{\pi}\int_1^t \frac{(t-1)^\gamma}{t(t-\alpha)^\gamma}\,dt, \tag{6}$$

as required.

When $x_1 \to \infty$, $b \to \infty$, $a\alpha^\gamma \to \infty$ and so $\alpha \to \infty$.

Now

$$z = \frac{b}{\pi\alpha^\gamma}\int_1^t \frac{(t-1)^\gamma}{t(t/\alpha - 1)^\gamma}\,dt$$

$$= \frac{a}{\pi}\int_1^t \frac{(t-1)^\gamma}{t(t/\alpha-1)^\gamma}\,dt$$

$$\to \frac{a}{\pi}\int_1^t \frac{(t-1)^\gamma}{t(-1)^\gamma}\,dt \text{ (as } \alpha\to\infty)$$

$$= \frac{a}{\pi}\int_1^t \frac{(1-t)^\gamma}{t}\,dt.$$

5.12 Vortex Rows

When a body moves slowly through a liquid rows of vortices are sometimes formed in its wake. These vortices can, when stable, be photographed. Here we consider infinite systems of parallel line vortices. Two-dimensional flow will be presumed throughout.

5.12.1 Single Infinite Row of Line Vortices

We first consider the case of a single infinite row of vortices each of strength κ at the points

$$z=0, \pm a, \pm 2a, \ldots, \pm na, \ldots$$

The complex velocity potential due to the $(2n+1)$ vortices nearest the origin is

$$\begin{aligned} w_n &= (i\kappa/2\pi)\{\log z+\log(z-a)+\log(z-2a)+\ldots+\log(z-na) \\ &\qquad +\log(z+a)+\log(z+2a)+\ldots+\log(z+na)\} \\ &= (i\kappa/2\pi)\log\{z(z^2-a^2)(z^2-4a^2)\ldots(z^2-n^2)^2\}. \\ &= \frac{i\kappa}{2\pi}\log\left\{\frac{\pi z}{a}\left(1-\frac{z^2}{a^2}\right)\left(1-\frac{z^2}{4a^2}\right)\ldots\left(1-\frac{z^2}{n^2a^2}\right)\right\}+\text{const.} \end{aligned}$$

The last term is but an immaterial constant. Ignoring this and letting $n\to\infty$, we obtain the complex velocity potential in the form

$$\begin{aligned} w &= \frac{i\kappa}{2\pi}\log\left\{\frac{\pi z}{a}\left(1-\frac{z^2}{a^2}\right)\left(1-\frac{z^2}{4a^2}\right)\ldots\left(1-\frac{z^2}{n^2a^2}\right)\ldots\right\} \\ &= (i\kappa/2\pi)\log\sin(\pi z/a), \end{aligned}$$

using the well-known infinite product form

$$\sin x = x\left(1-\frac{x^2}{\pi^2}\right)\left(1-\frac{x^2}{4\pi^2}\right)\ldots\left(1-\frac{x^2}{n^2\pi^2}\right)\ldots$$

The velocity components are found from

$$\begin{aligned} -u+iv &= \frac{dw}{dz} = \left(\frac{i\kappa}{2a}\right)\cot\left(\frac{\pi z}{a}\right) \\ &= \frac{i\kappa}{2a}\cdot\frac{\cos\{\pi(x+iy)/a\}\cdot\sin\{\pi(x-iy)/a\}}{\sin\{\pi(x+iy)/a\}\cdot\sin\{\pi(x-iy)/a\}}, \end{aligned}$$

whence

$$u = -\frac{\kappa}{2a}\cdot\frac{\operatorname{sh}(2\pi y/a)}{\operatorname{ch}(2\pi y/a)-\cos(2\pi x/a)},$$

$$v = \frac{\kappa}{2a}\cdot\frac{\sin(2\pi x/a)}{\operatorname{ch}(2\pi y/a)-\cos(2\pi x/a)}.$$

It is readily seen from these relations that all vortices remain at rest. Also the reader can easily show that as $y \to \infty$,

$$u \to \mp(\kappa/2a), \quad v \to 0,$$

the negative sign holding in the region $y>0$ and the positive in the region $y<0$. Thus above the row of vortices we have a uniform stream of speed $(\kappa/2a)$ in the negative x-direction and below it we have a uniform stream of speed $(\kappa/2a)$ in the positive x-direction. The motion is in some ways like two rigid blocks sliding over a layer of rollers separating them.

The above problem may also be solved using the conformal transformation $\zeta = \sin(\pi z/a)$. Let there be a line vortex of strength κ through $\zeta=0$ in the ζ-plane. Then the transformation maps the entire ζ-plane on to the strip $-\frac{1}{2}a < \text{Re}(z) \leqslant \frac{1}{2}a$ and also on to parallel strips in the z-plane. The line vortex through $\zeta=0$ maps into equal line vortices through $z=0, \pm a, \pm 2a, \ldots$ Also

$$w = \frac{i\kappa}{2\pi} \log \zeta = \frac{i\kappa}{2\pi} \log \sin\left(\frac{\pi z}{a}\right).$$

5.12.2 The Kármán Vortex Street

Now suppose we have two parallel rows, each containing an infinite number of line vortices, arranged as follows.

The first row consists of line vortices of strength κ at the points having Cartesian coordinates $(na, \frac{1}{2}b)$ $(n=0, \pm 1, \pm 2, \ldots)$.

The second row consists of line vortices of strength $-\kappa$ at the points $(\frac{1}{2} \cdot \overline{2n+1}a, -\frac{1}{2}b)$ $(n=0, \pm 1, \pm 2, \ldots)$. Such an arrangement is called a *Kármán vortex street*. The complex velocity potential may be found either from first principles or by using the result obtained for a single row. It is found to be

$$w = \frac{i\kappa}{2\pi} \log \sin \frac{\pi}{a}\left(z - \frac{ib}{2}\right) - \frac{i\kappa}{2\pi} \log \sin \frac{\pi}{a}\left(z - \frac{a}{2} + \frac{ib}{2}\right).$$

Neither row induces velocity in itself. The velocity of the vortex at $z = \frac{1}{2}a - \frac{1}{2}ib$ is given by

$$-u + iv = \frac{i\kappa}{2\pi} \frac{d}{dz}\left\{\log \sin \frac{\pi}{a}\left(z - \frac{ib}{2}\right)\right\}_{(a-ib)/2} = -\frac{\kappa}{2a} \operatorname{th}\left(\frac{\pi b}{a}\right).$$

Similarly we can show that any other vortex in the same row moves with the same velocity. This means that the entire lower row advances in the positive x-direction with velocity $(\kappa/2a)\operatorname{th}(\pi b/a)$. The entire upper row can be shown to advance with the same velocity, so that the vortex configuration remains unaltered at all times.

A Kármán vortex street is often realized when a flat plate moves broadside through a liquid.

EXERCISE 5

1. Infinite cylinders with boundaries $S_1, S_2, \ldots, S_n$ and a surrounding cylindrical surface S_{n+1} have their generators parallel to **k**. The intervening space is filled with ideal incompressible liquid in two-dimensional motion perpendicular to **k**. Show that, if there are no singularities in the liquid and no circulations round the cylinders, the motion is uniquely determined by the velocities of $S_1, S_2, \ldots, S_{n+1}$.

The space between two light concentric circular cylinders C_1, C_2 with radii $a, b (a<b)$ is filled with incompressible liquid of density ρ. The system is initially at rest. Impulses are applied simultaneously to C_1, C_2 perpendicular to their axes so that C_1 moves with velocity u and C_2 with velocity v perpendicular to it. Show that the kinetic energy of the liquid generated per unit length of the cylinders is $\frac{1}{2}\pi\rho(b^2+a^2)(b^2-a^2)^{-1}(a^2u^2+b^2v^2)$ and hence, or otherwise, determine the impulse per unit length of C_1.

(London Univ., B.Sc. Spec. Hons., 1960)

2. An incompressible, inviscid fluid flows past a stationary circular cylinder of radius a. The velocity at infinity is uniform, of magnitude U and perpendicular to the axis of the cylinder and there is no circulation. Obtain the velocity potential and the stream function of the flow. Show that the resultant force on the cylinder is zero, and that if the cylinder is divided into two parts by a plane through its axis inclined at an angle α to the stream, then the reaction per unit length between the two halves has components

$$2ap_0 - \tfrac{1}{3}a\rho U^2(5\cos^2\alpha + \sin^2\alpha)$$

and

$$\tfrac{8}{3}a\rho U^2 \sin\alpha\cos\alpha$$

respectively, perpendicular and parallel to the plane of separation. [p_0 is the pressure at infinity and ρ is the density of the fluid.]

(Bristol Univ., B.Sc. Spec. I, 1962)

3. Establish Bernoulli's equation for the irrotational motion of an inviscid fluid in the form

$$\int \frac{dp}{\rho} = \frac{\partial\varphi}{\partial t} - \Omega - \tfrac{1}{2}q^2 + g(t),$$

explaining the notation and the assumptions made.

An infinite circular cylinder of radius a moves from rest with acceleration f in a straight line perpendicular to its axis through an incompressible inviscid fluid initially at rest. There are no external forces. Show that immediately the motion starts, the fluid pressure at the rear of the cylinder is given by $p = p_0 - \rho fa$, where p_0 is the initial fluid pressure.

(Nottingham Univ., Maths. Hons. II, 1963)

4. An infinite rigid cylinder of radius a is at rest with its axis along the

z-axis. It is immersed in an incompressible inviscid fluid which has a uniform velocity $[-U, 0, 0]$ at infinity. Show that the equation of a streamline whose asymptote is a line at distance b from Ox is of the form

$$(r - a^2 r^{-1}) \sin \theta = b,$$

where (r, θ) are polar coordinates in the (x, y) plane. (It is to be assumed that the flow is irrotational, with no circulation around the cylinder.)

Show also that the time taken by a particle moving in this streamline to travel from the point $r = R$ upstream of the origin to the point where it crosses the plane $x = 0$ is

$$t(R) = \frac{1}{U} \int_{r_0}^{R} \frac{r^2\, dr}{[(r^2 - a^2)^2 - b^2 r^2]^{1/2}},$$

where r_0 is the positive root of $r^2 - br - a^2 = 0$.

When the fluid is at rest at infinity and the cylinder moves with velocity $[U, 0, 0]$, show that a fluid particle, originally at rest at distance b from the plane $y = 0$ is displaced by the passage of the cylinder a total distance

$$\delta = 2 \lim_{R \to \infty} [R - Ut(R)]$$

in the negative x direction. Explain why δ would be the same if U were time-dependent.

(Cambridge Univ., M.T. II, 1964)

5. If $w = \varphi + i\psi$ is an analytic function of a complex variable $z = x + iy$, prove that $\varphi(x, y)$ may be taken as the velocity potential for two-dimensional irrotational motion of an incompressible ideal fluid, and the curves $\psi(x, y) = \text{constant}$ as streamlines.

In such a two-dimensional motion, a source of strength m is placed at each of the points $(-1, 0)$ and $(1, 0)$, and a sink of strength $2m$ is placed at the origin. Show that the streamlines are the curves

$$(x^2 + y^2)^2 = x^2 - y^2 + kxy,$$

where k is a parameter.

(London Univ., Gen. B.Sc. II, 1962)

A two-dimensional motion of a liquid has the complex potential

$$w = U\left(z + \frac{a^2}{z}\right) + ik \log\left(\frac{z}{a}\right),$$

where the constants U, k, a are real and positive. Show that

(i) the velocity at infinity is U in the negative sense of the real axis;
(ii) the circle $|z| = a$ is a streamline;
(iii) there are, in general, two stagnation points;
(iv) there is a circulation $2\pi k$ about the circle.

(London Univ., B.Sc. Gen. II, 1958)

. A line source of strength m per unit length is at a distance c from the infinite plane which bounds a semi-infinite liquid of density ρ. Find the pressure at any point of this plane and hence show that the total force, on unit length of the plane parallel to the line source is $\pi\rho m^2/c$. What is its direction?

(Imperial College, B.Sc. Spec. Hons., 1956)

8. A circular cylinder of radius a moves with uniform velocity U perpendicular to its generators in an unbounded domain of incompressible inviscid fluid. Find the complex potential of the motion on the assumption that it is two-dimensional and irrotational and that there is a circulation κ round the cylinder. Show that if $\kappa > 4\pi a U$ there is just one point at which the fluid has the same velocity as the cylinder. Deduce that fluid in the neighbourhood of the cylinder is carried along with it.

(Durham Univ., Gen. B.Sc., 1962)

9. Inviscid liquid of constant density ρ is in irrotational motion in the space between two finite closed surfaces S_1, S_2. If the velocity potential φ is single-valued, prove that the kinetic energy of the liquid is

$$-\tfrac{1}{2}\rho\iint_{S_1}\varphi\frac{\partial\varphi}{\partial n}dS_1-\tfrac{1}{2}\rho\iint_{S_2}\varphi\frac{\partial\varphi}{\partial n}dS_2,$$

where $\partial/\partial n$ denotes differentiation along the normal drawn into the liquid.

Liquid is in two-dimensional irrotational motion with circulation κ in the space between two fixed infinite circular cylinders $r=a$, $r=b$, where $a<b$. Prove that the kinetic energy per unit length of the cylinders is $\tfrac{1}{4}(\rho\kappa^2/\pi)\log(b/a)$ and explain how to modify the first formula so that it can be applied to this case.

(London Univ., B.Sc. Spec. Hons., 1959)

10. The space between two concentric circular cylinders of radii $a, \sqrt{3}a$ is filled with liquid which flows irrotationally with circulation $4\pi Ua$ about the inner cylinder. The two cylinders are suddenly set in motion in opposite directions with speed U perpendicular to the axis. Introducing suitable coordinates, show that the complex potential of the resulting initial motion of the fluid is of the form

$$w=U(2z+3a^2z^{-1}+2ia\log z).$$

Find the kinetic energy of the liquid per unit length of the cylinders.

(London Univ., B.Sc. Spec. Hons., 1955)

11. The space on one side of an infinite plane wall, $y=0$, is filled with inviscid incompressible fluid, moving at infinity with velocity U in the direction of the x-axis. The motion of the fluid is steady and two-dimensional in the (x, y)-plane. A doublet of strength μ pointing at an angle α to the x-axis is situated at a distance a from the wall.

(i) Show that if $U=(2\mu\cos\alpha)/a^2$ there is a stagnation point at O, the foot of the perpendicular from the doublet on the wall.

(ii) If $\alpha=\pi/2$ and $U=2\mu/a^2$, calculate the pressure distribution at the wall, when the pressure at infinity is p_0.

(Bristol Univ., B.Sc. Spec., 1962)

12. Obtain the complex potential function appropriate to uniform streaming in two dimensions past a circular cylinder of radius a, the speed at great distance being U. If a marked particle of liquid is at distance b directly upstream of the axis of the cylinder O at time $t=0$, prove that at time t it will be at distance r from O where

$$Ut-b+r=\tfrac{1}{2}a\log\left\{\frac{(r+a)(b-a)}{(r-a)(b+a)}\right\}.$$

Use this result to show that if the marked particle moves a distance s while the liquid at infinity moves a distance d, then

$$s=\frac{b^2-a^2}{b+a\coth\{(d-s)/a\}}.$$

(Leeds Univ., B.Sc., 1962)

13. An inviscid incompressible fluid has a uniform velocity $\mathbf{U}$ at infinity and flows past a fixed elliptical cylinder of semi-axes $a, b, (a>b)$ whose major axis is parallel to the direction of motion at infinity. Verify that the flow at any point is given by the complex potential w where

$$w=Uc\exp\xi_0\cdot\cosh(\zeta-\xi_0),\quad z=c\cosh\zeta,$$

in the usual notation.

Deduce that at large distances from the body the effect of the cylinder on the flow is approximately the same as that of a circular cylinder of radius $\sqrt{[\frac{1}{2}b(a+b)]}$.

(Durham Univ., B.Sc. Hons., 1963)

14. Show that, for two-dimensional irrotational motion of an incompressible fluid, the complex velocity q may be expressed in terms of a regular function $w(z)$ by the equation

$$\bar{q}=-w'(z).$$

Describe, with the aid of a diagram, the flow pattern given by

$$w(z)=Uz-m\log(z-2a),$$

where U, m, a are all real constants.

If a circular cylinder of radius a is inserted with its centre at the origin, find the complex potential of the disturbed motion, there being no circulation about the cylinder.

(London Univ., Gen. B.Sc. II, 1959)

15. The complex potential of an unbounded, inviscid, incompressible fluid in irrotational two-dimensional motion in the complex z-plane is $f(z)$, where all the singularities of $f(z)$ are outside the circle $|z|=a$. If a circular cylinder with this circle as boundary is inserted in the fluid, show that the complex potential of the motion is then

$$w=f(z)+\bar{f}(a^2/z).$$

A line vortex of strength κ lies at a distance $b(>a)$ from the axis of a circular cylinder of radius a. If the vortex remains stationary find the circulation around the cylinder.

(Bristol Univ., Gen. B.Sc., 1963)

16 Liquid moves two-dimensionally without circulation in the region exterior to both the circles

$$|z-i|=\sqrt{2}, \quad |z+i|=\sqrt{2}.$$

Find the complex potential given that there is a unit source at $z=1$ and that the liquid is at rest at infinity. Find the velocity at the point $3i$.

(London Univ., B.Sc. Spec. Hons., 1964)

17. *Electrostatic form of the circle theorem*

Let $f(z)=\varphi(x,y)+i\psi(x,y)$ be the complex potential for a given two-dimensional distribution which is analytic in the region $|z|\geqslant a$. Suppose that the boundary $|z|=a$ is now introduced into the potential field and that $\varphi=0$ on this boundary. Show that the new complex potential for the region $|z|\geqslant a$ is now given by

$$w=f(z)-\bar{f}(a^2/z).$$

The reader conversant with electrostatic theory will recognize that this result gives the complex electrostatic potential when the earthed conducting sphere $|z|=a$ is introduced into a field whose complex potential was originally $f(z)$. Then $\text{Re}(w)$ gives the electrostatic potential of the field, the lines of force being given by $\text{Im}(w)=\text{const.}$

18. If $w=f(z)$ is the complex potential for a two-dimensional motion having no rigid boundaries and no singularities of flow within the circle $|z|=a$, show that on introducing the rigid boundary $|z|=a$ into the flow, the new complex potential is given by

$$w=f(z)+\bar{f}(a^2/z) \text{ for } |z|\geqslant a.$$

Show further, that if all singularities of flow are inside the circle $|z|=a$, then the above result still holds for $|z|\leqslant a$.

A source and a sink of equal strength are placed at the points $(a,0)$, $(-a,0)$ respectively, within a fixed circular boundary $|z|=2a$. Show that the streamlines are given by

$$16a^2y^2+\lambda y(r^2-4a^2)=(r^2-16a^2)(r^2-a^2)$$

where λ is a constant, and that the velocity at the point $(2a, \theta)$ is $20m \sin \theta / \{a(17 - 8 \cos 2\theta)\}$.

(Univ. of Aston in Birm., Gen. B.Sc., 1966)

19. Show that the equation for the stream function ψ in steady two-dimensional motion of a uniform incompressible inviscid fluid in which the vorticity is ω is

$$\frac{\partial^2 \psi}{\partial x^2} + \frac{\partial^2 \psi}{\partial y^2} = -\omega.$$

The fluid contained within an elliptic cylinder with semi-axes a, b is in steady two-dimensional motion with uniform vorticity ω. Show that the streamlines are similar ellipses, and that all elements of the fluid make one circuit in the time $2\pi(a^2 + b^2)/(\omega ab)$. Where does the fluid pressure have a maximum value?

(Cambridge Univ., M.T. II, 1957)

20. Show that the transformation

$$w = -Vz + Az^{1/2},$$

where A is a constant to be determined, gives the complex potential for the case of a homogeneous liquid streaming past the infinite parabolic cylinder

$$r^{1/2} \cos(\tfrac{1}{2}\theta) = a^{1/2},$$

the velocity at infinity being V in the positive direction of the x-axis.

(London Univ., Gen. B.Sc. II, 1963)

21. Verify that the two-dimensional irrotational motion of a liquid bounded by the lines $y = 0, y = 2a$ due to a source at the point $(0, a)$ is given by the complex potential

$$w = -m \log \cosh(\pi z / 2a),$$

where $2\pi m$ is the rate of emission from the source.

Give a rough sketch of the streamlines. Prove that the streamline which is asymptotic to $y = a + ka$, where $-1 < k < 1$, leaves the sources in the direction making an angle $\frac{1}{2}k\pi$ with the x-axis.

(London Univ., B.Sc. Spec. Hons., 1951)

22. Show that the complex potential

$$w = -V\left(\frac{\zeta - ib + c^2}{\zeta - ib}\right), \quad \zeta = \xi + i\eta,$$

gives irrotational streaming motion with speed V parallel to the ξ-axis past a fixed circular cylinder S of radius c with centre $0 + ib$.

If a is real and $z = \zeta + a^2/\zeta$, $z = x + iy$, show that

$$\frac{z - 2a}{z + 2a} = \frac{(\zeta - a)^2}{(\zeta + a)^2}$$

and, if $c^2 = a^2 + b^2$, determine the contour C in the z-plane corresponding to S.

If C represents the boundary fixed in a uniform stream with velocity V parallel to Ox, show that a circulation $-4\pi bV$ ensures that the velocity is finite at all points of C.

(London Univ., B.Sc. Spec. Hons., 1960)

23. If $\zeta = z + a^2/z, (a > 0)$, sketch the curve in the ζ-plane which is the transform of the circle

$$|z - b + a| = b \quad (b > 0)$$

in the z-plane, and prove that there is a cusp at the point $\zeta = -2a$.

If the curve in the ζ-plane is the profile of a rigid body surrounded by an infinite fluid which at great distances has velocity components $[-U \cos \alpha, -U \sin \alpha]$ along the ξ- and η-axes, and if the circulation round the body is such that the velocity is everywhere bounded, prove that the fluid flows past the cusp at a speed $(Ua/b) \cos \alpha$.

(Manchester Univ., Faculty of Technology, B.Sc. Tech. II, 1963)

24. Show that the transformation $t = \cosh(\pi z/b)$ can be used to map a semi-infinite strip of width b in the z-plane on to a half-plane in the t-plane.

A semi-infinite strip given by the boundaries $x = 0, y = b$ contains a source at $x = 0, y = 0$ sending liquid into the strip at the rate $2\pi m$ and a sink at $x = 0, y = b$ of equal strength. Show that the complex potential is $w = -4m \log \tanh(\pi z/2b)$.

(Liverpool Univ., B.Sc. Hons., 1963)

25. Verify that $w = ik \log\{(z - ia)/(z + ia)\}$ is the complex potential of a steady flow of liquid about a circular cylinder, the plane $y = 0$ being a rigid boundary.

Find the force exerted by the liquid on unit length of the cylinder.

(London Univ., B.Sc. Spec. Hons., 1955)

26. Find the resultant force on a fixed circular boundary $\mathscr{C}$ of radius a with centre at the origin if the fluid motion is caused entirely by a vortex of strength $2\pi\kappa$ at $2a + 0i$ and there is no circulation round $\mathscr{C}$.

(London Univ., B.Sc. Spec. Hons., 1960)

27. Show that the force exerted on a circular cylinder $|z| = a$ in the irrotational flow produced by a line source m at $z = 3a$ is $X = \rho m^2/(48\pi\alpha)$, $Y = 0$.

(Manchester Univ., Hons. Maths. I, 1963)

28. The complex potential of a steady two-dimensional fluid motion in the z-plane is

$$\sum_{1}^{\infty} b_n z^n \quad (b_n \text{ real constants}).$$

If the rigid cylindrical obstacle $|z| = a$ is inserted into the flow, prove that the cylinder will experience a thrust of magnitude

$$2\pi\rho\sum_{1}^{\infty} a^{2n} n(n+1) b_n b_{n+1}$$

per unit length. In which direction does the thrust act?

(Battersea College of Technology, Dip. Tech. Maths. I, 1962)

29. Inviscid liquid of constant density ρ flows without circulation past the fixed elliptic cylinder $x^2/a^2 + y^2/b^2 = 1$, the velocity of the undisturbed stream being $[-U\cos\alpha, -U\sin\alpha]$. Using the transformation

$$2z = (a+b)\zeta + (a-b)\zeta^{-1},$$

or otherwise, determine the complex potential w from the corresponding flow past a circular cylinder.

Show that, if $|z|$ is large,

$$w \simeq U[ze^{-i\alpha} + \tfrac{1}{2}(a+b)(b\cos\alpha + ia\sin\alpha)/z],$$

and hence find the couple M exerted on unit length of the cylinder.

(London Univ., B.Sc. Spec. Hons., 1958)

30. One boundary of a channel consists of the rectilinear segments

$$\begin{aligned} y&=0, \qquad 0 \leqslant x < \infty; \\ x&=0, \qquad 0 \leqslant y \leqslant a-b \quad (a>b); \\ y&=a-b, \qquad -\infty < x \leqslant 0, \end{aligned}$$

and the second boundary is $y=a$, $-\infty < x < \infty$. An incompressible liquid whose velocity components at $x=+\infty$ are $[-U, 0]$ flows along the channel which it fills completely. Show that the flow is given by

$$w = \varphi + i\psi = \frac{aU}{\pi}\log t, \quad z = \frac{a}{\pi}\int_{a^2/b^2}^{t} (t-1)^{1/2} t^{-1} (t - a^2/b^2)^{-1/2}\, dt,$$

and that along the curves of constant pressure

$$(1-\lambda)\exp\left(\frac{2\pi\varphi}{aU}\right) - 2\left(\frac{a^2}{b^2} - \lambda\right)\exp\left(\frac{\pi\varphi}{aU}\right)\cos\left(\frac{\pi\psi}{aU}\right) + \frac{a^4}{b^4} - \lambda = 0,$$

where λ is a parameter.

(London Univ., B.Sc. Spec. Hons., 1954)

31. Incompressible inviscid two-dimensional motion is taking place within a rectangle open at $y=+\infty$ and bounded by $y=0, x=a, x=-a$. This motion is due to a source m at $(a, 0)$ and an equal sink at $(-a, 0)$. Show that the complex potential w of the motion is given by $\sin(\pi z/2a) = \coth(w/4m)$.

Show further that the streamlines are given by $\cos(\pi x/2a) = \lambda\sinh(\pi y/2a)$, where λ is a constant.

(Nottingham Univ., Hons. Maths. II, 1962)

32. An incompressible liquid moves irrotationally in two dimensions within

an L-shaped region which has two rigid boundaries C_1 and C_2 defined as follows:

$$C_1 : x=0,\ 0<y<\infty; \quad y=0,\ 0<x<\infty;$$
$$C_2 : x=b,\ a<y<\infty; \quad y=a,\ b<x<\infty.$$

At $y=\infty$ the liquid is flowing in the negative y direction with speed U and it always remains in contact with the side walls. Show that, with the usual notation, the flow may be determined from the equations

$$z=A\int_0^t \frac{(\alpha-t)^{1/2}}{t^{1/2}(1-t)}\,dt, \quad w=B\log(t-1),$$

where α is real.

Prove that $A=b/\pi$, $B=bU/\pi$ and $\alpha=1+(a/b)^2$.

(London Univ., B.Sc. Spec. Hons., 1956)

33. The boundary $ABCDEF$ (described in a counter-clockwise direction) of an infinite liquid consists of a horizontal plane $AB(-\infty<x<\infty, y=h)$, two horizontal planes $CD(x\leqslant 0, y=h-k)$, $EF(x>0, y=0)$, and a vertical plane $ED(x=0, 0\leqslant y\leqslant h-k)$, where $h>k>0$. Liquid is in two-dimensional irrotational motion inside this region, the velocity for x large and positive being U parallel to AB or FE. Map the boundary on to the real axis of the plane of a complex variable t in such a way that A, B, C, D, E, F correspond to the values $-\infty, 0, 0, 1, a, \infty$ of t respectively.

Hence obtain the complex potential in the form

$$w=\frac{Uh}{\pi}\log\left(\frac{h^2-k^2t^2}{1-t^2}\right),$$

$$z=\frac{h}{\pi}\left\{\log\left(\frac{1+t}{1-t}\right)-\frac{k}{h}\log\left(\frac{h+kt}{h-kt}\right)\right\}.$$

(Imperial College, B.Sc. Spec., 1956).

34. In the notation of Section 5.5.4, p. 174, prove that the circulation round $\mathscr{C}$ may be expressed in the forms

$$\Gamma=\oint_{\mathscr{C}}(u\,dx+v\,dy)=-\mathrm{Re}\oint_{\mathscr{C}}\frac{dw}{dz}\,dz.$$

(The last integral is often evaluated using the residue calculus. The formula affords an effective means of answering Question 6(iv) of Exercise 5.)

6

ELEMENTS OF THERMODYNAMICS

6.1 The Equation of State of a Substance

The measurable quantities of a compressible substance (fluid or otherwise) are its pressure p, density ρ and temperature T. It is found that these quantities are connected through a functional relation of the form

$$F(p, \rho, T)=0, \tag{1}$$

where F is a single-valued function of the variables p, ρ, T. Such an equation is known as the *equation of state* of the substance. The form assumed by F depends on the nature of the substance. For certain kinds of gas, we shall see that it is of a very simple form. For other substances the form of F may be exceedingly complex.

We have seen in Chapter 3 that the pressure in a fluid is ascribable to the momentum change produced in its molecules on impinging against a small plane rigid surface. The density is proportional to the number of molecules in a prescribed volume. The notions derive from the molecular structure of matter. In addition, since the randomly moving molecules of a substance have a mean kinetic energy associated with them, this leads to the notion of temperature, being defined as a measure of this kinetic energy of random molecular motion.

For the moment we confine our attention to gaseous substances. In some cases, the molecules of a gas have but negligible volume and there are virtually no mutual attractions between the individual molecules. Such a gas is said to be a *perfect gas* and its equation of state assumes the simple form

$$p=R\rho T \tag{2}$$

where R is a constant for the particular gas under consideration. Physical chemists like to base the value of R on 1 gramme molecule of the gas; with this stipulation, R can be shown to be the same for all perfect gases and is then termed the universal gas constant. In this case $1/\rho$ is the volume of 1 gramme molecule of the gas. In this chapter, the chemists' convention will not be used; R will relate to any convenient unit mass of gas.

Such gases as air, oxygen, hydrogen, nitrogen, etc., are, for a wide range of values of p, ρ, T, very nearly perfect gases and for most cases the simple equation (1) may be taken as the equation of state. However, when allowance is made for the covolume (i.e. volume of the molecules) and the intermolecular attractions, a more accurate equation of state is

$$(p + av^{-2})(v - b) = RT \tag{3}$$

where $v = 1/\rho$ a = constant, b = covolume. The equation (3) is due to van der Waals.

Other equations of state have been proposed for real gases. Amongst the more common are the following due to Clausius and Dieterici:

$$\{p + (a/Tv^2)\}(v - b) = RT \quad \text{(Clausius)};$$
$$\{p + (a/v^n)\}(v - b) = RT \quad \text{(Dieterici)};$$
$$p(v - b) = RT \exp\{-c/(RTv)\} \quad \text{(Dieterici)}.$$

6.2 The First Law of Thermodynamics

The first law of thermodynamics asserts *mutual equivalence between heat and mechanical work.* Thus heat is held to be a form of energy and in fact treatises on the kinetic theory of gases indicate that heat energy is mechanical energy on the molecular scale.

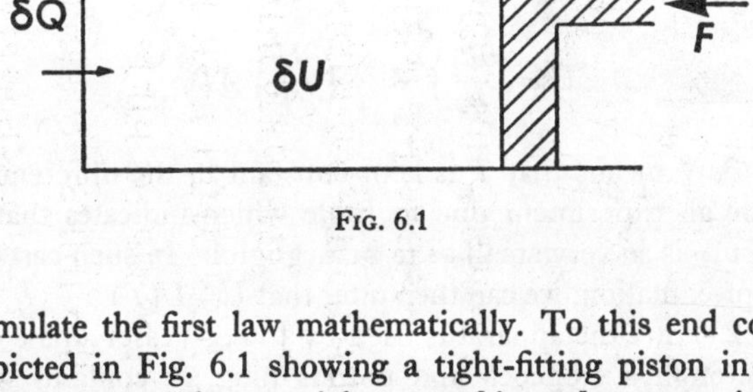

FIG. 6.1

We formulate the first law mathematically. To this end consider the model depicted in Fig. 6.1 showing a tight-fitting piston in a cylinder enclosing a compressible material, or *working substance* as it is usually called. The apparatus is thermally insulated from its surroundings. Suppose that a quantity δQ of heat (measured in mechanical units) is applied to the working substance from outside. A force F is applied to the piston. Then the heat supplied will produce two effects:

(i) it will increase the internal energy of the working substance by an amount δU;

(ii) it will perform useful external work δW against the applied force F.

In virtue of the first law of thermodynamics it follows that

$$\delta Q = \delta U + \delta W, \tag{1}$$

or in differential form

$$dQ = dU + dW. \tag{1'}$$

Equation (1′) is the mathematical formulation that is usually given of the first law. It is assumed that the piston moves slowly and without acceleration so that the external force F is always in equilibrium with the pressure forces exerted by the working substance.

If in the above process the movement of the piston is δx, then the work done against the external force F is $\delta W = F\delta x$. If δv denotes the volume expansion of the working substance and A the area of the piston, then, assuming equilibrium and no viscosity,

$$\delta W = F\delta x = pA\delta x = p\delta v,$$

where p is the pressure in the substance. Thus equation (1′) becomes

$$dQ = dU + p\,dv. \tag{2}$$

6.3 Internal Energy of a Gas

The internal energy of a gas may be expressed in terms of the measurable quantities p, ρ, T. Since these three quantities are related through the equation of state, the internal energy may be expressed as a function of any two of the three quantities. Let us suppose that U is expressed as a function of v, T, where $v = 1/\rho$, the volume of unit mass of gas. Then, when changes $\delta v, \delta T$ are made in v, T, the attendant change in U is

$$\delta U = \left(\frac{\partial U}{\partial v}\right)_T \delta v + \left(\frac{\partial U}{\partial T}\right)_v \delta T, \tag{1}$$

where $(\partial U/\partial v)_T$ means that T is kept constant in the differentiation, etc. We describe an experiment due to Joule which indicates that for many gases $(\partial U/\partial v)_T$ is so very small as to be negligible. In such cases, to a high order of approximation, we can then infer that $U = U(T)$.

Figure 6.2 shows the apparatus used by Joule. Vessel A is connected to vessel B through a stopcock and porous plug. A contains gas at high pressure and B is evacuated. When the stopcock is opened gas flows from A into B. The entire apparatus is thermally insulated and no temperature change can be detected as a result of the passage of gas from

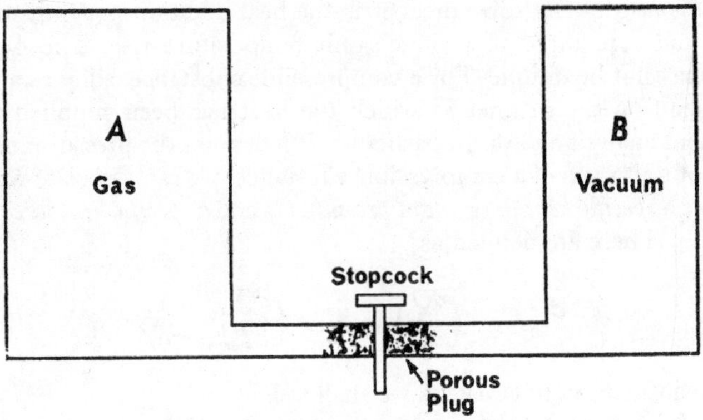

FIG. 6.2

A to B. If $U = U(v, T)$, then taking $\delta T = 0$, following the experimental evidence, (1) shows that

$$\delta U = \left(\frac{\partial U}{\partial v}\right)_T \delta v. \tag{2}$$

Now no heat is supplied from outside and no external work is done so that $\delta U = 0$. Thus (2) gives

$$0 = \left(\frac{\partial U}{\partial v}\right)_T \delta v. \tag{3}$$

Since $\delta v \neq 0$, (3) shows that $(\partial U/\partial v)_T = 0$ and hence that $U = U(T)$.

With actual gases, however, a small temperature rise can be detected with sufficient accuracy of measurement. The relation $U = U(T)$ is exactly true for a perfect gas but not for any other. Actually the internal energy U of a given quantity of gas may be written in the form

$$U = E + I,$$

where E is the kinetic energy due to the translatory motion of the gas molecules and I is the energy ascribable to all other causes such as molecular rotation and vibration, and also the potential energy of the molecules due to their intermolecular attractions. For a perfect gas the molecules are deemed to have no size and hence no rotational or vibrational energy and no mutual attractions and hence $U = E = E(T)$, in accordance with the experiment described. For real gases, however, $I = I(T, v) \neq 0$.

6.4 Specific Heats of a Gas

Let δQ be the amount of heat added to unit mass of a substance so as to produce a temperature increment δT. Then the rate of increase of heat added with temperature rise is $\partial Q/\partial T$. This defines a quantity known as the *specific heat* of the substance: it is the heat addition per unit mass of the substance required to produce unit temperature rise. The quantity $\partial Q/\partial T$ may not be unique. For a compressible substance, such as a gas, it will depend on the manner in which the heat has been supplied as the subsequent analysis shows. In particular, if p denotes the pressure and v the volume of unit mass of a compressible substance, we can associate with the substance a *specific heat at constant pressure* C_p and a *specific heat at constant volume* C_v. These are defined as

$$C_p = \left(\frac{\partial Q}{\partial T}\right)_p, \quad C_v = \left(\frac{\partial Q}{\partial T}\right)_v.$$

These quantities are unequal, as we shall see.

From the first law of thermodynamics,

$$dQ = p\,dv + dU$$

$$= p\,dv + \left(\frac{\partial U}{\partial T}\right)_v dT + \left(\frac{\partial U}{\partial v}\right)_T dv,$$

if we assume $U = U(v, T)$ for the internal energy of unit mass of the substance. Hence

$$C_v = \left(\frac{\partial Q}{\partial T}\right)_v = \left(\frac{\partial U}{\partial T}\right)_v. \tag{1}$$

To obtain C_p, we have

$$C_p = \left(\frac{\partial Q}{\partial T}\right)_p = \left(\frac{\partial U}{\partial T}\right)_v + \left\{p + \left(\frac{\partial U}{\partial v}\right)_T\right\}\left(\frac{\partial v}{\partial T}\right)_p.$$

Thus

$$C_p = C_v + \left\{p + \left(\frac{\partial U}{\partial v}\right)_T\right\}\left(\frac{\partial v}{\partial T}\right)_p. \tag{2}$$

Equations (1), (2) show that $C_p \neq C_v$.

The equations (1), (2) are applicable to any substance. Let us now apply them to the case of a perfect gas for which $U = U(T)$. Then (1) shows that

$$C_v = dU/dT, \tag{3}$$

and (2) gives

$$C_p = C_v + p\left(\frac{\partial v}{\partial T}\right)_p. \tag{4}$$

Further, the equation of state of a perfect gas is $pv = RT$, so that $(\partial v/\partial T)_p = R/p$ and (4) becomes

$$C_p = C_v + R. \tag{5}$$

Also, for a perfect gas, the kinetic theory shows that the internal energy U is directly proportional to the temperature T, and this leads to the result that C_v is constant. Equation (5) shows that C_p is then also constant and hence that

$$C_p/C_v=\gamma, \tag{6}$$

where γ is a constant termed the *adiabatic constant*. From (5), (6) we have at once

$$C_v=\frac{R}{\gamma-1}, \tag{7}$$

$$C_p=\frac{\gamma R}{\gamma-1}. \tag{8}$$

For air and for most diatomic gases (such as oxygen, hydrogen, nitrogen), $\gamma \simeq 1{\cdot}400$.

Observe that, for a perfect gas, (3) implies that

$$U(T)=C_vT$$

if the temperature scale is so chosen that $U(0)=0$. In these circumstances there is no internal energy when $T=0$. The temperature $T=0$ is called the *absolute zero* and the corresponding temperature scale is called the *absolute* or *Kelvin* scale.

6.5 Functions of State; Entropy

Let φ be some thermodynamic function of a substance expressible in terms of any two of the measurable quantities p, ρ, T. We have seen that this is always possible whenever the equation of state of the substance is known. It may happen that if we express φ as a function of, say, p and v $(v=1/\rho)$, then in changing from an initial state A to another state B the value of φ depends only on the conditions at A and B and not at all on the paths joining them. In these circumstances, φ is termed a *function of state*.

Suppose that the differential of φ is expressible in terms of dp and dv by the relation

$$d\varphi=M(p,v)\,dp+N(p,v)\,dv. \tag{1}$$

If $\varphi_B-\varphi_A$ is independent of the path taken in changing from the conditions at A to those at B, then $d\varphi$ must be an exact differential. The necessary and sufficient condition for this is

$$\left(\frac{\partial M}{\partial v}\right)_p=\left(\frac{\partial N}{\partial p}\right)_v. \tag{2}$$

If (2) holds, then $\varphi(p,v)$ is a function of state.

The mathematical formulation of the first law of thermodynamics gives the result

$$dQ = p\,dv + dU.$$

For a perfect gas, $dU = C_v\,dT = (C_v/R)\,d(pv)$, using the equation of state $pv = RT$. Thus for such a gas,

$$\begin{aligned} dQ &= \{1 + (C_v/R)\}p\,dv + (C_v/R)v\,dp \\ &= (C_p p\,dv + C_v v\,dp)/R. \end{aligned}$$

Identifying dQ with $d\varphi$, we have $M = (C_v/R)v$, $N = (C_p/R)p$.
Now

$$\left(\frac{\partial M}{\partial v}\right)_p = \frac{C_v}{R}; \quad \left(\frac{\partial N}{\partial p}\right)_v = \frac{C_p}{R}.$$

Since $C_p \neq C_v$, equation (2) is violated showing that dQ is not an exact differential. This means that the quantity of heat Q added to a unit mass of gas is not a function of state: it depends on the path chosen in the change from a given state A to a second state B.

Using $pv = RT$, the relation for dQ for a perfect gas may be written in the alternative form

$$dS = (C_p/v)\,dv + (C_v/p)\,dp, \tag{3}$$

where $dS = dQ/T$. Now

$$\frac{\partial}{\partial p}\left(\frac{C_p}{v}\right) = 0; \quad \frac{\partial}{\partial v}\left(\frac{C_v}{p}\right) = 0,$$

so that

$$\frac{\partial}{\partial p}\left(\frac{C_p}{v}\right) = \frac{\partial}{\partial v}\left(\frac{C_v}{p}\right).$$

This shows that dS is an exact differential. We may integrate this differential to give

$$S - S_0 = C_p \log v + C_v \log p, \tag{4}$$

or

$$\log (pv^\gamma) = (S - S_0)/C_v.$$

Thus

$$pv^\gamma = \exp\{(S - S_0)/C_v\}. \tag{5}$$

The quantity S is called the *entropy* per unit mass: dS is the entropy differential. Flows for which S is constant are called *isentropic*. Equation (5) shows that they are characterized by

$$pv^\gamma = \text{const.} \tag{6}$$

We return to this later (see Section 6.7).

6.6 Maxwell's Thermodynamic Relations

One form of the first law of thermodynamics applied to any working substance is

$$dQ = dU + p\,dv.$$

where dQ is the heat added to unit mass, dU is the internal energy gained, $p\,dv$ is the external work done and v is the volume per unit mass. Using the equation

$$dQ = T\,dS,$$

this first law may be expressed in the form

$$dU = T\,dS - p\,dv. \tag{1}$$

Now the internal energy of a given quantity of any substance is a function of state only. This is an experimental fact. Careful measurement shows that when a given amount of heat is supplied to a given mass of substance which is thermally insulated from its surroundings the temperature rise is always the same. The temperature of the substance, expressed in degrees absolute, is a monotonic single-valued function of the internal energy. Thus dU is an exact differential and so (1) implies

$$\left(\frac{\partial T}{\partial v}\right)_S = -\left(\frac{\partial p}{\partial S}\right)_v. \tag{1'}$$

Now define $H = U + pv$. Then H is called the *enthalpy* per unit mass. Thus H is a function of state. Then

$$\begin{aligned} dH &= dU + p\,dv + v\,dp \\ &= T\,dS + v\,dp, \end{aligned} \tag{2}$$

using equation (1). dH must be an exact differential. Thus

$$\left(\frac{\partial T}{\partial p}\right)_S = \left(\frac{\partial v}{\partial S}\right)_p. \tag{2'}$$

Now write $F = U - TS$. Then F is also a function of state. It is called the *free energy* per unit mass. Further,

$$\begin{aligned} dF &= dU - T\,dS - S\,dT \\ &= -p\,dv - S\,dT, \end{aligned} \tag{3}$$

using (1). Since dF is an exact differential,

$$\left(\frac{\partial p}{\partial T}\right)_v = \left(\frac{\partial S}{\partial v}\right)_T. \tag{3'}$$

Finally, write $G = U + pv - TS$. G is called the *Gibbs function* and it is a function of state. Further,

$$dG = dU + p\,dv + v\,dp - T\,dS - S\,dT$$

or

$$dG = v\,dp - S\,dT, \tag{4}$$

using (1). Since dG is an exact differential,

$$\left(\frac{\partial v}{\partial T}\right)_p = -\left(\frac{\partial S}{\partial p}\right)_T. \tag{4'}$$

The four equations (1′), (2′), (3′), (4′) are called *Maxwell's thermodynamic relations*. They are the necessary and sufficient conditions for the thermodynamic functions U, H, F, G to be functions of state.

EXAMPLE 1

Define the internal energy of a thermodynamical system and state the first law of thermodynamics. Give reasons for saying the internal energy is a function of state.

A simple substance is such that the internal energy u, the pressure p, and the volume v, of unit mass are related by $u=3pv$. Show that

$$c_v=3v\left(\frac{\partial p}{\partial T}\right)_v, \quad c_p=4p\left(\frac{\partial v}{\partial T}\right)_p.$$

Assuming that $\gamma(=c_p/c_v)$ is constant, show that

$$p^{3\gamma}v^4=F(T),$$

where $F(T)$ is a function of temperature T only.

(London Univ., B.Sc. Spec. I, 1962)

If dQ denotes the heat added to unit mass of the substance and du the internal energy gained, then since $p\,dv$ is the external work done, the first law of thermodynamics gives

$$dQ=p\,dv+du.$$

Taking $u=3pv$,

$$dQ=4p\,dv+3v\,dp.$$

Hence

$$c_v=\left(\frac{\partial Q}{\partial T}\right)_v=3v\left(\frac{\partial p}{\partial T}\right)_v.$$

$$c_p=\left(\frac{\partial Q}{\partial T}\right)_p=4p\left(\frac{\partial v}{\partial T}\right)_p.$$

Since $c_p=\gamma c_v$

$$4p\left(\frac{\partial v}{\partial T}\right)_p=3\gamma v\left(\frac{\partial p}{\partial T}\right)_v.$$

Now $(\partial v/\partial T)_p=1/(\partial T/\partial v)_p$; $(\partial p/\partial T)_v=1/(\partial T/\partial p)_v$. Hence

$$4p\left(\frac{\partial T}{\partial p}\right)_v=3\gamma v\left(\frac{\partial T}{\partial v}\right)_p. \tag{1}$$

Taking the equation of state in the form $T=f(p, v)$,

$$dT=\left(\frac{\partial T}{\partial p}\right)_v dp+\left(\frac{\partial T}{\partial v}\right)_p dv.$$

When $T=$const., $dT=0$ and so

$$\left(\frac{\partial T}{\partial v}\right)_p \bigg/ \left(\frac{\partial T}{\partial p}\right)_v = -\left(\frac{\partial p}{\partial v}\right)_T. \tag{2}$$

From (1), (2) we obtain, for $T=$const.,

$$-\left(\frac{\partial p}{\partial v}\right)_T = \frac{4p}{3\gamma v}$$

or $3\gamma(dp/p)+4(dv/v)=0$, subject to $T=$const. Integrating, $3\gamma \log p + 4 \log v = \psi(T)$, whence

$$p^{3\gamma}v^4=F(T) \quad (=\exp \psi(T)).$$

EXAMPLE 2

The pressure, volume, temperature, internal energy and entropy of a given mass of fluid in equilibrium are p, V, T, U and S respectively. Prove that if T and V are taken as independent variables, then

$$\frac{\partial U}{\partial T}=T\frac{\partial S}{\partial T},$$

$$\frac{\partial U}{\partial V}=T\frac{\partial S}{\partial V}-p=T\frac{\partial p}{\partial T}-p.$$

Using the last result, show that for a gas obeying van der Waals' equation

$$\{p+(a/V^2)\}(V-b)=RT,$$

where R, a, b are constants, the specific heat C_v is independent of V. Prove also that for small values of a and b, $C_p - C_v$ has the approximate value $R+2a/(VT)$.

(London Univ., B.Sc. Spec. I, 1963)

By the first law of thermodynamics,

$$dQ=dU+p\,dV$$

or

$$T\,dS=dU+p\,dV=\left(\frac{\partial U}{\partial V}\right)_T dV+\left(\frac{\partial U}{\partial T}\right)_V dT+p\,dV \tag{1}$$

Keeping V constant, $dV=0$ so that (1) gives

$$T\left(\frac{\partial S}{\partial T}\right)_V=\left(\frac{\partial U}{\partial T}\right)_V, \tag{2}$$

as required. Keeping T constant, $dT=0$ and (1) gives

$$T\left(\frac{\partial S}{\partial V}\right)_T=\left(\frac{\partial U}{\partial V}\right)_T+p$$

or

$$\left(\frac{\partial U}{\partial V}\right)_T=T\left(\frac{\partial S}{\partial V}\right)_T-p. \tag{3}$$

From (1),

$$d(U-TS)=-S\,dT-p\,dV. \tag{4}$$

But $U-TS$ is a function of state only so that $d(U-TS)$ is an exact differential. Hence, from (4)

$$\left(\frac{\partial S}{\partial V}\right)_T=\left(\frac{\partial p}{\partial T}\right)_V. \tag{5}$$

From (3), (5),

$$\left(\frac{\partial U}{\partial V}\right)_T=T\left(\frac{\partial p}{\partial T}\right)_V-p. \tag{6}$$

When $\{p+(a/V^2)\}(V-b)=RT$,

$$\left(\frac{\partial p}{\partial T}\right)_V=\frac{R}{V-b}.$$

Since $dQ=dU+p\,dV$,

$$C_v=\left(\frac{\partial Q}{\partial T}\right)_V=\left(\frac{\partial U}{\partial T}\right)_V,$$

and so

$$\frac{\partial C_v}{\partial V}=\frac{\partial^2 U}{\partial V\partial T}=\frac{\partial}{\partial T}\left(\frac{\partial U}{\partial V}\right)$$

$$=\frac{\partial}{\partial T}\left(\frac{RT}{V-b}-p\right)$$

$$=\frac{\partial}{\partial T}\left(\frac{a}{V^2}\right)=0,$$

showing that C_v is independent of V.

Now $dQ=dU+p\,dV$

$$=\left(\frac{\partial U}{\partial V}\right)_T dV+\left(\frac{\partial U}{\partial T}\right)_V dT+p\,dV,$$

so that

$$C_v=\left(\frac{\partial Q}{\partial T}\right)_V=\left(\frac{\partial U}{\partial T}\right)_V.$$

Hence

$$dQ=\left(\frac{\partial U}{\partial V}\right)_T dV+C_V\,dT+p\,dV,$$

and so

$$\begin{aligned}C_p&=\left(\frac{\partial Q}{\partial T}\right)_p\\&=C_v+\left\{\left(\frac{\partial U}{\partial V}\right)_T+p\right\}\left(\frac{\partial V}{\partial T}\right)_p\\&=C_v+T\left(\frac{\partial p}{\partial T}\right)_V\left(\frac{\partial V}{\partial T}\right)_p.\end{aligned}$$

But

$$\left(\frac{\partial p}{\partial T}\right)_V=\frac{R}{V-b}$$

and keeping p constant gives

$$\left(p+\frac{a}{V^2}\right)\left(\frac{\partial V}{\partial T}\right)_p-\frac{2a}{V^3}\left(\frac{\partial V}{\partial T}\right)_p(V-b)=R,$$

so that

$$\left(\frac{\partial V}{\partial T}\right)_p=\frac{R}{p+(a/V^2)-(2a/V^3)(V-b)}=\frac{R}{\{RT/(V-b)\}-(2a/V^3)(V-b)}.$$

Hence

$$\begin{aligned}C_p-C_v&=\frac{R^2T/(V-b)}{\{RT/(V-b)\}-(2a/V^3)(V-b)}\\&=R\{1-(2a/RTV^3)(V-b)^2\}^{-1}\\&=R\{1+(2a/RTV)+\ldots\}\\&\simeq R+(2a/TV).\end{aligned}$$

6.7 Isothermal, Adiabatic and Isentropic Processes

At this stage we introduce some important terminologies in thermodynamics. The change in a substance from a fixed state A to a fixed state B may be effected in infinitely many ways. For thermodynamic functions of state, we have seen that the mode of change from A to B is immaterial. However, there are thermodynamic functions, such as heat input Q, which do depend on the nature of the change as well as on the states, A, B.

A change from A to B in which the temperature T is kept constant is said to be *isothermal.* In the case of a perfect gas, since $p=R\rho T$ is the equation of state an isothermal change would be governed by Boyle's law:

$$p=k\rho \quad (k=\text{const.}).$$

An *adiabatic change* is one in which there is no heat exchange between the working substance and its surroundings. Even if there is viscosity in the working substance so that heat is dissipated by the frictional forces that are called into play, the change is still adiabatic if there is complete thermal insulation of the apparatus from its surroundings.

If a change is made so that the entropy of every single particle of the working substance stays constant, then such a change is termed *isentropic.* We have seen that isentropic changes in a perfect gas are characterized by equation (6) of Section 6.5. When the entropy of every single quantity of a substance of fixed mass is the same and stays constant in any change, the change is said to be *homentropic.* In the equation alluded to, the constant on the right-hand side is the same for each considered small quantity of gas in isentropic flow but a different constant attaches to each such quantity. For homentropic flow, however, the constant is the same throughout the entire volume of gas.

6.8 Heat Engines, Cycle of Changes, Reversibility

A *heat engine* is a device for obtaining mechanical work from heat. To this end the engine uses a working substance which may be solid, liquid or gaseous. In particular a steam engine utilizes steam as the working substance. In all engines an amount of heat Q_1 is drawn from a hot source and a lesser amount Q_2 is discharged to a cold sink, the difference (Q_1-Q_2) being given to the working substance as internal energy and as useful external work against an applied load, in accordance with the first law of thermodynamics. The *efficiency* η of such an engine is defined by the ratio

$$\eta=(Q_1-Q_2)/Q_1.$$

It may happen that in the functioning of a heat engine, the working substance undergoes a system of operations in which it is finally brought back to its initial state (i.e. to the same conditions of pressure, density and temperature), whilst having performed external work through the supply of heat from the source and the discharge of heat to the sink. Such a system of changes is called a *cycle.* It is often convenient to represent such a cycle on an *indicator diagram* which is the curve showing the dependence of the pressure p on the volume v of a fixed mass of the working substance. The indicator diagram of Fig. 6.3 shows a cycle of operations. The working substance is first taken from state A to state B via the path $\mathscr{C}_1$ and then returned to A via the path $\mathscr{C}_2$. The work done by the substance in expansion from A to B along $\mathscr{C}_1$ is represented by the area under the curve $\mathscr{C}_1$ and between the ordinates AA_1, BB_1. That done against the substance in passing from state B back to A along $\mathscr{C}_2$ is the area under the curve $\mathscr{C}_2$ and between AA_1, BB_1. Hence the total work done by the substance in the cycle is represented by the area of the shaded loop between the arcs $\mathscr{C}_1, \mathscr{C}_2$. At the end of the cycle the substance is in the same state A as at the beginning. Hence there is no change of internal energy of the working substance and the work done by the substance manifests itself entirely as useful external work performed.

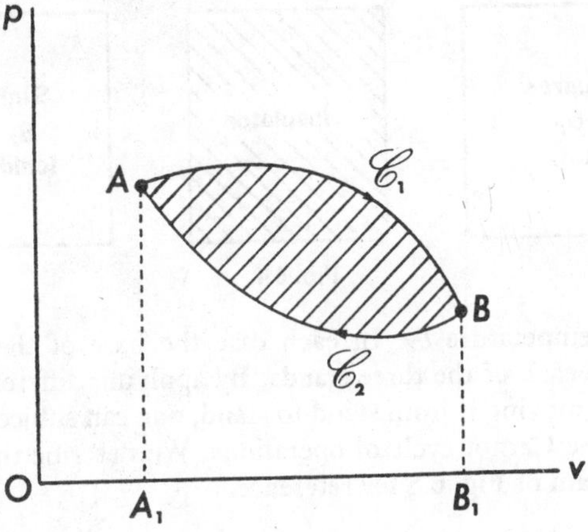

FIG. 6.3

If, in the above cycle, we can start at A, proceed to B via the path $\mathscr{C}_2$ reversed, then from B back to A along the path $\mathscr{C}_1$ reversed whilst taking exactly the same amount of heat from the sink as was taken from the source in the previous cycle and discharging to the source the same amount as was discharged to the sink in the previous cycle, then the cycle is said to be *reversible*. In such a reversible cycle exactly the same amount of work is done by the working substance whichever route is chosen. If these conditions are violated, the cycle is said to be *irreversible*. For real fluids there is always some dissipation of heat due to viscosity so that a truly reversible cycle can never be realized.

6.9 The Carnot Cycle

We now introduce an important reversible heat engine due originally to the French engineer Carnot. His name is given to the cycle of operations to which the working substance is subjected. The reader will refer to Fig. 6.4 to understand the functioning of the engine.

Figure 6.4 shows a piston enclosing a working substance in a cylinder, the base of which is a perfect conductor of heat, the walls and the piston being of perfect insulating material. The base of the cylinder may be placed on an insulating stand, or on a hot source at temperature θ_1 or on a

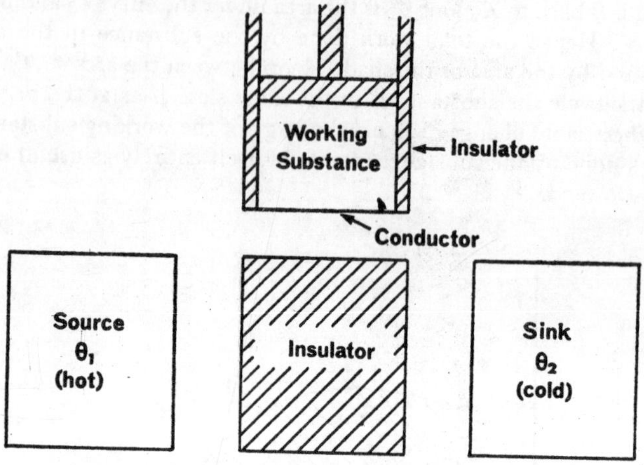

FIG. 6.4

cold sink at temperature θ_2. In each case the base of the cylinder fits perfectly on to each of the three stands. By applying different loadings to the piston and moving it from stand to stand, one can subject the working substance to the Carnot cycle of operations. We describe these using the indicator diagram of Fig. 6.5 for reference.

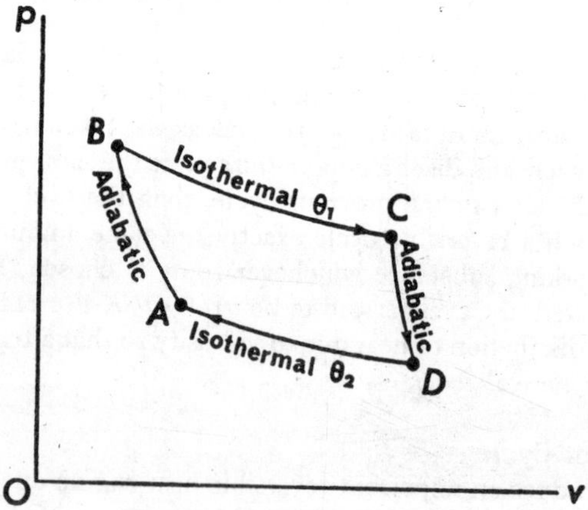

FIG. 6.5

The cylinder is first placed on the insulating stand. No heat can now enter or leave the working substance so that any change taking place is *adiabatic*. Let the working substance have a state represented by A in Fig. 6.5, its temperature being θ_2. The working substance is then compressed by driving the piston into the cylinder until the temperature of the working substance rises to $\theta_1(>\theta_2)$. Its state is now represented by B. The change from A to B has taken place along a curve called an *adiabatic* since the change has been effected entirely under adiabatic conditions.

With the working substance now at the state represented by B, the cylinder is removed from the non-conducting stand and placed with its base in contact with the hot source at temperature θ_1. Expansion now takes place under isothermal conditions, i.e. at the constant temperature θ_1, from the state represented by B to that represented by C. This expansion phase takes place along the curve BC which is termed an *isothermal*. Let Q_1 be the heat absorbed by the working substance along this isothermal. At C the expansion is complete.

The cylinder is now placed back on to the insulating stand and an adiabatic expansion is allowed to take place by pulling the piston out of the cylinder until the substance is brought back to the temperature θ_2, as represented by D. Thus CD is the adiabatic along which this expansion has taken place.

Finally the cylinder is removed from the insulator and placed with its base in contact with the sink at temperature θ_2. An isothermal compression is allowed to take place until the working substance is brought back to the state represented by A. In this process a certain amount of heat Q_2 is discharged to the sink.

Thus the working substance is taken through a complete cycle of operations comprising two adiabatics and two isothermals. Q_1 is the heat absorbed from the hot source and Q_2 that discharged to the cold sink. Thus (Q_1-Q_2) is the useful external work done by the working substance. Such a cycle is called a Carnot cycle and, as for any other complete cycle, the efficiency is $(Q_1-Q_2)/Q_1$.

The cycle of operations described above may be reversed. We describe briefly the reversal processes of the Carnot cycle.

(i) Starting from A, isothermal expansion at temperature θ_2 takes place along AD with the cylinder on the sink until D is reached. The heat absorbed from the sink is Q_2.

(ii) Adiabatic compression from D along DC takes place with the cylinder placed on the insulating stand.

(iii) Isothermal compression follows from C to A along CA with the cylinder placed on the hot source at temperature θ_1. The heat discharged to the hot source is Q_1.

(iv) Finally adiabatic expansion from B to A along BA takes place with the cylinder on the insulating stand.

In the reversed cycle, Q_1 is the heat discharged to the hot source and Q_2 that drawn from the cold sink. Thus (Q_1-Q_2) is the external work that has to be done on the substance. In the reversed cycle, the engine functions as a refrigeration machine inasmuch as the work done by the substance is making the cold sink cooler.

6.10 The Second Law of Thermodynamics

Stated briefly the second law of thermodynamics is: *Heat cannot pass from a body at a low temperature to one at a higher temperature without the assistance of an external agency.* Thus in the reversed Carnot cycle (refrigeration machine) to extract heat $Q_1(>Q_2)$ from the cold sink and discharge heat Q_2 to the hot source it was necessary to do work of amount (Q_1-Q_2) on the working substance. It is common experience that when two bodies at different temperatures are placed in contact, heat flows from the body at the higher temperature to that at the lower one; flow from the lower to the higher temperature cannot take place without some external assistance.

At the time of writing, the second law of thermodynamics is held by physicists everywhere to be a great universal truth. Its disproof would seriously undermine physics as we know it today. Clerk Maxwell, however, suggested a model which, if it could be made to work, would violate the law as literally stated. He imagined a box partitioned into separate compartments A and B, both of which contained gases at different temperatures. He supposed that A contained the gas at the higher temperature so that the mean velocity of the molecules in A was greater than that in B. Next he supposed that a frictionless and weightless trap-door was inserted in the partition separating A and B and that this trap-door could be operated by a tiny creature termed a demon. Maxwell further supposed that every time a fast moving molecule in B came towards the demon he opened the door and allowed the molecule to pass into A and then quickly shut the door. In the absence of frictional and gravitational forces, the demon could do no work on the system. Likewise, all incident slow-moving molecules were supposed to be transferred from A to B. In this way A became hotter and B colder, so that heat was made to flow from the colder body to the hotter one without doing any work. Actually the second law is statistical in kind. That is to say its enunciation applies to molecules taken on the average for given ensembles and not to individual molecules in the way described above.

A consequence of the second law of thermodynamics is that the entropy of a self-contained system tends to increase. This may be illustrated rather naïvely by the following model. Suppose A and B are two bodies at temperatures θ_1, θ_2 respectively, where $\theta_1>\theta_2$. When they are placed in contact, heat flows from A to B. Let δQ be the heat that has flowed from A to B in a time interval δt. Then in this time, the entropy lost by A is $\delta Q/\theta_1$ and that gained by B is $\delta Q/\theta_2$. Hence the net gain of entropy of the entire system is

$$\frac{\delta Q}{\theta_2}-\frac{\delta Q}{\theta_1}=\frac{\delta Q(\theta_1-\theta_2)}{\theta_1\theta_2}>0.$$

The second law of thermodynamics may be formulated in the following alternative and equivalent manner: *The entropy of an isolated system performing internal irreversible processes will increase.* The result may be established more rigorously and extensively than has been done for the simple model. Pushed to its logical conclusion the second law tells us that *the entropy of the universe is increasing* and this has led to many scientific and philosophical ramifications.

6.11 Carnot's Theorem

We now establish a theorem due to Carnot concerning the efficiency of a reversible engine. The theorem may be stated in two equivalent forms.

(i) *For prescribed temperature limits, no engine can be more efficient than a reversible engine working between those limits.*

(ii) *For prescribed temperature limits, all reversible engines working between the limits have the same efficiency.*

Proof:

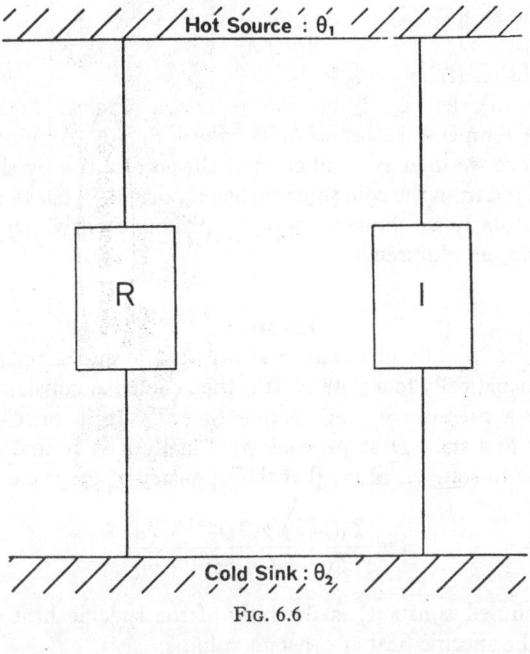

FIG. 6.6

Figure 6.6 shows two engines R (reversible) and I (irreversible) working between the same temperature limits θ_1, θ_2 where $\theta_1 > \theta_2$. To prove (i), we start by supposing that

$$\eta_I > \eta_R,$$

where η denotes efficiency, and show that this assumption contradicts the second law of thermodynamics. Let the amounts of working substance in each engine be so adjusted that each draws the same amount of heat Q_1

from the source at temperature θ_1. Let R discharge an amount of heat Q_2 to the sink and let I discharge Q_2' to it. Then

$$\eta_R=\frac{Q_1-Q_2}{Q_1};\quad \eta_I=\frac{Q_1-Q_2'}{Q_1}$$

Since $\eta_I>\eta_R$, by hypothesis, this leads to the result

$$Q_2>Q_2'.$$

Now let I drive R backwards, the two being coupled together. Then I draws Q_1 from the hot source, R discharges Q_1 to the hot source so that in total the hot source neither gains nor loses heat. The cold sink loses heat Q_2 to R and gains Q_2' from I. Thus the cold sink suffers a net loss of heat of $(Q_2-Q_2')>0$. This loss of heat has been effected by the coupled engines without external assistance. This contradicts the second law of thermodynamics and so

$$\eta_I \not> \eta_R,$$

which establishes (i).

To prove (ii), let R_1, R_2 be two reversible engines both operating between the temperature limits θ_1, θ_2 where $\theta_1>\theta_2$. Assume $\eta_{R_1}>\eta_{R_2}$. Then as before we infer contradiction of the second law by showing that heat is extracted from the cold source when R_1 drives R_2 backwards. Hence $\eta_{R_1}\not>\eta_{R_2}$. Similarly we show $\eta_{R_2}\not>\eta_{R_1}$ by making R_2 drive R_1 backwards. Thus $\eta_{R_1}=\eta_{R_2}$ as required.

Example

A perfect gas initially in a state A at pressure p_1 and temperature T_1 is expanded adiabatically to a state B. It is then cooled at constant volume to a state C at a pressure p_2 and temperature T_2. It is next compressed adiabatically to a state D at pressure p_1. Finally it is heated at constant pressure back to state A. Show that the efficiency of the cycle is

$$1-\frac{T_1(p_2T_1/p_1T_2)^{\gamma-1}-T_2}{\gamma\{T_1-T_2(p_1/p_2)^{(\gamma-1)/\gamma}\}},$$

where γ (assumed constant) is the ratio of the specific heat at constant pressure to the specific heat at constant volume.

(London Univ., B.Sc. Spec. I, 1963)

With reference to the indicator diagram of Fig. 6.7, the known quantities are p_1, T_1, p_2, T_2. Further, $v_1=RT_1/p_1$, $v_2=RT_2/p_2$. Now AB is an adiabatic and along this portion of the cycle no heat enters or leaves the system. Hence the change is homentropic and since the working substance is a perfect gas,

$$p_1v_1^{\gamma}=p_Bv_2^{\gamma}.$$

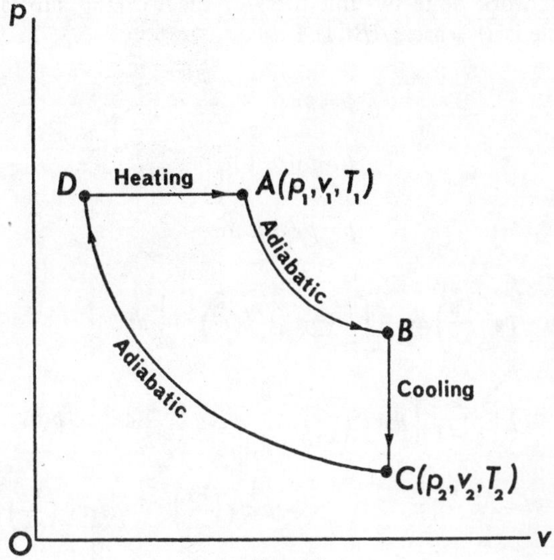

FIG. 6.7

Thus

$$\left\{\begin{aligned} p_B &= p_1\left(\frac{v_1}{v_2}\right)^{\gamma} = p_1\left(\frac{T_1p_2}{T_2p_1}\right)^{\gamma}, \\ v_B &= v_2 = \frac{RT_2}{p_2}, \\ T_B &= \frac{p_Bv_B}{R} = \left(\frac{T_1p_2}{T_2p_1}\right)^{\gamma}\frac{p_1v_2}{R} = \left(\frac{T_1p_2}{T_2p_1}\right)^{\gamma}\frac{p_1T_2}{p_2} = \left(\frac{p_1T_1}{p_1T_2}\right)^{\gamma-1}T_1. \end{aligned}\right.$$

Similarly

$$p_1v_D{}^{\gamma} = p_2v_2{}^{\gamma}$$

and

$$\left\{\begin{aligned} v_D &= \left(\frac{p_2}{p_1}\right)^{1/\gamma}v_2 = \left(\frac{p_2}{p_1}\right)^{1/\gamma}\frac{RT_2}{p_2}, \\ p_D &= p_1, \\ T_D &= \frac{p_Dv_D}{R} = \left(\frac{p_2}{p_1}\right)^{1/\gamma}\left(\frac{p_1T_2}{p_2}\right) = \left(\frac{p_1}{p_2}\right)^{(\gamma-1)/\gamma}T_2. \end{aligned}\right.$$

The heat per unit mass absorbed from the hot source along the constant pressure line DA is

$$Q = C_p(T_1 - T_D) = \left(\frac{\gamma R}{\gamma-1}\right)\left\{T_1 - \left(\frac{p_1}{p_2}\right)^{(\gamma-1)/\gamma}T_2\right\}.$$

The useful work done by unit mass of the working substance in the complete cycle is W=area $ABCDA$.

On DA,

$$p=p_1$$

On AB,

$$p=p_1(v_1/v)^\gamma.$$

On CD,

$$p=p_2(v_2/v)^\gamma.$$

Hence

$$W=\int_{v_D}^{v_1}\left(p_1-p_2\cdot\frac{v_2^\gamma}{v^\gamma}\right)dv+\int_{v_1}^{v_2}(p_1v_2^\gamma-p_2v_2^\gamma)\cdot\frac{1}{v^\gamma}dv$$

$$=p_1(v_1-v_D)+\frac{1}{\gamma-1}\left\{p_2v_2\left(\frac{v_2}{v_1}\right)^{\gamma-1}-p_2v_2\left(\frac{v_2}{v_D}\right)^{\gamma-1}+p_1v_1\right.$$

$$\left.-p_2v_2\left(\frac{v_2}{v_1}\right)^{\gamma-1}-p_1v_1\left(\frac{v_1}{v_2}\right)^{\gamma-1}+p_2v_2\right\}$$

$$=RT_1-R\left(\frac{p_2}{p_1}\right)^{1/\gamma}\left(\frac{p_1}{p_2}\right)T_2+\frac{R}{\gamma-1}\left\{-T_2\left(\frac{p_1}{p_2}\right)^{(\gamma-1)/\gamma}\right.$$

$$\left.+T_1-T_1\left(\frac{p_2T_1}{p_1T_2}\right)^{\gamma-1}+T_2\right\}$$

$$=R\left\{T_1-\left(\frac{p_1}{p_2}\right)^{1/\gamma}\left(\frac{p_1}{p_2}\right)T_2\right\}+\frac{R}{\gamma-1}\left\{-T_2\left(\frac{p_1}{p_2}\right)^{(\gamma-1)/\gamma}+T_1\right.$$

$$\left.-T_1\left(\frac{p_2T_1}{p_1T_2}\right)^{\gamma-1}+T_2\right\}$$

$$=R\left\{T_1-\left(\frac{p_1}{p_2}\right)^{(\gamma-1)/\gamma}T_2\right\}+\frac{R}{\gamma-1}\left\{T_2-T_1\left(\frac{p_2T_1}{p_1T_2}\right)^{\gamma-1}\right.$$

$$\left.+T_1-T_2\left(\frac{p_1}{p_2}\right)^{(\gamma-1)/\gamma}\right\}$$

$$=\frac{\gamma}{\gamma-1}R\left\{T_1-\left(\frac{p_1}{p_2}\right)^{(\gamma-1)/\gamma}T_2\right\}+\frac{R}{\gamma-1}\left\{T_2-T_1\left(\frac{p_2T_1}{p_1T_2}\right)^{\gamma-1}\right\}.$$

Thus

$$\eta=\frac{W}{Q}=1+\frac{T_2-T_1(p_2T_1/p_1T_2)^{\gamma-1}}{\gamma\{T_1-(p_1/p_2)^{(\gamma-1)/\gamma}T_2\}}.$$

(*N.B.* The indicator diagram of Fig. 6.7 is that for the diesel cycle.)

EXERCISE 6

1. The first law of thermodynamics can be written in the form

$$\Delta Q = dU + p\,dv.$$

Explain briefly the significance of the terms of this equation. Deduce that the entropy per unit mass of a perfect gas, of density ρ and at temperature T, is $c_v \log (T/\rho^{\gamma-1})$ assuming that its specific heats c_v, c_p are constant.

Two samples of such a gas having the same total mass m but at different temperatures T_1, T_2 are contained in equal but separate compartments of a box with adiabatic walls. They are separated by a thin diaphragm and initially the samples are at rest. The diaphragm is then broken, the samples mix and ultimately the gas comes to rest. Prove that the resulting increase of entropy in the universe is $mc_v \log \{(T_1+T_2)^2/(4T_1T_2)\}$.

(Durham Univ., B.Sc. Hons., 1962)

2. State the first law of thermodynamics.

Explain what is meant by a thermodynamic property, and by using the fact that entropy is a thermodynamic property prove that the specific internal energy u of a gas, obeying the van der Waals' equation of state satisfies the condition $(\partial u/\partial v)_T = a/v^2$.

Show that the specific heat of the gas at constant volume c_v is a function of temperature only and assuming c_v to be constant calculate u in terms of v and T.

An insulated cylinder contains a partition on each side of which is a unit mass of a van der Waals gas, for which c_v is constant. The temperatures in both sides have the common value T_0 but the pressures and volumes are unequal. The partition is suddenly removed and after the ensuing motion has ceased the temperature of the gas is T_1. Prove that $T_1 < T_0$.

(Bristol Univ., B.Sc. Spec. II, 1962)

3. Given that the equation of state of a certain gas is

$$pv = RT(1 + b/v),$$

prove that the internal energy E, expressed as a function of the entropy S and the volume v, is given by

$$E = f(S - R \log v + bR/v),$$

where f is an arbitrary function of its argument.

(Imperial College, B.Sc. Spec., 1949)

4. Neglecting second and higher powers of the constants a and b in van der Waals' equation of state

$$(p + a/v^2)(v - b) = RT,$$

obtain the following equation:

$$\left(\frac{\partial T}{\partial p}\right)_H = \frac{(2a/RT)-b}{C_p},$$

where C_p is the specific heat at constant pressure.

*Sketch the family of isenthalpic curves showing the temperature T as a function of the pressure p for constant heat content H.

(Imperial College, B.Sc. Spec., 1949)

5. Define the Helmholtz free energy and the Gibbs function of a gas and calculate the changes in these quantities when the system undergoes an infinitesimal reversible process from one equilibrium state to another.

Prove that

(i)

$$\left(\frac{\partial u}{\partial v}\right)_T + p = T\left(\frac{\partial p}{\partial T}\right)_v,$$

(ii)

$$c_p - c_v = T\left(\frac{\partial p}{\partial T}\right)_v \left(\frac{\partial v}{\partial T}\right)_p.$$

The equation of state of a gas is

$$pv/RT = 1 + B'p,$$

where B' is a function of temperature only. Calculate the specific entropy of the gas (apart from an arbitrary added function of the temperature) and show that in the case when $B' = b - a/T^2$, where a and b are absolute constants,

$$c_v = (c_v)_0 - 2bRp - Rp^2\{b + (a/T^2)\}^2,$$
$$c_p = (c_p)_0 + 2aRp/T^2,$$

where $(c_v)_0$ and $(c_p)_0$ are the respective values of these quantities at very low pressures.

(Bristol Univ., B.Sc. Spec., 1961)

6. A perfect gas, for which C_v and C_p are constants, is enclosed in an upright adiabatically insulating cylinder by the weight of a frictionless adiabatically insulating piston, and is in equilibrium with pressure p_1, volume V_1, and temperature T_1. A weight is placed on the piston, and the gas ultimately comes to a new equilibrium pressure p_2, volume V_2, and temperature T_2. Find the expression for the increase in internal energy of the gas and deduce that

$$\frac{T_2}{T_1} = \frac{1 + \lambda(\gamma - 1)}{\gamma},$$

where $\lambda = p_2/p_1$, and $\gamma = C_p/C_v$.

(London Univ., B.Sc. Spec. Hons. I, 1964)

7.* If U denotes the internal energy of a substance, S the entropy, $H(=U+pv)$ the enthalpy, $F(=U-TS)$ the free energy, $G(=U+pv-TS)$ the Gibbs function, establish the following results:

(i)

$$U=F-T\left(\frac{\partial F}{\partial T}\right)_v=H-p\left(\frac{\partial H}{\partial p}\right)_S;$$

(ii)

$$H=G-T\left(\frac{\partial G}{\partial T}\right)_p;$$

(iii)

$$\left(\frac{\partial H}{\partial p}\right)_T=v-T\left(\frac{\partial v}{\partial T}\right)_p.$$

8.* For a perfect gas prove that $\frac{\partial(T,S)}{\partial(p,v)}=1$, the independent variables being p, v.

9.* Show, that $C_p-C_v=-T\left(\frac{\partial v}{\partial T}\right)_p^2\left(\frac{\partial p}{\partial v}\right)_T$.

7

GAS DYNAMICS

7.1 Compressibility Effects in Real Fluids

When a body is moving at slow speeds through a medium such as air, the variation of the air density in most regions not too near the surface of the body is small and the methods of analysis of classical hydrodynamics for ideal fluids are applicable. Within the proximity of the surface, however, viscous effects are called into play. However, at very high speeds compressibility of the air takes effect and the methods used for ideal fluids are no longer applicable at any part of the field of air flow that is likely to be of interest. Thus for high-speed flow of a compressible fluid it is necessary that all equations expressing mass and energy conservation, momentum change, entropy change, etc., take into account the variable density of the medium. Such analyses often invoke the equation of state of the gas and, possibly, equations expressing isentropic or homentropic flow.

Another feature of compressible fluids is their ability to support wave motion. Thus we shall see in Section 7.3 that a small disturbance in a compressible fluid is propagated throughout the fluid as a wave motion and at a speed called the *speed of sound.* In an incompressible fluid these features are not realized: a small disturbance applied at any point of such a fluid is transmitted instantaneously throughout the whole field of the fluid. This was assumed in Section 3.11 where impulsive motion was studied.

7.2 The Elements of Wave Motion

Before investigating wave propagation in compressible fluids we first detail some of the more salient features of wave motion. We first treat wave motion in one dimension and then generalize the results to propagation in two or three dimensions.

7.2.1 The One-Dimensional Wave Equation

We start by considering the meaning of the function $\varphi(x, t)$ defined to be

$$\varphi(x, t) = f(x - ct), \tag{1}$$

where x signifies a distance associated with a time t, and c is a constant. From dimensional considerations, c must have the dimensions of speed if equation (1) is to have any physical significance. We note that in virtue of (1),

$$\varphi(x + cT, t + T) = f\{(x + cT) - c(t + T)\},$$

so that

$$\varphi(x + cT, t + T) = \varphi(x, t). \tag{2}$$

Equation (2) means that *the value of* φ *at distance* x *at time* t *is the same as that at distance* $(x + cT)$ *at time* $(t + T)$. In Fig. 7.1(i), φ is plotted against x at time t: an arbitrary shape is drawn. Figure 7.1(ii) shows the plot of φ against x at time $(t + T)$. Clearly the second figure gives a profile congruent to the first but shifted along the positive x-axis a distance cT. Since the movement takes place in time T, we infer that the profile moves with velocity c in the direction of the positive x-axis. Thus we have shown that a plausible physical interpretation of equation (1) is that φ *represents a disturbance moving in the positive* x*-direction with velocity* c*, its profile remaining undistorted.* This is a mathematical model of a one-dimensional wave propagation without distortion.

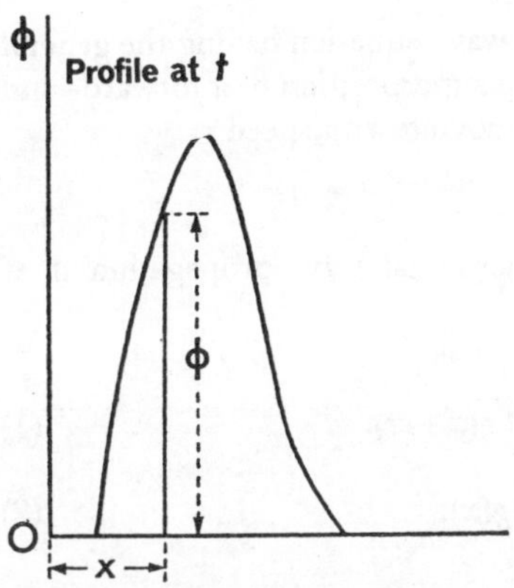

Fig. 7.1(i)

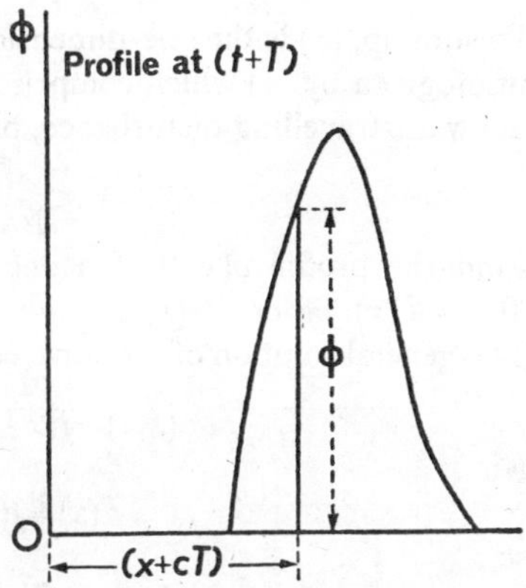

Fig. 7.1(ii)

In the same way we may show that

$$\varphi(x, t) = g(x + ct), \tag{3}$$

represents a disturbance moving without distortion in the negative x*-direction with speed* c*.*

Clearly the combination

$$\varphi(x, t) = f(x - ct) + g(x + ct) \tag{4}$$

represents the superposition of a forward- and a backward-travelling wave, each moving with speed c. By elementary partial differentiation, starting from (4), we can easily show that

$$\frac{\partial^2 \varphi}{\partial t^2} = \frac{c^2 \partial^2 \varphi}{\partial x^2}. \tag{5}$$

Equation (5) is known as the *one-dimensional wave equation*. The form of φ given by (4) is its general solution. This solution may be formally obtained by the following method due to d'Alembert. Put

$$\xi = x - ct, \quad \eta = x + ct.$$

Then it is easily shown that (5) reduces to

$$\partial^2 \varphi / \partial \xi \partial \eta = 0.$$

Thus

$$\partial \varphi / \partial \eta = \psi(\eta),$$

and so

$$\begin{aligned} \varphi &= \int \psi(\eta)\, d\eta + f(\xi) \\ &= g(\eta) + f(\xi) \\ &= g(x + ct) + f(x - ct). \end{aligned}$$

To sum up, (5) is the one-dimensional wave equation having the general solution given by (4) which comprises the superposition of a forward- and a backward-travelling disturbance, both moving with speed c.

Example

Find the profile $\varphi(x, t)$ of a one-dimensional wave propagation if at $t = 0$, $\varphi = F(x)$, $\partial \varphi / \partial t = G(x)$.

The general solution of the wave equation is

$$\varphi(x, t) = f(x - ct) + g(x + ct). \tag{1}$$

Then

$$F(x) = f(x) + g(x). \tag{2}$$

Also

$$\partial \varphi / \partial t = -cf'(x - ct) + cg'(x + ct),$$

so that

$$G(x) = -cf'(x) + cg'(x). \tag{3}$$

From (3)

$$\frac{1}{c} \int_0^x G(x)\, dx = -cf(x) + cg(x) + 2A. \tag{4}$$

Solving (2), (4) for $f(x), g(x)$ gives

$$f(x)=\tfrac{1}{2}F(x)-\frac{1}{2c}\int_0^x G(x)\,dx+A, \tag{5}$$

$$g(x)=\tfrac{1}{2}F(x)+\frac{1}{2c}\int_0^x G(x)\,dx-A. \tag{6}$$

From the last two relations

$$f(x-ct)=\tfrac{1}{2}F(x-ct)-\frac{1}{2c}\int_0^{x-ct} G(x)\,dx+A, \tag{7}$$

$$g(x+ct)=\tfrac{1}{2}F(x+ct)+\frac{1}{2c}\int_0^{x+ct} G(x)\,dx-A. \tag{8}$$

Substitution into (1) now gives the required solution:

$$\varphi(x,t)=\tfrac{1}{2}\{F(x-ct)+F(x+ct)\}+\frac{1}{2c}\int_{x-ct}^{x+ct} G(x)\,dx.$$

7.2.2 Wave Equations in Two and in Three Dimensions

If a disturbance takes place in three dimensions in such a way that the disturbance is constant over any plane perpendicular to the direction of propagation, then the wave is termed a *plane wave* and any such plane is called a *wave-front*. If such a wave is travelling with speed c in a direction specified by the unit vector $\mathbf{n}=[l,m,n]$, then the function $f(lx+my+nz-ct)$ satisfies these requirements since the wave fronts have equations $lx+my+nz=$const. at any considered time. Similarly $g(lx+my+nz+ct)$ would represent a disturbance travelling in the direction $-\mathbf{n}$ with the same speed. Hence the function

$$\varphi(x,y,z,t)=f(lx+my+nz-ct)+g(lx+my+nz+ct) \tag{1}$$

represents the superposition of plane waves travelling with speeds c in the directions $\pm\mathbf{n}$. By forming the second-order derivatives of φ w.r.t. x, y, z, t and using the fact that $l^2+m^2+n^2=1$ we arrive at the *wave equation in three dimensions*:

$$\nabla^2\varphi=\frac{\partial^2\varphi}{\partial x^2}+\frac{\partial^2\varphi}{\partial y^2}+\frac{\partial^2\varphi}{\partial z^2}=\frac{1}{c}\,\frac{\partial^2\varphi}{\partial t^2}. \tag{2}$$

Equation (1) gives the general solution of (2). The solution (1) may also be written in the form

$$\varphi=f(\mathbf{n}\cdot\mathbf{r}-ct)+g(\mathbf{n}\cdot\mathbf{r}+ct). \tag{1'}$$

In two dimensions the wave equation is

$$\frac{\partial^2\varphi}{\partial x^2}+\frac{\partial^2\varphi}{\partial y^2}=\frac{1}{c^2}\,\frac{\partial^2\varphi}{\partial t^2}$$

with general solution

$$\varphi = f(lx+my-ct)+g(lx+my+ct),$$

where $l^2+m^2=1$.

7.2.3 Spherical Waves

Let us now consider the three-dimensional wave equation

$$\nabla^2\varphi = \frac{1}{c^2}\frac{\partial^2\varphi}{\partial t^2}$$

in spherical polar coordinates (r, θ, ψ). Its form becomes

$$\frac{\partial^2\varphi}{\partial r^2}+\frac{2}{r}\frac{\partial\varphi}{\partial r}+\frac{1}{r^2\sin\theta}\frac{\partial}{\partial\theta}\left(\sin\theta\frac{\partial\varphi}{\partial\theta}\right)+\frac{1}{r^2\sin\theta}\frac{\partial^2\varphi}{\partial\psi^2}=\frac{1}{c^2}\frac{\partial^2\varphi}{\partial t^2}.$$

If there is spherical symmetry so that $\varphi=\varphi(r,t)$, this last equation reduces to

$$\frac{\partial^2\varphi}{\partial r^2}+\frac{2}{r}\frac{\partial\varphi}{\partial r}=\frac{1}{c^2}\frac{\partial^2\varphi}{\partial t^2}$$

or

$$\frac{\partial^2}{\partial r^2}(r\varphi)=\frac{1}{c^2}\frac{\partial^2}{\partial t^2}(r\varphi).$$

The general solution of the one-dimensional wave equation gives the solution of this last equation for $r\varphi$:

$$r\varphi = f(r-ct)+g(r+ct)$$

or

$$\underline{\varphi(r,t)=(1/r)\{f(r-ct)+g(r+ct)\}}.$$

The solution obtained here represents concentric spherical wave fronts, centre O, and having radii which increase or decrease at speed c. Here, the wave profiles change because of the factor $1/r$ in the solution.

7.2.4 Progressive and Stationary Waves

So far we have treated various types of wave equations whose solutions represent the superposition of wave-fronts travelling in opposite directions with speed c. In each case the wave profile remains unaltered, save in the case of spherical waves where it is diminished by the factor $1/r$. Such waves—plane or spherical—are called *progressive* because of their movement in some direction.

Reverting to the one-dimensional equation of Section 7.2.1, let us take the special forms

$$f(x-ct)=a\cos p(t-x/c),\quad g(x+ct)=a\cos p(t+x/c),$$

where a, p, c are constants. Then the functions $f(x-ct)$, $g(x+ct)$ correspond to *harmonic waves* and the equation (4) of that section gives

$$\varphi(x, t)=a \cos p(t-x/c)+a \cos p(t+x/c)=2a \cos (px/c) \cdot \cos (pt).$$

This type of disturbance is known as a *stationary* wave, because its profile does not move. Thus at all times, $\varphi=0$ at the fixed positions where $x=(k+\frac{1}{2})\pi c/p$ (k an integer). These determine the *nodes* or positions of zero displacement. The points where $x=kc\pi/p$ (k an integer) determine the *antinodes* or positions of maximum displacement. In both forms $f(x-ct)$, $g(x+ct)$, the *amplitude* is a. The *period* of each, and also of φ, is $2\pi/p$, the *angular frequency* is p and the *frequency in cycles per unit time* is $p/(2\pi)$. If, keeping t constant, we increase or decrease x by an amount $2\pi c/p$ or by whole-number multiples of this amount, then in both cases f, g, φ all remain unaltered. The quantity $2\pi c/p$ is called the *wavelength* of the harmonic wave or of the combination of harmonic waves which φ represents. We observe that

Velocity of harmonic wave = Frequency in cycles/sec. × Wavelength.

Also, we observe that the amplitude of φ is $2a \cos pt$, which varies with time.

The reader will readily see that if, in the forms for f and g, the cosines are replaced by sines, then similar results follow. Thus we may represent a progressive harmonic wave in one dimension by

$$f=a \begin{matrix}\cos \\ \sin\end{matrix}\, p(t \mp x/c),$$

where a is the amplitude, c the speed of propagation, $2\pi/p$ the frequency in cycles per unit time and $2\pi c/p$ the wavelength. When the negative sign is taken, the wave travels in the positive x-direction; when the positive sign is taken, it travels in the negative x-direction. We observe that the cosine and sine forms are the real and imaginary parts of the complex quantity

$$a \exp \{ip(t \mp x/c)\},$$

where $i=\sqrt{(-1)}$ and all other quantities are real. This affords a convenient way of representing a progressive harmonic wave. We make use of these forms in Section 9.12.2.

In physics periodic or harmonic wave motions are of two kinds: transverse and longitudinal. For *transverse waves*, the vibrations occur in planes at right-angles to the direction of propagation. Such cases occur in the propagation of electromagnetic waves, including the special case of light waves. When the vibrations occur in a specific plane, the wave is said to be *plane polarized*. For *longitudinal waves*, the vibrations occur in the direction of propagation. In the next section we study the propagation of sound waves, which are longitudinal. In Chapter 9 we meet both types.

7.3 The Speed of Sound in a Gas

Let us suppose that a small disturbance is created within a non-viscous gas such that:

(i) The motion is irrotational. Then there exists a velocity potential φ such that the fluid velocity at any point is $\mathbf{q} = -\nabla\varphi$.

(ii) The squares and products of all disturbances from the equilibrium state specified by p_0, ρ_0 can be neglected. Also q is so small that q^2 is negligible.

(iii) The isentropic law $p = k\rho^\gamma$ holds. This follows in consequence of assuming that changes take place so rapidly that heat exchanges and hence entropy changes are negligible.

We write $\rho = \rho_0(1+s)$, where s is called the *condensation*. Then the equation of continuity $(\partial\rho/\partial t) + \nabla\cdot(\rho\mathbf{q}) = 0$ becomes

$$(\partial s/\partial t) - \nabla\cdot\{(1+s)\nabla\varphi\} = 0,$$

or, if we assume the velocity $-\nabla\varphi$ to be so small in magnitude that $s\nabla\varphi$ is negligible,

$$\partial s/\partial t = \nabla^2\varphi. \tag{1}$$

Under zero body force the equation of motion becomes

$$\int\frac{dp}{\rho} - \frac{\partial\varphi}{\partial t} = \text{const.}, \tag{2}$$

neglecting $\mathbf{q}^2$. Taking

$$p = (p_0/\rho_0{}^\gamma)\rho^\gamma,$$

$$\frac{dp}{d\rho} = \frac{\gamma p_0}{\rho_0{}^\gamma}\rho^{\gamma-1} = \frac{\gamma p_0}{\rho_0}\left(\frac{\rho}{\rho_0}\right)^{\gamma-1} = \frac{\gamma p_0}{\rho_0}(1+s)^{\gamma+1} \simeq a_0^2,$$

where

$$a_0^2 = \gamma p_0/\rho_0. \tag{3}$$

Hence

$$\int\frac{dp}{\rho} = \int\left(\frac{dp}{d\rho}\cdot\frac{d\rho}{\rho}\right) \simeq a_0^2\int\frac{d\rho}{\rho} = a_0^2\log\rho + \text{const.} \tag{4}$$

From (2), (4),

$$\begin{aligned}\partial\varphi/\partial t &= a_0^2\log\rho + \text{const.}\\ &= a_0^2\{\log\rho_0 + \log(1+s)\} + \text{const.}\\ &\simeq a_0^2 s + \text{const.}, \quad \text{to the first order.}\end{aligned}$$

Thus, absorbing the constant into s gives

$$\partial\varphi/\partial t = a_0^2 s. \tag{5}$$

Eliminating s from (1), (5) now gives

$$\partial^2\varphi/\partial t^2 = a_0^2\nabla^2\varphi. \tag{6}$$

Equation (6) is a wave type of equation. It shows that small disturbances are propagated in the gas with speed

$$a_0 = \left(\frac{dp}{d\rho}\right)^{1/2} = \left(\frac{\gamma p_0}{\rho_0}\right)^{1/2}. \tag{7}$$

This speed is called the *speed of sound* in the gas. A vibrating tuning fork would produce disturbances propagating with such a velocity, though *any* small disturbance—audible or not—would travel through the gas at this speed.

The equation (7) was devised under isentropic conditions. When we wish to emphasize this we write

$$a^2 = (\partial p/\partial \rho)_S.$$

7.4 Equations of Motion of a Gas

In Chapter 3 we obtained the equation of continuity

$$(\partial\rho/\partial t) + \nabla \cdot (\rho \mathbf{q}) = 0$$

for a moving compressible fluid and also the equation of motion

$$\frac{d\mathbf{q}}{dt} = \mathbf{F} - \left(\frac{1}{\rho}\right) \nabla p.$$

In the case of steady motion under no body forces, these give

$$\nabla \cdot (\rho \mathbf{q}) = 0, \tag{1}$$

$$(\mathbf{q} \cdot \nabla)\mathbf{q} = -(1/\rho)\nabla p. \tag{2}$$

Also, for such flow, Bernoulli's equation becomes

$$\tfrac{1}{2}\mathbf{q}^2 + \int \frac{dp}{\rho} = \text{const.} \tag{3}$$

In the special case of isentropic flow—for which the entropy of each particle stays constant along any streamline— $p = k\rho^\gamma$ and (3) reduces to

$$\tfrac{1}{2}\mathbf{q}^2 + a^2/(\gamma - 1) = \text{const.} \tag{3'}$$

In (3′) the constant is the same along any streamline, but unless the flow is homentropic it will vary from one streamline to another. The condition for constant entropy of a fluid particle in its motion along a streamline is expressed by the equation

$$(\mathbf{q} \cdot \nabla)S = 0 \tag{4}$$

since for steady flow $\mathbf{q} \cdot \nabla$ is the operation of time differentiation following the fluid particle. In addition the equation of state may be taken in either of the equivalent forms

$$f(p, \rho, T) = 0 \tag{5}$$

or

$$p = F(\rho, S), \tag{5'}$$

where the forms of f, F are known. Equation (5′) is more convenient for discussing cases of isentropic and homentropic flows.

In the above, equations (1), (2), (4) and (5) or (5′) are distinct equations for finding $p, \rho, \mathbf{q}, S$. Bernoulli's equation is really derived from the equation of motion, but the forms (3), (3′) are very useful, as we shall see. Thus the problem of gas flow is determinable.

Since $(\mathbf{q}\cdot\nabla)\mathbf{q}=\nabla(\frac{1}{2}\mathbf{q}^2)-\mathbf{q}\wedge\boldsymbol{\zeta}$, scalar multiplying (2) through by $\mathbf{q}$ gives

$$\mathbf{q}\cdot\nabla(\tfrac{1}{2}\mathbf{q}^2)=-(1/\rho)\mathbf{q}\cdot\nabla p.$$

Using the equation of state in the form (5′), we have

$$\begin{aligned}\nabla p&=\textstyle\sum\mathbf{i}(\partial p/\partial x)\\&=\sum\mathbf{i}\left\{\left(\frac{\partial p}{\partial\rho}\right)_S\frac{\partial\rho}{\partial x}+\left(\frac{\partial p}{\partial S}\right)_\rho\frac{\partial S}{\partial x}\right\}\\&=a^2\nabla\rho+(\partial p/\partial S)_\rho\nabla S.\end{aligned}$$

For homentropic flow, $\nabla p=a^2\nabla\rho$ and (2) becomes

$$\mathbf{q}\cdot\nabla(\tfrac{1}{2}\mathbf{q}^2)=-(a^2/\rho)\,(\nabla\rho)\cdot\mathbf{q}$$

From (1),

$$\rho\nabla\cdot\mathbf{q}+(\nabla\rho)\cdot\mathbf{q}=0$$

and so

$$\mathbf{q}\cdot\nabla(\tfrac{1}{2}\mathbf{q}^2)=a^2\nabla\cdot\mathbf{q}. \tag{6}$$

Equation (6) is an important result to which we shall return later.

7.5 Subsonic, Sonic and Supersonic Flows

Let q be the speed of a gas at a certain location and let a be the local speed of sound, where $a=(dp/d\rho)^{1/2}=(\gamma p/\rho)^{1/2}$. Then the local *Mach number* M is the dimensionless parameter defined by the relation

$$M=q/a.$$

When $q=a$, $M=1$ and the flow is said to be *sonic*, since the speed of flow and the speed of sound are the same. When $M<1$, $q<a$ and the flow is *subsonic*, i.e. the speed of flow of the gas is less than the local speed of sound in it. When $M>1$, $q>a$ and the flow is *supersonic*, i.e. the speed of flow of the gas is greater than the local speed of sound.

Subsonic and supersonic flows have many different physical features. It is thus important to know the Mach number to discern what type of flow pattern is realized. We examine these physical features by discussing the nature of spherical sound waves in a moving stream of gas.

First consider a source O emitting sound waves in a gas at rest (Fig. 7.2(i)). Spherical wave fronts centred on O travel outwards from O, and at time t after starting from O the disturbance is spread uniformly over the surface S of the sphere centre O and radius at.

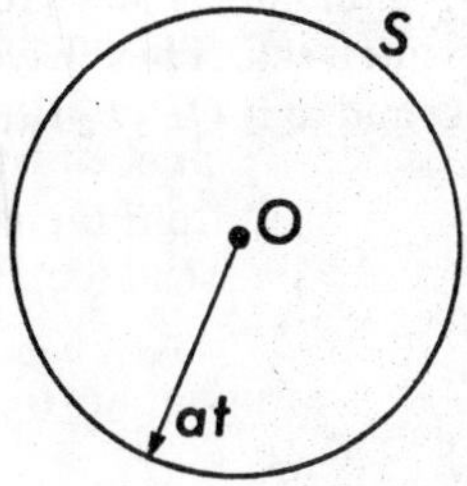

FIG. 7.2(i)

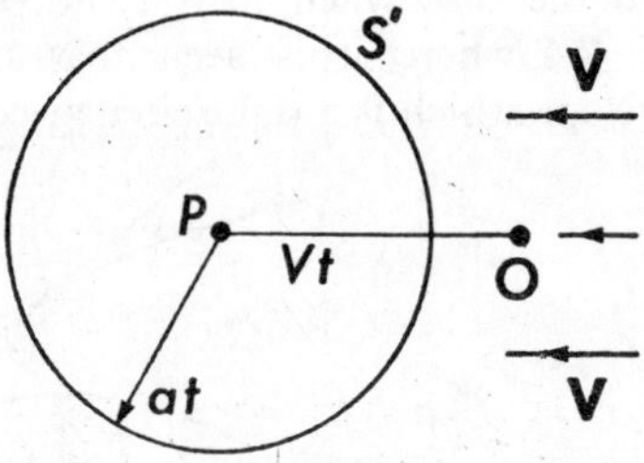

FIG. 7.2(ii)

Now suppose the gas flows with uniform velocity $\mathbf{V}$ past the source. Then at time t, every particle of S is displaced a distance $\mathbf{V}t$ relative to S' and the disturbance which was initially at O is now on the surface S' of a sphere centre P and radius at where $\overline{OP} \equiv \mathbf{V}t$ (Fig. 7.2(ii)). When $M<1$, $V<a$ and O lies within S'. When $M>1$, $V>a$ and O lies outside S'. Let us consider each of these in turn.

When $M<1$, $P_1, P_2, P_3, \ldots$ denote the centres of the spherical disturbance at times $t, 2t, 3t \ldots$, the radii of the corresponding spheres being $at, 2at, 3at, \ldots$. Further, $\overline{OP_1} \equiv \mathbf{V}t$, $\overline{OP_2} \equiv 2\mathbf{V}t$, $\overline{OP_3} \equiv 3\mathbf{V}t, \ldots$. Figure 7.3 illustrates the case. It is seen that the disturbances at times $t, 2t, 3t, \ldots$ are on the boundaries of non-intersecting spheres.

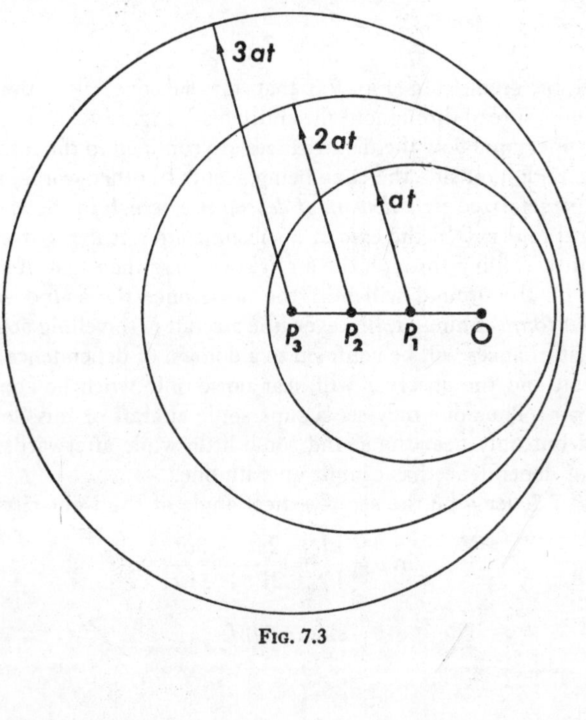

FIG. 7.3

In the case when $M>1$, the corresponding situation is depicted in Fig. 7.4 where it is seen that the spheres intersect. They have an envelope which is a right circular cone, vertex O and axis $OP_1P_2P_3 \ldots$.

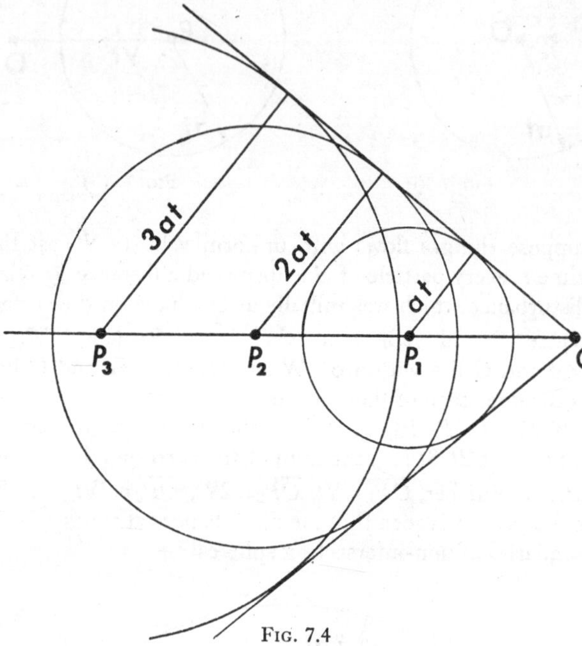

FIG. 7.4

It will be seen from Fig. 7.3 that for subsonic flow the spherical disturbances spread throughout the entire field. Figure 7.4 shows, however, that for supersonic flow the disturbances are confined to the interior of the cone, the region outside the cone being silent. In other words, supersonic flow is characterized by a *domain of dependence*, which in the above case is the conical interior. In the case of a subsonic aircraft flying overhead (i.e. an aircraft travelling through the air at speed less than that of sound) any observer on the ground will hear the noise once the sound waves have spread out to meet him. If, however, the aircraft is travelling at supersonic speed, disturbances will be confined to a domain of dependence relative to the aircraft and the observer will hear noise only when he comes within this domain. Thus one may see a supersonic aircraft or missile travelling overhead but only hear the sound some little while afterwards when the domain of dependence has caught up with one.

In Fig. 7.3, let μ be the semi-vertical angle of the right-circular cone. Then

$$\sin \mu = \frac{at}{Vt} = \frac{2at}{2Vt} = \frac{3at}{3Vt} \ldots$$

or

$$\sin \mu = 1/M.$$

The angle μ is called the *Mach angle*; it is real only when $M \geqslant 1$ and non-existent for subsonic flow. The cone is termed a *Mach cone.*

In two dimensions, the spheres in the above models become circles and the cone becomes a pair of *Mach lines* or *Mach waves.* When uniform supersonic flow takes place parallel to a plane wall with a small excrescence, as may occur in a supersonic wind tunnel, a Mach wave forms at the excrescence as shown in Fig. 7.5. Such a wave may be discerned by special photographic processes. The Mach wave is set at the Mach angle μ to the stream.

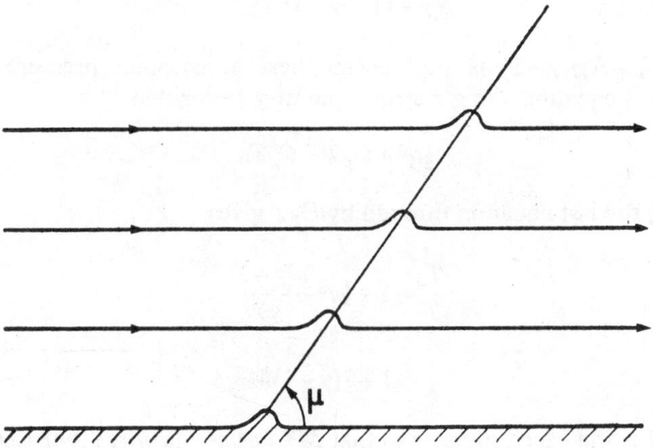

FIG. 7.5

The term *transonic flow* features in aerodynamics and may be taken somewhat arbitrarily to cover the range of Mach numbers $0{\cdot}8 < M < 1{\cdot}2$. *Hypersonic flow* takes place at very high Mach numbers, say $M > 5$.

7.6 Isentropic Gas Flow

In Section 7.4 we saw that for isentropic flow of a gas

$$\tfrac{1}{2}\mathbf{q}^2 + a^2/(\gamma - 1) = \text{const. along a streamline.}$$

The L.H.S. of this equation shows that the maximum value of q is attained whenever $a = 0$: denote this value by q_{max}. This is entirely a theoretical value which cannot obtain in practice. Also we introduce the *critical speed of sound* a_* defined to be the value of a when $q = a$. Then Bernoulli's equation of isentropic flow along a streamline may be written in the forms

$$\tfrac{1}{2}\mathbf{q}^2 + \frac{a^2}{\gamma - 1} = \tfrac{1}{2}q_{max}^2 = \frac{\gamma + 1}{2(\gamma - 1)}a_*^2 = \frac{a_0^2}{\gamma - 1},$$

where a_0 is the value of a on the streamline when $q=0$, i.e. it is the *stagnation speed of sound.* Another form of this equation is

$$\tfrac{1}{2}\mathbf{q}^2+\frac{\gamma p}{\rho(\gamma-1)}=\frac{\gamma p_0}{\rho_0(\gamma-1)},$$

the suffix 0 referring to stagnation conditions.

Now, for a perfect gas,

$$\frac{\gamma p}{\rho(\gamma-1)}=\frac{\gamma RT}{\gamma-1}=C_pT,$$

where $C_p=R\gamma/(\gamma-1)$ is the specific heat at constant pressure. Thus Bernoulli's equation along a streamline may be written

$$\tfrac{1}{2}\mathbf{q}^2+C_pT=C_pT_0.$$

Dividing the last equation through by C_pT gives

$$\frac{T_0}{T}=1+\tfrac{1}{2}q^2\frac{\rho(\gamma-1)}{\gamma p}$$

$$=1+\tfrac{1}{2}(\gamma-1)M^2,$$

where M is the Mach number so that $M^2=q^2/a^2=(q^2\rho/\gamma p)$. For isentropic flow we have the relations

$$\frac{a_0^2}{a^2}=\left(\frac{\gamma p_0}{\rho_0}\right)\div\left(\frac{\gamma p}{\rho}\right)=\frac{T_0}{T}=\left(\frac{p_0}{p}\right)^{(\gamma-1)/\gamma}=\left(\frac{\rho_0}{\rho}\right)^{\gamma-1}$$

so that

$$p_0/p=\{1+\tfrac{1}{2}(\gamma-1)M^2\}^{\gamma/(\gamma-1)},$$
$$\rho_0/\rho=\{1+\tfrac{1}{2}(\gamma-1)M^2\}^{1/(\gamma-1)}.$$

Other isentropic relations, which are easily found from Bernoulli's equation, are

$$\frac{T}{T_0}=1-\tfrac{1}{2}(\gamma-1)\left(\frac{u}{a_0}\right)^2,$$

$$\frac{T}{T_0}=1-\frac{\gamma-1}{\gamma+1}\left(\frac{u}{a_*}\right)^2,$$

which yield the following relations for p/p_0 and ρ/ρ_0:

$$\frac{p}{p_0}=\left\{1-\tfrac{1}{2}(\gamma-1)\left(\frac{u}{a_0}\right)^2\right\}^{\gamma/(\gamma-1)},$$

$$\frac{\rho}{p_0}=\left\{1-\tfrac{1}{2}(\gamma-1)\left(\frac{u}{a_0}\right)^2\right\}^{1/(\gamma-1)}$$

$$\frac{p}{p_0}=\left\{1-\frac{\gamma-1}{\gamma+1}\left(\frac{u}{a_*}\right)^2\right\}^{\gamma/(\gamma-1)},$$

$$\frac{\rho}{\rho_0}=\left\{1-\frac{\gamma-1}{\gamma+1}\left(\frac{u}{a_*}\right)^2\right\}^{1/(\gamma-1)}.$$

At sonic or critical speeds ($M=1$), $p=p_*$, $\rho=\rho_*$, $T=T_*$ and so

$$\frac{p_*}{p_0}=\left(\frac{2}{\gamma+1}\right)^{\gamma/(\gamma-1)}\simeq 0{\cdot}528 \text{ for air } (\gamma\simeq 1{\cdot}400);$$

$$\frac{\rho_*}{\rho_0}=\left(\frac{2}{\gamma+1}\right)^{1/(\gamma-1)}\simeq 0{\cdot}630 \text{ for air};$$

$$\frac{T_*}{T_0}=\frac{2}{\gamma+1}\simeq 0{\cdot}833 \text{ for air.}$$

7.7 Reservoir Discharge through a Channel of Varying Section

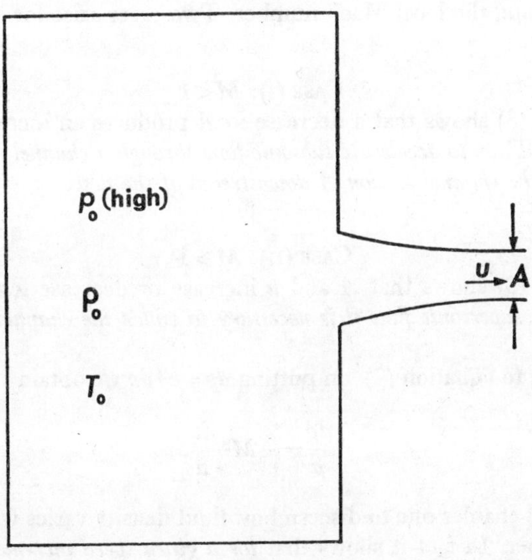

FIG. 7.6

Figure 7.6 depicts a reservoir containing stationary gas at high pressure p_0, density ρ_0, temperature T_0. An open-ended axially-symmetric channel is fitted to the reservoir and we assume that the gas discharges steadily and isentropically into the air at the open section where the pressure is less than p_0. Let the section of the channel vary so slowly that to a first order of approximation the velocity is constant across any section (though, of course, varying from section to section). Then to all intents and purposes the flow is one-dimensional.

Let p be the pressure, ρ the density, u the gas velocity at a location of the channel of section A. For steady flow the equation of continuity is $\rho uA = \text{const.}$ and, on differentiating this logarithmically,

$$\frac{d\rho}{\rho}+\frac{du}{u}+\frac{dA}{A}=0. \tag{1}$$

Bernoulli's equation,

$$\tfrac{1}{2}u^2+\int\frac{dp}{\rho}=\text{const.},$$

in differential form is

$$u\,du+(dp/\rho)=0. \tag{2}$$

Putting $dp=a^2\,d\rho$ in (2) and eliminating $d\rho/\rho$ from (1), (2) gives

$$(1-M^2)\frac{du}{u}=-\frac{dA}{A}, \tag{3}$$

where $M=u/a$, the local Mach number. Two cases arise for discussion.

CASE (I): $M<1$

Equation (3) shows that a decrease in A produces an increase in u and conversely. Thus *to accelerate subsonic flow through a channel it is necessary to decrease the channel section A downstream of the flow.*

CASE (II): $M>1$

Equation (3) shows that A and u increase or decrease together. Thus *to accelerate supersonic flow it is necessary to widen the channel downstream of the flow.*

Reverting to equation (2), on putting $dp=a^2\,d\rho$ we obtain

$$\frac{d\rho}{\rho}=-M^2\frac{du}{u}. \tag{4}$$

Equation (4) enables one to discern how fluid density varies with changing Mach number. In fact it shows that *for a given speed increment there is a density drop whose magnitude increases with increasing Mach number*

Further, for $M \geqslant 1$ the drop in density is so great that the channel must expand to satisfy continuity requirements.

From equation (3), if $dA/A=0$, then either $M=1$ or $du=0$. The case $du=0$ is realized in incompressible flow where the speed of the flow reaches a maximum at the stage when the channel section attains a minimum area of cross-section. For compressible fluids, M may be unity when the section A is a minimum.

To summarize the above results we may say that *if, starting with subsonic flow in the channel, we decrease the section downstream, then the flow is accelerated until the section has attained a certain minimum at which the Mach number is unity. If beyond this minimum section we now widen the channel, then the flow can be accelerated downstream of the section to produce supersonic flow.* This illustrates the principle of *flow through a nozzle.* The minimum section is termed the *throat.* (See Fig. 7.7.)

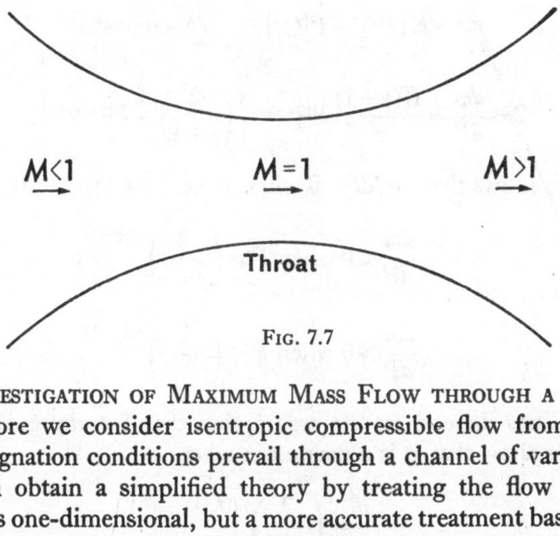

FIG. 7.7

7.7.1 INVESTIGATION OF MAXIMUM MASS FLOW THROUGH A NOZZLE

As before we consider isentropic compressible flow from a reservoir where stagnation conditions prevail through a channel of varying section. We again obtain a simplified theory by treating the flow through the channel as one-dimensional, but a more accurate treatment based on axially symmetric flows that are supersonic can be made by the methods of Section 7.9.5. The channel is assumed to taper steadily to a minimum section at the outlet.

Let A be the section at the outlet of the channel where the velocity is u, pressure p, density ρ. Then applying Bernoulli's equation along a streamline from the reservoir to the section A,

$$\tfrac{1}{2}u^2+\left(\frac{\gamma}{\gamma-1}\right)\frac{p}{\rho}=\left(\frac{\gamma}{\gamma-1}\right)\frac{p_0}{\rho_0},$$

so that

$$u=\left\{\frac{2\gamma}{\gamma-1}\cdot\frac{p_0}{\rho_0}\left(1-\frac{p}{p_0}\cdot\frac{\rho_0}{\rho}\right)\right\}^{1/2}.$$

Since $\rho_0/\rho=(p_0/p)^{1/\gamma}$,

$$u=\left[\frac{2\gamma}{\gamma-1}\cdot\frac{p_0}{\rho_0}\left\{1-\left(\frac{p}{p_0}\right)^{(\gamma-1)/\gamma}\right\}\right]^{1/2}. \tag{1}$$

The mass flux across the section A per unit time is

$$m=\rho uA=\rho_0 uA(p/p_0)^{1/\gamma}. \tag{2}$$

Eliminating u from (1), (2) gives

$$m=A\left[\frac{2\gamma}{\gamma-1}p_0\rho_0\left(\frac{p}{p_0}\right)^{2/\gamma}\left\{1-\left(\frac{p}{p_0}\right)^{(\gamma-1)/\gamma}\right\}\right]^{1/2}. \tag{3}$$

To find the stationary values of m for fixed A and variable p/p_0, write $P=p/p_0$ so that

$$m^2=K(P^{2/\gamma}-P^{(\gamma+1)/\gamma}). \quad (K=\text{const.}). \tag{4}$$

From (4),

$$2m\frac{dm}{dP}=\frac{K(\gamma+1)}{\gamma}P^{(2-\gamma)/\gamma}\left\{\left(\frac{2}{\gamma+1}\right)-P^{(\gamma-1)/\gamma}\right\}. \tag{5}$$

Equation (5) shows that $dm/dP=0$ when $P=\{2/(\gamma+1)\}^{\gamma/(\gamma-1)}$;

$$\frac{dm}{dP}<0 \text{ when } P>\left(\frac{2}{\gamma+1}\right)^{\gamma/(\gamma-1)};$$

$$\frac{dm}{dP}>0 \text{ when } P<\left(\frac{2}{\gamma+1}\right)^{\gamma/(\gamma-1)}.$$

The inequalities depend on the assumption that for any gas $\gamma>1$. The above results show that m is a maximum when

$$\frac{p}{p_0}=\left(\frac{2}{\gamma+1}\right)^{\gamma/(\gamma-1)} \tag{6}$$

and

$$m_{\max}=A\left\{\left(\frac{2\gamma}{\gamma+1}\right)p_0\rho_0\left(\frac{2}{\gamma+1}\right)^{2/(\gamma-1)}\right\}^{1/2}. \tag{7}$$

Equation (6) means that $p=p_*$ (see end of Section 7.6) and hence that *for maximum isentropic mass flow, conditions at the exit plane are sonic.*

From (3), (7),

$$\frac{m}{m_{\max}}=\left(\frac{\gamma+1}{\gamma-1}\right)^{1/2}\left\{\tfrac{1}{2}(\gamma+1)^{1/(\gamma-1)}\left(\frac{p}{p_0}\right)^{1/\gamma}\right\}\left\{1-\left(\frac{p}{p_0}\right)^{(\gamma-1)/\gamma}\right\}^{1/2} \tag{8}$$

which enables the curve of Fig. 7.8 to be sketched. Figure 7.8 shows that as p/p_0 is decreased from 1 to p_*/p_0, the ratio m/m_{max} increases from zero to a maximum of unity corresponding to sonic conditions at the outlet. Mathematically, m/m_{max} then decreases as p/p_0 is further decreased, but this is not realized physically. It is obvious that any reduction of outlet pressure can never result in a decrease of mass flow. Hence we infer that the basic assumptions on which the above analysis has been developed break down. In fact for $p<p_*$ the flow through the channel no longer remains isentropic. For such a case the pressure must be p_* at some location between the reservoir and the outlet. At such a point $M=1$. This can occur only at the outlet where the section is of minimum area, the flow being accelerated from subsonic to sonic at the outlet by tapering the channel. What happens is that at the outlet a plane discontinuity or *shock wave* is formed, immediately upstream of which the pressure is p_* and immediately downstream of which $p<p_*$. Across the shock wave there is an entropy gain and the pressure, density, temperature, etc., are discontinuous across it. These features will appear in the more detailed study of shock waves which will be made in Section 7.8.

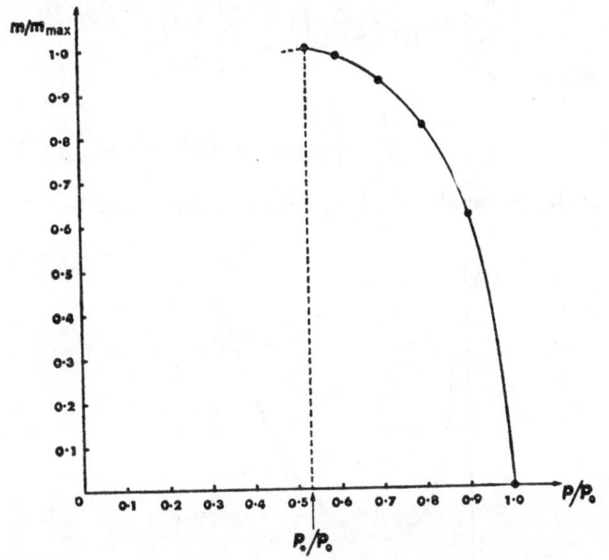

FIG. 7.8

The above treats the case where the channel tapers to a minimum section at the outlet. Now suppose that a diverging channel is fitted on to this to produce a nozzle of the type shown in Fig. 7.7. The former outlet section is now the throat of the nozzle and we have sonic conditions at the throat when the rate of mass flow is a maximum. Further, upstream of the throat the flow is subsonic and downstream of it the flow is supersonic. Let m_{max}

denote the maximum rate of mass flow. In fact equation (3) of this section gives the area of any section where the pressure is p in the form

$$A=\frac{m_{\max}}{\left[\frac{2\gamma}{\gamma-1}p_0\rho_0\left(\frac{p}{p_0}\right)^{2/\gamma}\left\{1-\left(\frac{p}{p_0}\right)^{(\gamma-1)/\gamma}\right\}\right]^{1/2}}. \tag{9}$$

Thus if A_* denotes the area of the throat section where conditions are sonic, we find

$$\left(\frac{A}{A_*}\right)^2=\frac{(p_*/p_0)^{2/\gamma}\{1-(p_*/p_0)^{(\gamma-1)/\gamma}\}}{(p/p_0)^{2/\gamma}\{1-(p/p_0)^{(\gamma-1)/\gamma}\}}.$$

Using

$$p/p_0=\{1+\tfrac{1}{2}(\gamma-1)M^2\}^{-\gamma/(\gamma-1)},$$

and

$$p_*/p_0=\{2/(\gamma+1)\}^{\gamma/(\gamma-1)},$$

we find on reduction

$$\left(\frac{A}{A_*}\right)^2=\frac{1}{M^2}\left\{\frac{2}{\gamma+1}\left(1+\frac{\gamma-1}{2}M^2\right)\right\}^{(\gamma+1)/(\gamma-1)} \tag{10}$$

For air with $\gamma=1{\cdot}4$

$$\frac{A}{A_*}=\left(\frac{1}{M}\right)(\tfrac{5}{6}+\tfrac{1}{6}M^2)^3. \tag{10'}$$

Figure 7.9 shows the plot of A/A_* versus M for $\gamma=1{\cdot}4$.

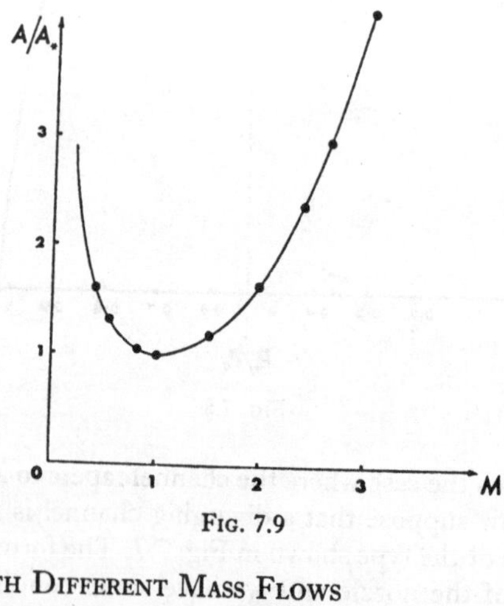

FIG. 7.9

7.7.2 Nozzle with Different Mass Flows

In the foregoing treatment we studied the maximum mass flow through a nozzle having minimum section at the throat at which conditions are sonic. Now suppose the mass flow is *subcritical* or less than the permissible maximum for isentropic conditions. Let A_m be the area of the throat section. Then for a prescribed subcritical rate of mass flow, we could design another nozzle of smaller throat section for which the flow would be a maximum if forced through this second nozzle. Let A_* denote the area of the throat of the second nozzle; at this section conditions would be sonic for the prescribed rate of mass flow. Thus $A_m/A_* > 1$ and the flow in the nozzle whose throat is of section A_m is everywhere subsonic. In this nozzle, the flow is first accelerated subsonically until it reaches the throat (where it does not become sonic) and then it is decelerated downstream of the throat. The pressure will decrease to a minimum at the throat and then increase again downstream of the throat. This is evident from the isentropic formula

$$p/p_0 = \{1 + \tfrac{1}{2}(\gamma - 1)M^2\}^{-\gamma/(\gamma-1)}.$$

When the flow is *critical* (i.e. the maximum mass rate of flow occurs consistent with the flow being isentropic throughout) the flow downstream of the throat (at which conditions are now sonic) may be either subsonic or supersonic, as illustrated in Figs. 7.10(i), (ii). In both cases the flow is isentropic throughout.

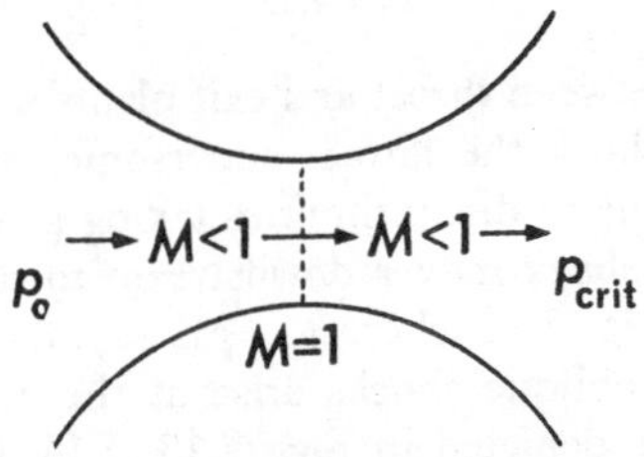

FIG. 7.10(i)

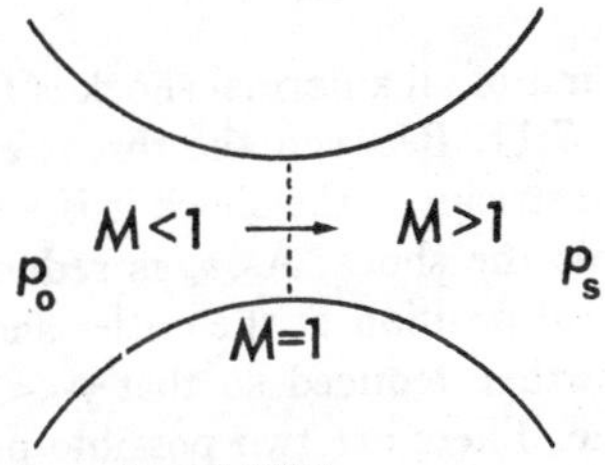

FIG. 7.10(ii)

The reason why both configurations are possible for critical flow may be seen from Fig. 7.9. Downstream of the throat let a section A be considered. Then $A/A_* > 1$. Figure 7.9 shows that for any such ratio A/A_*, there are two values of M, one subsonic, the other supersonic, and both corresponding to isentropic conditions. Downstream of the throat in the case of Fig. 7.10(i) the subsonic flow is decelerated isentropically by the widening nozzle section, but in Fig. 7.10(ii) the supersonic flow is accelerated isentropically by the widening section. The pressure at the exit plane of the nozzle (downstream of the throat) may be either p_{crit} in the case of subsonic flow downstream of the throat or p_s in the case of supersonic flow downstream of the throat. Then $p_s < p_{crit}$. In the case of supersonic flow, at a section A downstream of the throat, Fig. 7.9 shows there is but one value of $M > 1$ and corresponding to this there will be one pressure given by the isentropic formula $p/p_0 = \{1 + \tfrac{1}{2}(\gamma - 1)M^2\}^{-\gamma/(\gamma-1)}$.

The same remarks hold for the section in which $M<1$. Thus in the case of critical mass flow illustrated in Fig. 7.10(ii), there is only one velocity and one pressure at each section so long as the flow remains isentropic throughout. For subcritical isentropic flow, however, a continuous range of subsonic speeds may be produced at any section.

Let us next consider the case when the exit pressure p_e lies between p_s and p_{crit} so that $p_s<p_e<p_{crit}$. In this case the exit pressure cannot be reached isentropically. A plane shock wave will be formed normal to the direction of flow and situated between the throat and the exit plane, though it may be situated critically at the exit plane. These cases are depicted in Figs. 7.11, 7.12.

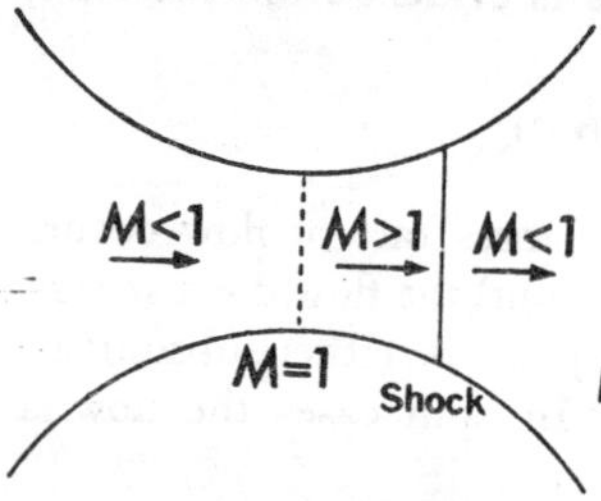

Fig. 7.11

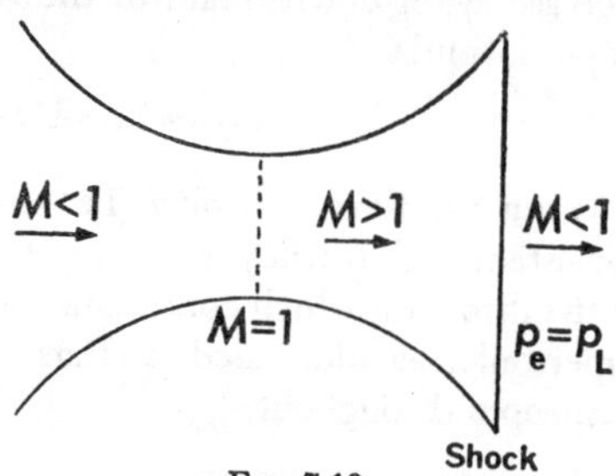

Fig. 7.12

First of all a normal shock is formed between throat and exit plane as in Fig. 7.11. Between the throat and the shock the flow is supersonic, but downstream of the shock it is subsonic due to discontinuities taking place across the shock. As p_e is reduced, this shock moves downstream to the critical position at the outlet shown in Fig. 7.12 where $p_e=p_L$, say. As p is further reduced so that $p_S<p_e<p_L$, oblique shocks arise at the exit plane. There are two possible patterns as depicted in Figs. 7.13, 7.14. In both cases reflections occur. Figure 7.13 shows the case of *simple reflection*: Fig. 7.14 that of *Mach reflection* in which a plane normal shock joins the oblique ones in the manner shown. We shall discuss reflection of shocks in Section 7.8.4.

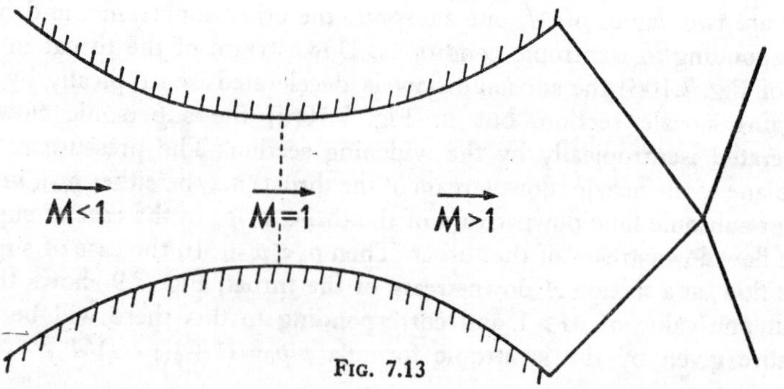

Fig. 7.13

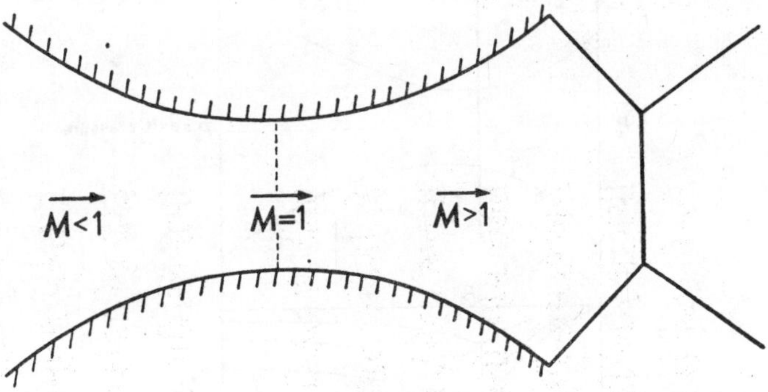

FIG. 7.14

When $p_e = p_s$ the flow is entirely supersonic in the nozzle downstream of the throat and isentropic as we have seen. The nozzle is shock-free.

Finally, for $p_e < p_s$ the flow is completely supersonic in the nozzle downstream of the throat and a train of expansion waves is set up at the exit plane. These expansion waves turn the flow away from the axis (Fig. 7.15).

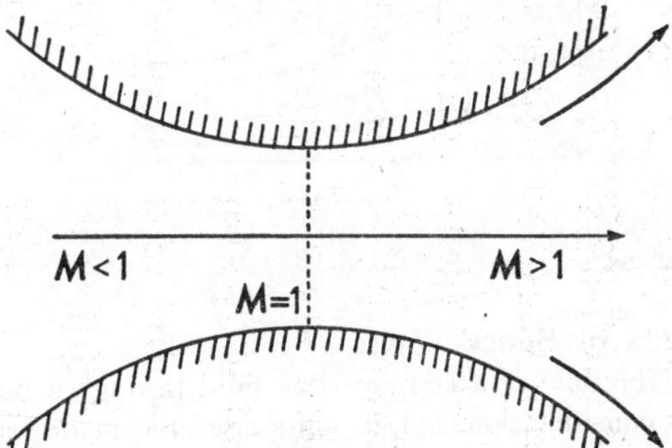

FIG. 7.15

Thus the complete discussion of flow through a nozzle is complicated. The diagram in 7.16 is an attempt to summarize the cardinal features of such flows. It shows the variations of pressure and Mach number along the axis of the nozzle for different mass flows. For subcritical flows, $M<1$ throughout and $p>p_*$ throughout. For maximum mass flow the isentrope (i.e. line of constant entropy) divides at $p=p_*$, $M=1$, the line joining p_* to p_{crit} being a subsonic isentrope, that joining p_* to p_S being a supersonic one. In between the two, jumps occur in the curves due to shock formation in the nozzle.

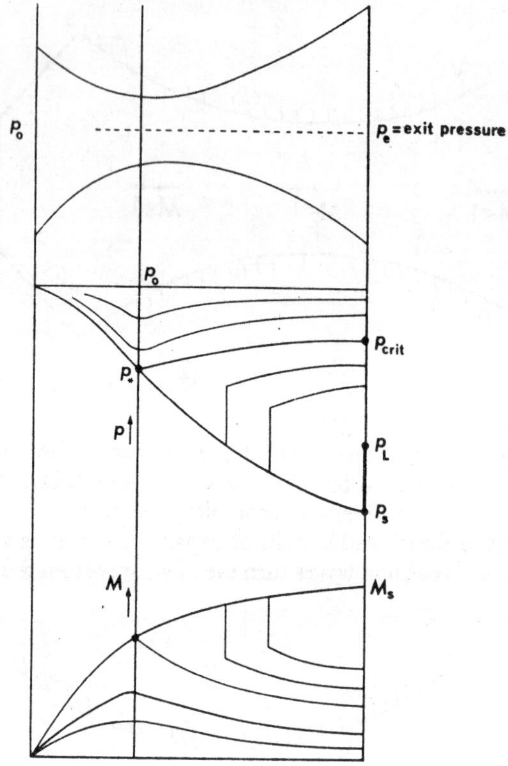

FIG. 7.16

7.8 Shock Waves

7.8.1 Formation of Shock Waves

In studying the flow of a compressible fluid through a nozzle we saw that continuity considerations led, in some cases, to the formation of plane shock waves across which there were discontinuities in pressure, density, temperature, etc. The principles of conservations of mass, momentum and energy still apply across these plane discontinuities, but we shall see that there is an entropy gain. Before embarking on detailed analysis of shock waves we instance some further cases of their formation. It should perhaps be pointed out that shock waves are not really waves in the sense outlined in Section 7.2. They are plane discontinuities, pulse-like in nature and are sometimes more appropriately called *shock fronts.*

Figure 7.17 shows a piston being driven with uniform velocity u into a long open tube—a so-called *shock tube.* The gas particles within the vicinity of the piston acquire the uniform velocity u but those some way ahead of the piston are at rest. A plane normal discontinuity or shock front travels forwards with velocity $U(>u)$ into the virgin gas as the piston advances into the tube. The shock is the mechanism by which the gas between it and the piston acquires the velocity u. The existence of the shock can be detected experimentally by certain delicate kinds of photographic methods (Schlieren, shadowgraph, etc.). The velocity $U>a$, the local speed of sound in the fluid. A simple physical explanation of the shock formation in this case is the following. Suppose we approximate the continuous motion of the piston by a set of forward-moving pulses, each of short duration. When the piston makes the first short movement forward,

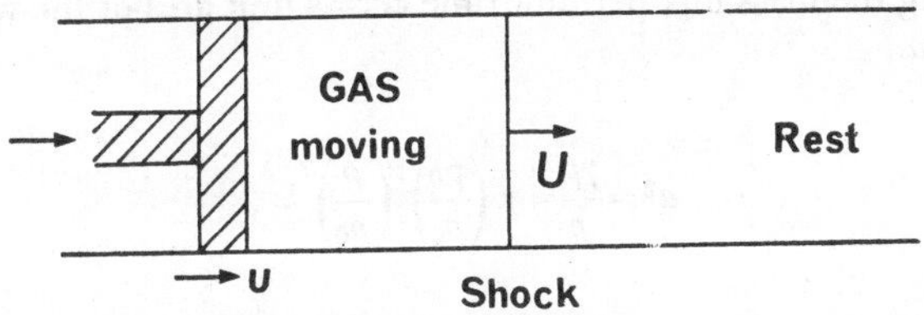

FIG. 7.17

a small disturbance is propagated forwards into the gas at the speed of sound. This small amplitude wave (or sound wave) heats the gas slightly and since $a \propto \sqrt{T}$, where T is the temperature, the second pulse will be propagated as another sound wave at a speed slightly in excess of the first one. Similarly the third pulse will be propagated at a speed slightly in excess of the second and so on. Thus the discreet pulses cause a train of sound waves of ever increasing velocity to be propagated through the gas. The tendency is for the faster moving rearmost waves to catch up with the slower moving foremost ones. In so doing the sound waves coalesce to form a more powerful shock front moving at a speed U in excess of the local speed of sound.

The above is a simplified model explaining shock formation in the tube when the piston is actuated with constant velocity. It is a helpful way of understanding the formation of the shock front but is open to the objection that the piston's motion is approximated to a system of short discrete pulses. Let us, then, look at the problem rather more analytically.

First of all consider a sound wave of velocity a moving into a gas at rest (Fig. 7.18(i)). The pressure ahead of the wave is p, the density ρ and the particle velocity zero. The pressure immediately behind the wave is $p+\delta p$, the density $\rho+\delta\rho$ and the particle velocity δu, the quantities $\delta p, \delta\rho, \delta u$ being of small order since the disturbance is weak. Figure 7.18(ii) shows the equivalent model obtained when the sound wave is brought to rest by imposing a backward velocity of a on the entire system.

Considering Fig. 7.18(ii), if we apply the equation of continuity across the stationary wave, we obtain

$$\rho a = (\rho + \delta\rho)\,(a - \delta u),$$

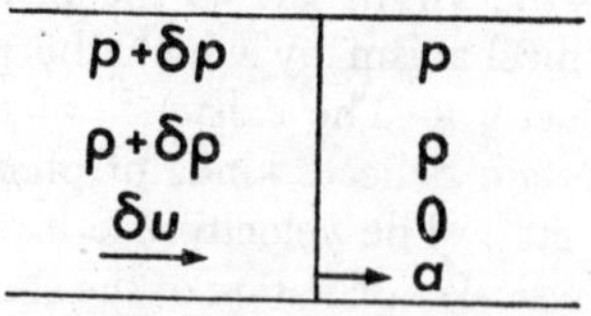

FIG. 7.18(i)

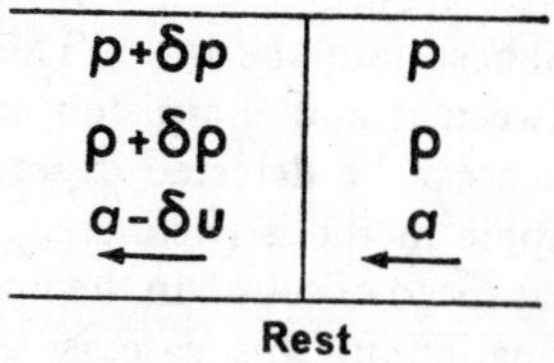

FIG. 7.18(ii)

considering the mass flux per unit time across unit area of the wave. Thus $d\rho/\rho = du/a$.

Now

$$a^2 = \frac{\gamma p}{\rho} = \gamma \left(\frac{p_0}{\rho_0}\right)\left(\frac{\rho}{\rho_0}\right)^{\gamma-1}$$

Thus

$$du = a\, d\rho/\rho = \gamma^{1/2} p_0{}^{1/2} \rho_0{}^{-\gamma/2} \rho^{(\gamma-3)/2}\, d\rho,$$

and so

$$u = \int_{\rho_0}^{\rho} \frac{du}{d\rho}\, d\rho = \frac{2}{\gamma-1} \cdot \frac{\gamma^{1/2} p_0^{1/2}}{\rho_0^{\gamma/2}} \left(\rho^{(\gamma-1)/2} - \rho_0^{(\gamma-1)/2}\right)$$

$$= \left(\frac{2a_0}{\gamma-1}\right)\left\{\left(\frac{\rho}{\rho_0}\right)^{(\gamma-1)/2} - 1\right\}.$$

Hence

$$(\rho/\rho_0)^{(\gamma-1)/2} = 1 + \tfrac{1}{2}(\gamma-1)\,(u/a_0), \tag{1}$$

where a_0 is the speed of sound in the undisturbed gas. Now

$$\frac{a^2}{a_0^2} = \left(\frac{\gamma p}{\rho}\right)\left(\frac{\rho_0}{\gamma p_0}\right) = \left(\frac{\rho_0}{\rho}\right)\left(\frac{p}{p_0}\right) = \left(\frac{\rho}{\rho_0}\right)^{\gamma-1}.$$

Hence

$$a = a_0(\rho/\rho_0)^{(\gamma-1)/2} = a_0 + \tfrac{1}{2}(\gamma-1)u, \tag{2}$$

using (1). Now each small disturbance propagates itself at a velocity equal to the local speed of sound *relative to the fluid.* Thus if the fluid moves with velocity u,

Velocity of propagation of disturbance $= u + a = a_0 + \tfrac{1}{2}(\gamma+1)u$.

Hence in a short time interval, τ, the distance moved by the disturbance is

$$(u+a)\tau = \{a_0 + \tfrac{1}{2}(\gamma+1)u\}\tau. \tag{3}$$

Now suppose the velocity profile of the disturbance generated by the moving piston is as shown in Fig. 7.19(i).

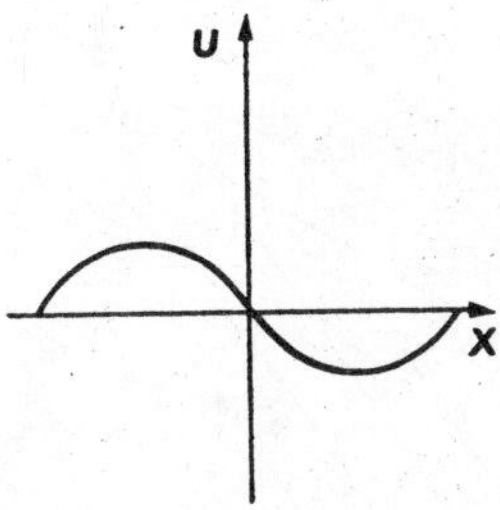

FIG. 7.19(i)

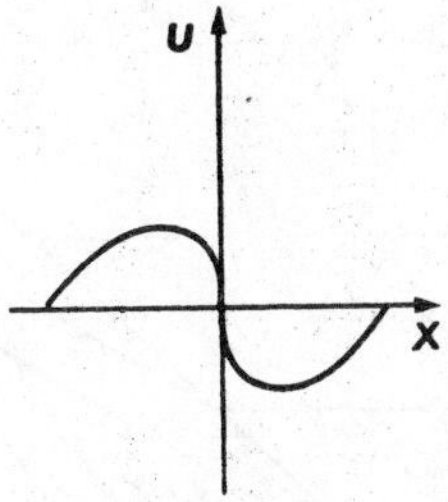

FIG. 7.19(ii)

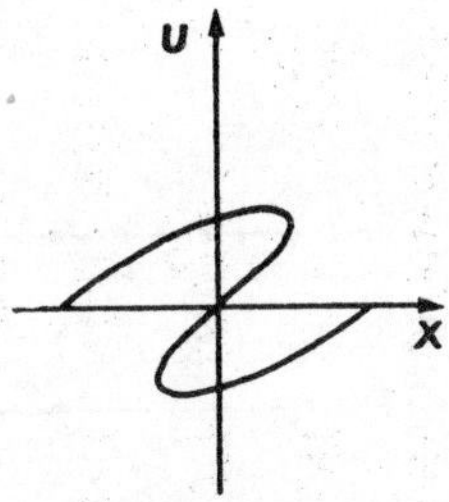

FIG. 7.19(iii)

Equation (3) shows that in a given interval τ, *the points of high velocity move farther to the right than those of low velocity*. Thus in Fig. 7.19(i) the crest has moved farther to the right than the trough and the profile is disturbed as in Fig. 7.19(ii). The steep-fronted compressive part of the wave will eventually become vertical. Further distortion takes place with lapse of time until the configuration depicted in Fig. 7.19(iii) is obtained, if our assumptions of isentropic flow are correct. This would make u a multi-valued function of x, which is physically impossible. Thus we expect the flow to cease to be isentropic and a shock wave to form.

The type of shock wave just considered is a *normal shock*—so-called because it is perpendicular to the incident gas stream. We now consider *oblique shocks* which are shocks inclined at oblique angles to the direction of flow. An oblique shock would result in superposing everywhere on a normal shock pattern a uniform velocity parallel to the shock front, so that velocity directions fore and aft of the front become obliquely inclined to the shock. An example of oblique shock formation is afforded by placing a wedge symmetrically in a uniform supersonic stream as in Fig. 7.20. Oblique shock waves are formed at the vertex. These turn the uniform supersonic incident flow in Region 1 so that downstream of each shock the flows in Regions 2, 3 are parallel to the faces of the wedge, the flows in these regions being uniform. The pattern of Fig. 7.20 can be observed very clearly by Schlieren photography.

It should be emphasized that for gases shock waves are compressive, i.e. in the passage of gas across a shock compression takes effect. This result will be established in Section 7.8.2. Save in the case of wet steam, expansion shocks are not realized for gaseous materials.* (For solid materials, however, compressive and expansive shocks are realized when the stresses enter the plastic range.)

Neglecting the viscosity of a fluid leads to the result that shock waves are plane discontinuities across which there are sudden jumps in pressure, density, temperature, etc. Actually it has been shown by taking account of

* See W. G. PENNEY and H. H. M. PIKE, *Shock Waves and the Propagation of Finite Pulses in Fluids*. The Physical Society Reports on Progress in Physics, Vol. XIII (1950).

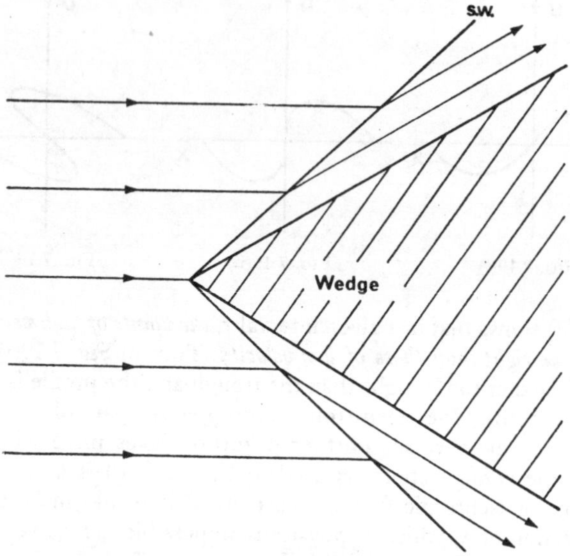

FIG. 7.20

the finite viscosity and thermal conductivity of the gas that in fact shocks are thin régimes through which the changes take place continuously but rapidly. The finite thickness of a shock zone is apparent from Schlieren photographs. It is of the order of a few mean free paths of the gas molecules. In this chapter we shall neglect viscous and thermal effects, but they will be taken into account in Section 9.13.

7.8.2 ELEMENTARY ANALYSIS OF NORMAL SHOCK WAVES

Here we treat the case of a stationary normal shock. If, as in the model of the piston advancing into the shock tube, the shock is not stationary, the problem can always be treated by reducing the shock to rest by imposing an equal and opposite velocity on to the entire system.

Figure 7.21 shows such a stationary shock front separating two uniform gaseous regions 1, 2. In Region 1, the pressure, density, temperature, particle velocity are respectively p_1, ρ_1, T_1, u_1 and a similar notation is used for the corresponding quantities in Region 2. We presume the elements in Region 1 are known, the problem being to find those in Region 2. To this end considerations of continuity, momentum and energy are invoked.

In Fig. 7.21, the mass flux of gas per unit time per unit area crossing the wave from Region 1 is $\rho_1 u_1$. That entering Region 2 is $\rho_2 u_2$. Since no gas is created or destroyed across the front

$$\rho_1 u_1 = \rho_2 u_2 = m, \quad \text{say.} \tag{1}$$

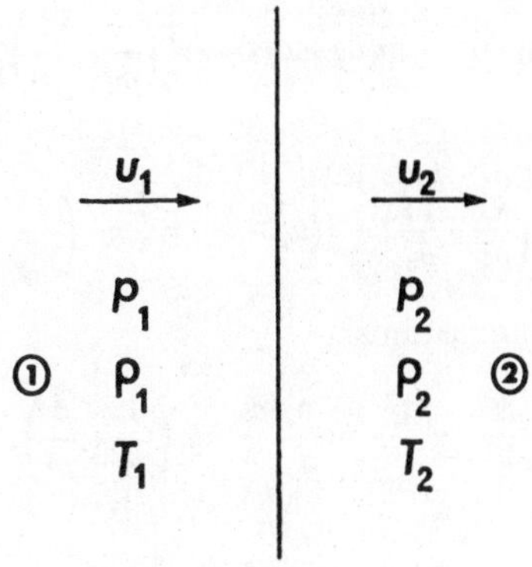

FIG. 7.21

Since the net force on unit area of the shock from 1 to 2 is $p_1 - p_2$ and the momentum gain per unit time across unit area from 1 to 2 is $m(u_2 - u_1)$,

$$p_1 - p_2 = m(u_2 - u_1)$$

or

$$p_1 + mu_1 = p_2 + mu_2. \tag{2}$$

Now the rate at which the pressures do work on the mass flux m is $p_1u_1 - p_2u_2$. This is equal to the total rate of gain of energy of m. Now the total energy per unit mass of gas in Region 1 is due to two causes: kinetic and internal. These give the respective contributions $\frac{1}{2}u_1^2$ and $C_vT_1 = RT_1/(\gamma-1) = p_1/\{\rho_1(\gamma-1)\}$. Thus the total change of energy per unit mass in crossing from 1 to 2 is

$$\left(\tfrac{1}{2}u_2^2 + \frac{p_2}{(\gamma-1)\rho_2}\right) - \left(\tfrac{1}{2}u_1^2 + \frac{p_1}{(\gamma-1)\rho_1}\right),$$

and so the total rate of change applied to the mass flux m is m times the above. Hence, by conservation of energy,

$$p_1u_1 - p_2u_2 = m\left(\tfrac{1}{2}u_2^2 + \frac{p_2}{(\gamma-1)\rho_2}\right) - m\left(\tfrac{1}{2}u_1^2 + \frac{p_1}{(\gamma-1)\rho_1}\right).$$

Since $m = \rho_1u_1 = \rho_2u_2$, this last equation can be written in the symmetrical form

$$\tfrac{1}{2}u_1^2 + \frac{\gamma}{\gamma-1}\frac{p_1}{\rho_1} = \tfrac{1}{2}u_2^2 + \frac{\gamma}{\gamma-1}\frac{p_2}{\rho_2}. \tag{3}$$

The equations (1)–(3) suffice to find p_2, ρ_2, u_2 in terms of p_1, ρ_1, u_1. The temperature T_2 could then be found from the equation of state $p_2 = R\rho_2T_2$. From (1), (2),

$$p_1 - p_2 = m(u_2 - u_1) = m^2\left(\frac{1}{\rho_2} - \frac{1}{\rho_1}\right).$$

From (1), (3),

$$\frac{\gamma}{\gamma-1}\left(\frac{p_1}{\rho_1} - \frac{p_2}{\rho_2}\right) - \tfrac{1}{2}(u_2^2 - u_1^2) = \tfrac{1}{2}m^2\left(\frac{1}{\rho_2^2} - \frac{1}{\rho_1^2}\right).$$

Dividing the last two equations,

$$\frac{\gamma}{\gamma-1}\cdot\frac{p_1/\rho_1 - p_2/\rho_2}{p_1 - p_2} = \tfrac{1}{2}\left(\frac{1}{\rho_2} + \frac{1}{\rho_1}\right),$$

whence

$$\frac{p_2}{p_1} = \frac{(\gamma-1)\rho_1 - (\gamma+1)\rho_2}{(\gamma-1)\rho_2 - (\gamma+1)\rho_1}, \tag{4}$$

or

$$\frac{\rho_2}{\rho_1} = \frac{(\gamma-1)p_1 + (\gamma+1)p_2}{(\gamma+1)p_1 + (\gamma-1)p_2}. \tag{5}$$

Equations (4), (5) are called the *Rankine–Hugoniot relations.* We shall assume the gas to be compressed across a shock. This will be established later from entropy considerations and the second law of thermodynamics. For $\gamma = 1{\cdot}4$, equation (4) gives

$$\frac{p_2}{p_1} = \frac{1 - 6(\rho_2/\rho_1)}{(\rho_2/\rho_1) - 6}$$

and Fig. 7.22 shows the plot of p_2/p_1 versus ρ_2/ρ_1 for this case, together with the plot of the isentropic equation $p_2/p_1 = (\rho_2/\rho_1)^{1{\cdot}4}$ for $1 \leqslant \rho_2/\rho_1 \leqslant 6$. Clearly the gas cannot be compressed by a normal shock to a density more than six times that in the incident stream.

The equation (3) may be written in the form

$$\tfrac{1}{2}u_1^2 + a_1^2/(\gamma-1) = \tfrac{1}{2}u_2^2 + a_2^2/(\gamma-1). \tag{3'}$$

The form of this equation (3′) is exactly the same as Bernoulli's equation for compressible isentropic flow when there is no shock wave. Thus the critical speed a_* is the same on both sides of the shock and each side of equation (3′) is equal to $\{(\gamma+1)a_*^2/2(\gamma-1)\}$. Hence the same is also true of equation (3) and so this yields

$$\begin{cases} \dfrac{p_1}{\rho_1} = \dfrac{\gamma+1}{2\gamma}a_*^2 - \dfrac{\gamma-1}{2\gamma}u_1^2, \\[2ex] \dfrac{p_2}{\rho_2} = \dfrac{\gamma+1}{2\gamma}a_*^2 - \dfrac{\gamma-1}{2\gamma}u_2^2. \end{cases}$$

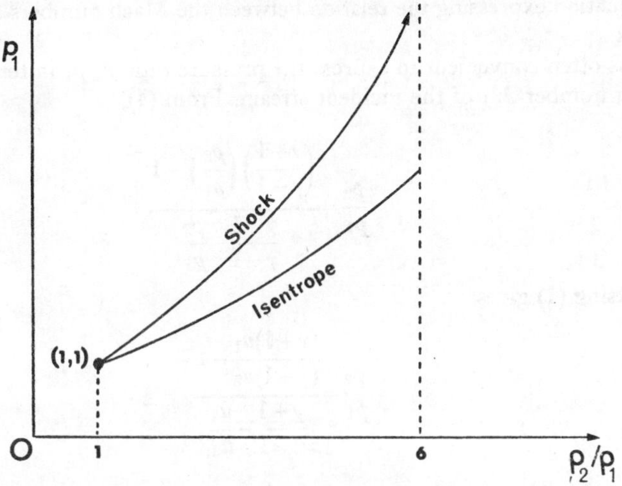

FIG. 7.22

From equation (2),

$$u_2-u_1=\frac{p_1-p_2}{m}=\frac{p_1}{\rho_1 u_1}-\frac{p_2}{\rho_2 u_2}$$

$$=\frac{\gamma+1}{2\gamma}\,a_*^2\left(\frac{1}{u_1}-\frac{1}{u_2}\right)+\frac{\gamma-1}{2\gamma}\,(u_2^2-u_1^2).$$

If we assume $u_2\neq u_1$, then, on simplification, this yields *Prandtl's relation*:

$$u_1u_2=a_*^2. \tag{6}$$

From $\frac{1}{2}u_1^2+\{a_1^2/(\gamma-1)\}=\{(\gamma+1)a_*^2/2(\gamma-1)\}$,
we have

$$u_1^2\left\{\tfrac{1}{2}+\frac{1}{(\gamma-1)M_1^2}\right\}=\frac{\gamma+1}{2(\gamma-1)}\,a_*^2\quad(M_1=u_1/a_1),$$

or

$$\left(\frac{u_1}{a_*}\right)^2=\frac{\gamma+1}{(\gamma-1)+(2/M_1^2)}.$$

Similarly

$$\left(\frac{u_2}{a_*}\right)^2=\frac{\gamma+1}{(\gamma-1)+(2/M_2^2)}\quad(M_2=u_2/a_2)$$

Multiplying together the last two equations and using (6), we obtain on simplification

$$M_2^2=\frac{1+\frac{1}{2}(\gamma-1)M_1^2}{\gamma M_1^2-\frac{1}{2}(\gamma-1)}, \tag{7}$$

an equation expressing the relation between the Mach numbers across the shock.

It is often convenient to express the pressure ratio p_2/p_1 in terms of the Mach number M_1 of the incident stream. From (4),

$$\frac{p_2}{p_1}=\frac{\left(\frac{\gamma+1}{\gamma-1}\right)\left(\frac{\rho_2}{\rho_1}\right)-1}{\frac{\gamma+1}{\gamma-1}-\frac{\rho_2}{\rho_1}}$$

and using (1) gives

$$\frac{p_2}{p_1}=\frac{\frac{(\gamma+1)u_1}{(\gamma-1)u_2}-1}{\frac{\gamma+1}{\gamma-1}-\frac{u_1}{u_2}}$$

$$=\frac{\left(\frac{\gamma+1}{\gamma-1}\right)\frac{u_1^2}{a_*^2}-1}{\frac{\gamma+1}{\gamma-1}-\frac{u_1^2}{a_*^2}}.$$

Using

$$\left(\frac{u_1}{a_*}\right)^2=\frac{\gamma+1}{(\gamma-1)+(2/M_1^2)},$$

on simplification we obtain

$$\frac{p_2}{p_1}=\frac{2\gamma M_1^2}{\gamma+1}-\frac{\gamma-1}{\gamma+1}. \tag{8}$$

Using (5), (8) there results

$$\frac{\rho_2}{\rho_1}=\frac{2+(\gamma-1)M_1^2}{(\gamma+1)M_1^2}. \tag{9}$$

We now discuss further implications of Prandtl's relation (6). The relation implies that

(i) if $u_1>a_*$, then $u_2<a_*$;
(ii) if $u_1<a_*$, then $u_2>a_*$.

It is necessary to decide which, if either, of these alternatives is valid. The Rankine–Hugoniot relations alone are insufficient to settle the matter. To this end it is necessary to use the fact that there must be entropy gain across a shock in virtue of the second law of thermodynamics. Let S_1, S_2

be the specific entropies per unit mass on each side of the shock. Then from the definition of Section 6.5 it is easily shown that

$$S_2 - S_1 = C_v \log\left\{\left(\frac{p_2}{p_1}\right)\cdot\left(\frac{\rho_1}{\rho_2}\right)^{\gamma}\right\}.$$

Using equations (8), (9), we find

$$\frac{S_2 - S_1}{C_v} = \log\left[\left\{\frac{2\gamma M_1^2 - (\gamma-1)}{\gamma+1}\right\}\left\{\frac{(\gamma-1)M_1^2+2}{(\gamma+1)M_1^2}\right\}^{\gamma}\right].$$

Put $M_1^2 = 1 + m$, where m is small. Then

$$\frac{S_2 - S_1}{C_v} = \log\left\{\left(1+\frac{2\gamma m}{\gamma+1}\right)\left(1+\frac{\gamma-1}{\gamma+1}m\right)^{\gamma}\left(1+m\right)^{-\gamma}\right\}$$

$$= \log\left(1+\frac{2\gamma m}{\gamma+1}\right) + \gamma\log\left(1+m\frac{\gamma-1}{\gamma+1}\right) - \gamma\log(1+m)$$

$$= \left(\frac{2\gamma}{\gamma+1}\right)m - \tfrac{1}{2}\left(\frac{2\gamma}{\gamma+1}\right)^2 m^2 + \tfrac{1}{3}\left(\frac{2\gamma}{\gamma+1}\right)^3 m^3 - \ldots$$

$$+\gamma\left\{\frac{\gamma-1}{\gamma+1}m - \tfrac{1}{2}\left(\frac{\gamma-1}{\gamma+1}\right)^2 m^2 + \tfrac{1}{3}\left(\frac{\gamma-1}{\gamma+1}\right)^3 m^3 - \ldots\right\}$$

$$-\gamma(m - \tfrac{1}{2}m^2 + \tfrac{1}{3}m^3 - \ldots)$$

$$= \tfrac{2}{3}\gamma(\gamma-1)(\gamma+1)^{-2}m^3 + \ldots.$$

Since $(S_2 - S_1)/C_v > 0$, $\gamma > 1$, this implies $m > 0$ and hence $M_1 > 1$. Hence

$$\left(\frac{u_1}{a_*}\right)^2 = \frac{\gamma+1}{(\gamma-1)+(2/M_1^2)} > \frac{\gamma+1}{(\gamma-1)+2} = 1$$

or $u_1 > a_*$, and so $u_2 < a_*$.

Thus case (i) holds and case (ii) is false. It is similarly shown that $M_2 < 1$. In other words we have *shown that for a stationary normal shock, the upstream flow is supersonic but passage through the shock reduces it to subsonic.*

With $M_1 > 1$, (8) shows $p_2 > p_1$ and (9) shows $\rho_2 > \rho_1$. Hence *in passing through a shock, the gas is compressed.*

Example

Normal reflection of a plane shock.

A plane normal shock is advancing with uniform velocity U_1 towards a parallel rigid wall in a gaseous medium. Ahead of the shock the gas is at rest, the pressure is p_0 and the density ρ_0. Behind the shock the gas moves with uniform velocity u_1 in the direction of the shock, the pressure is p_1 and the density is ρ_1. After reflection from the wall the shock moves in the opposite direction, leaving in its wake gas at pressure p_2 and density ρ_2. Show that

$$\frac{p_2-p_0}{p_1-p_0}=\frac{(3\gamma-1)p_1+(\gamma+1)p_0}{(\gamma-1)p_1+(\gamma+1)p_0}.$$

Discuss the value of this ratio for a weak shock for which $p_1/p_0 \sim 1$, and for a strong shock for which $p_1/p_0 \sim \infty$.

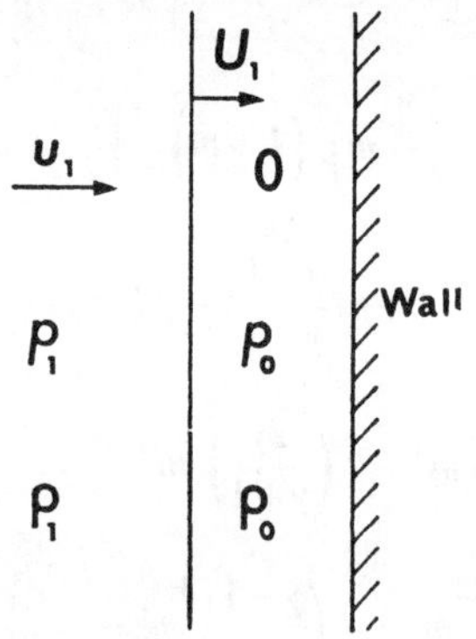

FIG. 7.23(i)

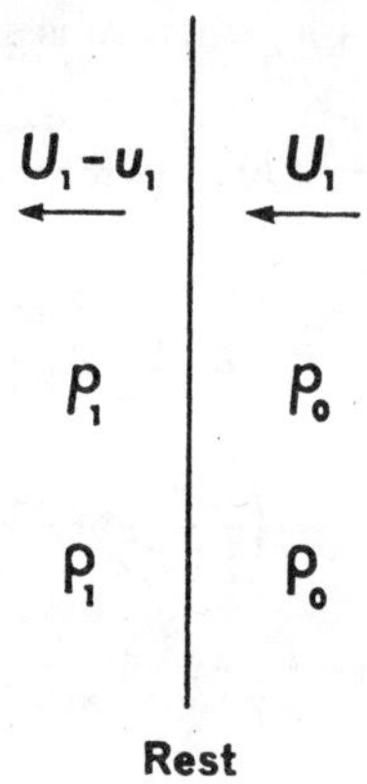

FIG. 7.23(ii)

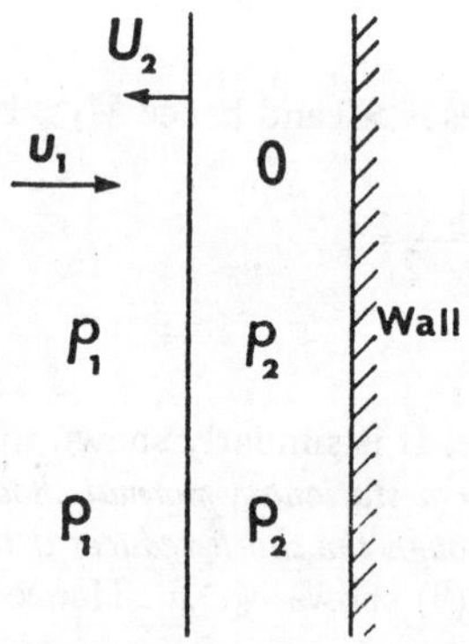

FIG. 7.24(i)

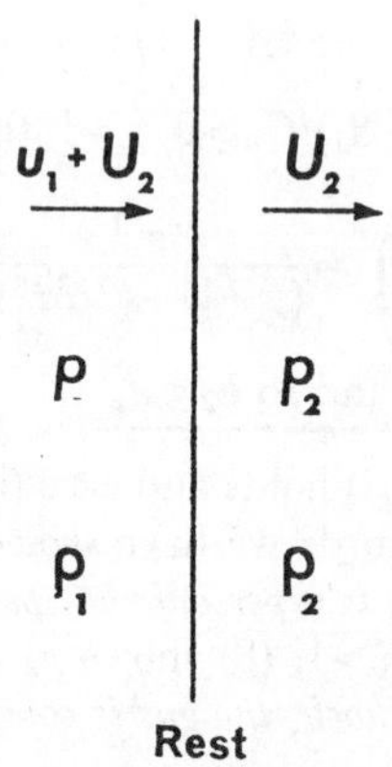

FIG. 7.24(ii)

Figure 7.23(i) shows the shock advancing towards the wall. Figure 7.23(ii) is the same configuration but with the shock reduced to rest by superposing a backward velocity U_1 on the entire system. Figure 7.24(i) shows the reflected shock receding from the wall with velocity U_2 and leaving behind a region at rest. Figure 7.24(ii) is the same configuration with a backward velocity U_2 superposed on the system to bring the shock to rest. The problem is analysed by writing down the pertinent conditions across the stationary shocks in Figs. 7.23(ii), 7.24(ii).

Energy considerations applied to the stationary shocks give

$$\tfrac{1}{2}U_1^2+\frac{a_0^2}{\gamma-1}=\tfrac{1}{2}(U_1-u_1)^2+\frac{a_1^2}{\gamma-1}=\tfrac{1}{2}\left(\frac{\gamma+1}{\gamma-1}\right)u_*^2, \tag{1}$$

$$\tfrac{1}{2}U_2^2+\frac{a_2^2}{\gamma-1}=\tfrac{1}{2}(U_2+u_1)^2+\frac{a_1^2}{\gamma-1}=\tfrac{1}{2}\left(\frac{\gamma+1}{\gamma-1}\right)a_*^2. \tag{2}$$

Using Prandtl's relations

$$U_1(U_1-u_1)=a_*^2=U_2(U_2+u_1), \tag{3}$$

(1), (2) become

$$\tfrac{1}{2}(U_1-u_1)^2+\frac{a_1^2}{\gamma-1}=\tfrac{1}{2}\left(\frac{\gamma+1}{\nu-1}\right)U_1(U_1-u_1), \tag{1'}$$

$$\tfrac{1}{2}(U_2+u_1)^2+\frac{a_1^2}{\gamma-1}=\tfrac{1}{2}\left(\frac{\gamma+1}{\gamma-1}\right)U_2(U_2+u_1). \tag{2'}$$

Writing

$$M_1=\frac{U_1-u_1}{a_1},\quad M_2=\frac{U_2+u_1}{a_1}, \tag{4}$$

equation (1′) becomes

$$\tfrac{1}{2}M_1^2+\frac{1}{\gamma-1}=\tfrac{1}{2}\left(\frac{\gamma+1}{\gamma-1}\right)M_1\left(M_1+\frac{u_1}{a_1}\right),$$

or

$$M_1^2+\tfrac{1}{2}(\gamma+1)\,(u_1/a_1)M_1-1=0. \tag{5}$$

Similarly (2′) becomes

$$M_2^2-\tfrac{1}{2}(\gamma+1)\,(u_1/a_1)M_2-1=0. \tag{6}$$

Comparing (5), (6), we see that M_1 and $-M_2$ are the roots of the quadratic equation in M:

$$M^2+\tfrac{1}{2}(\gamma+1)\,(u_1/a_1)M-1=0. \tag{7}$$

Hence, by considering the products of the roots of (7),

$$M_1M_2=1. \tag{8}$$

From Fig. 7.23(ii),

$$p_0/p_1=\{2\gamma M_1^2-(\gamma-1)\}/(\gamma+1), \tag{9}$$

and from Fig. 7.24(ii),

$$\frac{p_2}{p_1}=\frac{2\gamma M_2^2-(\gamma-1)}{\gamma+1}. \tag{10}$$

Using (8), (9), (10),

$$\left(\frac{p_0}{p_1}+\frac{\gamma-1}{\gamma+1}\right)\left(\frac{p_2}{p_1}+\frac{\gamma-1}{\gamma+1}\right)=\frac{4\gamma^2}{(\gamma+1)^2}M_1M_2=\frac{4\gamma^2}{(\gamma+1)^2},$$

whence

$$\frac{p_2}{p_1}=\frac{(3\gamma-1)p_1-(\gamma-1)p_0}{(\gamma+1)p_0+(\gamma-1)p_1}. \tag{11}$$

It is easily shown that the excess pressure ratio $(p_2-p_0)/(p_1-p_0)$ is

$$\frac{p_2-p_0}{p_1-p_0}=\frac{(3\gamma-1)(p_1/p_0)+(\gamma+1)}{(\gamma-1)(p_1/p_0)+(\gamma+1)}. \tag{12}$$

Then for a *weak shock*, $p_1/p_0\sim 1$ and so

$$(p_2-p_0)/(p_1-p_0)\sim 2.$$

(For a sound wave, $p_1=p_0$ and the ratio is exactly 2.) For a *strong shock*, $p_1/p_0\sim\infty$ and so

$$\frac{p_2-p_0}{p_1-p_0}\sim\frac{3\gamma-1}{\gamma-1}=8 \text{ for } \gamma=1{\cdot}4.$$

7.8.3 Elementary Analysis of Oblique Shock Waves

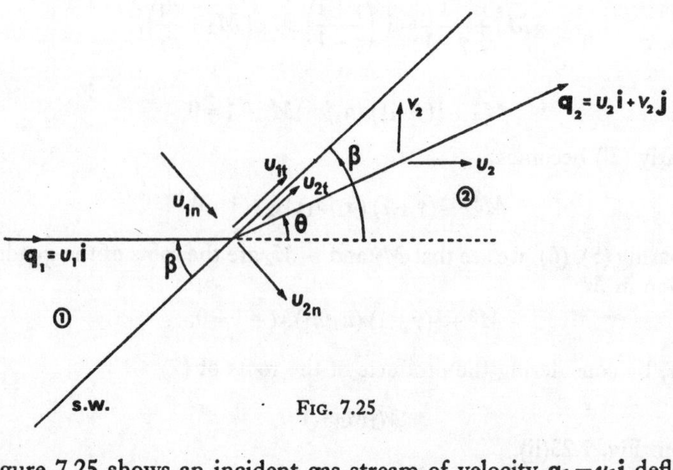

Fig. 7.25

Figure 7.25 shows an incident gas stream of velocity $\mathbf{q}_1=u_1\mathbf{i}$ deflected through an angle θ by an oblique shock wave set at angle β to the direction of incident flow specified by the unit vector **i**. After deflection the new velocity is $\mathbf{q}_2 = u_2\mathbf{i} + v_2\mathbf{j}$ where **j** is the unit vector at right-angles to **i** in the plane of flow. The shock wave divides the field into Regions 1, 2. It is desired to find the flow in Region 2 in terms of that in Region 1. The components of $\mathbf{q}_1$ and $\mathbf{q}_2$ normal and tangential to the shock wave are specified by $[u_{1\mathrm{n}}, u_{1\mathrm{t}}]$, $[u_{2\mathrm{n}}, u_{2\mathrm{t}}]$ in the senses shown.

The equation of continuity applied across the shock is

$$\rho_1 u_{1\mathrm{n}} = \rho_2 u_{2\mathrm{n}} = m, \quad \text{say,} \tag{1}$$

m being the mass flux per unit time across unit area of the shock. Since the normal force per unit area of the shock driving gas from Region 1 to Region 2 is $(p_1 - p_2)$ and the rate of change of momentum in the normal direction from 1 to 2 is $m(u_{2\mathrm{n}} - u_{1\mathrm{n}})$, we have, on equating the two,

$$p_1 + m u_{1\mathrm{n}} = p_2 + m u_{2\mathrm{n}}. \tag{2}$$

As there is no tangential force, the momentum equation applied tangentially across the shock is

$$m u_{1\mathrm{t}} = m u_{2\mathrm{t}},$$

or

$$u_{1\mathrm{t}} = u_{2\mathrm{t}}. \tag{3}$$

By arguments similar to those used in the treatment of normal shocks, the energy equation is

$$\tfrac{1}{2}q_1^2 + \frac{\gamma p_1}{(\gamma-1)\rho_1} = \tfrac{1}{2}q_2^2 + \frac{\gamma p_2}{(\gamma-1)\rho_2} = \tfrac{1}{2}\left(\frac{\gamma+1}{\gamma-1}\right)a_*^2 \tag{4}$$

or

$$\tfrac{1}{2}(u_{1\mathrm{n}}^2 + u_{1\mathrm{t}}^2) + \frac{\gamma}{\gamma-1}\frac{p_1}{\rho_1} = \frac{(\gamma+1)a_*^2}{2(\gamma-1)} = \tfrac{1}{2}(u_{2\mathrm{t}}^2 + u_{2\mathrm{n}}^2) + \frac{\gamma}{\gamma-1}\frac{p_2}{\rho_2}.$$

Substituting the values of p_1, p_2 found from (4) in terms of a_* into (2) and using (1) gives on simplification

$$u_{2\mathrm{n}}\left\{a_*^2 + u_{1\mathrm{n}}^2 - \left(\frac{\gamma-1}{\gamma+1}\right)u_{1\mathrm{t}}^2\right\} = u_{1\mathrm{n}}\left\{a_*^2 + u_{2\mathrm{n}}^2 - \left(\frac{\gamma-1}{\gamma+1}\right)u_{2\mathrm{t}}^2\right\}.$$

Since $u_{1\mathrm{t}} = u_{2\mathrm{t}}$,

$$u_{2\mathrm{n}}\left\{a_*^2 + u_{1\mathrm{n}}^2 - \left(\frac{\gamma-1}{\gamma+1}\right)u_{1\mathrm{t}}u_{2\mathrm{t}}\right\} = u_{1\mathrm{n}}\left\{a_*^2 + u_{2\mathrm{n}}^2 - \left(\frac{\gamma-1}{\gamma+1}\right)u_{1\mathrm{t}}u_{2\mathrm{t}}\right\},$$

or

$$(u_{2\mathrm{n}} - u_{1\mathrm{n}})\left\{a_*^2 - u_{1\mathrm{n}}u_{2\mathrm{n}} - \left(\frac{\gamma-1}{\gamma+1}\right)u_{1\mathrm{t}}u_{2\mathrm{t}}\right\} = 0.$$

If $u_{2n}=u_{1n}$, then since $u_{2t}=u_{1t}$, the flows in Regions 1 and 2 are identical. Hence $u_{2n}\neq u_{1n}$, and so

$$u_{1n}u_{2n}=a_*^2-\left(\frac{\gamma-1}{\gamma+1}\right)u_{2t}^2. \tag{5}$$

Now $u_{1n}=u_1\sin\beta$, $u_{2n}=u_2\sin\beta-v_2\cos\beta$, $u_{2t}=u_2\cos\beta+v_2\sin\beta$. Hence substituting in (5) we obtain, on reduction,

$$u_1\tan\beta(u_2\tan\beta-v_2)=a_*^2\sec^2\beta-(\gamma-1)(\gamma+1)^{-1}(u_2+v_2\tan\beta). \tag{6}$$

Since $u_{1t}=u_{2t}$, $u_1\cos\beta=u_2\cos\beta+v_2\sin\beta$, so that

$$\tan\beta=(u_1-u_2)/v_2. \tag{7}$$

Substituting $\tan\beta$ from (7) into (6) we obtain, on simplification,

$$v_2^2=(u_1-u_2)^2\cdot\frac{u_1u_2-a_*^2}{\dfrac{2u_1^2}{\gamma+1}-u_1u_2+a_*^2}. \tag{8}$$

If we introduce the dimensionless variables

$$\bar{u}_1=\frac{u_1}{a_*},\quad \bar{u}_2=\frac{u_2}{a_*},\quad \bar{v}_2=\frac{v_2}{a_*},$$

then

$$\bar{v}_2^2=(\bar{u}_1-\bar{u}_2)^2\cdot\frac{\bar{u}_1\bar{u}_2-1}{\dfrac{2\bar{u}_1^2}{\gamma+1}-\bar{u}_1\bar{u}_2+1}. \tag{9}$$

When $\bar{u}_1$ is fixed we can plot $\bar{v}_2$ against $\bar{u}_2$. The curve so obtained is called a *shock-polar*. Figure 7.26 displays such a curve. It is a Descartes folium.

From (9), $\bar{v}_2=0$ when $\bar{u}_2=\bar{u}_1$ and when $\bar{u}_2=1/\bar{u}_1$. Thus there are two waves giving zero deflection of flow. These are characterized by the points A, B which are inverse in the circle centre $(0, 0)$ and of unit radius. This circle divides the plane into two regions; outside the circle the flow is everywhere supersonic and within the circle it is subsonic. On the circle the flow is sonic. Hence the unit circle is termed the *sonic circle*.

A radius vector through O inclined at angle θ to OBA may cut the shock polar in the three points D, C, E. The velocity corresponding to E is not realizable since it is greater than that corresponding to A, the velocity in the incident stream. OC gives a smaller decrease in speed than does OD. Hence C corresponds to a weak shock and D to a strong one. Experiment seems to show that in most cases the weaker shock occurs for a given stream deflection θ. It is thought, but perhaps not yet confirmed, that the stronger

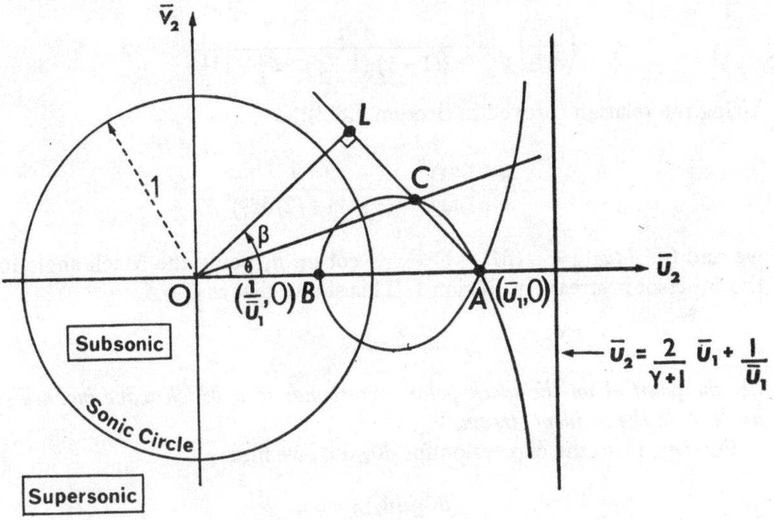

FIG. 7.26

shock is unstable. Taking C to have the coordinates $(\bar{u}_2, \bar{v}_2)$ corresponding to the dimensionless velocities in the Region 2, we see that

$$\text{gradient of } AC = \bar{v}_2/(\bar{u}_2 - \bar{u}_1),$$

so that if OL is drawn perpendicularly to AC as in Fig. 7.25,

$$\text{gradient of } OL = (\bar{u}_1 - \bar{u}_2)/\bar{v}_2.$$

From equation (7), $\tan\beta = (u_1 - u_2)/v_2 = (\bar{u}_1 - \bar{u}_2)/\bar{v}_2$.
Hence

$$\underline{\angle LOA = \beta.}$$

Thus it follows that *if a radius vector drawn through O making an angle θ with OBA cuts the shock polar in D, C, E as in Fig. 7.26, then* $\overline{OC}$ *represents the dimensionless velocity downstream of the shock for a given deflection θ of the flow, and the angle between OBA and the perpendicular OL drawn to AC gives the angle β at which the oblique shock is set to the direction of the incident stream in Region 1.*

From equation (9) we find

$$\frac{d\bar{v}_2}{d\bar{u}_2} = \frac{(\gamma+1)^{-1}\bar{u}_1^4 - 3(\gamma+1)^{-1}\bar{u}_1^3\bar{u}_2 + \bar{u}_1^2\bar{u}_2^2 + 2(\gamma+1)^{-1}\bar{u}_1^2 - 2\bar{u}_1\bar{u}_2 + 1}{(\bar{u}_1\bar{u}_2 - 1)^{1/2}\{2(\gamma+1)^{-1}\bar{u}_1^2 - \bar{u}_1\bar{u}_2 + 1\}^{3/2}}.$$

At A, $\bar{u}_2 = \bar{u}_1$, $\bar{v}_2 = 0$, and so there is no change of velocity at A and we expect A to correspond to a very weak wave. Also at A we find

$$\left(\frac{d\bar{v}_2}{d\bar{u}_2}\right)_A = -\frac{(\bar{u}_1^2-1)^{1/2}}{\{(1-\gamma)(1+\gamma)^{-1}\bar{u}_1^2+1\}^{1/2}}.$$

Using the relation (proved in Section 7.8.2)

$$\bar{u}_1^2 = \left(\frac{u_1}{a_*}\right)^2 = \frac{\gamma+1}{(\gamma-1)+(2/M_1^2)},$$

we find $(d\bar{v}_2/d\bar{u}_2)_A = -(M_1^2-1)^{1/2} = -\cot\mu_1$, μ_1 being the Mach angle for the supersonic stream in Region 1. This shows that as $C \to A$,

$$\beta \to \mu_1,$$

i.e. *the point A on the shock polar corresponds to a Mach wave inclined at angle μ_1 to the incident stream.*

Further, from the expression for $d\bar{v}_2/d\bar{u}_2$, we find

$$(d\bar{v}_2/d\bar{u}_2)_B = \infty,$$

so that as $C \to B$, $\beta \to 0$. *Thus the point B on the shock polar corresponds to a normal shock.*

Let us now examine the case when the points C, D coincide at T on the shock polar. This will give a maximum flow deflection θ_{max} (Fig. 7.27). For $\theta < \theta_{max}$ and for given $\bar{u}_1$, there are two possible plane oblique shocks, one strong, the other weak. For $\theta > \theta_{max}$, the type of solution considered apparently breaks down.

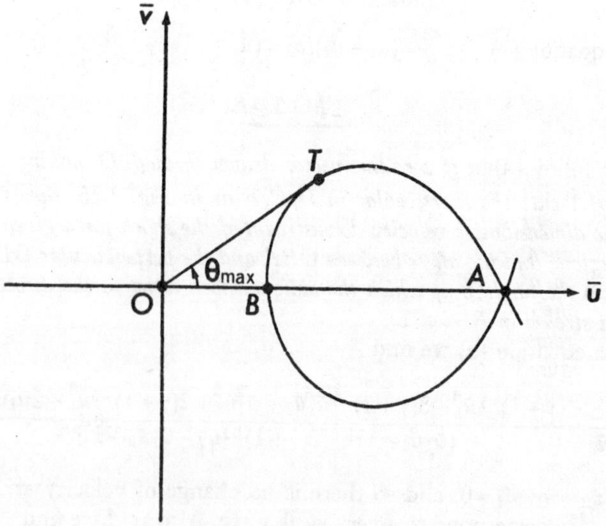

FIG. 7.27

To discuss the cases $\theta<\theta_{max}$, $\theta>\theta_{max}$ in greater detail, let us consider flow past a symmetrically-positioned wedge having vertex angle 2θ and placed in a uniform supersonic stream, for which $\bar{u}_1$ is the dimensionless velocity. For $\theta<\theta_{max}$, a pair of plane shocks emanate from the vertex, each at angle β to the direction of the incident stream, as in Fig. 7.28(i). Their directions are readily found from the shock polar appropriate to the incident flow. When, however, $\theta>\theta_{max}$, plane shocks are no longer possible and experiment shows that a *bow wave*, detached from the vertex, is formed as shown in Fig. 7.28(ii). This bow wave is a curved shock wave and makes an angle $\pi/2$ with the axis of symmetry at their intersection, but the inclination of the tangent to it tends to the Mach angle μ_1 at great distances from the axis. The strength of the bow wave decreases with distance from the axis and it approaches a Mach wave in form.

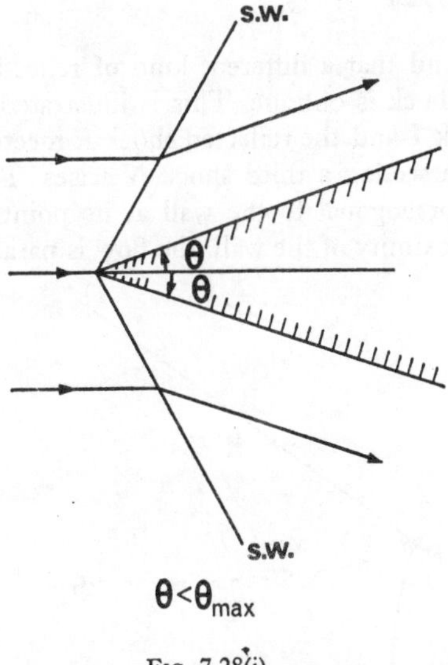

FIG. 7.28(i)

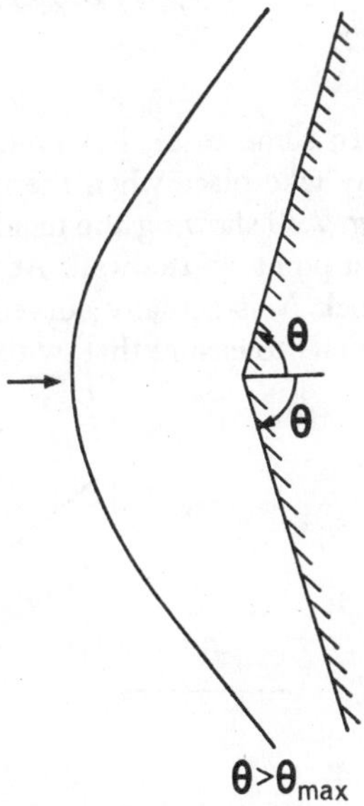

FIG. 7.28(ii)

7.8.4 Formation of Reflected Oblique Shocks

Figure 7.29 shows a wedge placed in a uniform supersonic stream of gas initially flowing parallel to a straight wall. Shock waves arise from the apex A, and the wave I which is incident on the wall at O suffers reflection from O so that a reflected shock wave is formed. The flow in Region 1 is parallel to the wall. The incident wave I causes deflection of the flow in Region 2 as shown. As the flow must be restored to its original direction parallel to the wall in the Region 3, this necessitates the formation of the reflected wave R. The flows in the three regions are all uniform, but different one from another. The case in which the reflected shock is attached to the wall is known as *simple reflection* and is obtained in the case of weak shock waves.

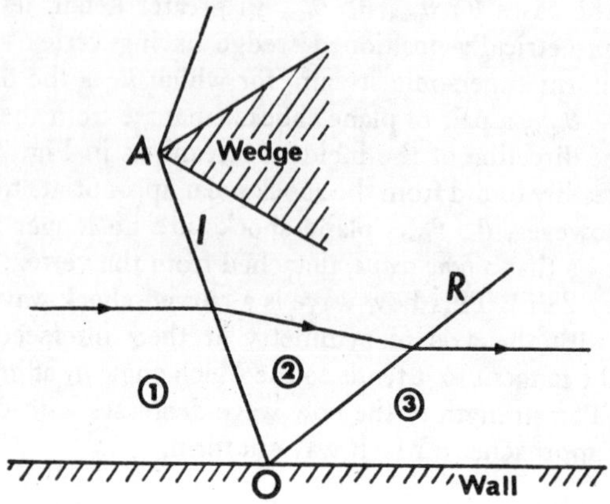

FIG. 7.29.

In some cases, however, it is found that a different kind of reflection may take place when the incident shock is oblique. This is illustrated in Fig. 7.30 showing the incident shock I and the reflected shock R meeting at a point off the wall. At their intersection a third shock N arises. This shock N is actually curved but is orthogonal to the wall at its point of contact to ensure that, within the proximity of the wall, the flow is parallel

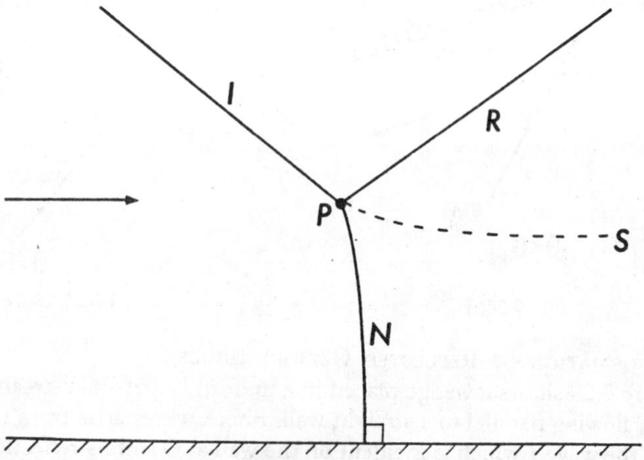

FIG. 7.30

to the wall. In analysis, N is often approximated to a plane shock. N is called a *Mach wave*, but is not to be confused with the type of Mach wave introduced previously. At the triple point P there arises a surface S along which the tangential component of velocity varies and across which it is discontinuous, Also the pressure is continuous and there is no transport of fluid across S. S is called a *slip stream* or *vortex sheet* since it is characterized by a trail of vortices. Such a phenomenon is realized when the incident shock is strong and the flow deflection large. This type of reflection is called *Mach reflection.*

Mach reflection is realized when an atomic bomb bursts above ground. Figure 7.31 shows such a bomb bursting at O, a point above ground level. A train of spherical waves emanates from O and on reaching the ground these waves suffer reflection. Those parts of the ground within a certain radial distance from O will be subject to simple shock reflection since the deflection of the incident air stream is not very great. At large distances from O, however, Mach reflection takes effect. The powerful Mach waves travel outwards as shown. This is why such an explosion over a large metropolis would cause widespread havoc over areas remote from the centre of explosion.

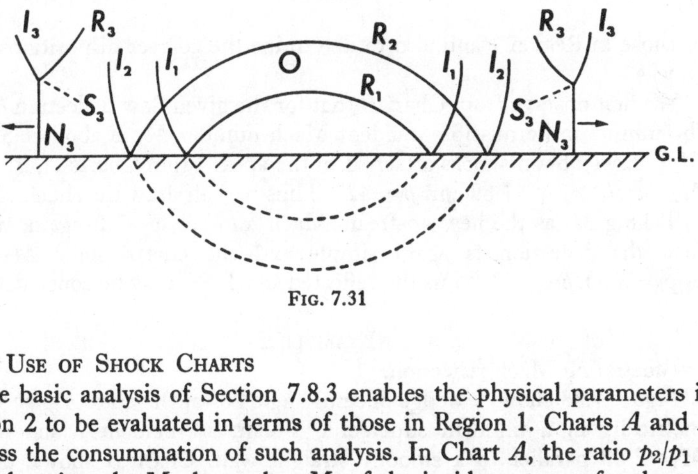

FIG. 7.31

7.8.5 USE OF SHOCK CHARTS

The basic analysis of Section 7.8.3 enables the physical parameters in Region 2 to be evaluated in terms of those in Region 1. Charts A and B express the consummation of such analysis. In Chart A, the ratio p_2/p_1 is plotted against the Mach number M_1 of the incident stream for the two cases in which the flow deflection θ and the shock angle β are constant. Chart B shows a similar plot of the downstream Mach number M_2 against M_1. Both charts have been plotted for $\gamma = 1{\cdot}4$. Interpolation between the curves is very easy. We illustrate applications of the charts with two examples.

EXAMPLE 1

Simple reflection at a wall.

Figure 7.32 shows a wedge of semi-angle 12° at the apex placed symmetrically in a uniform supersonic stream at incident Mach number $M_1 = 2{\cdot}00$ parallel to a smooth straight wall. AB is the shock arising from the apex A and BC its reflection from the straight wall. It is required to evaluate the pressures and Mach numbers in the Regions 2 and 3 in terms

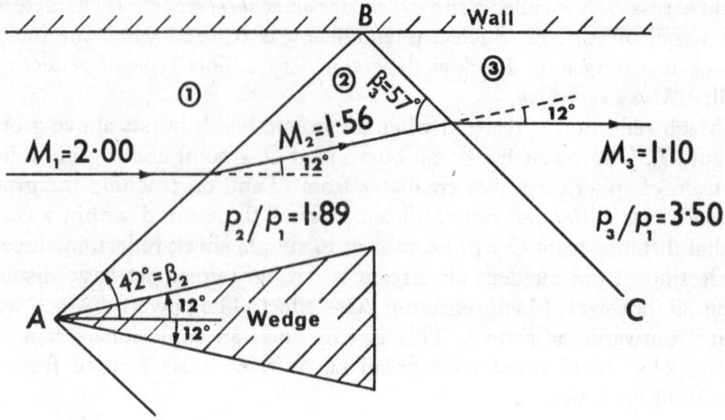

FIG. 7.32

of those in Region 1 and also to determine the correct dispositions of the shocks.

We first observe from Chart *B* that for the given flow deflection $\theta = 12°$, the minimum permissible incident Mach number M_1 is about 1·5. Since $M_1 = 2$, we obtain shocks attached to the apex. From Charts *A*, *B* we find $M_2 = 1{\cdot}56$, $p_2/p_1 = 1{\cdot}89$ and $\beta_2 = 42°$. Thus we can draw the shock *AB*.

Taking M_2 as the new upstream Mach number, $\theta = 12°$ again, we confirm the reflection is again simple and the charts show $M_3 = 1{\cdot}10$, $p_3/p_2 = 3{\cdot}50$, $\beta_3 = 57°$. Thus the reflected shock *BC* may be constructed.

EXAMPLE 2

Illustrating Mach reflection.

Figure 7.33 shows a wedge of semi-angle 20° at the apex *A* placed symmetrically in a uniform supersonic stream at incident Mach number $M_1 = 2{\cdot}00$ parallel to a smooth straight wall. Chart *B* shows that for $\theta = 20°$, the shock waves will be attached to *A*, since the incident Mach number for critical attachment is less than 2.

With $M_1 = 2$, $\theta_{12} = 20°$, the charts show $M_2 = 1{\cdot}21$, $p_2/p_1 = 2{\cdot}85$, $\beta_{12} = 53{\cdot}5°$ in Region 2. Thus we can construct the shock wave *AO*. It is found, taking the new incident Mach number as 1·21, that the original direction of flow, requiring a deflection back through 20°, cannot be obtained by simple reflection. (Chart *B* shows that the incident Mach number for critical simple reflection is greater than 1·21.) Thus Mach reflection takes effect.

To construct the Mach reflection, proceed as follows. First choose an arbitrary point *O* on the shock wave *AO*. Its actual position on the wave depends on downstream conditions. Knowing that the total deflection δ from Regions 1 to 4 is the same as that from Regions 1 to 3 and that $p_4/p_1 = p_3/p_1$, the required pressure ratio and total deflection δ may be

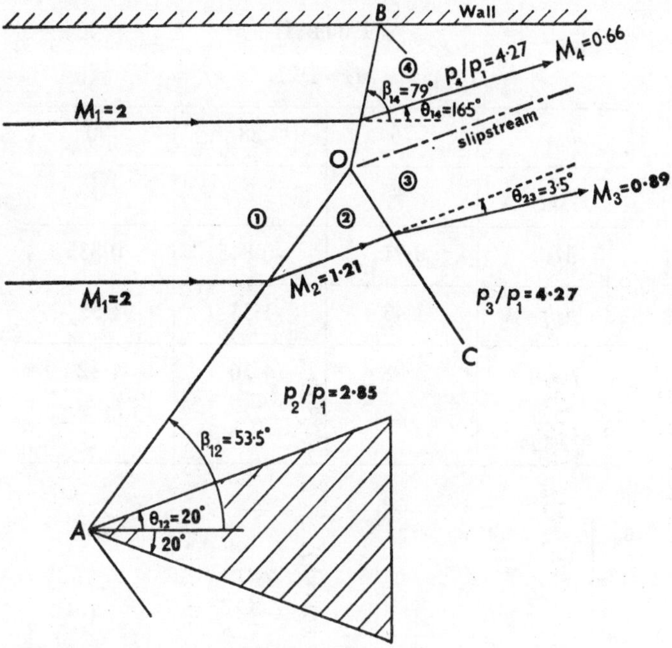

FIG. 7.33

found by plotting p_4/p_1 and p_3/p_1 against values of δ on the same axes as in Fig. 7.33. The point of intersection Y gives the required value of pressure ratio and total deflection satisfying the above conditions. The information needed to plot the graph is obtained by reading off from the unstable part of the charts (corresponding to the strong shocks) values of pressure ratio for given δ's. The appropriate information is contained in Tables I, II below.

TABLE I

$M_1 = 2$

β_{14}	75°	80°	85°	90°
θ_{14}	19°	15°	6°	0
M_4	0·72	0·64	0·595	0·585
p_4/p_1	4·2	4·3	4·45	4·5

TABLE II

$M_2 = 1{\cdot}21$

β_{23}	76	88	90
θ_{23}	4	2	0
M_3	0·91	0·845	0·835
p_3/p_2	1·43	1·53	1·55
$p_3/p_1 = 2{\cdot}85\, p_3/p_2$	4·08	4·36	4·42

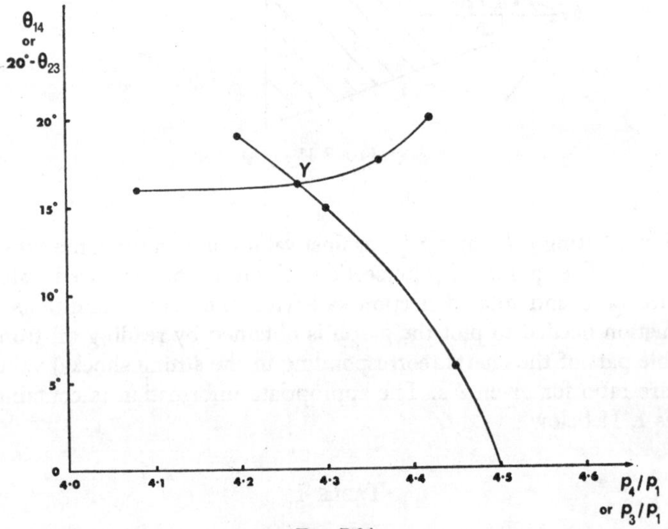

FIG. 7.34

Having M_1, θ_{14}, M_2, θ_{23}, it is now possible to read off M_4, β_{14}, M_3, β_{23} using the strong shock solution of Chart B and draw in OB, OC and the slip stream OE.

(*N.B.* In actual fact the wave OB is curved and normal to the wall at B to prevent any deflection of the flow there, but the direction $\overline{OB}$ is that of the tangent to the wave at O. However, the above construction is thought to be a good approximation to the required solution.)

7.9 The Method of Characteristics applied to Supersonic Homentropic Irrotational Gas Flows

In Section 7.4 we obtained the equation tor homentropic gas flow in the form

$$\mathbf{q} \cdot \nabla(\tfrac{1}{2}\mathbf{q}^2) = a^2 \nabla \cdot \mathbf{q}, \tag{1}$$

where $\mathbf{q}$ is the local particle velocity of the gas and a is the speed of sound. Now suppose the flow is irrotational. Then there exists a velocity potential φ such that $\mathbf{q} = -\nabla\varphi$ and (1) becomes

$$a^2\nabla^2\varphi + \mathbf{q} \cdot \nabla(\tfrac{1}{2}\mathbf{q}^2) = 0. \tag{2}$$

Now

$$\mathbf{q} \cdot \nabla(\tfrac{1}{2}\mathbf{q}^2) = \left(u\frac{\partial}{\partial x} + v\frac{\partial}{\partial y} + w\frac{\partial}{\partial z}\right)(u^2+v^2+w^2)$$

$$= u\left(u\frac{\partial u}{\partial x} + v\frac{\partial v}{\partial x} + w\frac{\partial w}{\partial x}\right)$$

$$+ v\left(u\frac{\partial u}{\partial y} + v\frac{\partial v}{\partial y} + w\frac{\partial w}{\partial y}\right)$$

$$+ w\left(u\frac{\partial u}{\partial z} + v\frac{\partial v}{\partial z} + w\frac{\partial w}{\partial z}\right)$$

$$= -(u^2\varphi_{xx} + v^2\varphi_{yy} + w^2\varphi_{zz} + 2vw\varphi_{yz} + 2wu\varphi_{zx} + 2uv\varphi_{xy}),$$

where $\varphi_x = \partial\varphi/\partial x$, etc. Hence (2) becomes

$$(a^2-u^2)\varphi_{xx} + (a^2-v^2)\varphi_{yy} + (a^2-w^2)\varphi_{zz} - 2vw\varphi_{yz} - 2wu\varphi_{zx} - 2uv\varphi_{xy} = 0. \tag{3}$$

Equation (3) is a second-order partial differential equation for φ. It is nonlinear save for the case $a = \infty$, corresponding to incompressible flow (which latter reduces to $\nabla^2\varphi = 0$). We consider special cases of (3) leading to solutions using the *method of characteristics*—a technique outlined in the Appendix.

7.9.1 The Method of Characteristics for Two-Dimensional, Homentropic, Irrotational Flow

If such flow takes place in planes parallel to that containing the x- and y-axes, then the second-order partial differential equation for φ becomes

$$(a^2-u^2)\varphi_{xx} - 2uv\varphi_{xy} + (a^2-v^2)\varphi_{yy} = 0. \tag{1}$$

Since $u = -\varphi_x$, $v = -\varphi_y$,

$$dx\varphi_{xx} + dy\varphi_{xy} = -du, \tag{2}$$

$$dx\varphi_{xy} + dy\varphi_{yy} = -dv. \tag{3}$$

Equations (1), (2), (3) permit of unique solutions for the second-order derivatives of φ (all other quantities being supposed known) provided that $\Delta \neq 0$, where

$$\Delta = \begin{vmatrix} a^2 - u^2 & -2uv & a^2 - v^2 \\ dx & dy & 0 \\ 0 & dx & dy \end{vmatrix}. \tag{4}$$

The equation of the characteristics is $\Delta = 0$, or

$$(a^2 - u^2)\left(\frac{dy}{dx}\right)^2 + 2uv\frac{dy}{dx} + (a^2 - v^2) = 0. \tag{5}$$

Thus the characteristic directions are given by

$$\frac{dy}{dx} = \frac{-uv \pm a\sqrt{(u^2 + v^2 - a^2)}}{a^2 - u^2} = \frac{-uv \pm a\sqrt{(q^2 - a^2)}}{a^2 - u^2}. \tag{6}$$

These are real and distinct if $q > a$, i.e. if the flow is ***supersonic***: the equation (1) is then ***hyperbolic***. If $q = a$, the flow is ***sonic*** and (1) is ***parabolic***: there is only one family of characteristics in this case. If $q < a$, the flow is ***subsonic*** and (1) is ***elliptic***: the characteristics are not real in this case.

Along the characteristics the following condition prevails.

$$\begin{vmatrix} a^2 - u^2 & 0 & a^2 - v^2 \\ dx & -du & 0 \\ 0 & -dv & dy \end{vmatrix} = 0$$

or

$$\frac{dv}{du} = -\frac{a^2 - u^2}{a^2 - v^2}\frac{dy}{dx}. \tag{7}$$

From (6), (7), taking

$$\left(\frac{dy}{dx}\right)_{\text{I}} = \frac{-uv + a\sqrt{(q^2 - a^2)}}{a^2 - u^2},$$

gives

$$\left(\frac{dv}{du}\right)_{\text{I}} = \frac{uv - a\sqrt{(q^2 - a^2)}}{a^2 - v^2}.$$

Taking

$$\left(\frac{dy}{dx}\right)_{\text{II}} = \frac{-uv - a\sqrt{(q^2 - a^2)}}{a^2 - u^2},$$

gives

$$\left(\frac{dv}{du}\right)_{\text{II}} = \frac{uv + a\sqrt{(q^2 - a^2)}}{a^2 - v^2}.$$

Thus

$$\left(\frac{dy}{dx}\right)_{\text{I}}\left(\frac{dv}{du}\right)_{\text{II}}=\frac{a^2(u^2+v^2-a^2)-u^2v^2}{(a^2-u^2)(a^2-v^2)}=-1. \tag{8.1}$$

Similarly

$$\left(\frac{dy}{dx}\right)_{\text{II}}\left(\frac{dv}{du}\right)_{\text{I}}=-1. \tag{8.2}$$

Now suppose that at each point (x, y) of the *physical plane*, or plane of flow, we draw the two characteristic directions (supposing the flow to be supersonic). If $[u, v]$ denote the velocity components at each point (x, y), let the *hodograph plane* be drawn in which v is plotted against u and in such a way that the x- and u-axes are parallel and so are the y- and v-axes. At each point (u, v) of the hodograph plane draw the two characteristic directions dv/du given by (7). Then the equations (8.1), (8.2) show *that characteristics of the first kind in the physical plane are at right-angles to those of the second kind in the hodograph plane, and characteristics of the second kind in the physical plane are at right-angles to those of the first kind in the hodograph plane.*

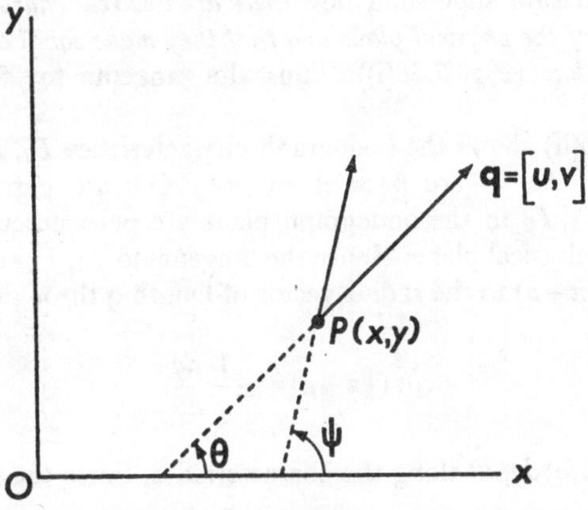

FIG. 7.35

To find the characteristic directions at a point $P(x, y)$ of the physical plane, consider Fig. 7.35 showing the velocity $\mathbf{q}$ at P making angle θ with $\overline{Ox}$ and one of the characteristics making angle ψ with this direction so that $dy/du = \tan\psi$ is the gradient of the characteristic at P. Then, from (6),

$$\tan\psi=\frac{dy}{dx}=\frac{-uv\pm a\sqrt{(q^2-a^2)}}{a^2-u^2}.$$

Since the flow is supersonic, $a=q\sin\mu$, where μ is the Mach angle at P. Also, $u=q\cos\theta$, $v=q\sin\theta$. Hence

$$\tan\psi=\frac{-\cos\theta\sin\theta\pm\sin\mu\cos\mu}{\sin^2\mu-\cos^2\theta}$$

$$=\frac{\sin 2\theta\mp\sin 2\mu}{\cos 2\theta+\cos 2\mu}$$

$$=\frac{\sin 2\theta+\sin(\mp 2\mu)}{\cos 2\theta+\cos(\mp 2\mu)}$$

$$=\frac{2\sin(\theta\mp\mu)\cos(\theta\pm\mu)}{2\cos(\theta\mp\mu)\cos(\theta\pm\mu)},$$

$$=\tan(\theta\mp\mu).$$

Thus

$$\psi=\theta\mp\mu.$$

This shows that for supersonic flow *there are two real characteristics $\mathscr{C}_1, \mathscr{C}_2$ at each point of the physical plane and that they make equal angles μ with the direction of flow* (Fig. 7.36(i)). Thus the tangents to $\mathscr{C}_1, \mathscr{C}_2$ at P are Mach lines.

Figure 7.36(ii) shows the hodograph characteristics Γ_1, Γ_2 in the (u, v)-plane, where Ox, Ou are parallel and $O'y, O'v$ are parallel. Then the tangents to Γ_1, Γ_2 in the hodograph plane are perpendicular to those to $\mathscr{C}_2, \mathscr{C}_1$ in the physical plane. Hence the tangents to Γ_1, Γ_2 are each inclined at an angle $(\frac{1}{2}\pi-\mu)$ to the radius vector of length $\mathbf{q}$ through O'. Thus

$$\cot(\tfrac{1}{2}\pi-\mu)=\pm\frac{1}{q}\frac{dq}{d\theta},$$

or $\tan\mu=\pm(1/q)\,dq/d\theta$ along the characteristics. Since the flow is homentropic,

$$\tfrac{1}{2}q^2+\frac{a^2}{\gamma-1}=\tfrac{1}{2}\left(\frac{\gamma+1}{\gamma-1}\right)a_*^2$$

whence

$$q^2(\gamma-1+2\sin^2\mu)=\text{const}.$$

Differentiating logarithmically gives

$$\frac{dq}{q}+\frac{2\sin\mu\cos\mu\,d\mu}{\gamma-1+2\sin^2\mu}=0.$$

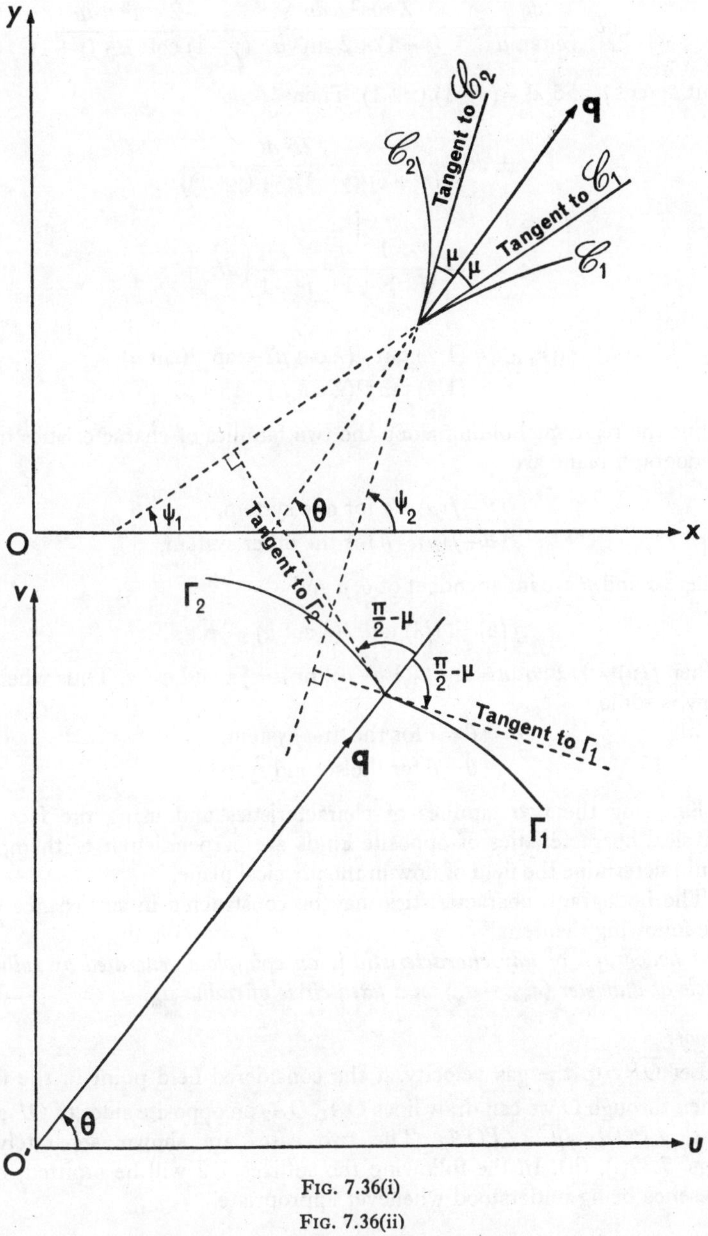

FIG. 7.36(i)

FIG. 7.36(ii)

Thus

$$\pm d\theta = \frac{dq}{q \tan \mu} = -\frac{2 \cos^2 \mu \, d\mu}{(\gamma-1)+2 \sin^2 \mu} = \frac{-2 \cot^2 \mu \, d\mu}{(\gamma-1) \cot^2 \mu + (\gamma+1)}.$$

Put $t = \cot \mu$ and $\lambda^2 = (\gamma-1)/(\gamma+1)$. Then

$$\pm \, d\theta = \frac{2t^2 \, dt}{(t^2+1)\{(\gamma-1)t^2+(\gamma+1)\}}$$

$$= \left(\frac{1}{(\lambda t)^2+1} - \frac{1}{t^2+1}\right) dt,$$

so that

$$\begin{aligned} \text{const.} \pm \theta &= (1/\lambda) \tan^{-1}(\lambda \cot \mu) - \tan^{-1}(\cot \mu) \\ &= (1/\lambda) \tan^{-1}(\lambda \cot \mu) - \tfrac{1}{2}\pi + \mu. \end{aligned}$$

Thus the relations holding along the two families of characteristics in the hodograph plane are

$$\begin{cases} \theta - f(\mu) = \alpha \text{ for one system,} \\ \theta + f(\mu) = \beta \text{ for the other system,} \end{cases}$$

where α and β are independent of θ, μ and

$$f(\mu) = (1/\lambda) \tan^{-1}(\lambda \cot \mu) - \tfrac{1}{2}\pi + \mu.$$

When $f(\mu)=0$, $\lambda \cot \mu = \tan \{\lambda(\frac{1}{2}\pi - \mu)\}$ or $\mu = \frac{1}{2}\pi$ and $q=a$. Thus when the flow is sonic,

$$\begin{cases} \theta = \alpha \text{ for the first system,} \\ \theta = \beta \text{ for the second system.} \end{cases}$$

Knowing the two families of characteristics and using the fact that physical characteristics of opposite kinds are perpendicular to them, one could determine the field of flow in the physical plane.

The hodograph characteristics may be constructed in accordance with the following theorem *

The hodograph of any characteristic is an epicycloid generated by rolling a circle of diameter $(q_{max} - a_*)$ on a fixed circle of radius a_*.

Proof:

Let $\overline{OP} \equiv \mathbf{q}$, the gas velocity at the considered field point in the flow. Then through O we can draw lines OA_1, OA_2 on opposite sides of OP such that $\angle POA_1 = \mu = \angle POA_2$. The two cases are shown separately in Figs. 7.37(i), (ii). In the following the suffixes 1, 2 will be omitted, their presence being understood whenever appropriate.

* G. Temple, 'The method of characteristics for supersonic flow', *R. and M.*, No. 2091 (1944).

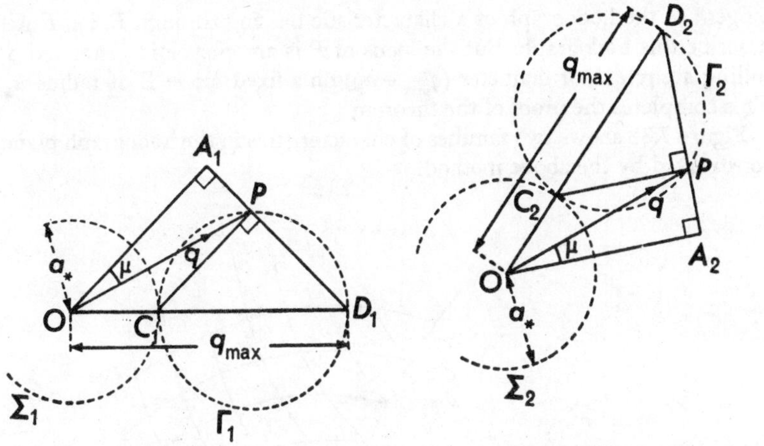

FIG. 7.37(i) FIG. 7.37(ii)

On OP as hypotenuse construct a right-angled triangle OAP such that $\angle OAP=\frac{1}{2}\pi$, $\angle AOP=\mu$. Then $AP=q\sin\mu=a$. Produce $\overline{AP}$ to D where $OD=q_{max}$, and $\frac{1}{2}q^2+a^2/(\gamma-1)=\frac{1}{2}(\gamma+1)(\gamma-1)^{-1}a_*^2=\frac{1}{2}q_{max}^2$. Draw $\overline{PC}\|\overline{AO}$ to meet OD in C. Then

$$\begin{aligned}OA^2&=q^2-a^2\\&=q^2-\tfrac{1}{2}(\gamma-1)(q_{max}^2-q^2)\\&=\tfrac{1}{2}(\gamma+1)q^2-\tfrac{1}{2}(\gamma-1)q_{max}^2.\end{aligned}$$

Thus

$$AD^2=q_{max}^2-OA^2=\tfrac{1}{2}(\gamma+1)(q_{max}^2-q^2).$$

Also

$$AP^2=a^2=\tfrac{1}{2}(\gamma-1)(q_{max}^2-q^2).$$

Thus

$$\frac{AP}{AD}=\left(\frac{\gamma-1}{\gamma+1}\right)^{1/2}=a_*$$

and so $CD=q_{max}-a_*$.

With centre O draw the circle Σ of radius a_* and also draw the circle Γ on CD as diameter. Since $\angle CPD=\frac{1}{2}\pi$, Γ passes through P and it touches Σ at C.

Keeping Σ fixed, allow the circle Γ of diameter $(q_{max}-a_*)$ to roll without slipping on Σ. Then C will be the instantaneous centre of rotation of the motion of Γ. Thus the direction of motion of the point P, regarded as a point fixed on the circumference of Γ, will be along APD, since this line is perpendicular to CP. But APD is at right-angles to OA which latter is in the direction of a characteristic in the physical plane since $\angle POA=\mu$. Hence APD is tangential at P to the hodograph characteristic of the opposite kind through P. Hence P will always move in the direction of the tangent to the hodograph of a characteristic passing through P, i.e. P will describe this hodograph. But the locus of P is an epicycloid generated by rolling a circle Γ of diameter $(q_{max} - a_*)$ on a fixed circle Σ of radius a_*. This completes the proof of the theorem.

Figure 7.38 shows two families of characteristics in the hodograph plane, constructed by the above method.

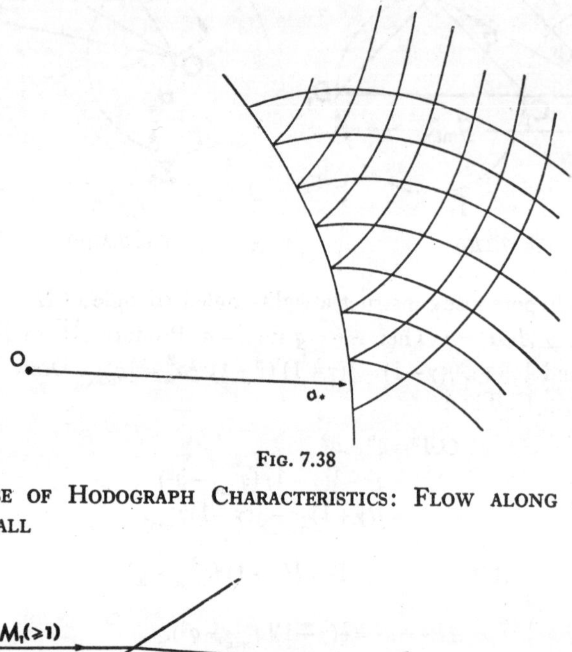

FIG. 7.38

7.9.2 USE OF HODOGRAPH CHARACTERISTICS: FLOW ALONG A CONVEX WALL

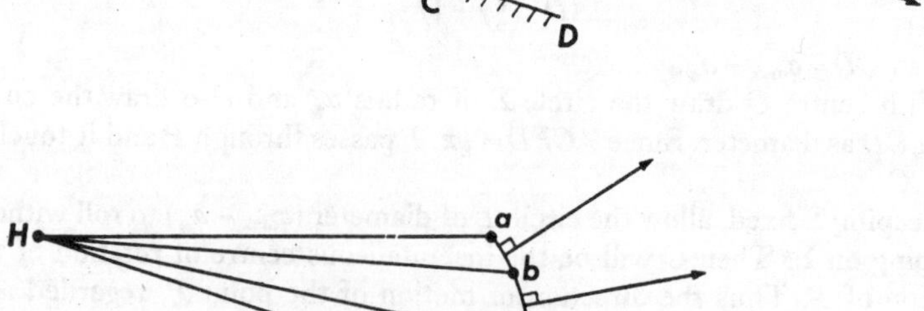

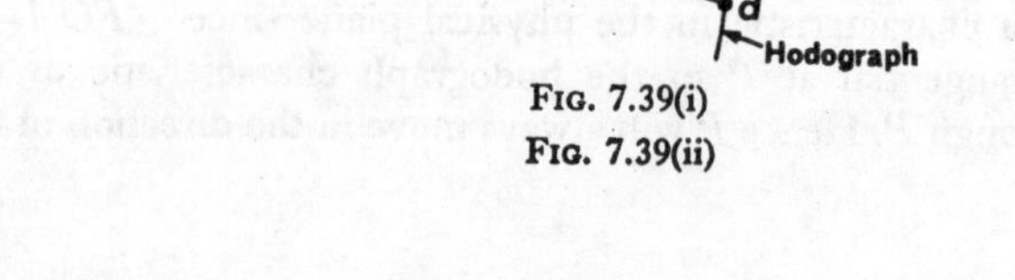

FIG. 7.39(i)
FIG. 7.39(ii)

Suppose it is desired to analyse the flow of a gas (at supersonic speeds) round a curved wall which is convex to the region of flow. First, let the wall be approximated by a series of straight-line segments $OA, AB, BC, CD, \ldots$ taken sufficiently close together and such that the angles between successive segments are small (say 2°–5°). Then the field of supersonic flow is parallel to each segment $\overline{OA}, \overline{AB}, \overline{BC}, \ldots$, its appropriate direction at each of the points $A, B, C, \ldots$ being achieved by a Mach wave or characteristic centred on $A, B, C, \ldots$ It is required to construct each of these Mach waves and to find the velocities in the uniform compartments into which the Mach waves divide the physical plane.

Figure 7.39(ii) shows the appropriate hodograph characteristic. $\overline{Ha}$ represents the velocity vector for the uniform region in which flow is parallel to $\overline{OA}$ and known. We draw $\overline{Ha} \| \overline{OA}$ and similarly, $\overline{Hb} \| \overline{AB}$, $\overline{Hc} \| \overline{BC}$, ... Then $ab, bc, cd, \ldots$ are approximately straight-line segments of the hodograph. Draw the outward normals to the midpoints of these segments. These normals are parallel to corresponding characteristics in the physical plane. In Fig. 7.39(i), draw lines through $A, B, C, \ldots$ parallel to the normals through the midpoints of $ab, bc, cd, \ldots$ These give, to high order of accuracy, the directions of the Mach waves through $A, B, C, \ldots$ Also the vectors $\overline{Hb}, \overline{Hc}, \overline{Hd}, \ldots$ give the gas velocities in the regions of the physical plane parallel to $\overline{AB}, \overline{BC}, \overline{CD}, \ldots$ Thus the pattern of flow in the physical plane is determined.

7.9.3 Use of Characteristic Coordinates

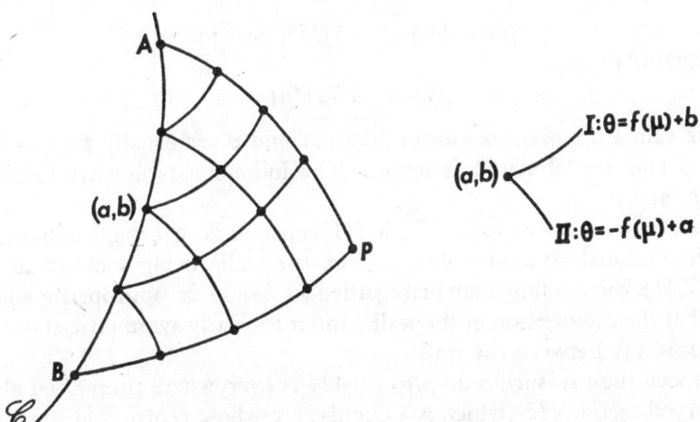

Fig. 7.40(i) Fig. 7.40(ii)

Let AB be an arc of a curve $\mathscr{C}$ in a field of completely supersonic flow and such that at no point of this arc is the tangent a Mach line (Fig. 7.40(i)). Then at each point of AB a pair of real characteristics may be drawn. Suppose the physical data are prescribed along AB, i.e. $p, \rho, T, \mathbf{q}$ are all known. Dissect the arc AB by a number of points sufficiently close together. At each such point of AB we can draw a pair of characteristics I, II as in Fig. 7.40(ii). In this way a lattice, bounded by the sides of the curvilinear triangle PAB, is formed. In practice the joins of neighbouring points of the lattice will be approximated by straight-line segments. Within this region PAB all physical quantities may be evaluated at the lattice points using the 'method of characteristics'.

The essence of the method is as follows. Along an upward-going characteristic I, we have

$$\theta=f(\mu)+b, \quad b=\text{const.}, \tag{1}$$

and along a downward-giving characteristic II,

$$\theta=-f(\mu)+a, \quad a=\text{const.}, \tag{2}$$

where

$$f(\mu)=\mu+(1/\lambda)\tan^{-1}(\lambda\cot\mu)-\tfrac{1}{2}\pi \tag{3}$$

and $\lambda=\sqrt{\{(\gamma-1)/(\gamma+1)\}}$. At the intersection of I, II we thus have

$$\theta=\tfrac{1}{2}(a+b), \tag{4}$$

$$f(\mu)=\tfrac{1}{2}(a-b). \tag{5}$$

The intersection of I,II is said to have *characteristic coordinates* (a, b), but in general the system is not orthogonal. Thus $\frac{1}{2}(a+b)$ determines the *direction* of the stream at (a, b). From $\frac{1}{2}(a-b)$ we can find μ, hence M and hence the pressure using

$$p/p_0=[1+\tfrac{1}{2}(\gamma-1)M^2]^{-\gamma/(\gamma-1)}.$$

The quantity

$$N=a-b=2f(\mu) \tag{6}$$

is thus called the *pressure number*. Both N and θ are usually expressed in degrees and are tabulated functions. The following table cites values of N, p/p_0 and μ.

As an example of the use of Table III, consider flow through a divergent two-dimensional straight-walled nozzle, the walls being inclined at 20° (Fig. 7.41). Such a flow may be regarded as due to an appropriate source placed at the intersection of the walls, and it is clearly symmetric about the plane midway between the walls.

Suppose the pressure ratio $p/p_0=0{\cdot}424$ is everywhere prescribed along the curved section AF (which is a circular arc whose centre is at the intersection of the walls). Then along AF, $N=6°$, $\mu=58°\ 18'$, $M=\operatorname{cosec}\mu=1{\cdot}1753$. Along the lower wall, take $\theta=0$ and along the upper wall take $\theta=20°$. Divide the arc AF by equally spaced points B, C, D, E. Then along $ABCDEF$, $N=6°$ everywhere and using

$$\theta=\tfrac{1}{2}(a+b), \quad N=a-b,$$

TABLE III

N	p/p_0	μ	N	p/p_0	μ	N	p/p_0	μ
0°	0·527	90° 00′	40°	0·180	34° 14′	80°	0·0545	23° 07′
2°	0·479	67° 43′	42°	0·171	33° 29′	82°	0·0509	22° 42′
4°	0·449	62° 00′	44°	0·162	32° 46′	84°	0·0475	22° 18′
6°	0·424	58° 10′	46°	0·153	32° 5′	86°	0·0443	21° 54′
8°	0·402	55° 12′	48°	0·145	31° 24′	88°	0·0413	21° 30′
10°	0·382	52° 43′	50°	0·137	30° 47′	90°	0·0383	21° 07′
12°	0·363	50° 36′	52°	0·130	30° 10′	92°	0·0357	20° 44′
14°	0·345	48° 43′	54°	0·122	29° 34′	94°	0·0331	20° 22′
16°	0·329	47° 03′	56°	0·116	28° 59′	96°	0·0307	20° 01′
18°	0·313	45° 32′	58°	0·109	28° 25′	98°	0·0284	19° 39′
20°	0·298	44° 09′	60°	0·103	27° 52′	100°	0·0265	19° 20′
22°	0·284	42° 51′	62°	0·097	27° 21′	102°	0·0243	18° 57′
24°	0·270	41° 39′	64°	0·0912	26° 49′	104°	0·0224	18° 36′
26°	0·257	40° 32′	66°	0·0858	26° 20′	106°	0·0207	18° 16′
28°	0·245	39° 29′	68°	0·0806	25° 50′			
30°	0·233	38° 30′	70°	0·0757	25° 21′			
32°	0·221	37° 34′	72°	0·0711	24° 53′			
34°	0·210	36° 40′	74°	0·0666	24° 26′			
36°	0·200	35° 49′	76°	0·0623	23° 59′			
38°	0·190	35° 01′	78°	0·0583	23° 33′			

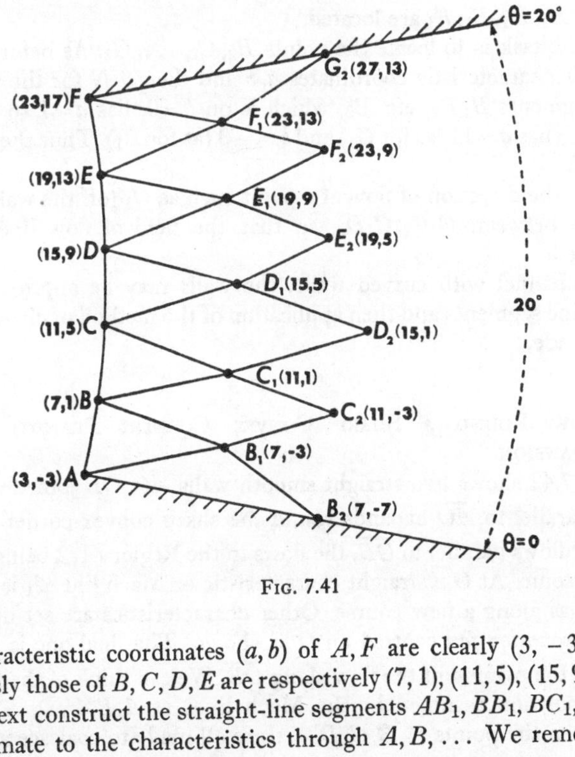

FIG. 7.41

the characteristic coordinates (a, b) of A, F are clearly $(3, -3)$, $(23, 17)$. Obviously those of B, C, D, E are respectively $(7, 1)$, $(11, 5)$, $(15, 9)$, $(19, 13)$.

We next construct the straight-line segments $AB_1, BB_1, BC_1, \ldots$ which approximate to the characteristics through $A, B, \ldots$ We remember that along an upward-going characteristic, b is constant and along a downward going one, a is constant. Thus the characteristic coordinates of B_1, C_1, D_1, E_1, F_1 are respectively $(7, -3)$, $(11, 1)$, $(15, 5)$, $(19, 9)$, $(23, 13)$. To determine C_1, for instance, proceed as follows. The characteristic coordinates of B, C_1 are $(7, 1)$, $(11, 1)$, so that $(9, 1)$ are those of the midpoint of BC_1 (to a high order of accuracy). Hence

$$\text{mean } \theta \text{ for } BC_1 = \tfrac{1}{2}(9° + 1°) = 5°,$$
$$\text{mean } N \text{ for } BC_1 = 9° - 1° = 8°,$$

and so mean μ for $BC_1 = 55° 12'$.
Thus the angle which BC_1 makes with the wall $\theta = 0$ is $55° 12' + 5° = 60° 12'$. The characteristic coordinates of the midpoint of CC_1 are $(11, 3)$ and

$$\text{mean } \theta \text{ for } CC_1 = \tfrac{1}{2}(11° + 3°) = 7°,$$
$$\text{mean } N \text{ for } CC_1 = 11° - 3° = 8°,$$

and so mean μ for $CC_1 = 55° 12'$.
Thus the angle CC_1 makes with $\theta = 0$ is $55° 12' - 7° = 48° 12'$. Hence by drawing BC_1, CC_1 at angles $60° 12'$ and $48° 12'$ to $\theta = 0$ we locate C_1. Similarly B_1, D_1, E_1, F_1 are located.

The next task is to locate the points $B_2, C_2, \ldots, G_2$. As before one first finds their characteristic coordinates a, b and then θ, N for the midpoints of line segments $B_1 B_2$, etc. B_2, which is on $\theta = 0$, has $a = 7$ so that B_2 is $(7, -7)$. C_2 has $a = 11$ (as for C_1) and $b = -3$ (as for B_1). Thus the process is continued.

Finally, the direction of flow at a point such as C_1 (off the walls) bisects the angle between C_1C_2, C_1D_2, so that the field of flow lines may be constructed.

For a channel with curved walls, the walls may be approximated by straight-line segments and then application of the method of characteristics may be made.

7.9.4 Flow Round a Sharp Convex Corner: Prandtl–Meyer Expansion

Figure 7.42 shows two straight smooth walls AO, OB joined at O. Gas flowing parallel to $\overline{AO}$ expands round the sharp convex corner at O and ultimately flows parallel to $\overline{OB}$, the flows in the Regions 1, 2 being uniform and supersonic. At O a straight characteristic or Mach line $\mathscr{C}_1$ is set up to turn the gas along a new course. Other characteristics are set up at O to form an *expansion fan* centred on O as shown. The final one is $\mathscr{C}_2$ which completes the turning of the flow along $\overline{OB}$. We can think of the fan as the ultimate position assumed by the Mach lines at $A, B, C, D, \ldots$ in Fig. 7.38(i) when the points $A, B, C, D, \ldots$ are all made to coalesce at a single

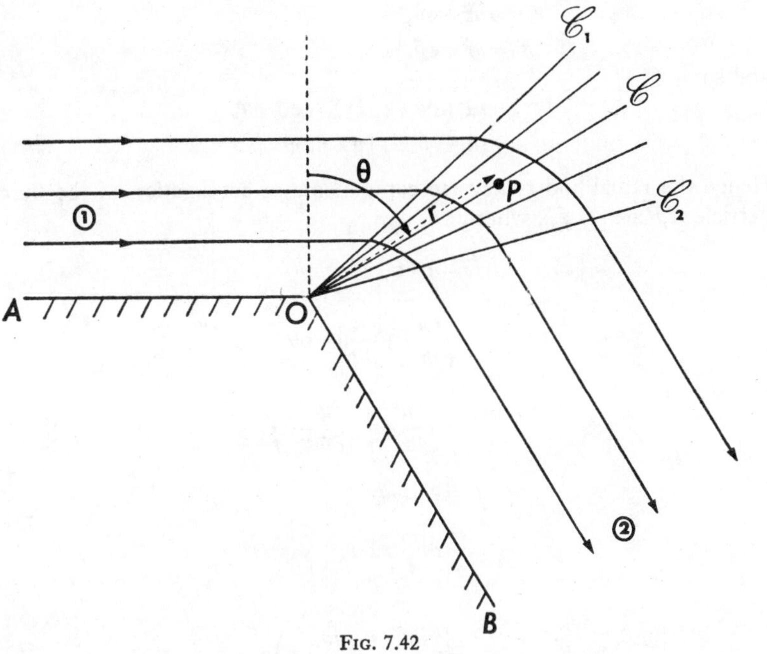

FIG. 7.42

point O, so that the gradual bend of Fig. 7.39(i) has become the sharp one of Fig. 7.42. Note that in Region 1 of Fig. 7.42 all Mach lines are parallel to $\mathscr{C}_1$ and in Region 2 they are all parallel to $\mathscr{C}_2$. Also $\mathscr{C}_1, \mathscr{C}_2$ make angles with the flow directions in Regions 1, 2 equal to the Mach angles μ_1, μ_2 for those Regions. Thus the gas is expanded by such a process which is termed a *Prandtl–Meyer expansion.*

In order to analyse the flow within the expansion fan, let P be a point on a characteristic $\mathscr{C}$ through O where $OP=r$ and the angular distance of P from the normal to AO through O is θ. Let u, v be the radial and transverse velocity components at P so that $u=\dot{r}$, $v=r\dot{\theta}$. Then the vectorial velocity of the gas at P is $\mathbf{q}$ where

$$\mathbf{q}=u\hat{\mathbf{r}}+v\hat{\boldsymbol{\theta}}.$$

If we define the unit vector $\hat{\mathbf{z}}$ by $\hat{\mathbf{z}}=\hat{\mathbf{r}}\wedge\hat{\boldsymbol{\theta}}$, then the vector angular velocity of the frame in which the polar coordinates are measured is $\dot{\theta}\hat{\mathbf{z}}$. Hence using standard theory of rotating axes the total acceleration of a fluid particle at P is $\mathbf{f}$ where

$$\mathbf{f}=\mathbf{f}_0+(\dot{\theta}\hat{\mathbf{z}})\wedge\mathbf{q},$$

and where $\mathbf{f}_0$ denotes the acceleration of the particle relative to the frame. (See F. CHORLTON, *Textbook of Dynamics*, p. 42, Van Nostrand (1963).) Keeping the axes fixed, since

$$\mathbf{q} = u\hat{\mathbf{r}} + v\hat{\boldsymbol{\theta}},$$

$$\mathbf{f}_0 = \dot{u}\hat{\mathbf{r}} + \dot{v}\hat{\boldsymbol{\theta}},$$

and so

$$\mathbf{f} = \dot{u}\hat{\mathbf{r}} + \dot{v}\hat{\boldsymbol{\theta}} + (\dot{\theta}\hat{\mathbf{z}}) \wedge (u\hat{\mathbf{r}} + v\hat{\boldsymbol{\theta}})$$
$$= (\dot{u} - v\dot{\theta})\hat{\mathbf{r}} + (\dot{v} + u\dot{\theta})\hat{\boldsymbol{\theta}}.$$

Hence the radial and transverse components of acceleration of the fluid particle at P are $[f_r, f_\theta]$, where

$$f_r = \dot{u} - v\dot{\theta}$$

$$= \frac{\partial u}{\partial r}\dot{r} + \frac{\partial u}{\partial \theta}\dot{\theta} - v\dot{\theta}$$

$$= u\frac{\partial u}{\partial r} + \frac{v}{r}\left(\frac{\partial u}{\partial \theta} - v\right);$$

$$f_\theta = \dot{v} + u\dot{\theta}$$

$$= \frac{\partial v}{\partial r}\dot{r} + \frac{\partial v}{\partial \theta}\dot{\theta} + u\dot{\theta}$$

$$= u\frac{\partial v}{\partial r} + \frac{v}{r}\left(\frac{\partial v}{\partial \theta} + u\right).$$

The equation of motion of the fluid, in the absence of body forces, is

$$\mathbf{f} = -\frac{1}{\rho}\nabla p = -\frac{1}{\rho}\left(\frac{\partial p}{\partial r}\hat{\mathbf{r}} + \frac{1}{r}\frac{\partial p}{\partial \theta}\hat{\boldsymbol{\theta}}\right).$$

Thus equating components gives

$$u\frac{\partial u}{\partial r} + \frac{v}{r}\frac{\partial u}{\partial \theta} - \frac{v^2}{r} = -\frac{1}{\rho}\frac{\partial p}{\partial r} \tag{1}$$

$$u\frac{\partial v}{\partial r} + \frac{v}{r}\frac{\partial v}{\partial \theta} + \frac{uv}{r} = -\frac{1}{\rho r}\frac{\partial p}{\partial \theta}. \tag{2}$$

The equation of continuity, $\nabla \cdot (\rho\mathbf{q}) = 0$ for steady flow, assumes the form

$$\frac{1}{r}\left\{\frac{\partial}{\partial r}(\rho r u) + \frac{\partial}{\partial \theta}(\rho v)\right\} = 0$$

or

$$\frac{\partial}{\partial r}(\rho r u) + \frac{\partial}{\partial \theta}(\rho v) = 0. \tag{3}$$

Along any Mach line $\mathscr{C}$ conditions are uniform so that all physical quantities in equations (1), (2), (3) are independent of r. Then these equations simplify to the forms

$$v = du/d\theta, \tag{1'}$$

$$v\left(u + \frac{dv}{d\theta}\right) = -\frac{1}{\rho}\frac{dp}{d\theta}, \tag{2'}$$

$$\rho u + (d/d\theta)(\rho v) = 0. \tag{3'}$$

From (3′),

$$\rho\left(u + \frac{dv}{d\theta}\right) + v\frac{d\rho}{d\theta} = 0, \tag{4}$$

and eliminating $u + dv/d\theta$ from (2′), (4) gives

$$-\frac{1}{v}\frac{dp}{d\theta} + v\frac{d\rho}{d\theta} = 0$$

or

$$v^2 = dp/d\rho = a^2.$$

Thus

$$v = du/d\theta = a, \tag{5}$$

where a is the speed of sound at P. Equation (5) shows that $q \geqslant a$ so that *at no point of the field is the flow subsonic.*

From (1′), (2′),

$$v\frac{dv}{d\theta} + u\frac{du}{d\theta} = -\frac{1}{\rho}\frac{dp}{d\theta},$$

or

$$\tfrac{1}{2}(u^2 + v^2) + \int \frac{dp}{\rho} = \text{const.} = \tfrac{1}{2}q_{max}^2. \tag{6}$$

Equation (6), is, of course, Bernoulli's equation recovered. If we assume the flow to be isentropic so that $p = k\rho^\gamma$, then (6) gives

$$u^2 + (\gamma+1)(\gamma-1)^{-1}v^2 = q_{max}^2. \tag{7}$$

From (1′), (7),

$$v = du/d\theta = \lambda\sqrt{(q_{max}^2 - u^2)}, \tag{8}$$

where $\lambda^2 = (\gamma-1)/(\gamma+1)$. Integrating (8),

$$\left.\begin{aligned} u &= q_{max}\sin\lambda(\theta-\theta_0), \\ v &= a = du/d\theta = \lambda q_{max}\cos\lambda(\theta-\theta_0), \end{aligned}\right\} \tag{9}$$

where θ_0 is an integration constant. When $\theta = \theta_0$, $u = 0$, $v = a$.

Let us now suppose the flow in Region 1 is sonic so that $u = 0$, $v = a$, $\theta_0 = 0$, the characteristic $\mathscr{C}_1$ being normal to OA at O. Then the Prandtl–Meyer expansion will turn sonic flow into supersonic and throughout the expansion fan,

$$\left.\begin{aligned} u &= q_{max} \sin \lambda\theta, \\ v = a &= \lambda q_{max} \cos \lambda\theta \end{aligned}\right\}.$$

Further,

$$\cot \mu = \frac{u}{v} = \frac{1}{\lambda} \tan \lambda\theta.$$

Now

$$\begin{aligned} p/p_0 &= \{1 + \tfrac{1}{2}(\gamma - 1)M^2\}^{-\gamma/(\gamma-1)} \\ &= \{1 + \tfrac{1}{2}(\gamma - 1) \operatorname{cosec}^2 \mu\}^{-\gamma/(\gamma-1)} \\ &= \{(2 \cos^2 \lambda\theta)/(\gamma + 1)\}^{\gamma/(\gamma-1)}, \quad \text{on simplification.} \end{aligned}$$

The last relationship shows that $p = 0$ when

$$\lambda\theta = \tfrac{1}{2}\pi,$$

or

$$\theta = \pi/(2\lambda).$$

At any point $P(r, \theta)$ in the fan, a diagram shows that the angle which the direction of flow makes with the initial direction in Region 1 is φ where

$$\begin{aligned} \varphi &= \theta + \mu - \tfrac{1}{2}\pi \\ &= \theta + \tan^{-1}(\lambda \cot \lambda\theta) - \tfrac{1}{2}\pi. \end{aligned}$$

When $\theta = \pi/(2\lambda)$,

$$\varphi = \pi/(2\lambda) - \tfrac{1}{2}\pi \simeq 129°{\cdot}3 \text{ for } \gamma = 1{\cdot}400.$$

This shows that *the maximum angle through which uniform sonic flow can be turned by the mechanism of Prandtl–Meyer expansion round a sharp bend is approximately* 129°·3. If the angle of bend exceeds this critical amount, then the flow in Region 2 separates from OB to form a void as shown in Fig. 7.43.

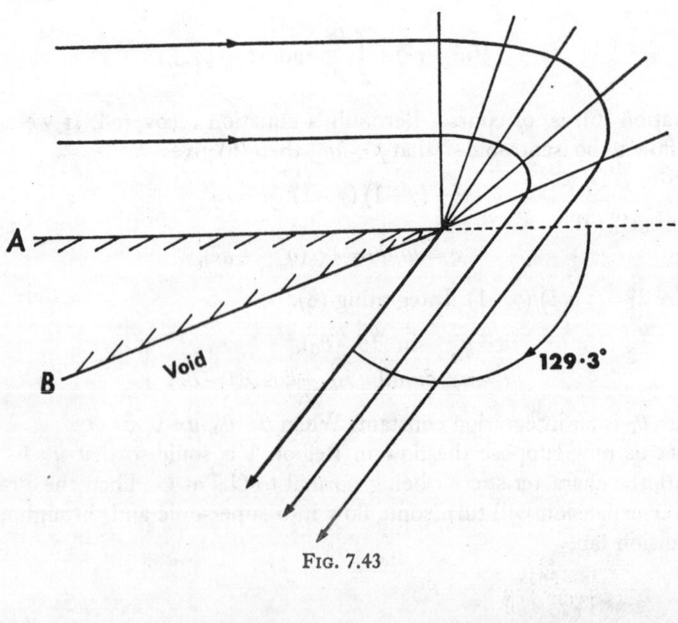

Fig. 7.43

7.9.5 Axially-Symmetric Flows in Three Dimensions

Now suppose we have a field of irrotational homentropic supersonic flow symmetrical about an axis Ox. Let (x, R, θ) be the cylindrical polar co-ordinates at any point P in the field, $R=0$ specifying the axis of symmetry. Then all physical quantities are functions of x, R only. Starting from the equation of motion in the form

$$\mathbf{q}\cdot\nabla(\tfrac{1}{2}\mathbf{q}^2)=a^2\nabla\cdot\mathbf{q},$$

and taking

$$\mathbf{q}=u\mathbf{i}+v\hat{\mathbf{R}}=-\nabla\varphi,$$
$$u=-\varphi_x,\ v=-\varphi_R,$$

we have

$$\nabla\equiv\mathbf{i}\frac{\partial}{\partial x}+\hat{\mathbf{R}}\frac{\partial}{\partial R}+\hat{\boldsymbol{\theta}}\left(\frac{1}{R}\frac{\partial}{\partial\theta}\right),$$

$$\mathbf{q}\cdot\nabla\equiv u\frac{\partial}{\partial x}+v\frac{\partial}{\partial R},$$

$$(\mathbf{q}\cdot\nabla)(\tfrac{1}{2}\mathbf{q}^2)=\tfrac{1}{2}\left(u\frac{\partial}{\partial x}+v\frac{\partial}{\partial R}\right)(u^2+v^2)$$

$$=u^2\frac{\partial u}{\partial x}+uv\frac{\partial v}{\partial x}+vu\frac{\partial u}{\partial R}+v^2\frac{\partial v}{\partial R}$$

$$=-u^2\varphi_{xx}-2uv\varphi_{xR}-v^2\varphi_{RR}.$$

Also,

$$\nabla\cdot\mathbf{q}=-\nabla^2\varphi$$

$$=-\frac{1}{R}\left\{\frac{\partial}{\partial x}\left(R\frac{\partial\varphi}{\partial x}\right)+\frac{\partial}{\partial R}\left(R\frac{\partial\varphi}{\partial R}\right)+\frac{\partial}{\partial\theta}(0)\right\}$$

$$=-\varphi_{xx}-\varphi_{RR}-(1/R)\varphi_R.$$

Thus the equation of fluid motion becomes

$$(a^2-u^2)\varphi_{xx}-2uv\varphi_{xR}+(a^2-v^2)\varphi_{RR}+(a^2/R)\varphi_R=0. \tag{1}$$

From the relation $\mathbf{q}=-\nabla\varphi(x, R)$, we have

$$dx\varphi_{xx}+dR\varphi_{xR}=-du, \tag{2}$$

$$dx\varphi_{xR}+dR\varphi_{RR}=-dv. \tag{3}$$

Equations (1), (2), (3) show that the characteristic directions are given by

$$\begin{vmatrix}(a^2-u^2) & -2uv & (a^2-v^2)\\ dx & dR & 0\\ 0 & dx & dR\end{vmatrix}=0$$

or

$$\frac{dR}{dx}=\frac{-uv\pm a\sqrt{(u^2+v^2-a^2)}}{a^2-u^2}. \tag{4}$$

The equation (4) shows that real distinct characteristics are obtained whenever $a<q$, i.e. when the flow is supersonic. Thus, as for the case of plane flow, the method of characteristics is available for supersonic irrotational homentropic flow.

Along the characteristics the following relationship obtains:

$$\begin{vmatrix} (a^2-u^2) & (a^2/R)v & (a^2-v^2) \\ dx & -du & 0 \\ 0 & -dv & dR \end{vmatrix}=0,$$

or

$$\frac{dv}{du}=-\frac{a^2-u^2}{a^2-v^2}\frac{dR}{dx}-\frac{a^2v}{R(a^2-v^2)}\frac{dR}{du}. \tag{5}$$

From (4), (5),

$$\frac{dv}{du}=\frac{uv\mp a\sqrt{(q^2-a^2)}}{a^2-v^2}-\frac{a^2v}{a^2-v^2}\cdot\frac{1}{R}\cdot\frac{dR}{du}. \tag{6}$$

For purposes of computation (4) and (6) are expressed in more convenient forms by putting

$$u=q\cos\theta,\ v=q\sin\theta.$$

(Note that θ no longer has the significance of a cylindrical polar coordinate element.) Then, writing $a=q\sin\mu$, equation (4) gives

$$\frac{dR}{dx}=\frac{-\sin\theta\cos\theta\pm\sin\mu\cos\mu}{\sin^2\mu-\cos^2\theta}=\frac{\sin 2\theta\mp\sin 2\mu}{\cos 2\theta+\cos 2\mu}$$

$$=\frac{\sin 2\theta+\sin(\mp 2\mu)}{\cos 2\theta+\cos(\mp 2\mu)}=\frac{2\sin(\theta\mp\mu)\cos(\theta\pm\mu)}{2\cos(\theta\mp\mu)\cos(\theta\pm\mu)},$$

so that

$$dR/dx=\tan(\theta\mp\mu). \tag{7}$$

Equation (6) becomes

$$\frac{dv}{du}=\frac{\cos\theta\sin\theta\mp\sin\mu\cos\mu}{\sin^2\mu-\sin^2\theta}-\frac{q\sin^2\mu\sin\theta}{\sin^2\mu-\sin^2\theta}\,\frac{1}{R}\,\frac{dR}{du}.$$

Writing $z=(1/q)\,dq/d\theta$

$$\frac{dv}{du}=\frac{dv/d\theta}{du/d\theta}=\frac{z\sin\theta+\cos\theta}{z\cos\theta-\sin\theta}.$$

Also,

$$\frac{dR}{du}=\frac{dR}{d\theta}\bigg/\frac{du}{d\theta}=\frac{dR/d\theta}{(dq/d\theta)\cos\theta-q\sin\theta}.$$

Thus

$$\frac{z\sin\theta+\cos\theta}{z\cos\theta-\sin\theta}=\frac{\cos\theta\sin\theta\mp\cos\mu\sin\mu}{\sin^2\mu-\sin^2\theta}$$
$$-\frac{\sin^2\mu\sin\theta}{\sin^2\mu-\sin^2\theta}\cdot\frac{1}{z\cos\theta-\sin\theta}\frac{1}{R}\frac{dR}{d\theta}.$$

Now

$$\frac{\cos\theta\sin\theta\mp\cos\mu\sin\mu}{\sin^2\mu-\sin^2\theta}=\frac{\sin 2\theta+\sin(\mp 2\mu)}{\cos 2\theta-\cos(\mp 2\mu)}=\frac{2\sin(\theta\mp\mu)\cos(\theta\pm\mu)}{-2\sin(\theta\mp\mu)\sin(\theta\pm\mu)}$$
$$=-\cot(\theta\pm\mu).$$

Also,

$$\frac{\sin^2\mu\sin\theta}{\sin^2\mu-\sin^2\theta}=\frac{2\sin^2\mu\sin\theta}{\cos 2\theta-\cos(\mp 2\mu)}=\frac{2\sin^2\mu\sin\theta}{-2\sin(\theta\mp\mu)\sin(\theta\pm\mu)}.$$

Thus we have

$$\frac{z\sin\theta+\cos\theta}{z\cos\theta-\sin\theta}$$
$$=-\cot(\theta\pm\mu)+\frac{\sin^2\mu\sin\theta}{\sin(\theta\mp\mu)\sin(\theta\pm\mu)}\cdot\frac{1}{z\cos\theta-\sin\theta}\cdot\frac{1}{R}\frac{dR}{d\theta},$$

or

$$z\{\sin\theta\sin(\theta\pm\mu)+\cos\theta\cos(\theta\pm\mu)\}+\{\cos\theta\sin(\theta\pm\mu)-\sin\theta\cos(\theta\pm\mu)\}$$
$$=\frac{\sin^2\mu\sin\theta}{\sin(\theta\mp\mu)}\cdot\frac{1}{R}\cdot\frac{dR}{d\theta}.$$

Hence

$$z\cos\mu\pm\sin\mu=\frac{\sin^2\mu\sin\theta}{\sin(\theta\mp\mu)}\cdot\frac{1}{R}\cdot\frac{dR}{d\theta},$$

or

$$\frac{1}{q}\frac{dq}{d\theta}=\mp\tan\mu+\frac{\sin\mu\tan\mu\sin\theta}{\sin(\theta\mp\mu)}\frac{1}{R}\frac{dR}{d\theta}. \tag{8}$$

Equations (7) and (8) form the basis of a suitable numerical technique for constructing the characteristics and determining the velocity increments at the network points.

EXERCISE 7

1. A gas flows adiabatically under no body forces in a thin tube whose

cross-section is σ at any point at a distance x from a fixed cross-section. Show that the speed q of the gas satisfies the equation

$$\frac{d}{dx}(\log \sigma)+\left(1-\frac{q^2}{c^2}\right)\frac{d}{dx}(\log q)=0$$

where c is the speed of sound at the point considered.

The speed of the gas in the section $\frac{1}{4}\pi \leqslant x \leqslant \frac{3}{4}\pi$ is given by $q=a\sin x$, $(a<c)$ and $\sigma=\alpha$ when $x=\frac{1}{4}\pi$. Assuming the adiabatic law to hold in this region, find the cross-section of the tube at any point in the section.

(Durham Univ., Gen. B.Sc., 1963)

2. Derive the equation of motion of a perfect fluid. For steady irrotational flow under a conservative force $-\text{grad } V$, show that

$$\int\frac{dp}{\rho}+\tfrac{1}{2}q^2+V=C$$

where p is a function of the density ρ only, where q is the speed of the gas and where C is a constant.

Show that if a compressible fluid flows steadily through a tube of slowly varying cross-section A, then, neglecting external forces,

$$\frac{1}{A}\frac{dA}{dl}=\frac{q^2-a^2}{a^2q}\frac{dq}{dl}$$

where l is the distance along the tube and a is the local speed of sound. Discuss briefly the Laval nozzle.

Further show that if the flow is adiabatic,

$$q^{\gamma-1}\{c_0^2-\tfrac{1}{2}(\gamma-1)q^2\}=\lambda A^{1-\gamma}$$

where c_0 is the speed of sound at a stagnation point, γ is the ratio of the specific heats and λ is a constant.

(Liverpool Univ., B.Sc. Hons., 1963)

3. Derive Bernoulli's equation for the steady motion of a compressible fluid in the form

$$2(c^2-c_0^2)=(\gamma-1)(U^2-q^2),$$

where c is the local speed of sound, q is the local speed of the fluid, γ the adiabatic constant, and c_0, U constants. Show that the local density of the fluid is related to q by the equation

$$\rho=\rho_0\{1+\tfrac{1}{2}(\gamma-1)(U^2-q^2)/c_0^2\}^{1/(\gamma-1)}$$

where ρ_0 is a constant.

If the flow is three-dimensional and purely radial, show that along a line of flow

$$\rho q r^2=\text{constant},$$

where r is the radial distance. Deduce that if the motion is subsonic then q must steadily decrease with r, and that if $\rho \to \rho_0 > 0$, then $q \to 0$.

(London Univ., B.Sc. Spec. Hons., 1959)

4. A polytropic gas with pressure p_0, density ρ_0, and sound speed a_0, occupies the space $x>0$ of a shock tube bounded by a piston at $x=0$. A plane shock wave is formed in the gas by suddenly moving the piston forward with speed U. Show that the speed of the shock s is given by

$$s=\tfrac{1}{2}\{U+\sqrt{[U^2+4(1-\lambda^2)^2a_0^2]}\}/(1-\lambda^2), \quad \text{where } \lambda^2=(\gamma-1)/(\gamma+1).$$

When $U/a_0 \gg 1$ show that the pressure p_1 and density ρ_1 behind the shock are given approximately by

$$\frac{\rho_1}{\rho_0} \simeq \frac{1}{\lambda^2}, \quad \frac{p}{p_0} \simeq \frac{(1+\lambda^2)U^2}{(1-\lambda^2)^2 a_0^2}.$$

(London Univ., B.Sc. Spec. Hons., 1962)

5. A wedge is introduced symmetrically into a uniform stream of sufficiently high Mach number for there to be a pair of oblique shocks attached to the vertex of the wedge. Formulate the equations of the system, and show that the semi-angle of the wedge must not exceed $\operatorname{cosec}^{-1} \gamma$ if this régime is to be possible.

(Imperial College, B.Sc. Spec., 1964)

6. State the Rankine–Hugoniot relations connecting flow conditions on the two sides of a stationary normal shock in a perfect gas.

The Mach number and density upstream are M_1, ρ_1, respectively and the density downstream is ρ_2. Prove that

$$\frac{\rho_2}{\rho_1}=\frac{(\gamma+1)M_1^2}{2+(\gamma-1)M_1^2},$$

where γ is the ratio of the specific heats of the gas.

Suppose now that the shock is inclined at an angle θ to the upstream direction where the flow is hypersonic ($M_1 \gg 1$). Show that the flow deflection φ caused by the shock satisfies

$$\tan \varphi = \sin 2\theta/(\gamma+\cos 2\theta)$$

and find its maximum value.

(Cambridge Univ., M.T. II, 1964)

7. Describe in detail the connections between the velocities and between the variables of state on the two sides of a shock wave.

Show that in the case of a stationary normal shock in a perfect gas, in which it can be assumed that the coefficients of specific heat c_p, c_v are constants ($\gamma=c_p/c_v$), the rise in temperature across the shock is

$$T_2-T_1=2\gamma p_1(\gamma M_1^2+1)(M_1^2-1)/\{\rho_1 M_1^2 c_p(\gamma+1)^2\}$$

where p_1, ρ_1, M_1 are the pressure, the density and the Mach number upstream from the shock.

Explain how the results for a normal shock may be modified to give the results for a stationary oblique shock. Prove that, if the suffix 2 refers to downstream conditions, $\{\rho_2(p_2-p_1)\}/\{p_1(\rho_2-\rho_1)\}$ depends only on the speed of the upstream flow and the angle it makes with the shock wave.

(London Univ., B.Sc. Spec. Hons., 1964)

8. Derive the general equation of motion of steady irrotational homentropic compressible flow in the form

$$a^2\nabla^2\varphi+\mathbf{q}\cdot\nabla(\tfrac{1}{2}\mathbf{q}^2)=0,$$

where φ is the velocity potential, $\mathbf{q}=-\nabla\varphi$ is the fluid velocity and a is the local speed of sound in the fluid.

Show that if such a flow is two-dimensional and specified by polar coordinates (r, θ) in the plane of flow, then the equation may be written in the form

$$(a^2-u^2)\varphi_{rr}-2\frac{uv}{r}\varphi_{r\theta}+(a^2-v^2)\frac{\varphi_{\theta\theta}}{r^2}=(a^2+v^2)\frac{u}{r},$$

$[u, v]$ being the radial and transverse velocity components, the subscripts r, θ denoting partial differentiation with respect to the specified variables. Show that the directions of the characteristics in the physical plane are given by the roots of the quadratic equation

$$(a^2-v^2)\left(\frac{dr}{d\theta}\right)^2+2uvr\frac{dr}{d\theta}+r^2(a^2-u^2)=0$$

and obtain the condition for the characteristics to be real and distinct. Write down (in determinantal form) the appropriate conditions which obtain along the characteristics.

(Univ. of Aston in Birmingham, Tech. Maths. IV, 1965)

9. Derive the equation for the velocity potential φ in steady, homentropic, irrotational, plane flow of a compressible fluid,

$$(a^2-u^2)\frac{\partial^2\varphi}{\partial x^2}-2uv\frac{\partial^2\varphi}{\partial x\partial y}+(a^2-v^2)\frac{\partial^2\varphi}{\partial y^2}=0$$

where $u=\partial\varphi/\partial x, v=\partial\varphi/\partial y$ and a is the speed of sound.

Show that the transformation in which u and v are taken as independent variables, and

$$\Phi=ux+vy-\varphi,$$

implies that $\partial\Phi/\partial u=x$, $\partial\Phi/\partial v=y$ and leads to the equation

$$(a^2-u^2)\frac{\partial^2\Phi}{\partial v^2}+2uv\frac{\partial^2\Phi}{\partial u\partial v}+(a^2-v^2)\frac{\partial^2\Phi}{\partial v^2}=0.$$

By considering the variations of x and y in a direction $dv/du = \lambda$ in the hodograph plane, λ being a parameter, show that there are two characteristic directions, given by the roots λ_1, λ_2 of

$$(a^2 - u^2) - 2uv\lambda + (a^2 - v^2)\lambda^2 = 0,$$

which are real where the flow is supersonic, and along which the conditions

$$dx + \lambda_2\, dy = 0, \quad dx + \lambda_1\, dy = 0$$

respectively hold.

(Imperial College, B.Sc. Spec., 1963)

8

VISCOUS FLOW

8.1 Stress Components in a Real Fluid

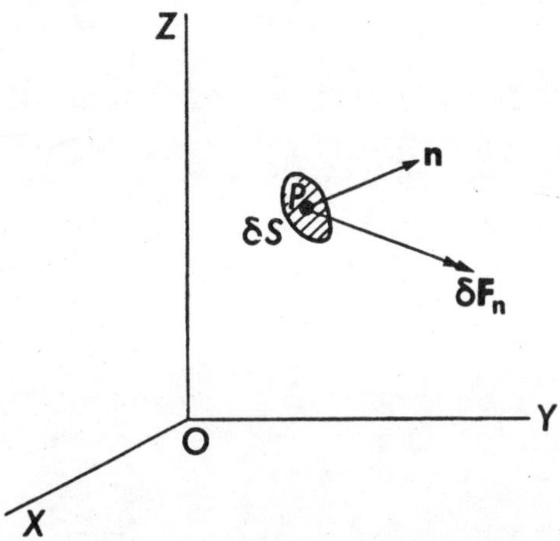

FIG. 8.1

Let δS be a small rigid plane area inserted at a point P in a viscous fluid. Cartesian coordinates (x, y, z) are referred to a set of fixed axes OX, OY, OZ. Suppose that $\delta \mathbf{F}_n$ is the force exerted by the moving fluid on one side of δS, the unit vector $\mathbf{n}$ being taken to specify the normal at P to δS on this side. We know that in the case of an inviscid fluid, $\delta \mathbf{F}_n$ is aligned with $\mathbf{n}$. For a viscous fluid, however, frictional forces are called into play between the fluid and the surface so that $\delta \mathbf{F}_n$ will also have a component tangential to δS. We suppose the Cartesian components of $\delta \mathbf{F}_n$ to be $[\delta F_{nx}, \delta F_{ny}, \delta F_{nz}]$, so that

$$\delta \mathbf{F}_n = \delta F_{nx}\mathbf{i} + \delta F_{ny}\mathbf{j} + \delta F_{nk}\mathbf{k}.$$

Then the *components of stress* parallel to the axes are defined to be p_{nx}, p_{ny}, p_{nz}, where

$$p_{nx} = \lim_{\delta S \to 0} \left(\frac{\delta F_{nx}}{\delta S} \right) = \frac{dF_{nx}}{dS},$$

$$p_{ny} = \lim_{\delta S \to 0} \left(\frac{\delta F_{ny}}{\delta S} \right) = \frac{dF_{ny}}{dS},$$

$$p_{nz} = \lim_{\delta S \to 0} \left(\frac{\delta F_{nz}}{\delta S} \right) = \frac{dF_{nz}}{dS}.$$

In the components p_{nx}, p_{ny}, p_{nz}, the first suffix n denotes the direction of the normal to the elemental plane δS: the second suffix, x, y, or z, denotes the direction in which the component is measured.

If we identify $\mathbf{n}$ in turn with the unit vectors $\mathbf{i}, \mathbf{j}, \mathbf{k}$ in $\overline{OX}, \overline{OY}, \overline{OZ}$ (which is achieved by suitably re-orientating δS), we obtain the following three sets of stress components:

$$\begin{array}{lll} p_{xx}, & p_{xy}, & p_{xz}; \\ p_{yx}, & p_{yy}, & p_{yz}; \\ p_{zx}, & p_{zy}, & p_{zz}. \end{array}$$

The diagonal elements p_{xx}, p_{yy}, p_{zz} of this array are called ***normal*** or ***direct stresses***. The remaining six elements are called ***shearing stresses***. For an inviscid fluid,

$$p_{xx} = p_{yy} = p_{zz} = -p;$$
$$p_{xy} = p_{xz} = \ldots = 0.$$

Here, we regard the direct stresses as positive when they are tensile and negative when compressive, so that p is the hydrostatic pressure.

The matrix

$$\begin{bmatrix} p_{xx} & p_{xy} & p_{xz} \\ p_{yx} & p_{yy} & p_{yz} \\ p_{zx} & p_{zy} & p_{zz} \end{bmatrix}$$

is called the ***stress matrix***. Knowledge of its components enables us to calculate the total forces on any area at any chosen point. The quantities $p_{ij} (i, j = x, y, z)$ are called the components of the ***stress tensor*** whose matrix is of the above form. Clearly p_{ij} is a second-order tensor.

8.2 Relations between Cartesian Components of Stress

We consider the motion of a small rectangular parallelepiped of viscous fluid, its centre being $P(x, y, z)$ and its edges of lengths $\delta x, \delta y, \delta z$, parallel to fixed Cartesian axes (Fig. 8.2). The mass $\rho \delta x \delta y \delta z$ of the fluid element remains constant and the element is presumed to move along with the fluid. In the diagram the points P_1, P_2 have coordinates $(x - \frac{1}{2}\delta x, y, z)$, $(x + \frac{1}{2}\delta x, y, z)$.

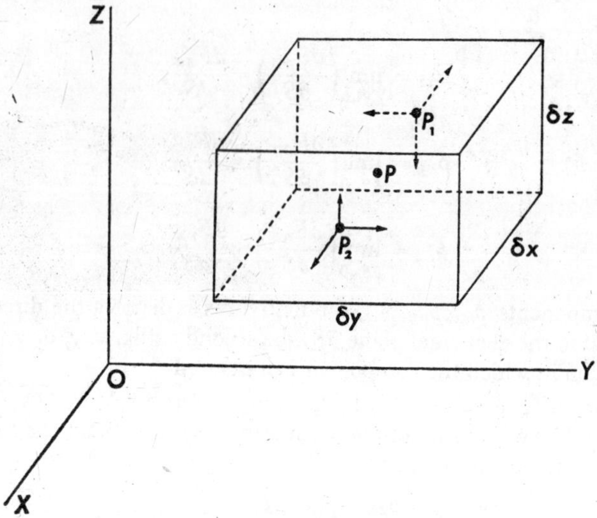

FIG. 8.2

At $P(x, y, z)$, the force components parallel to $\overline{OX}, \overline{OY}, \overline{OZ}$ on the surface of area $\delta y \times \delta z$ through P and having $\mathbf{i}$ as unit normal are

$$[p_{xx}\delta y\delta z, \quad p_{xy}\delta y\delta z, \quad p_{xz}\delta y\delta z].$$

At $P_2(x+\frac{1}{2}\delta x, y, z)$, since $\mathbf{i}$ is the unit normal measured outwards from the fluid, the corresponding force components across the parallel plane of area $\delta y \times \delta z$ are

$$\left[\left\{p_{xx}+\tfrac{1}{2}\delta x\left(\frac{\partial p_{xx}}{\partial x}\right)\right\}\delta y\delta z, \quad \left\{p_{xy}+\tfrac{1}{2}\delta x\left(\frac{\partial p_{xy}}{\partial x}\right)\right\}\delta y\delta z,\right.$$
$$\left.\left\{p_{xz}+\tfrac{1}{2}\delta x\left(\frac{\partial p_{xz}}{\partial x}\right)\right\}\delta y\delta z\right].$$

For the parallel plane through $P_1(x-\frac{1}{2}\delta x, y, z)$, since $-\mathbf{i}$ is the unit normal drawn outwards from the fluid element, the corresponding components are

$$\left[-\left\{p_{xx}-\tfrac{1}{2}\delta x\left(\frac{\partial p_{xx}}{\partial x}\right)\right\}\delta y\delta z, \quad -\left\{p_{xy}-\tfrac{1}{2}\delta x\left(\frac{\partial p_{xy}}{\partial x}\right)\right\}\delta y\delta z,\right.$$
$$\left.-\left\{p_{xz}-\tfrac{1}{2}\delta x\left(\frac{\partial p_{xz}}{\partial x}\right)\right\}\delta y\delta z\right].$$

The forces on the parallel planes through P_1 and P_2 are equivalent to a single force at P with components

$$\left[\frac{\partial p_{xx}}{\partial x}, \quad \frac{\partial p_{xy}}{\partial x}, \quad \frac{\partial p_{xz}}{\partial x}\right]\delta x\delta y\delta z$$

together with couples whose moments (to the third order) are

$$\begin{cases} -p_{xz}\delta x\delta y\delta z \text{ about } Oy; \\ +p_{xy}\delta x\delta y\delta z \text{ about } Oz. \end{cases}$$

Similarly the pair of faces perpendicular to the y-axis give a force at P having components

$$\left[\frac{\partial p_{yx}}{\partial y}, \quad \frac{\partial p_{yy}}{\partial y}, \quad \frac{\partial p_{yz}}{\partial y}\right]\delta x\delta y\delta z$$

together with couples of moments

$$\begin{cases} -p_{yx}\delta x\delta y\delta z \text{ about } Oz; \\ +p_{yz}\delta x\delta y\delta z \text{ about } Ox. \end{cases}$$

The pair of faces perpendicular to the z-axis give a force at P having components

$$\left[\frac{\partial p_{zx}}{\partial z}, \quad \frac{\partial p_{zy}}{\partial z}, \quad \frac{\partial p_{zz}}{\partial z}\right]\delta x\delta y\delta z$$

together with couples of moments

$$\begin{cases} -p_{zy}\delta x\delta y\delta z \text{ about } Ox; \\ +p_{zx}\delta x\delta y\delta z \text{ about } Oy. \end{cases}$$

Taking into account the surface forces on all six faces of the cuboid, we see that they reduce to a single force at P having components

$$\left[\left(\frac{\partial p_{xx}}{\partial x}+\frac{\partial p_{yx}}{\partial y}+\frac{\partial p_{zx}}{\partial z}\right), \quad \left(\frac{\partial p_{xy}}{\partial x}+\frac{\partial p_{yy}}{\partial y}+\frac{\partial p_{zy}}{\partial z}\right), \quad \left(\frac{\partial p_{xz}}{\partial x}+\frac{\partial p_{yz}}{\partial y}+\frac{\partial p_{zz}}{\partial z}\right)\right]\delta x\delta y\delta z,$$

together with a vector couple having Cartesian components

$$[(p_{yz}-p_{zy}), \quad (p_{zx}-p_{xz}), \quad (p_{xy}-p_{yx})]\delta x\delta y\delta z.$$

Now suppose the external body forces acting at P are $[X, Y, Z]$ per unit mass, so that the total body force on the element has components $[X, Y, Z]\,\rho\delta x\delta y\delta z$. Take moments about the **i**-direction through P. Then

Total moment of forces = Moment of inertia about axis × Angular acceleration,

i.e.

$$(p_{yz}-p_{zy})\delta x\delta y\delta z+0_4=0_5.$$

Here 0_n signifies a quantity of the nth order of smallness in δx, δy, δz. Thus, to the third order of smallness,

$$(p_{yz}-p_{zy})\delta x\delta y\delta z=0,$$

so that as the element becomes vanishingly small, we obtain

$$p_{yz}=p_{zy}.$$

Similarly,

$$p_{zx}=p_{xz}, \quad p_{xy}=p_{yx}.$$

Thus the stress matrix is diagonally symmetric and contains only six unknowns.

8.3 Translational Motion of Fluid Element

Taking account of surface forces and body forces, the total force component in the **i**-direction acting on the element in the last section is

$$\left(\frac{\partial p_{xx}}{\partial x}+\frac{\partial p_{yx}}{\partial y}+\frac{\partial p_{zx}}{\partial z}\right)\delta x\delta y\delta z+X\rho\delta x\delta y\delta z.$$

As the mass $\rho\delta x\delta y\delta z$ is considered fixed, if the velocity of P is $\mathbf{q}=[u, v, w]$ at time t, then the equation of motion in the **i**-direction is

$$\left(\frac{\partial p_{xx}}{\partial x}+\frac{\partial p_{yx}}{\partial y}+\frac{\partial p_{zx}}{\partial z}\right)\delta x\delta y\delta z+\rho X\delta x\delta y\delta z=(\rho\delta x\delta y\delta z)\frac{du}{dt}$$

or

$$\frac{\partial p_{xx}}{\partial x}+\frac{\partial p_{yx}}{\partial y}+\frac{\partial p_{zx}}{\partial z}+\rho X=\rho\frac{du}{dt}.$$

Now $u=u(x, y, z, t)$ so that

$$\frac{du}{dt}=\frac{\partial u}{\partial t}+u\frac{\partial u}{\partial x}+v\frac{\partial u}{\partial y}+w\frac{\partial u}{\partial z}.$$

Thus by cyclic permutation we obtain the three equations of motion in the **i**, **j**, **k** directions:

$$\begin{cases}\dfrac{\partial u}{\partial t}+u\dfrac{\partial u}{\partial x}+v\dfrac{\partial u}{\partial y}+w\dfrac{\partial u}{\partial z}=X+\dfrac{1}{\rho}\left(\dfrac{\partial p_{xx}}{\partial x}+\dfrac{\partial p_{yx}}{\partial y}+\dfrac{\partial p_{zx}}{\partial z}\right),\\[2ex] \dfrac{\partial v}{\partial t}+u\dfrac{\partial v}{\partial x}+v\dfrac{\partial v}{\partial y}+w\dfrac{\partial v}{\partial z}=Y+\dfrac{1}{\rho}\left(\dfrac{\partial p_{xy}}{\partial x}+\dfrac{\partial p_{yy}}{\partial y}+\dfrac{\partial p_{zy}}{\partial z}\right),\\[2ex] \dfrac{\partial w}{\partial t}+u\dfrac{\partial w}{\partial x}+v\dfrac{\partial w}{\partial y}+w\dfrac{\partial w}{\partial z}=Z+\dfrac{1}{\rho}\left(\dfrac{\partial p_{xz}}{\partial x}+\dfrac{\partial p_{yz}}{\partial y}+\dfrac{\partial p_{zz}}{\partial z}\right).\end{cases}$$

In tensor form, if the coordinates are (x_i), the velocity components (u_i), the body force components (X_i) $(i=1,2,3)$, the equations of motion may be written

$$\frac{\partial u_i}{\partial t}+u_j u_{i,j}=X_i+\frac{1}{\rho}p_{ji,j} \quad (i,j=1,2,3).$$

8.4 The Rate of Strain Quadric and Principal Stresses

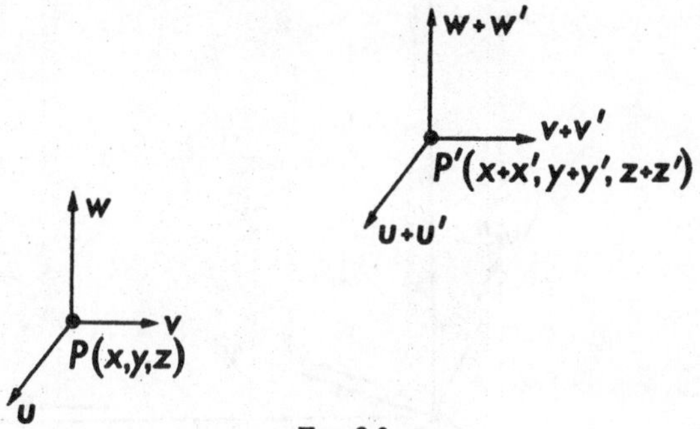

FIG. 8.3

In Fig. 8.3, $P(x,y,z)$, $P'(x+x',y+y',z+z')$ are neighbouring points in a moving fluid at which the corresponding velocities have Cartesian components $[u,v,w]$, $[u+u',v+v',w+w']$, all primed quantities being small. Then, from Section 2.11, the rate of strain quadric for P is

$$ax'^2+by'^2+cz'^2+2fy'z'+2gz'x'+2hx'y'=\text{const.},$$

where $a=\partial u/\partial x, \ldots, 2f=(\partial w/\partial y)+(\partial v/\partial z), \ldots$. Moreover the velocity increments due to pure strain are

$$[(ax'+hy'+gz'), \quad (hx'+by'+fz'), \quad (gx'+fy'+cz')].$$

When the rate of strain quadric is referred to its principal axes, it assumes the form

$$Ax''^2+By''^2+Cz''^2=\text{const.},$$

and the velocity increments due to pure strain referred to the principal axes are $[Ax'', By'', Cz'']$. The stresses in the fluid element are independent of its translation and rotation and depend solely on its distortion which is due to the rates of strain set up. At each point P of the moving fluid there is a unique rate of strain quadric whose axes are in the directions in which the lines joining P to neighbouring particles on the axes are undergoing elongations along the axes. Thus we conclude that the stresses across the axial or *principal planes* of the rate of strain quadric are normal to these planes. Such stresses are called *principal stresses* and will be denoted by p_1, p_2, p_3.

8.5 Some Further Properties of the Rate of Strain Quadric

Before proceeding further with the analysis of viscous fluid motion it is necessary to consider some further properties of the rate of strain quadric.

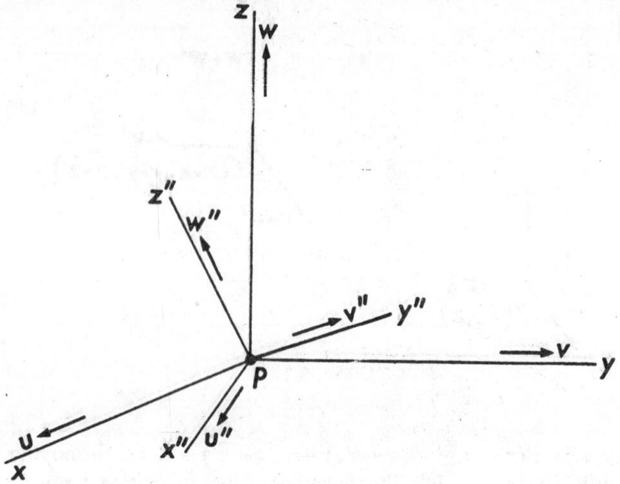

FIG. 8.4

Let P be a point in the fluid moving with velocity $\mathbf{q}$ and let Px, Py, Pz be a set of axes through P parallel to a given set of fixed Cartesian axes. Let $[u, v, w]$ be the components of $\mathbf{q}$ with respect to these axes. Then, in accordance with Section 2.11, the rate of strain quadric with centre at P has equation

$$ax^2+by^2+cz^2+2fyz+2gzx+2hxy=\text{const.}, \tag{1}$$

where $a=(\partial u/\partial x)_P, \ldots, 2f=(\partial w/\partial y)_P+(\partial v/\partial z)_P, \ldots$. Here the point (x, y, z) is a neighbouring point to P. Now let Px'', Py'', Pz'' be the principal axes for this quadric surface and let $[u'', v'', w'']$ be the components of $\mathbf{q}$ along the directions of these principal axes. Then the equation of the rate of strain quadric referred to its principal axes is

$$Ax''^2+By''^2+Cz''^2=\text{const.}, \tag{2}$$

where $A=(\partial u''/\partial x'')_P, \ldots$.

If we suppose Px'', Py'', Pz'' to have D.C.'s $[l_1, m_1, n_1]$, $[l_2, m_2, n_2]$, $[l_3, m_3, n_3]$ with respect to the axes Px, Py, Pz, then Px has D.C.'s $[l_1, l_2, l_3]$, etc., with respect to the principal axes. These facts are best seen from the following table.

TABLE OF DIRECTION COSINES

	Px	Py	Pz
Px''	l_1	m_1	n_1
Py''	l_2	m_2	n_2
Pz''	l_3	m_3	n_3

Clearly the following relations hold:

$$\begin{cases} l_1^2+m_1^2+n_1^2=1, \text{ etc.,} \\ l_1^2+l_2^2+l_3^2=1, \text{ etc.,} \\ l_1m_1+l_2m_2+l_3m_3=0, \text{ etc.,} \\ l_1l_2+m_1m_2+n_1n_2=0, \text{ etc.} \end{cases}$$

(In all there are 12 such equations.)

We use the above properties to evaluate a, b, c, f, g, h in terms of A, B, C and the D.C.'s. Since

$$\frac{\partial}{\partial x} \equiv \frac{\partial x''}{\partial x}\frac{\partial}{\partial x''}+\frac{\partial y''}{\partial x}\frac{\partial}{\partial y''}+\frac{\partial z''}{\partial x}\frac{\partial}{\partial z''}$$

$$\equiv l_1\frac{\partial}{\partial x''}+l_2\frac{\partial}{\partial y''}+l_3\frac{\partial}{\partial z''}$$

and

$$u=l_1u''+l_2v''+l_3w'',$$

$$a=\frac{\partial u}{\partial x}=l_1^2\frac{\partial u''}{\partial x''}+l_2^2\frac{\partial v''}{\partial y''}+l_3^2\frac{\partial w''}{\partial z''},$$

since $(\partial v''/\partial z'')+(\partial w''/\partial y'')=0$, etc. Similar relations hold for b, c so that

$$\left.\begin{aligned} a&=l_1^2A+l_2^2B+l_3^2C, \\ b&=m_1^2A+m_2^2B+m_3^2C, \\ c&=n_1^2A+n_2^2B+n_3^2C. \end{aligned}\right\} \tag{3}$$

Adding together equations (3) gives

$$a+b+c=A+B+C,$$

showing that $a+b+c$ is an invariant at any particular point for all orientations of axes. This is obvious since

$$a+b+c=\text{div }\mathbf{q}.$$

We further observe that

$$2f=\frac{\partial w}{\partial y}+\frac{\partial v}{\partial z}$$

$$=\left(m_1\frac{\partial}{\partial x'}+m_2\frac{\partial}{\partial y'}+m_3\frac{\partial}{\partial z'}\right)(n_1u'+n_2v'+n_3w')$$

$$+\left(n_1\frac{\partial}{\partial x'}+n_2\frac{\partial}{\partial y'}+n_3\frac{\partial}{\partial z'}\right)(m_1u'+m_2v'+m_3w')$$

$$=2\left(m_1n_1\frac{\partial u'}{\partial x'}+m_2n_2\frac{\partial v'}{\partial y'}+m_3n_3\frac{\partial w'}{\partial z'}\right).$$

By cyclic permutation we find two similar relations so that

$$\left.\begin{aligned}f&=m_1n_1A+m_2n_2B+m_3n_3C,\\ g&=n_1l_1A+n_2l_2B+n_3l_3C,\\ h&=l_1m_1A+l_2m_2B+l_3m_3C.\end{aligned}\right\}\tag{4}$$

8.6 Stress Analysis in Fluid Motion

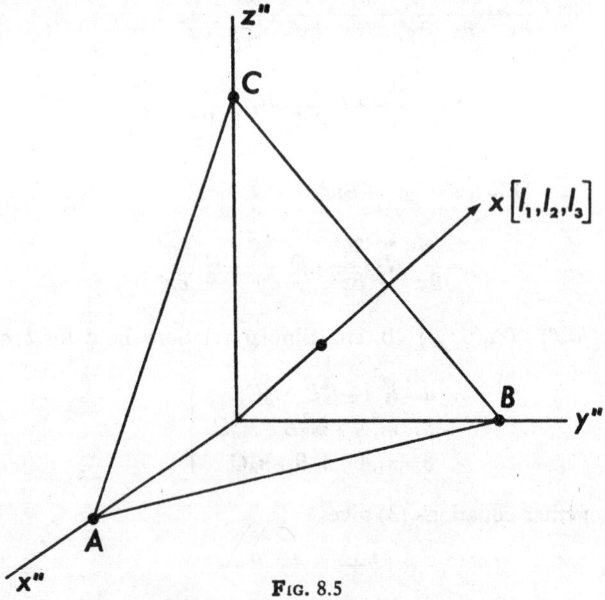

FIG. 8.5

With the notation of Section 8.5, a plane perpendicular to Px cuts the principal axes Px'', Py'', Pz'' of the rate of strain quadric in A, B, C to form a small tetrahedron of fluid $PABC$ (Fig. 8.5). If δA denotes this area of the face ABC, then $l_1\delta A$, $l_2\delta A$, $l_3\delta A$ are the areas of the faces PBC, PCA, PAB. Since these last three are principal planes, using the notation of Section 9.4 it follows that the only forces on them are the normal forces $p_1l_1\delta A, p_2l_2\delta A, p_3l_3\delta A$. The forces on the face ABC are $p_{xx}\delta A, p_{xy}\delta A, p_{xz}\delta A$ in the x-, y-, z-directions.

The equation of motion in the x-direction (assuming the element to be a particle of fixed mass moving with the fluid) is

$$p_{xx}\delta A - l_1(p_1l_1\delta A) - l_2(p_2l_2\delta A) - l_3(p_3l_3\delta A) = 0_3,$$

where 0_3 is a term of the third order. Thus in the limit as the volume of the element tends to zero, we obtain

$$\left.\begin{aligned} p_{xx} &= l_1^2p_1 + l_2^2p_2 + l_3^2p_3, \\ \text{and similarly}\quad p_{yy} &= m_1^2p_1 + m_2^2p_2 + m_3^2p_3, \\ p_{zz} &= n_1^2p_1 + n_2^2p_2 + n_3^2p_3. \end{aligned}\right\} \qquad (1)$$

Adding the equations (1) together we find

$$p_{xx} + p_{yy} + p_{zz} = p_1 + p_2 + p_3 \quad (= -3p, \text{ say}).$$

Thus the sum of the normal stresses across any three perpendicular planes at a point is an invariant. We denote this sum by $-3p$, so that p now denotes the mean pressure at the point.

Resolving in the direction Py and neglecting third order terms,

$$\left.\begin{aligned} p_{xy} &= l_1m_1p_1 + l_2m_2p_2 + l_3m_3p_3, \\ \text{and similarly}\quad p_{yz} &= m_1n_1p_1 + m_2n_2p_2 + m_3n_3p_3, \\ p_{zx} &= n_1l_1p_1 + n_2l_2p_2 + n_3l_3p_3. \end{aligned}\right\} \qquad (2)$$

Thus equations (1) and (2) express the six distinct components of the stress matrix in terms of the principal stresses.

8.7 Relations between Stress and Rate of Strain

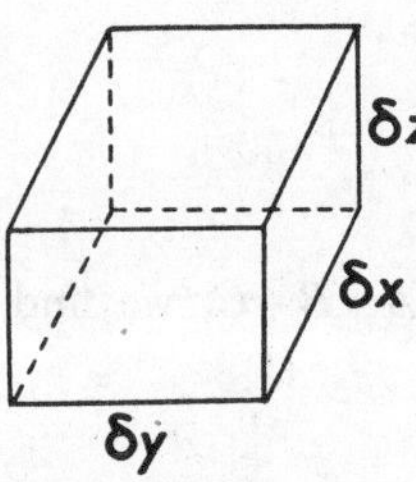

FIG. 8.6(i)

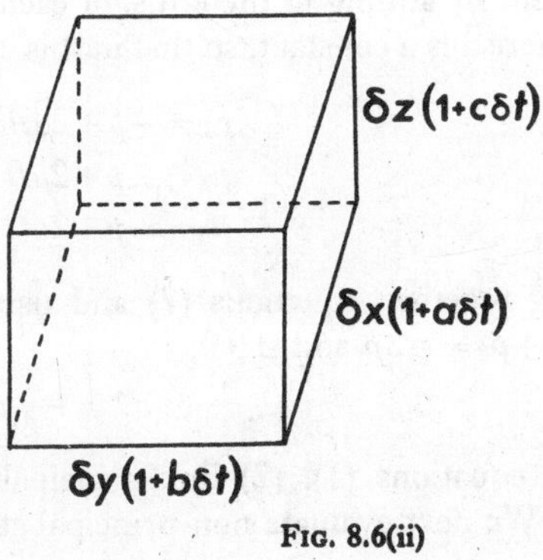

FIG. 8.6(ii)

Using the analyses of the foregoing sections, we now link together the stresses and rates of strain in a viscous fluid in motion. To this end suppose Fig. 8.6(i) represents a particle of fluid at time t in the shape of a rectangular parallelepiped of edges $\delta x, \delta y, \delta z$, parallel to fixed Cartesian axes. At time t the velocity component in the x-direction at the corner (x, y, z) of the box is u and so that at the corner $(x+\delta x, y, z)$ is $u+(\partial u/\partial x)\delta x$ or $u+a\delta x$. Thus at time $(t+\delta t)$, the edge δx has grown to length $\delta x+a\delta x\delta t$, since $a\delta x$ is the relative velocity increase between its two ends. Similarly, the edges $\delta y, \delta z$ have grown to lengths $\delta y(1+b\delta t)$, $\delta z(1+c\delta t)$, respectively. Thus the volumetric increment in the interval δt is

$$\delta x\delta y\delta z(1+a\delta t)(1+b\delta t)(1+c\delta t)-\delta x\delta y\delta z \simeq (a+b+c)\delta x\delta y\delta z\delta t,$$

which gives a dilatation or volumetric strain in time δt of $(a+b+c)\delta t$. Hence at time t, the *rate of dilatation* is Δ where

$$\Delta=a+b+c=\frac{\partial u}{\partial x}+\frac{\partial v}{\partial y}+\frac{\partial w}{\partial z}=\text{div }\mathbf{q}.$$

This quantity has been seen to be invariant at each point of the fluid: its value is also $A+B+C$, in terms of principal rates of strain. Further, we know that the equation of continuity for an incompressible fluid is $\Delta=0$, but for a compressible one $\Delta\neq 0$. We discuss the two cases separately.

In the case of an *incompressible fluid*, we suppose that the principal stresses p_1, p_2, p_3 differ from their mean value $-p$ by quantities proportional to the rates of distortion A, B, C in the principal directions. Thus we write

$$\left.\begin{aligned} p_1&=-p+2\mu A,\\ p_2&=-p+2\mu B,\\ p_3&=-p+2\mu C,\end{aligned}\right\} \tag{1}$$

where μ is a constant.

In the case of a *compressible fluid*, we have the additional effect of the rate of dilatation Δ manifesting itself equally in all directions. This effect we represent by adding to the R.H.S. of each of the equations (1) the quantity $\lambda\Delta$, where λ is a constant, so that in this case

$$\left.\begin{aligned} p_1&=-p+2\mu A+\lambda\Delta,\\ p_2&=-p+2\mu B+\lambda\Delta,\\ p_3&=-p+2\mu C+\lambda\Delta.\end{aligned}\right\} \tag{2}$$

Adding together equations (2) and using $\Delta=A+B+C$, we find, since $p_1+p_2+p_3=-3p$ and $\Delta\neq 0$,

$$\lambda=-\tfrac{2}{3}\mu. \tag{3}$$

The equations (1), (2) link principal stresses with principal rates of strain. We next evaluate non-principal stresses $p_{xx}, \ldots, p_{yz}, \ldots$ in terms of non-principal rates of strain. In Section 8.6, equations (1), we showed that

$$p_{xx}=l_1^2p_1+l_2^2p_2+l_3^2p_3.$$

Using the equations (2) of this Section, we obtain

$$\begin{aligned} p_{xx} &= -p(l_1^2+l_2^2+l_3^2)+2\mu(l_1^2A+l_2^2B+l_3^2C)+\lambda\Delta(l_1^2+l_2^2+l_3^2) \\ &= -p+2\mu a+\lambda\Delta \text{ (from Section 8.5)} \\ &= -p+2\mu(\partial u/\partial x)+\lambda\Delta. \end{aligned}$$

Permuting the symbols gives the three equations

$$\left.\begin{aligned} p_{xx} &= -p+2\mu(\partial u/\partial x)+\lambda\Delta, \\ p_{yy} &= -p+2\mu(\partial v/\partial y)+\lambda\Delta, \\ p_{zz} &= -p+2\mu(\partial w/\partial z)+\lambda\Delta, \end{aligned}\right\} \qquad (3')$$

where $\Delta=\text{div}\,\mathbf{q}$, $\Delta=0$ for incompressible flow and $\lambda=-\frac{2}{3}\mu$ for compressible flow.

Further, from Section 8.6, equations (2),

$$\begin{aligned} p_{yz} &= m_1n_1p_1+m_2n_2p_2+m_3n_3p_3 \\ &= (-p+\lambda\Delta)\,(m_1n_1+m_2n_2+m_3n_3)+2\mu(m_1n_1A+m_2n_2B+m_3n_3C) \\ &= 2\mu f \quad \text{(from Section 8.5)} \\ &= \mu\left(\frac{\partial w}{\partial y}+\frac{\partial v}{\partial z}\right). \end{aligned}$$

Thus we obtain the three equations

$$\left.\begin{aligned} p_{yz}=p_{zy} &= \mu\left(\frac{\partial w}{\partial y}+\frac{\partial v}{\partial z}\right), \\ p_{zx}=p_{xz} &= \mu\left(\frac{\partial u}{\partial z}+\frac{\partial w}{\partial x}\right), \\ p_{xy}=p_{yx} &= \mu\left(\frac{\partial v}{\partial x}+\frac{\partial u}{\partial y}\right), \end{aligned}\right\} \qquad (4)$$

which are true for compressible and incompressible fluids.

The equations (3), (4) may be conveniently combined in tensorial form. Thus if (x_i) denote the Cartesian coordinates, (u_i) the velocity components $(i=1, 2, 3)$, then both sets of equations may be written

$$p_{ij}=(\lambda\Delta-p)\delta_{ij}+\mu(u_{i,j}+u_{j,i}) \quad (i,j=1,2,3), \qquad (5)$$

where $\Delta=u_{j,i}$, $p=-\frac{1}{3}p_{i,i}$, $\Delta=0$ for incompressible flow, $\lambda=-\frac{2}{3}\mu$ for compressible flow.

8.8 The Coefficient of Viscosity and Laminar Flow

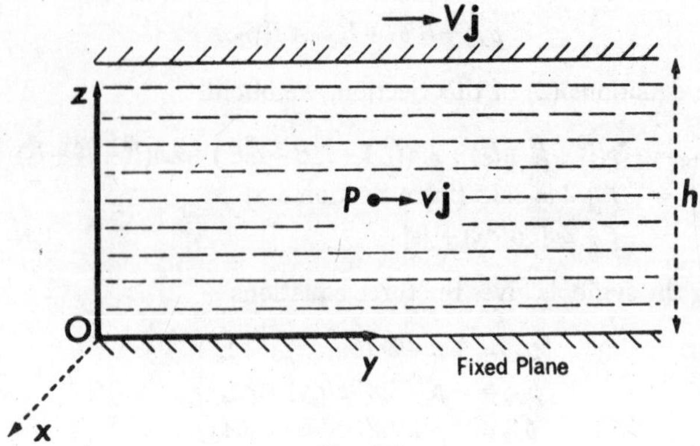

FIG. 8.7

Figure 8.7 shows two parallel planes $z=0, z=h$ a small distance h apart, the space between containing a thin film of viscous fluid. The plane $z=0$ is held fixed whilst the upper plane is given a constant velocity rightwards of amount $V\mathbf{j}$. Then provided V is not excessively large, the layers of liquid in contact with $z=0$ are at rest whilst those in contact with $z=h$ are moving with velocity $V\mathbf{j}$, i.e. there is no slip between fluid and either surface. A velocity gradient is set up in the fluid between the planes. At some point $P(x, y, z)$ in between the planes the fluid velocity will be $v\mathbf{j}$ where $0<v<V$ and v is independent of x, y. Thus when z is fixed, v is fixed, i.e. the fluid moves in layers parallel to the two planes. Such flow is termed *laminar*. Due to the viscosity of the fluid there is friction between these parallel layers.

Experimental work shows that the shearing stress on the moving plane is proportional to V/h when h is sufficiently small. Thus we write this stress in the form $\mu' V/h$, where μ' is a constant called the *coefficient of viscosity*. Now suppose $h \to 0$. Then the stress on the fixed plane is

$$p_{zy} = \mu' \lim_{h\to 0} \left(\frac{V}{h}\right) = \mu' \frac{dv}{dz}. \tag{1}$$

In the equations (4) of Section 8.7 we put $u=0, w=0, v=v(z)$, then we obtain

$$p_{zy} = \mu(dv/dz), \quad p_{xz}=0, \quad p_{yx}=0. \tag{2}$$

The equations (1), (2) above clearly show that

$$\mu' = \mu, \tag{3}$$

i.e. the constant μ of Section 8.7 is the coefficient of viscosity.

From (1) above we readily find the dimensions of μ. Thus

$$[\mu]=\frac{[p_{zy}]}{[dv/dz]}=\frac{(MLT^{-2})/L^2}{(LT^{-1})L^{-1}}=ML^{-1}T^{-1}, \tag{4}$$

where M, L, T signify mass, length and time.

In aerodynamics, a rather more important quantity is the ***kinematic coefficient of viscosity*** ν defined by

$$\nu=\mu/\rho. \tag{5}$$

Thus

$$[\nu]=L^2T^{-1}. \tag{6}$$

For most fluids μ depends on the pressure and temperature. For gases, according to the kinetic theory, μ is independent of the pressure but decreases with the temperature.

8.9 The Navier-Stokes Equations of Motion of a Viscous Fluid

In Section 8.3 we obtained the translational equation of motion in the form

$$\frac{du}{dt}=X+\frac{1}{\rho}\left(\frac{\partial p_{xx}}{\partial x}+\frac{\partial p_{yx}}{\partial y}+\frac{\partial p_{zx}}{\partial z}\right).$$

On substituting

$$p_{xx}=-p+2\mu(\partial u/\partial x)+\lambda\Delta,$$

$$p_{yx}=\mu\left(\frac{\partial v}{\partial x}+\frac{\partial u}{\partial y}\right),$$

$$p_{zx}=\mu\left(\frac{\partial u}{\partial z}+\frac{\partial w}{\partial x}\right),$$

we obtain

$$\frac{du}{dt}=X-\frac{1}{\rho}\frac{\partial p}{\partial x}+\nu\nabla^2u+\left(\nu+\frac{\lambda}{\rho}\right)\frac{\partial\Delta}{\partial x}.$$

Since $\lambda=-\frac{2}{3}\mu$ for a compressible fluid and since $\Delta=0$ for an incompressible fluid this equation may be written unambiguously for the two cases in the form

$$\frac{du}{dt}=X-\frac{1}{\rho}\frac{\partial p}{\partial x}+\nu\nabla^2u+\frac{1}{3}\nu\frac{\partial\Delta}{\partial x}.$$

Thus the equations of motion in the three directions may be written

$$\left.\begin{aligned}\frac{du}{dt}&=X-\frac{1}{\rho}\frac{\partial p}{\partial x}+\nu\nabla^2 u+\frac{1}{3}\nu\frac{\partial\Delta}{\partial x},\\ \frac{dv}{dt}&=Y-\frac{1}{\rho}\frac{\partial p}{\partial y}+\nu\nabla^2 v+\frac{1}{3}\nu\frac{\partial\Delta}{\partial y},\\ \frac{dw}{dt}&=Z-\frac{1}{\rho}\frac{\partial p}{\partial z}+\nu\nabla^2 w+\frac{1}{3}\nu\frac{\partial\Delta}{\partial z}.\end{aligned}\right\} \tag{1}$$

The tensor form of these equations is

$$\frac{du_i}{dt}=X_i-\frac{1}{\rho}p_{,i}+\nu u_{i,jj}+\tfrac{1}{3}\nu\Delta_{,i}. \tag{1'}$$

Writing $\mathbf{q}=[x, y, z]$, $\mathbf{F}=[X, Y, Z]$ the vectorial form of equations (1) is clearly

$$\frac{d\mathbf{q}}{dt}=\mathbf{F}-\nabla\int\frac{dp}{\rho}+\nu\nabla^2\mathbf{q}+\tfrac{1}{3}\nu\nabla(\nabla\cdot\mathbf{q}). \tag{2}$$

Now $d\mathbf{q}/dt=(\partial\mathbf{q}/\partial t)+(\mathbf{q}\cdot\nabla)\mathbf{q}=(\partial\mathbf{q}/\partial t)+\nabla(\tfrac{1}{2}\mathbf{q}^2)-\mathbf{q}\wedge(\nabla\wedge\mathbf{q})$,

$$\nabla\wedge(\nabla\wedge\mathbf{q})=\nabla(\nabla\cdot\mathbf{q})-\nabla^2\mathbf{q},$$

so that (2) may also be written

$$\frac{\partial\mathbf{q}}{\partial t}+\nabla(\tfrac{1}{2}\mathbf{q}^2)-\mathbf{q}\wedge(\nabla\wedge\mathbf{q})=\mathbf{F}-\nabla\int\frac{dp}{\rho}+\frac{4}{3}\nu\nabla(\nabla\cdot\mathbf{q})-\nu\nabla\wedge(\nabla\wedge\mathbf{q}). \tag{3}$$

Any of the forms (1), (2), (3) are called the *Navier–Stokes equations of motion.*

For incompressible flow, the forms (2), (3) give

$$\begin{aligned}d\mathbf{q}/dt&=\mathbf{F}-(1/\rho)\nabla p+\nu\nabla^2\mathbf{q}\\ &=\mathbf{F}-(1/\rho)\nabla p-\nu\nabla\wedge(\nabla\wedge\mathbf{q})\end{aligned} \tag{4}$$

where, as before, $d\mathbf{q}/dt$ may be developed in the form on the L.H.S. of (3). The equation (4) shows that for incompressible flow the equation of motion differs from Euler's equation of motion in inviscid flow by the term $-\nu\nabla\wedge(\nabla\wedge\mathbf{q})$. This term, due to viscosity, increases the complexity by increasing the order of the differential equation. Thus an additional boundary condition is required. This is furnished by the condition that there must be no slip between a viscous fluid and its boundary. For this reason, we cannot obtain the solution to the corresponding inviscid flow problem by solving (4) and then letting $\nu\to 0$.

8.10 Some Solvable Problems in Viscous Flow

There is no general solution to the Navier–Stokes equations. Nevertheless there are some special problems which can be solved. We consider a few here. The problems treated all relate to incompressible fluids.

8.10.1 Steady Motion between Parallel Planes

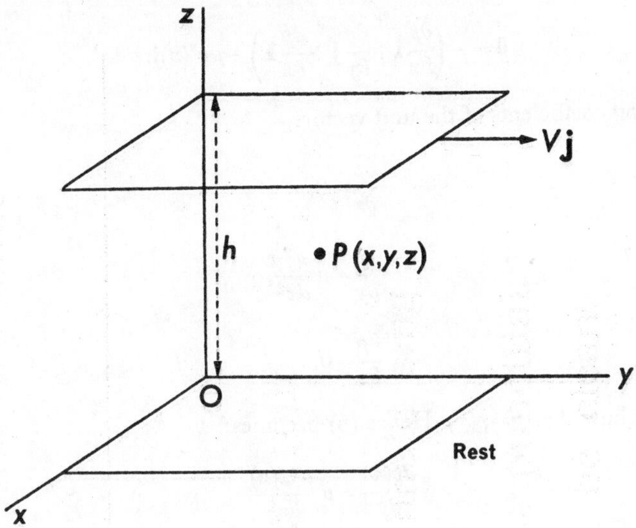

Fig. 8.8

The region $0 \leqslant z \leqslant h$ between the planes $z=0$, $z=h$ is filled with incompressible viscous fluid (Fig. 8.8). The plane $z=0$ is held at rest and the plane $z=h$ moves with constant velocity $V\mathbf{j}$. It is required to determine the nature of the flow when conditions are steady, assuming there is no slip between the fluid and either boundary, neglecting body forces.

Let $P(x, y, z)$ be any point within the fluid. Then the velocity $\mathbf{q}$ at P will be of the form

$$\mathbf{q}=v(y, z)\mathbf{j}. \tag{1}$$

The equation of continuity, $\nabla \cdot \mathbf{q}=0$, gives

$$\partial v/\partial y=0 \tag{2}$$

and from (1), (2) we infer that

$$\mathbf{q}=v(z)\mathbf{j}. \tag{3}$$

With no body forces, the Navier–Stokes vector equation of motion may be taken in the form

$$\frac{\partial \mathbf{q}}{\partial t}+(\mathbf{q}\cdot\nabla)\mathbf{q}=-\frac{1}{\rho}\nabla p+\nu\nabla^2\mathbf{q}. \tag{4}$$

The form (4) is suitable here since we are dealing with axes *fixed in space*. Since the flow is steady, $\partial\mathbf{q}/\partial t=\mathbf{0}$. Also

$$(\mathbf{q}\cdot\nabla)\mathbf{q}=(v\partial/\partial y)v(z)\mathbf{j}=\mathbf{0};$$

and

$$\nabla^2\mathbf{q}=v''(z)\mathbf{j}.$$

Hence (4) gives

$$\mathbf{0}=-\left(\frac{\partial p}{\partial x}\mathbf{i}+\frac{\partial p}{\partial y}\mathbf{j}+\frac{\partial p}{\partial z}\mathbf{k}\right)+\mu v''(z)\mathbf{j}.$$

Equating coefficients of the unit vectors,

$$\frac{\partial p}{\partial x}=0, \tag{5}$$

$$\frac{\partial p}{\partial y}=\mu\frac{d^2v}{dz^2}, \tag{6}$$

$$\frac{\partial p}{\partial z}=0. \tag{7}$$

(5), (7) show that $p=p(y)$. Hence (6) becomes

$$\frac{dp(y)}{dy}=\mu\frac{d^2v(z)}{dz^2}. \tag{8}$$

The L.H.S. of (8) is a function of y only: the R.H.S. is a function of z only. Hence each is a constant. As the fluid is moving in the positive y-direction, the pressure $p(y)$ should decrease as y increases. Hence $dp/dy<0$ and so we take

$$\frac{dp(y)}{dy}=\frac{\mu d^2v(z)}{dz^2}=-P,$$

where $P>0$. Solving for v gives

$$v(z)=A+Bz-(P/2\mu)z^2. \tag{9}$$

When $z=0$, $v=0$ and when $z=h$, $v=V$. Hence we find

$$v(z)=\left(\frac{V}{h}+\frac{Ph}{2\mu}\right)z-\frac{P}{2\mu}z^2. \tag{10}$$

---

(10) shows that the velocity profile between the plates is *parabolic*.

The total flow per unit breadth across a plane perpendicular to Oy is

$$\int_0^h v(z)\,dz=\tfrac{1}{2}Vh+\frac{1}{12}\frac{Ph^3}{\mu},$$

and the mean velocity across such a section is

$$\frac{1}{h}\int_0^h v(z)\,dz = \tfrac{1}{2}V + \frac{1}{12}\,\frac{Ph^2}{\mu}.$$

The tangential stress at any point $P(x, y, z)$ is

$$\mu\frac{dv}{dz} = \frac{V}{h} + \frac{Ph}{2\mu} - \frac{Pz}{\mu}.$$

Thus the drag per unit area on the lower plane is $(V/h)+(Ph/2\mu)$ and that on the upper plane is $(V/h)-(Ph/2\mu)$.

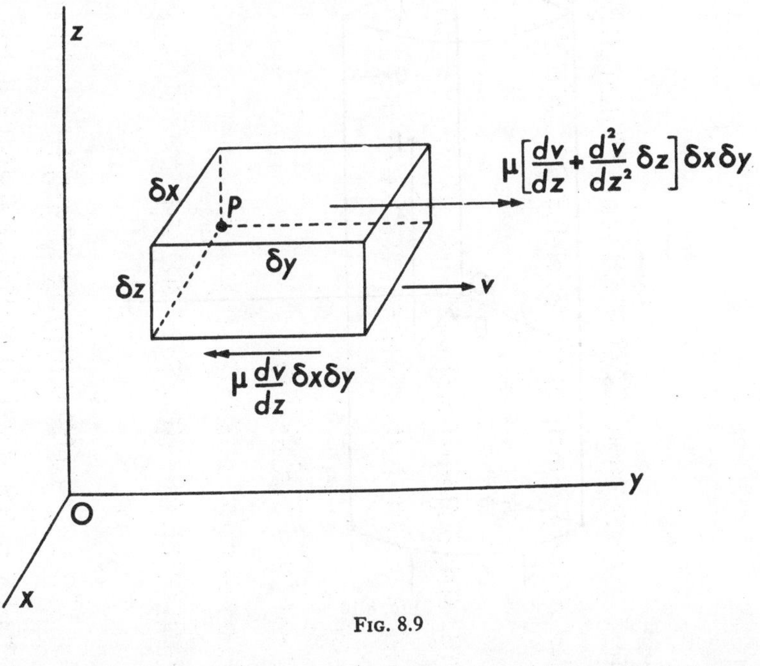

FIG. 8.9

The problem may be treated without using the Navier–Stokes equations. Figure 8.9 shows a rectangular element $\delta x\times\delta y\times\delta z$ of the fluid, the point P being the point (x, y, z). The viscous force on the lower plane, whose velocity is $v(z)\mathbf{j}$, is $-\mu(dv/dz)\delta x\delta y\mathbf{j}$ and that on the upper plane is $\mu\{(dv/dz)+(d^2v/dz^2)\delta z\}\delta x\delta y\mathbf{j}$. The pressure forces on the three faces at P normal to $\mathbf{i}, \mathbf{j}, \mathbf{k}$ are $p\delta y\delta z\mathbf{i}$, $p\delta z\delta x\mathbf{j}$, $p\delta x\delta y\mathbf{k}$: those on the opposite faces of the element are $-\{p+(\partial p/\partial x)\delta x\}\delta y\delta z\mathbf{i}$, $-\{p+(\partial p/\partial y)\delta y\}\delta z\delta x\mathbf{j}$, $-\{p+(\partial p/\partial z)\delta z\}\delta x\delta y\mathbf{k}$. Since the element is unaccelerated, we find, resolving in the three directions.

$$\begin{cases} -\partial p/\partial x = 0, \\ \mu(d^2v/dz^2) - \partial p/\partial y = 0, \\ -\partial p/\partial z = 0, \end{cases}$$

as before.

8.10.2 Steady Flow through Tube of Uniform Circular Cross-Section (Poiseuille Flow)

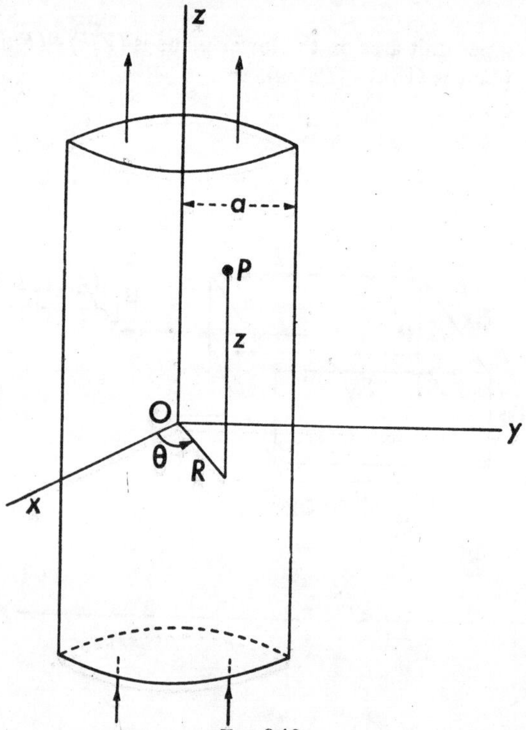

Fig. 8.10

Figure 8.10 illustrates the steady flow of an inviscid incompressible fluid through a circular tube of radius a. P is a point in the fluid having cylindrical polar coordinates (R, θ, z), referred to the origin O on the axis of the tube which is taken as the z-axis. We assume there are no body forces. Then continuity considerations applied to an annular-shaped element of radii $R, R+\delta R$ of the fluid indicate that the fluid velocity is of the form

$$\mathbf{q} = w(R)\mathbf{k}. \tag{1}$$

Let us take the Navier–Stokes vector equation in the form

$$\frac{\partial \mathbf{q}}{\partial t}+(\mathbf{q}\cdot\nabla)\mathbf{q}=-\frac{1}{\rho}\nabla p-\nu\nabla\wedge(\nabla\wedge\mathbf{q}). \tag{2}$$

Here this form is adopted with the last term $-\nu\nabla\wedge(\nabla\wedge\mathbf{q})$, and not $+\nu\nabla^2\mathbf{q}$ as in the previous case, because the axes are *not fixed in space*. For the same reason the form given on the L.H.S. is more useful than $d\mathbf{q}/dt$. Then we have

$$\partial\mathbf{q}/\partial t=\mathbf{0} \quad \text{(steady conditions)},$$

$$(\mathbf{q}\cdot\nabla)\mathbf{q}=(w\partial/\partial z)\,[w(R)\mathbf{k}]=\mathbf{0},$$

$$\nabla p=\frac{\partial p}{\partial R}\,\hat{\mathbf{R}}+\frac{1}{R}\,\frac{\partial p}{\partial\theta}\,\hat{\boldsymbol{\theta}}+\frac{\partial p}{\partial z}\,\mathbf{k}.$$

[*N.B.* In the above, $(\mathbf{q}\cdot\nabla)\mathbf{q}$ is easy to evaluate since the operand involves the constant unit vector $\mathbf{k}$.]

Also

$$\nabla\wedge\mathbf{q}=\frac{1}{R}\begin{vmatrix}\hat{\mathbf{R}} & R\hat{\boldsymbol{\theta}} & \mathbf{k}\\ \dfrac{\partial}{\partial R} & \dfrac{\partial}{\partial\theta} & \dfrac{\partial}{\partial z}\\ 0 & 0 & w(R)\end{vmatrix}=-w'(R)\hat{\boldsymbol{\theta}}.$$

Hence

$$\nabla\wedge(\nabla\wedge\mathbf{q})=\frac{1}{R}\begin{vmatrix}\hat{\mathbf{R}} & R\hat{\boldsymbol{\theta}} & \mathbf{k}\\ \dfrac{\partial}{\partial R} & \dfrac{\partial}{\partial\theta} & \dfrac{\partial}{\partial z}\\ 0 & -Rw'(R) & 0\end{vmatrix}=-\frac{1}{R}\,\frac{d}{dR}(Rw')\mathbf{k}.$$

Thus the equation (2) becomes

$$\mathbf{0}=-\frac{1}{\rho}\left[\frac{\partial p}{\partial R}\,\hat{\mathbf{R}}+\frac{\partial p}{R\partial\theta}\,\hat{\boldsymbol{\theta}}+\frac{\partial p}{\partial z}\,\mathbf{k}\right]+\frac{\nu}{R}\,\frac{d(Rw')}{dR}\,\mathbf{k}.$$

Equating coefficients of the unit vectors gives

$$\frac{\partial p}{\partial R}=0, \tag{3}$$

$$\frac{\partial p}{\partial\theta}=0, \tag{4}$$

$$\frac{\partial p}{\partial z}=\frac{\mu}{R}\,\frac{d(Rw')}{dR}. \tag{5}$$

(3), (4) show that $p=p(z)$ so that (5) becomes

$$\frac{dp(z)}{dz}=\frac{\mu}{R}\,\frac{d}{dR}\,[Rw'(R)]. \tag{6}$$

The L.H.S. of (6) is a function of z only: the R.H.S. is a function of R only. Hence each is constant. As flow is supposed to occur in the positive direction, we suppose $dp/dz<0$. Take each side of (6) to be $-P$, where P is a positive constant. Then

$$\frac{d}{dR}(Rw') = -\frac{PR}{\mu},$$

or

$$\frac{Rdw}{dR} = A - \frac{PR^2}{2\mu}.$$

Hence

$$\frac{dw}{dR} = \frac{A}{R} - \frac{1}{2}\frac{PR}{\mu}$$

and so

$$w(R) = B + A\log R - \tfrac{1}{4}(PR^2/\mu). \tag{7}$$

Now w is finite on $R=0$. Thus we require $A=0$. Also on $R=a$, $w=0$ since there is no slip. Then $B=\frac{1}{4}(Pa^2/\mu)$. Hence

$$\underline{w(R) = \tfrac{1}{4}(P/\mu)(a^2 - R^2).} \tag{8}$$

The form (8) shows that the velocity profile is parabolic, i.e. the plot of w against R from $R=0$ to a is of parabolic shape.

The volume of fluid discharged over any section per unit time is

$$Q = \int_0^a w(R)\cdot 2\pi R\,dR = \frac{\pi P a^4}{8\mu}.$$

If p denotes the pressure difference at two points on the axis of the tube distant l apart, then $P=p/l$.

8.10.3 Steady Flow between Concentric Rotating Cylinders

Figure 8.11 shows two concentric infinite cylinders of radii $a, b(b>a)$ with viscous liquid in between. The inner cylinder is held at rest whilst the outer is rotated with constant angular velocity Ω. Let (R, θ, z) be the cylindrical polar coordinates at a point P in the liquid referred to some point O on the axis of symmetry. Then when the motion is steady the velocity at P will be of the form

$$\mathbf{q} = R\omega(R)\hat{\boldsymbol{\theta}}, \tag{1}$$

where $\omega(R)$ is the angular velocity of the liquid at radial distance R. The pressure at P will clearly be of the form

$$p = p(R). \tag{2}$$

This time we take the Navier–Stokes vector equation in the form

$$\frac{\partial \mathbf{q}}{\partial t}+\nabla(\tfrac{1}{2}\mathbf{q}^2)-\mathbf{q}\wedge(\nabla\wedge\mathbf{q})=-\frac{1}{\rho}\nabla p-\nu\nabla\wedge(\nabla\wedge\mathbf{q}). \tag{3}$$

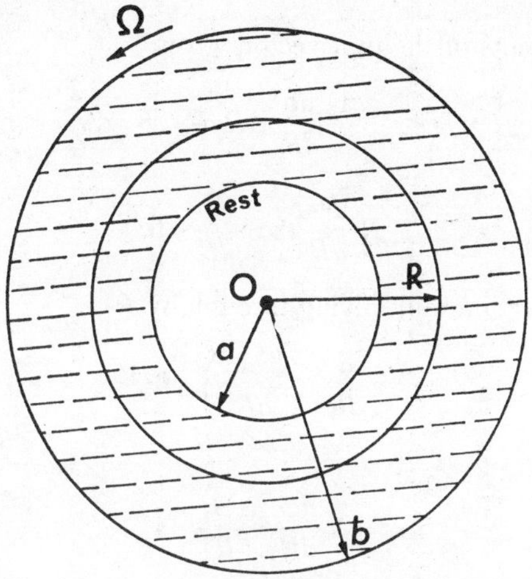

FIG. 8.11

The form of the L.H.S. chosen here is preferred to that of the last example because the form $(\mathbf{q}\cdot\nabla)\mathbf{q}$ is difficult to evaluate directly as the unit vector $\hat{\boldsymbol{\theta}}$ featured in (1) is not constant. For steady flow $\partial\mathbf{q}/\partial t=\mathbf{0}$. Also, since $\nabla\equiv\hat{\mathbf{R}}(\partial/\partial R)+\hat{\boldsymbol{\theta}}(\partial/R\partial\theta)+\mathbf{k}(\partial/\partial z)$,

$$\nabla(\tfrac{1}{2}\mathbf{q}^2)=\hat{\mathbf{R}}\frac{\partial}{\partial R}(\tfrac{1}{2}R^2\omega^2)=\left\{R^2\omega\frac{d\omega}{dR}+R\omega^2\right\}\hat{\mathbf{R}}.$$

We have

$$\nabla\wedge\mathbf{q}=\frac{1}{R}\begin{vmatrix}\hat{\mathbf{R}} & R\hat{\boldsymbol{\theta}} & \mathbf{k}\\ \dfrac{\partial}{\partial R} & \dfrac{\partial}{\partial\theta} & \dfrac{\partial}{\partial z}\\ 0 & R^2\omega & 0\end{vmatrix}=\left(R\frac{d\omega}{dR}+2\omega\right)\mathbf{k};$$

$$\mathbf{q}\wedge(\nabla\wedge\mathbf{q})=R\omega\hat{\boldsymbol{\theta}}\wedge\left(R\frac{d\omega}{dR}+2\omega\right)\mathbf{k}=R\omega\left(R\frac{d\omega}{dR}+2\omega\right)\hat{\mathbf{R}};$$

$$\nabla\wedge(\nabla\wedge\mathbf{q})=\frac{1}{R}\begin{vmatrix}\hat{\mathbf{R}} & R\hat{\boldsymbol{\theta}} & \mathbf{k}\\ \dfrac{\partial}{\partial R} & \dfrac{\partial}{\partial\theta} & \dfrac{\partial}{\partial z}\\ 0 & 0 & R\dfrac{d\omega}{dR}+2\omega\end{vmatrix}=-\left(R\frac{d^2\omega}{dR^2}+3\frac{d\omega}{dR}\right)\hat{\boldsymbol{\theta}}.$$

Thus (3) becomes

$$-R\omega^2\hat{\mathbf{R}} = -\frac{1}{\rho}\frac{dp}{dR}\hat{\mathbf{R}} + \nu\left(R\frac{d^2\omega}{dR^2} + 3\frac{d\omega}{dR}\right)\hat{\boldsymbol{\theta}}.$$

Equating coefficients of the unit vectors gives

$$\frac{1}{\rho}\frac{dp}{dR} = R\omega^2, \tag{4}$$

$$R\frac{d^2\omega}{dR^2} + 3\frac{d\omega}{dR} = 0. \tag{5}$$

From (5), we have, on multiplying through by R^2,

$$\frac{d}{dR}\left(\frac{R^3 d\omega}{dR}\right) = 0$$

whence

$$\frac{d\omega}{dR} = \frac{A}{R^3},$$

$$\omega = B - (A/2R^2).$$

Taking $\omega=0$ when $R=a$, $\omega=\Omega$ when $R=b$, we find

$$\omega(R) = \Omega b^2(1 - a^2R^{-2})(b^2 - a^2)^{-1}. \tag{6}$$

The fluid velocity at radial distance R has magnitude

$$q = R\omega(R) = \Omega b^2(R - a^2R^{-1})(b^2 - a^2)^{-1}.$$

Thus the viscous stress at this location is

$$\mu\frac{dq}{dR} = \mu\Omega b^2(1 + a^2R^{-2})(b^2 - a^2)^{-1}.$$

Hence the couple per unit length on the inner cylinder due to viscous drag has moment

$$(2\pi a \times 1) \times a \times \{2\mu\Omega b^2(b^2 - a^2)^{-1}\} = 4\pi\mu a^2 b^2 \Omega (b^2 - a^2)^{-1}.$$

This result forms the basis of measuring μ for some liquids using a rotation viscometer.

8.11 Steady Viscous Flow in Tubes of Uniform Cross-Section

Before leaving the subject of solvable viscous flow problems, we treat the case of steady incompressible unaccelerated flow through a tube of any uniform cross-section. We neglect body forces. Thus with $d\mathbf{q}/dt = \mathbf{0}$, $\mathbf{F} = \mathbf{0}$, the Navier–Stokes vector equation becomes

$$(1/\rho)\nabla p - \nu\nabla^2\mathbf{q} = \mathbf{0}. \tag{1}$$

Let us work with fixed coordinate axes Ox, Oy, Oz, with Oz taken parallel to the flow so that $\mathbf{q}=w(x,y)\mathbf{k}$. Then (1) gives

$$\nabla p=\mu\left(\frac{\partial^2 w}{\partial x^2}+\frac{\partial^2 w}{\partial y^2}\right)\mathbf{k}. \tag{2}$$

Equating coefficients of the unit vectors,

$$\frac{\partial p}{\partial x}=0, \tag{3}$$

$$\frac{\partial p}{\partial y}=0, \tag{4}$$

$$\frac{\partial p}{\partial z}=\mu\left(\frac{\partial^2 w}{\partial x^2}+\frac{\partial^2 w}{\partial y^2}\right). \tag{5}$$

Equations (3), (4) show $p=p(z)$ so that (5) gives

$$\frac{dp(z)}{dz}=\mu\left(\frac{\partial^2 w(x,y)}{\partial x^2}+\frac{\partial^2 w(x,y)}{\partial y^2}\right). \tag{6}$$

The L.H.S. of (6) is a function of z only: the R.H.S. is a function of x, y only. Thus each is a constant, say $-P$, the minus sign being taken as we expect p to decrease as z increases. Then the problem reduces to the solution of the partial differential equation

$$\frac{\partial^2 w(x,y)}{\partial x^2}+\frac{\partial^2 w(x,y)}{\partial y^2}=-\frac{P}{\mu}, \tag{7}$$

subject to w vanishing on the walls of the tube.

8.11.1 A Uniqueness Theorem

To obtain solutions of (7) which are practical, it is desirable to establish a uniqueness theorem. A form which is a little more general than that required here is the following:

If $\partial^2 w/\partial x^2+\partial^2 w/\partial y^2=f(x,y)$ at all points (x,y) of a region S in the plane Ox, Oy bounded by a closed curve $\mathscr{C}$ and if f is prescribed at each point (x,y) of S and w at each point of $\mathscr{C}$, then any solution $w=w(x,y)$ satisfying these conditions is unique.

Proof:

Let $w=w_1(x,y)$, $w=w_2(x,y)$ be two solutions satisfying

$$\frac{\partial^2 w}{\partial x^2}+\frac{\partial^2 w}{\partial y^2}=f(x,y)$$

in S together with the prescribed boundary conditions on $\mathscr{C}$. We prove that $w_1 \equiv w_2$. To this end, write

$$W(x, y) = w_1(x, y) - w_2(x, y).$$

Then $\partial^2 W/\partial x^2 + \partial^2 W/\partial y^2 = f - f = 0$. Also, on $\mathscr{C}$, $W = 0$, since $w_1 = w_2$ on $\mathscr{C}$.

Now consider

$$I = \iint_S \left\{ \left(\frac{\partial W}{\partial x}\right)^2 + \left(\frac{\partial W}{\partial y}\right)^2 \right\} dx\, dy = \iint_S \left\{ \left(\frac{\partial W}{\partial x}\right)^2 + \left(\frac{\partial W}{\partial y}\right)^2 + W\left(\frac{\partial^2 W}{\partial x^2} + \frac{\partial^2 W}{\partial y^2}\right) \right\} dx\, dy$$

$$= \iint_S \left\{ \frac{\partial}{\partial x}\left(W \frac{\partial W}{\partial x}\right) + \frac{\partial}{\partial y}\left(W \frac{\partial W}{\partial y}\right) \right\} dx\, dy$$

$$= \oint_{\mathscr{C}} \left(W \frac{\partial W}{\partial x} dy - W \frac{\partial W}{\partial y} dx \right) = 0, \quad \text{since } W = 0 \text{ on } \mathscr{C}.$$

Now $(\partial W/\partial x)^2 + (\partial W/\partial y)^2 \geqslant 0$. Since $I = 0$, $(\partial W/\partial x)^2 + (\partial W/\partial y)^2 = 0$ in S which will be true only if

$$\partial W/\partial x = 0 = \partial W/\partial y \text{ at each point of } S.$$

This shows that $W = \text{const.}$ in S. Since $W = 0$ on $\mathscr{C}$, we infer from the continuity of W that $W \equiv 0$ throughout S. This establishes the result.

8.11.2 Tube having Uniform Elliptic Cross-Section

Suppose the cross-section of the tube has equation

$$\frac{x^2}{a^2} + \frac{y^2}{b^2} = 1.$$

Then we must solve

$$\frac{\partial^2 w}{\partial x^2} + \frac{\partial^2 w}{\partial y^2} = -\frac{P}{\mu}$$

subject to $w = 0$ on this section.

We first observe that the function

$$w = k\left(1 - \frac{x^2}{a^2} - \frac{y^2}{b^2}\right)$$

is appropriately zero on the boundary. Regarding k as constant, on substituting w into the partial differential equation we have

$$k\left(-\frac{2}{a^2}-\frac{2}{b^2}\right)=-\frac{P}{\mu}$$

or

$$k=Pa^2b^2/\{2\mu(a^2+b^2)\}.$$

Thus

$$\underline{w(x,y)=Pa^2b^2(1-x^2a^{-2}-y^2b^{-2})\{2\mu(a^2+b^2)\}^{-1}.}$$

The uniqueness theorem shows that this is the required solution.

The volume discharged through the tube per unit time is

$$Q=\int_S w(x,y)\,dS.$$

This can readily be evaluated by double integration over the section. The following expedient is shorter, however. We observe that w is constant over the boundary of the ellipse

$$x=\lambda a\cos\theta,\ y=\lambda b\sin\theta\quad(0\leqslant\lambda\leqslant 1)$$

where, in fact, $w=Pa^2b^2(1-\lambda^2)/\{2\mu(a^2+b^2)\}$. The area between this ellipse and the ellipse

$$x=(\lambda+\delta\lambda)a\cos\theta,\ y=(\lambda+\delta\lambda)b\sin\theta$$

is $\pi(\lambda+\delta\lambda)^2ab-\pi\lambda^2ab\simeq 2\pi ab\lambda\delta\lambda$, to the first order. Thus the volume flowing per unit time between the two elliptic cross-sections is

$$\pi Pa^3b^3(1-\lambda^2)\lambda\delta\lambda/\{\mu(a^2+b^2)\},$$

so that the total volume discharged per unit time is

$$Q=\frac{\pi Pa^3b^3}{\mu(a^2+b^2)}\int_0^1(1-\lambda^2)\lambda\,d\lambda=\frac{\pi Pa^3b^3}{4\mu(a^2+b^2)}.$$

On taking $b=a$, we recover the formula for Q for the circular section of Section 8.10.2.

8.11.3 Tube having Equilateral Triangular Cross-Section

Suppose the section of the tube is the equilateral triangle bounded by the lines

$$x=a,\ y=\pm 3^{-1/2}x.$$

Then if we take

$$w=k(x-a)\left(y^2-\tfrac{1}{3}x^2\right),$$

$w=0$ appropriately on the boundary. Substituting for w in

$$\frac{\partial^2 w}{\partial x^2}+\frac{\partial^2 w}{\partial y^2}=-\frac{P}{\mu}$$

we find

$$k = 3P/(4\mu a).$$

Thus

$$\underline{w = 3P(x-a)\,(y^2 - \tfrac{1}{3}x^2)/(4\mu a).}$$

This solution is unique in virtue of the uniqueness theorem.

The volume discharged per unit time is

$$Q = \int_S w\,dS = 2\int_0^a dx \int_0^{3^{-1/2}x} w\,dy = \frac{Pa^4}{60 \cdot 3^{1/2} \cdot \mu}.$$

8.11.4 Use of Harmonic Functions

It is sometimes possible to see a particular integral $w = w_1(x, y)$ for

$$\frac{\partial^2 w}{\partial x^2} + \frac{\partial^2 w}{\partial y^2} = -\frac{P}{\mu} \text{ in } S$$

which may not satisfy $w = 0$ over the entire boundary $\mathscr{C}$. In these circumstances one may be able to proceed by looking for a suitable harmonic function $w = w_2(x, y)$ satisfying Laplace's equation

$$\frac{\partial^2 w}{\partial x^2} + \frac{\partial^2 w}{\partial y^2} = 0 \text{ in } S$$

so that

$$w = w_1 + kw_2$$

is a solution of the original Poisson-type equation. If one can determine k, and any other constants to satisfy $w = 0$ over the entire boundary, then $w = w_1 + kw_2$ is the unique solution to the problem. It is, therefore, desirable to discuss some solutions of the harmonic equation.

By substituting $w = X(x)Y(y)$ into Laplace's equation and separating the variables we easily show that typical harmonic functions are

$$w(x, y) = \begin{pmatrix} \text{sh} \\ \text{ch} \end{pmatrix} nx \begin{pmatrix} \sin \\ \cos \end{pmatrix} ny \text{ or } e^{\pm nx} \begin{pmatrix} \sin \\ \cos \end{pmatrix} ny;$$

$$w(x, y) = \begin{pmatrix} \sin \\ \cos \end{pmatrix} nx \begin{pmatrix} \text{sh} \\ \text{ch} \end{pmatrix} ny \text{ or } e^{\pm ny} \begin{pmatrix} \sin \\ \cos \end{pmatrix} nx.$$

As the partial differential equation is linear, suitable combinations of these may be combined with any admissible particular integral to form the required solution. Such a procedure usually leads to Fourier series expansions. We illustrate this.

Example

Steady flow through a channel of uniform rectangular cross-section.

Suppose the cross-section is bounded by the planes $x=\pm a, y=\pm b$. Then the problem is to solve

$$\frac{\partial^2 w}{\partial x^2}+\frac{\partial^2 w}{\partial y^2}=-\frac{P}{\mu}\quad (|x|<a,\ |y|<b)$$

subject to $w=0$ on $x=\pm a$ and on $y=\pm b$.

A particular integral satisfying $w=0$ on $x=\pm a$ is

$$w_1(x,y)=P(a^2-x^2)/(2\mu).$$

This does not, however, vanish on $y=\pm b$.

We now seek a suitable solution $w=w_2(x,y)$ satisfying $\nabla^2 w=0$ so that $w=w_1+w_2$ satisfies the given partial differential equation and all boundary conditions.

From symmetry we obviously require that w must be the same at the four points (x,y), $(x,-y)$, $(-x,y)$, $(-x,-y)$. We first choose w_2 to satisfy the same boundary conditions as w_1. A suitable form for w_2 would be

$$w_2=k_n\cos\frac{(2n+1)\pi x}{2a}\,\mathrm{ch}\,\frac{(2n+1)\pi y}{2a},$$

where n is an integer, since $\cos\{(2n+1)\pi/2\}=0$. By superposition

$$w(x,y)=\frac{P(a^2-x^2)}{2\mu}+\sum_{n=0}^{\infty}\left\{k_n\cos\frac{(2n+1)\pi x}{2a}\,\mathrm{ch}\,\frac{(2n+1)\pi y}{2a}\right\}$$

satisfies the given partial differential equation and secures that $w=0$ on $x=\pm a$. We now determine the k_n so that $w=0$ on $y=\pm b$. This requires that

$$-\frac{P}{2\mu}(a^2-x^2)=\sum_{n=0}^{\infty}\left[k_n\,\mathrm{ch}\left\{\frac{(2n+1)\pi b}{2a}\right\}\cos\left\{\frac{(2n+1)\pi x}{2a}\right\}\right],$$

for $-a\leqslant x\leqslant a$. The R.H.S. is a Fourier series expansion for the function $-(P/2\mu)(a^2-x^2)$. Multiplying both sides by $\cos\{(2n+1)\pi x/2a\}$ and integrating from $x=-a$ to $x=a$, we find

$$k_n=\frac{(-1)^{n+1}32a^3}{\pi^3(2n+1)^3\,\mathrm{ch}\,\{(2n+1)\pi b/2a\}}\,\frac{P}{2\mu},$$

yielding the solution

$$w(x,y)=\frac{Pa^2}{2\mu}\left\{1-\frac{x^2}{a^2}-\frac{32}{\pi^3}\sum_{n=0}^{\infty}\frac{(-1)^n\,\mathrm{ch}\,\{(2n+1)\pi y/2a\}\cos\{(2n+1)\pi x/2a\}}{(2n+1)^3\,\mathrm{ch}\,\{(2n+1)\pi b/2a\}}\right\}.$$

8.12 Diffusion of Vorticity

Consider the Navier–Stokes vector equation for an incompressible viscous fluid in the form

$$(\partial \mathbf{q}/\partial t)+\nabla(\tfrac{1}{2}\mathbf{q}^2)-\mathbf{q}\wedge\boldsymbol{\zeta}=\mathbf{F}-\nabla\int\frac{dp}{\rho}+\nu\nabla^2\mathbf{q}.$$

If the body forces are conservative, so that $\nabla\wedge\mathbf{F}=\mathbf{0}$, on taking the curl of both sides of the equation, we find

$$\nabla\wedge(\partial\mathbf{q}/\partial t)-\nabla\wedge(\mathbf{q}\wedge\boldsymbol{\zeta})=\nu\nabla\wedge(\nabla^2\mathbf{q}). \tag{1}$$

Now $\nabla\wedge(\partial\mathbf{q}/\partial t)=(\partial/\partial t)(\nabla\wedge\mathbf{q})=\partial\boldsymbol{\zeta}/\partial t$; $\nabla\wedge(\mathbf{q}\wedge\boldsymbol{\zeta})=(\boldsymbol{\zeta}\cdot\nabla)\mathbf{q}-(\mathbf{q}\cdot\nabla)\boldsymbol{\zeta}$, using $\nabla\cdot\boldsymbol{\zeta}=\nabla\cdot(\nabla\wedge\mathbf{q})=0$ and $\nabla\cdot\mathbf{q}=0$. Further, $\nabla\wedge(\nabla^2\mathbf{q})=\nabla^2(\nabla\wedge\mathbf{q})=\nabla^2\boldsymbol{\zeta}$. Thus (1) becomes

$$\partial\boldsymbol{\zeta}/\partial t+(\mathbf{q}\cdot\nabla)\boldsymbol{\zeta}-(\boldsymbol{\zeta}\cdot\nabla)\mathbf{q}=\nu\nabla^2\boldsymbol{\zeta},$$

or

$$d\boldsymbol{\zeta}/dt=(\boldsymbol{\zeta}\cdot\nabla)\mathbf{q}+\nu\nabla^2\boldsymbol{\zeta}. \tag{2}$$

The first term on the right-hand side of (2) represents the rate at which $\boldsymbol{\zeta}$ varies for a given particle when the vortex lines move with the fluid, the strengths of the vortices remaining constant. For slow motion, this term is negligible and (2) approximates to

$$d\boldsymbol{\zeta}/dt=\nu\nabla^2\boldsymbol{\zeta}. \tag{3}$$

In the case of *two-dimensional motion*, if we take

$$\mathbf{q}=u(x,y)\mathbf{i}+v(x,y)\mathbf{j}\quad\text{(fixed axes)},$$

then

$$\boldsymbol{\zeta}=\nabla\wedge\mathbf{q}=\left(\frac{\partial v}{\partial x}-\frac{\partial u}{\partial y}\right)\mathbf{k},$$

and

$$(\boldsymbol{\zeta}\cdot\nabla)\mathbf{q}=(\partial v/\partial x-\partial u/\partial y)(\partial/\partial z)\mathbf{q}=\mathbf{0},$$

so that (2) reduces to (3) precisely in this case.

Thus for slow three-dimensional motion, or for two-dimensional motion, we may regard (3) as an equation describing the way in which vorticity is transmitted throughout a viscous fluid. This equation, when broken down into its components, yields scalar equations which are of the same form as the equation of heat conduction in a liquid. Thus we conclude that vorticity diffuses through a liquid in much the same manner as heat does. By analogy, then, we assert that vorticity cannot be generated within the interior of a viscous fluid. It is in fact transmitted from the boundaries into the fluid. By way of illustration one may recall that a sailing ship will generate vortices in its wake, arising from the hull which is a moving boundary. With the passage of time the disturbance is soon damped out as the vortices diffuse through the water.

In the special case when the motion is two-dimensional and in circles with centres on the z-axis, at any point having cylindrical polar coordinates (R, θ, z), when the flow is unsteady,

$$q = q(R, t)\hat{\theta},$$

so that

$$\zeta = \frac{1}{R}\begin{vmatrix} \hat{\mathbf{R}} & \hat{\mathbf{R}}\hat{\theta} & \mathbf{k} \\ \dfrac{\partial}{\partial R} & \dfrac{\partial}{\partial \theta} & \dfrac{\partial}{\partial z} \\ 0 & Rq & 0 \end{vmatrix} = \frac{1}{R}\,\frac{\partial}{\partial R}(Rq)\,\mathbf{k}.$$

Since $\nabla \equiv \hat{\mathbf{R}}(\partial/\partial R) + \hat{\theta}(\partial/R\partial\theta) + \mathbf{k}(\partial/\partial z)$,

$$(\mathbf{q}\cdot\nabla)\zeta = \mathbf{k}\,\frac{q}{R^2}\,\frac{\partial^2(Rq)}{\partial\theta\,\partial R} = \mathbf{0}.$$

Thus

$$d\zeta/dt = \partial\zeta/\partial t.$$

Since $\zeta = \zeta(R, t)\mathbf{k}$,

$$\frac{\partial \zeta}{\partial t} = \frac{\partial \zeta}{\partial t}\mathbf{k}, \quad \nabla^2\zeta = (\nabla^2\zeta)\,\mathbf{k}.$$

Further, $\nabla^2\zeta = (1/R)\{(\partial/\partial R)\,(R\partial\zeta/\partial R)\} = (\partial^2\zeta/\partial R^2) + (1/R)\,(\partial\zeta/\partial R)$. Thus equation (3) assumes the form

$$\frac{\partial \zeta}{\partial t} = \nu\left(\frac{\partial^2\zeta}{\partial R^2} + \frac{1}{R}\,\frac{\partial\zeta}{\partial R}\right). \tag{4}$$

From standard theory on heat conduction,* the solution of (4) is

$$\zeta = \frac{\kappa}{8\pi\nu t}\exp\left(\frac{-R^2}{4\nu t}\right) \tag{5}$$

where κ is a constant. This shows that ζ decays rapidly with time.

8.13 Energy Dissipation due to Viscosity

If we follow a particle of viscous fluid of fixed mass $\rho\delta v$ and moving at any time t with velocity $\mathbf{q}$, then its kinetic energy is $\frac{1}{2}(\rho\delta v)\mathbf{q}^2$. Thus the rate of gain of kinetic energy at time t as we follow the particle is

$$\frac{d}{dt}\{\tfrac{1}{2}(\rho\delta v)\mathbf{q}^2\} = \rho\delta v\mathbf{q}\cdot\frac{d\mathbf{q}}{dt}.$$

Hence the total rate of gain of kinetic energy of the entire fluid (of volume V) is

* See CARSLAW, *Conduction of Heat*, p. 152 (1921).

$$\int_V \rho \mathbf{q} \cdot \left(\frac{d\mathbf{q}}{dt}\right) dv = \rho \int_V \mathbf{q} \cdot \left(\frac{d\mathbf{q}}{dt}\right) dv,$$

for an incompressible fluid. Now for such a fluid, the Navier–Stokes vector equation may be taken in the form

$$\frac{d\mathbf{q}}{dt} = \mathbf{F} - \nabla \int \frac{dp}{\rho} - \nu \nabla \wedge (\nabla \wedge \mathbf{q}).$$

Hence the rate of energy dissipation due to viscosity is W, where

$$W = \mu \int_V \mathbf{q} \cdot \{\nabla \wedge (\nabla \wedge \mathbf{q})\} \, dv.$$

Now $\nabla \cdot \{\mathbf{q} \wedge \text{curl}\, \mathbf{q}\} = (\text{curl}\, \mathbf{q})^2 - \mathbf{q} \cdot \{\nabla \wedge (\nabla \wedge \mathbf{q})\}$ by the methods of Chapter 1. Hence

$$W = \mu \int_V (\nabla \wedge \mathbf{q})^2 \, dv - \mu \int_V \nabla \cdot (\mathbf{q} \wedge \text{curl}\, \mathbf{q}) \, dv$$

$$= \mu \int_V (\nabla \wedge \mathbf{q})^2 \, dv - \mu \int_S \mathbf{n} \cdot (\mathbf{q} \wedge \text{curl}\, \mathbf{q}) \, dS,$$

where S is the total surface enclosing the volume V.

In the case when the boundary S is at rest and there is no slip between fluid and boundary, $\mathbf{q} = \mathbf{0}$ on S so that

$$W = \mu \int_V \zeta^2 \, dv.$$

8.14 Steady Flow past a Fixed Sphere

Suppose a solid sphere of radius a is held with its centre at the origin in a uniform stream of viscous incompressible liquid having undisturbed velocity $\mathbf{U} = -U\mathbf{i}$. Here U is supposed small. Then in the neighbourhood of the sphere the fluid velocity $\mathbf{q}$ will differ from the ambient velocity $\mathbf{U}$ but will still be of small magnitude. The Navier–Stokes vector equation, in the usual notation, is

$$\frac{\partial \mathbf{q}}{\partial t} + \mathbf{q} \cdot \nabla \mathbf{q} = \mathbf{F} - \frac{1}{\rho} \nabla p - \nu \nabla \wedge (\nabla \wedge \mathbf{q}).$$

When the flow is steady, $\partial \mathbf{q}/\partial t = \mathbf{0}$. Since $\mathbf{q}$ is everywhere small $\mathbf{q} \cdot \nabla \mathbf{q}$ may be supposed small compared with $|\nabla \wedge (\nabla \wedge \mathbf{q})|$. Thus we shall assume the approximation

$$\mathbf{F} - (1/\rho) \nabla p - \nu \nabla \wedge (\nabla \wedge \mathbf{q}) = \mathbf{0} \qquad (1)$$

for slow steady flow, where $\mathbf{q} \to -U\mathbf{i}$ at infinity and $\mathbf{q} = \mathbf{0}$ on the surface

$r=a$. We shall further assume that the body forces are conservative, so that $\nabla\wedge\mathbf{F}=\mathbf{0}$. Then on taking the curl of both sides of (1), we find

$$\text{curl curl curl } \mathbf{q}=0. \tag{2}$$

The equation of continuity is $\nabla\cdot\mathbf{q}=0$. This is satisfied by a vector function $\mathbf{A}$ (a vector potential) where

$$\mathbf{q}=\nabla\wedge\mathbf{A}. \tag{3}$$

From (2), (3),

$$\text{curl curl curl curl } \mathbf{A}=\mathbf{0}. \tag{4}$$

$$\begin{aligned}\nabla\wedge(\mathbf{U}\wedge\mathbf{r})&=\sum\{\mathbf{i}\wedge(\partial/\partial x)(\mathbf{U}\wedge\mathbf{r})\}=\sum\{\mathbf{i}\wedge(\mathbf{U}\wedge\partial\mathbf{r}/\partial x)\}\\&=\sum\{\mathbf{i}\wedge(\mathbf{U}\wedge\mathbf{i})\}=\mathbf{U}(\sum\mathbf{i}\cdot\mathbf{i})-\sum(\mathbf{U}\cdot\mathbf{i})\,\mathbf{i}\\&=3\mathbf{U}-\mathbf{U}=2\mathbf{U}.\end{aligned}$$

Hence when $r\to\infty$, $\mathbf{U}\to\frac{1}{2}\nabla\wedge(\mathbf{U}\wedge\mathbf{r})$ which is satisfied by $\mathbf{A}\to\frac{1}{2}(\mathbf{U}\wedge\mathbf{r})$.

Let us try to find solutions of (4) in the form

$$\mathbf{A}=r^{-n}\mathbf{U}\wedge\mathbf{r}. \tag{5}$$

Here n is an unknown. Clearly we expect $n=0$ to render one solution to satisfy conditions at infinity. We have

$$\begin{aligned}\nabla\wedge\mathbf{A}&=\sum\{\mathbf{i}\wedge\{\partial/\partial x)\,(r^{-n}\mathbf{U}\wedge\mathbf{r})\}\\&=\sum\mathbf{i}\wedge\{-nr^{-n-1}(\partial r/\partial x)\mathbf{U}\wedge\mathbf{r}+r^{-n}\mathbf{U}\wedge\{\partial\mathbf{r}/\partial x)\}\\&=-nr^{-n-2}\sum\{x\mathbf{i}\wedge(\mathbf{U}\wedge\mathbf{r})\}+r^{-n}\sum\{\mathbf{i}\wedge(\mathbf{U}\wedge\mathbf{i})\}\\&=r^{-n}\{\sum\mathbf{U}-\sum(\mathbf{U}\cdot\mathbf{i})\mathbf{i}\}-nr^{-n-2}\mathbf{r}\wedge(\mathbf{U}\wedge\mathbf{r})\\&=2r^{-n}\mathbf{U}-nr^{-n-2}(r^2\mathbf{U}-(\mathbf{U}\cdot\mathbf{r})\mathbf{r})\\&=(2-n)r^{-n}\mathbf{U}+nr^{-n-2}(\mathbf{U}\cdot\mathbf{r})\mathbf{r}.\end{aligned}$$

Thus

$$\begin{aligned}\nabla\wedge(\nabla\wedge\mathbf{A})&=(2-n)\sum\{\mathbf{i}\wedge(\partial/\partial x)\,(r^{-n}\mathbf{U})\}+n\sum[\mathbf{i}\wedge(\partial/\partial x)\{r^{-n-2}(\mathbf{U}\cdot\mathbf{r})\mathbf{r}\}]\\&=(2-n)\sum\{\mathbf{i}\wedge(-nr^{-n-1}x\mathbf{U}/r)\}\\&\quad+n\sum[\mathbf{i}\wedge\{-(n+2)r^{-n-3}(x/r)\,(U\cdot\mathbf{r})\mathbf{r}+r^{-n-2}(\mathbf{U}\cdot\mathbf{i})\mathbf{r}+r^{-n-2}(\mathbf{U}\cdot\mathbf{r})\mathbf{i}\}]\\&=-n(2-n)r^{-n-2}(\mathbf{r}\wedge\mathbf{U})+nr^{-n-2}(\mathbf{U}\wedge\mathbf{r})=n(3-n)r^{-n-2}(\mathbf{U}\wedge\mathbf{r}).\end{aligned}$$

Hence, since $\mathbf{A}=r^{-n}\mathbf{U}\wedge\mathbf{r}$,

$$\begin{aligned}\text{curl curl curl curl } \mathbf{A}&=n(3-n)\nabla\wedge[\nabla\wedge\{r^{-n-2}(\mathbf{U}\wedge\mathbf{r})\}]\\&=n(3-n)\,(n+2)\,(1-n)\,r^{-n-4}(\mathbf{U}\wedge\mathbf{r}).\end{aligned}$$

Thus (4) is satisfied by a function of the form (5) for $n=-2, 0, 1$ or 3. By superposition we may take as a more general solution

$$\mathbf{A}=(\mathbf{U}\wedge\mathbf{r})\left(k_1r^2+k_2+\frac{k_3}{r}+\frac{k_4}{r^3}\right). \tag{6}$$

Since $\mathbf{A}\to\frac{1}{2}(\mathbf{U}\wedge\mathbf{r})$ as $r\to\infty$, we require

$$k_1=0,\ k_2=\tfrac{1}{2},$$

so that

$$\mathbf{A}=(\mathbf{U}\wedge\mathbf{r})\,[\tfrac{1}{2}+(k_3/r)+(k_4/r^3)]. \tag{7}$$

Now

$$\mathbf{q}=\nabla\wedge\mathbf{A}$$

$$=\tfrac{1}{2}\nabla\wedge(\mathbf{U}\wedge\mathbf{r})+k_3\nabla\wedge(r^{-1}\mathbf{U}\wedge\mathbf{r})+k_4\nabla\wedge(r^{-3}\mathbf{U}\wedge\mathbf{r})$$

and, as we have shown,

$$\nabla\wedge(r^{-n}\mathbf{U}\wedge\mathbf{r})=(2-n)r^{-n}\mathbf{U}+nr^{-n-2}(\mathbf{U}\cdot\mathbf{r})\mathbf{r},$$

$$\mathbf{q}=\mathbf{U}+k_3\{r^{-1}\mathbf{U}+r^{-3}(\mathbf{U}\cdot\mathbf{r})\mathbf{r}\}+k_4\{-r^{-3}\mathbf{U}+3r^{-5}(\mathbf{U}\cdot\mathbf{r})\mathbf{r}\}$$
$$=(1+k_3r^{-1}-k_4r^{-3})\mathbf{U}+(k_3r^{-3}+3k_4r^{-5})\,(\mathbf{U}\cdot\mathbf{r})\mathbf{r}. \tag{8}$$

On $r=a$, $\mathbf{q}=\mathbf{0}$ everywhere, so that

$$1+k_3a^{-1}-k_4a^{-3}=0=k_3a^{-3}+3k_4a^{-5}.$$

We find $k_3=-\frac{3}{4}a$, $k_4=\frac{1}{4}a^3$ so that (8) gives the fluid velocity

$$\mathbf{q}=\left\{1-\frac{3}{4}\frac{a}{r}-\frac{1}{4}\left(\frac{a}{r}\right)^3\right\}\mathbf{U}+\frac{3}{4}\left(\frac{a^3}{r^5}-\frac{a}{r^3}\right)(\mathbf{U}\cdot\mathbf{r})\mathbf{r}, \tag{9}$$

and the vector potential is

$$\mathbf{A}=\left(\frac{1}{2}-\frac{3}{4}\frac{a}{r}+\frac{1}{4}\frac{a^3}{r^3}\right)(\mathbf{U}\wedge\mathbf{r}). \tag{10}$$

The vorticity vector is

$$\boldsymbol{\zeta}=\nabla\wedge\mathbf{q}=\nabla\wedge(\nabla\wedge\mathbf{A}).$$

Since we have shown that $\nabla\wedge[\nabla\wedge\{r^{-n}\mathbf{U}\wedge\mathbf{r}\}]=n(3-n)r^{-n-2}(\mathbf{U}\wedge\mathbf{r})$,

$$\boldsymbol{\zeta}=-\tfrac{3}{2}ar^{-3}\,\mathbf{U}\wedge\mathbf{r}. \tag{11}$$

Thus at the point in the fluid having spherical polar coordinates (r, θ, ψ), the origin being the centre of the sphere, and the line through it in the direction $\mathbf{i}$ specifying $\theta=0$, the magnitude of the vorticity is

$$\zeta=\tfrac{3}{2}ar^{-2}U\sin\theta. \tag{12}$$

From Section 8.13, the rate of energy dissipation due to vorticity, since the boundary of the sphere is at rest, is

$$W=\mu\int_V \zeta^2\,dv,$$

where V is the entire volume of fluid outside the sphere. Thus

$$W=\mu\int_{\theta=0}^{\pi}\int_{r=a}^{\infty}2\pi r^2\sin\theta\cdot\zeta^2\,dr$$

$$=\frac{9}{2}\mu\pi U^2a^2\int_a^{\infty}r^{-2}\,dr\cdot\int_0^{\pi}\sin^3\theta\,d\theta$$

$$=6\mu\pi U^2a.$$

If D denotes the total drag on the sphere, then $W=DU$. Thus

$$\underline{D=6\pi\mu aU.} \tag{13}$$

Clearly if the sphere moves with constant velocity $U\mathbf{i}$ in the fluid at rest, the same formula holds for the viscous drag.

In the special case when a solid sphere of radius a and density σ falls vertically through liquid of density $\rho(<\sigma)$, the viscous drag produces a terminal velocity U, say. When this is attained we have, equating the weight of the sphere to the upthrust plus drag,

$$\tfrac{4}{3}\pi a^3\sigma g=\tfrac{4}{3}\pi a^3\rho g+6\pi a\mu U,$$

or

$$\underline{U=\tfrac{2}{9}(\sigma-\rho)a^2g/\mu.} \tag{14}$$

8.15 Dimensional Analysis, Reynolds Number

So far we have mainly investigated those viscous flow problems which are analytically tractable, though we saw in the last Section that approximations had to be made to determine the drag on a solid sphere fixed in a uniform stream. In practical cases, such as arise in the design of ships, aircraft, underwater projectiles, etc., it is usually necessary to carry out experiments on models and to relate their behaviour to that of the full-scale body. Generally speaking, this requires not only that the model and the full-scale body should be geometrically similar, but also, as far as possible that one should ensure that the two possess *dynamical similarity* in the sense that the ratios of corresponding dynamical quantities at corresponding stations should be equal. Thus, for example, if we wish to simulate the motion of a ship, it may be desirable to secure that near the hull the ratio of viscous to inertial forces shall be the same for the full-scale body and for the model. We now develop criteria which enable conditions of dynamical similarity to be obtained between the full-scale body and the model. The method used involves constructing universal or dimensionless equations and forms part of the science of *dimensional analysis.*

In the notation of Section 8.9, the Navier–Stokes equation of motion of a viscous incompressible fluid in the x-direction is

$$\frac{\partial u}{\partial t}+u\frac{\partial u}{\partial x}+v\frac{\partial u}{\partial y}+w\frac{\partial u}{\partial z}=X-\frac{1}{\rho}\frac{\partial p}{\partial x}+\nu\left(\frac{\partial^2 u}{\partial x^2}+\frac{\partial^2 u}{\partial y^2}+\frac{\partial^2 u}{\partial z^2}\right). \tag{1}$$

Suppose L, V, P denote a characteristic length, velocity and pressure respectively. Then the length, velocities and pressures in (1) may be expressed in terms of these standards. Thus we write

$$\left.\begin{aligned} x&=Lx',\\ y&=Ly',\\ z&=Lz';\end{aligned}\right\} \tag{2}$$

$$\left.\begin{aligned} u&=Vu',\\ v&=Vv',\\ w&=Vw';\end{aligned}\right\} \tag{3}$$

$$p=Pp' \tag{4}$$

where all primed quantities are pure numbers having no dimensions. Then, since L/V is the characteristic time,

$$\frac{\partial u}{\partial t}=\frac{\partial(Vu')}{\partial(LV^{-1}t')}=\frac{V^2}{L}\frac{\partial v'}{\partial t'};$$

$$u\frac{\partial u}{\partial x}=(Vu')\frac{\partial(Vu')}{\partial(Lx')}=\frac{V^2}{L}u'\frac{\partial u'}{\partial x'},\quad \text{etc.};$$

$$\frac{1}{\rho}\frac{\partial p}{\partial x}=\frac{1}{\rho}\frac{\partial(Pp')}{\partial(Lx')}=\frac{P}{\rho L}\frac{\partial p'}{\partial x'};$$

$$\frac{\partial^2 u}{\partial x^2}=\frac{\partial^2(Vu')}{\partial(Lx')^2}=\frac{V}{L^2}\frac{\partial^2 u'}{\partial x'^2}.$$

Substituting these results into (1) and simplifying gives

$$\frac{\partial u'}{\partial t'}+u'\frac{\partial u'}{\partial x'}+v'\frac{\partial u'}{\partial y'}+w'\frac{\partial u'}{\partial z'}=\frac{LX}{V^2}-\frac{P}{\rho V^2}\frac{\partial p'}{\partial x'}+\frac{\nu}{VL}\left(\frac{\partial^2 u'}{\partial x'^2}+\frac{\partial^2 u'}{\partial y'^2}+\frac{\partial^2 u'}{\partial z'^2}\right). \tag{5}$$

In equation (5), the L.H.S. is entirely dimensionless. Hence the R.H.S. must likewise be so. It follows that the three quantities

$$\frac{LX}{V^2},\ \frac{P}{\rho V^2},\ \frac{\nu}{VL}$$

must be dimensionless quantities. *In order to produce a faithful model of a given incompressible viscous flow it is essential to keep these three numbers constant.*

The first number (LX/V^2) tells us how to scale the body forces.

The second number $(P/\rho V^2)$ ensures dynamical similarity in the two flows at points where viscosity is unimportant. Such points would occur at stations remote from the boundaries.

The third number ensures dynamical similarity at corresponding points near the boundaries where viscous effects supervene. Its reciprocal is called the *Reynolds number* and is denoted by R so that

$$R = VL/\nu. \tag{6}$$

It should be pointed out, however, that in actual practice, when simulating a given flow, it is seldom, if ever, possible to keep all three numbers constant and also preserve geometrical similarity. In such circumstances one has to decide beforehand what are to be the main purposes of the simulation. If, for example, the study is primarily that of the viscous drag on sailing ships, then it is most important to keep the Reynolds number constant when modelling the full-scale. If it is not also possible to keep $(P/\rho V^2)$ constant, then it means that the model will not be a faithful reproduction of the prototype at stations remote from the boundary where viscous effects are unimportant. This would not matter from the point of view of the purpose in hand.

The Reynolds number is useful for discerning important features of a given flow. Thus a large value for R indicates that the fluid is lightly viscous. A small R indicates very high viscosity. Water and air are fluids of small viscosity and in motion they usually give a high Reynolds number. Also it is known from experiment that if the value of VL/ν exceeds a certain critical value the flow ceases to be laminar near boundaries. In these circumstances, the flow becomes irregular in that eddies or vortices form just off the boundary: such flow is called *turbulent*. The theory developed in this chapter is not at all applicable to turbulent flow.

8.16 Prandtl's Boundary Layer

Now suppose we have laminar flow of a fluid at high Reynolds number over a smooth solid boundary such as a flat plate or an aerofoil. If the boundary is at rest, then the fluid in contact with it will also be at rest. As we move outwards along the normal, the velocity of the fluid will gradually increase until eventually the full stream velocity V is attained. Strictly speaking this is approached asymptotically, but the greatest rates of change of velocity occur within a thin layer of fluid in contact with the boundary. This is called the *boundary layer*. It is a region in which the viscous forces and the inertial forces are of comparable magnitude. Outside this layer the viscous forces are very much smaller than the inertial ones. Figure 8.12 shows laminar flow over a plane surface. Ox is taken in the surface, Oy normal to it and the normal is drawn at P on the surface. At various points on this normal, vectors are drawn parallel to the surface and of appropriate lengths to represent the fluid velocities at those points.

We may define the thickness h of the boundary layer as the region in which the x-component of velocity varies from 0 to $0{\cdot}99U$. The upper bound has been fixed arbitrarily; another bound of similar magnitude would give a comparable value for h. Prandtl's theory assumes viscous forces are important in the region $0<y\leqslant h$ but negligible for $y>h$. Such assumptions permit approximate calculations to be made for h.

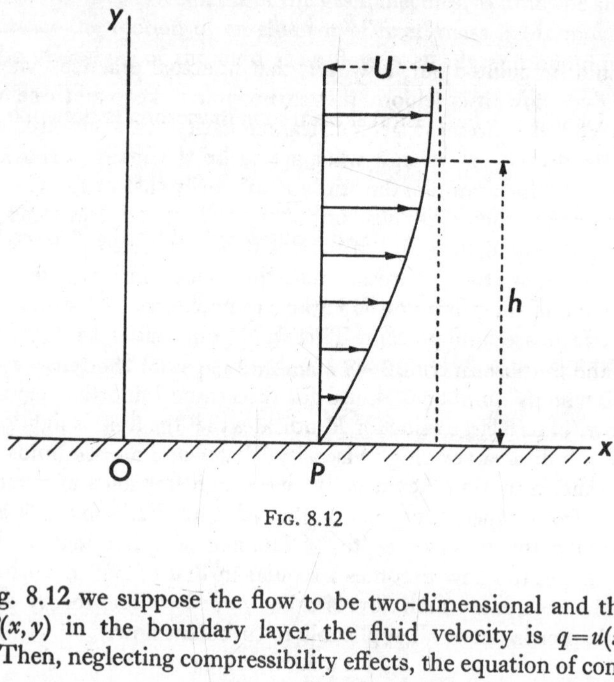

FIG. 8.12

In Fig. 8.12 we suppose the flow to be two-dimensional and that at a point $P(x,y)$ in the boundary layer the fluid velocity is $q=u(x,y)\mathbf{i}+v(x,y)\mathbf{j}$. Then, neglecting compressibility effects, the equation of continuity is

$$\frac{\partial u}{\partial x}+\frac{\partial v}{\partial y}=0 \tag{1}$$

and, in the absence of body forces, the Navier–Stokes equations are

$$\frac{\partial u}{\partial t}+u\frac{\partial u}{\partial x}+v\frac{\partial u}{\partial y}=-\frac{1}{\rho}\frac{\partial p}{\partial x}+\nu\left(\frac{\partial^2 u}{\partial x^2}+\frac{\partial^2 u}{\partial y^2}\right), \tag{2}$$

$$\frac{\partial v}{\partial t}+u\frac{\partial v}{\partial x}+v\frac{\partial v}{\partial y}=-\frac{1}{\rho}\frac{\partial p}{\partial y}+\nu\left(\frac{\partial^2 v}{\partial x^2}+\frac{\partial^2 v}{\partial y^2}\right). \tag{3}$$

Let L be a characteristic length in the direction Ox. The equations (1)–(3) may be rendered universal by putting

$$\bar{u}=\frac{u}{U}, \quad \bar{v}=\frac{Lv}{hU},$$

$$\bar{x}=\frac{x}{L}, \quad \bar{y}=\frac{y}{h},$$

$$\bar{t}=\frac{Ut}{L} \quad \bar{p}=\frac{p}{\rho U^2}.$$

Then, on simplification, we obtain

$$\frac{\partial \bar{u}}{\partial \bar{x}}+\frac{\partial \bar{v}}{\partial \bar{y}}=0, \tag{1'}$$

$$\frac{\partial \bar{u}}{\partial \bar{t}}+\bar{u}\frac{\partial \bar{u}}{\partial \bar{x}}+\bar{v}\frac{\partial \bar{u}}{\partial \bar{y}}=-\frac{\partial \bar{p}}{\partial \bar{x}}+\frac{1}{R}\left\{\frac{\partial^2 \bar{u}}{\partial \bar{x}^2}+\left(\frac{L}{h}\right)^2\frac{\partial^2 \bar{u}}{\partial \bar{y}^2}\right\} \tag{2'}$$

$$\frac{\partial \bar{v}}{\partial \bar{t}}+\bar{u}\frac{\partial \bar{v}}{\partial \bar{x}}+\bar{v}\frac{\partial \bar{v}}{\partial \bar{y}}=-\left(\frac{L}{h}\right)^2\frac{\partial \bar{p}}{\partial \bar{y}}+\frac{1}{R}\left\{\frac{\partial^2 \bar{v}}{\partial \bar{x}^2}+\left(\frac{L}{h}\right)^2\frac{\partial^2 \bar{v}}{\partial \bar{y}^2}\right\} \tag{3'}$$

where R is the Reynolds number UL/ν.

In (2′), since $(L/h)^2 \gg 1$, we may make the approximation

$$\frac{\partial \bar{u}}{\partial \bar{t}}+\bar{u}\frac{\partial \bar{u}}{\partial \bar{x}}+\bar{v}\frac{\partial \bar{u}}{\partial \bar{y}}=-\frac{\partial \bar{p}}{\partial \bar{x}}+\frac{L^2}{Rh^2}\frac{\partial^2 \bar{u}}{\partial \bar{y}^2}.$$

At a point in the boundary layer but not on the surface, it seems reasonable to assume that

$$\bar{u}=0(1), \quad \bar{x}=0(1), \quad \bar{y}=0(1), \quad \bar{t}=0(1), \quad \bar{p}=0(1).$$

Hence

$$(\partial \bar{u}/\partial \bar{t})=0(1), \quad (\partial \bar{u}/\partial \bar{x})=0(1), \quad (\partial \bar{p}/\partial \bar{x})=0(1).$$

Then

$$\bar{v}\frac{\partial \bar{u}}{\partial \bar{y}} \sim \left(\frac{Lv}{hU}\right)\left(\frac{uh}{Uy}\right)=0(1).$$

Also

$$\bar{u}(\partial \bar{u}/\partial \bar{x})=0(1), \quad (\partial^2 \bar{u}/\partial \bar{y}^2)=0(1),$$

so that (4) gives

$$0(1)+0(1)+0(1)=0(1)+(L^2/Rh^2)0(1),$$

showing that,

$$h=0(LR^{-1/2})=0(\nu L/U)^{1/2}. \tag{5}$$

For air if we take as a typical value $\nu=1{\cdot}56\times10^{-4}$ ft^2 sec^{-1}, and $L=4$ ft, $U=200$ ft sec^{-1}, then (5) suggests that

$$h \sim 1{\cdot}77\times10^{-3}\,\text{ft} \simeq \underline{2{\cdot}1\times10^{-2}\,\text{in.}}$$

It often happens that for sufficiently large stream velocities, the boundary layer formed on a solid surface may break away leaving a region of highly turbulent flow downstream of the point at which break away occurs. Such a feature is often observed when air flows over an aerofoil. In the boundary layer, because of the high velocity gradient normal to the boundary, vorticity will form. In fact in the above two-dimensional case, the magnitude of the vorticity is approximately $\partial u/\partial y$, since $\partial v/\partial x$ is small. This vorticity is swept downstream and also diffuses outwards from the boundary like heat. (We have seen that heat conduction and vorticity satisfy the same type of equation.) This has the effect of causing the boundary layer to thicken downstream. In addition, it may happen that the flow outside the boundary layer is being retarded with a consequent increase of pressure. This pressure gradient will be transmitted to the boundary layer and in the case of subsonic flow it can be transmitted upstream in the boundary layer. The subsequent pressure build-up in the boundary layer together with energy loss due to vorticity may result in the fluid reversing its velocity at some point on the boundary. At such a point the boundary layer breaks away from the surface leaving a trail of eddies in its wake. This frequently happens on the upper surface of an aerofoil leading to a considerable loss of lift or *stall*. The eddies may be removed by means of suction ducts in the aerofoil. Figure 8.13 shows flow over the upper surface of an aerofoil. Along the normals through A, B, C velocity profiles are drawn at various points. Break away of the boundary layer occurs at B.

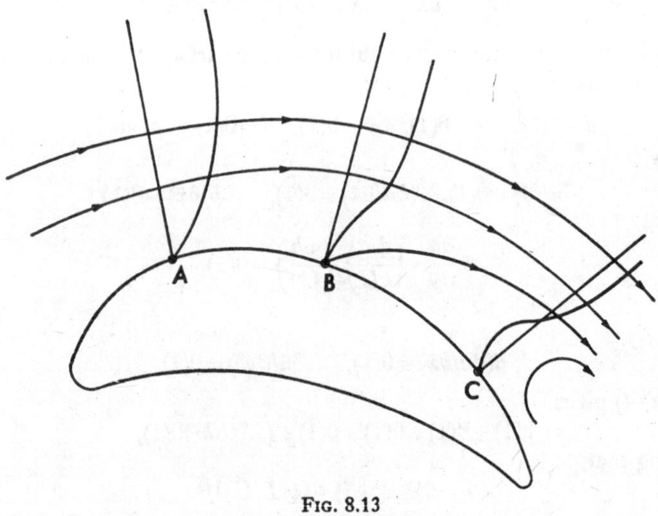

FIG. 8.13

8.16.1 KÁRMÁN'S INTEGRAL EQUATION

Referring back to equations (1) and (2) of Section 8.16, in which ρ is assumed constant in the boundary layer so that ν is also constant there, we observe that

$$u\frac{\partial u}{\partial x}+v\frac{\partial u}{\partial y}=u\frac{\partial u}{\partial x}+\frac{\partial(uv)}{\partial y}-u\frac{\partial v}{\partial y}=2u\frac{\partial u}{\partial x}+\frac{\partial(uv)}{\partial y},$$

using equation (1). Hence equation (2) becomes

$$\frac{\partial u}{\partial t}+\frac{\partial(u^2)}{\partial x}+\frac{\partial(uv)}{\partial y}=-\frac{1}{\rho}\frac{\partial p}{\partial x}+\nu\frac{\partial^2 u}{\partial y^2},$$

if, in accordance with the foregoing analysis, we neglect $\partial^2 u/\partial x^2$. If the boundary layer is confined to the region $0\leqslant y\leqslant h$, then integrating the last equation from $y=0$ to $y=h$ gives

$$\int_0^h \frac{\partial u}{\partial t}dy+\int_0^h \frac{\partial(u^2)}{\partial x}dy+[uv]_0^h=-\frac{1}{\rho}\frac{\partial p}{\partial x}h+\nu\left\{\left(\frac{\partial u}{\partial y}\right)_h-\left(\frac{\partial u}{\partial y}\right)_0\right\}.$$

But if $u=U$ when $y=h$,

$$[uv]_0^h=Uv_h=U\int_0^h \frac{\partial v}{\partial y}dy=-U\int_0^h \frac{\partial u}{\partial x}dy.$$

Thus, using $\partial u/\partial y=0$ when $y=h$,

$$\int_0^h \frac{\partial u}{\partial t}dy+\frac{\partial}{\partial x}\int_0^h u^2\,dy-U\frac{\partial}{\partial x}\int_0^h u\,dy=-\frac{h}{\rho}\frac{\partial p}{\partial x}-\left(\frac{\partial u}{\partial y}\right)_0.$$

This integral equation is due to Kármán.

EXERCISE 8

1. Show that the Navier–Stokes equation for steady, viscous incompressible flow under conservative body forces may be developed in the form

$$\mathbf{q}\wedge\boldsymbol{\zeta}=\nabla(\Omega+\tfrac{1}{2}\mathbf{q}^2+p/\rho)+\nu\,\text{curl}\,\boldsymbol{\zeta},$$

where $\boldsymbol{\zeta}=\text{curl}\,\mathbf{q}$, the vorticity vector. Show further that for two-dimensional flow in the (x, y)-plane, on taking the curl of both sides of this equation

$$\frac{\partial(\nabla^2\psi,\psi)}{\partial(x,y)}=-\nu\nabla^4\psi,$$

where ψ is the stream function such that

$$\mathbf{q}=-(\partial\psi/\partial y)\mathbf{i}+(\partial\psi/\partial x)\mathbf{j}.$$

2. With the usual notation, show that the rate of energy dissipation in a viscous liquid due to viscosity alone is

$$\mu\int_v \mathbf{q}\cdot(\nabla\wedge\nabla\wedge\mathbf{q})\,dv=\mu\int_v(\nabla\wedge\mathbf{q})^2\,dv-\mu\int_S n\cdot(\mathbf{q}\wedge\nabla\wedge\mathbf{q})\,dS$$

where S is the total closed surface containing v.

If the liquid is contained within a vessel at rest, what is the rate of dissipation of energy?

If the vessel is a surface of revolution about which it is rotating with angular velocity ω, taking $\mathbf{q}=[u, v, w]$ in Cartesians, the z-axis being the axis of rotation, show that there is an additional term in the rate of energy dissipation of

$$2\mu\omega^2\int_S(lx+my)\,dS$$

where $[l, m, n]$ are the direction cosines of the inward-drawn normal at the element dS of the surface of the vessel.

3. The equation governing the motion of a viscous fluid of constant density ρ is

$$\rho D\mathbf{q}/Dt=\rho\mathbf{F}-\nabla p+\tfrac{1}{3}\mu\nabla(\nabla\cdot\mathbf{q})+\mu\nabla^2\mathbf{q},$$

with the usual notation. Show that the vorticity vector ζ of an incompressible viscous fluid moving under no external forces satisfies the differential equation

$$D\zeta/Dt=(\zeta\cdot\nabla)\mathbf{q}+\nu\nabla^2\zeta,$$

ν being the coefficient of kinematic viscosity.

Motion is due to an infinite line vortex of strength k concentrated along the z-axis at time $t=0$ in an infinite viscous liquid. Show that at time t the vorticity ζ at a distance r from the z-axis satisfies

$$\frac{\partial\zeta}{\partial t}=\nu\left(\frac{\partial^2\zeta}{\partial r^2}+\frac{1}{r}\frac{\partial\zeta}{\partial r}\right).$$

Verify that the solution is

$$\zeta=\frac{k}{2\nu t}\exp\left(\frac{-r^2}{4\nu t}\right),$$

and show that the circulation within an infinite circular cylinder of radius a whose axis is the z-axis decreases from $2\pi k$ to zero as t increases from zero to infinity.

(Nottingham Univ., Maths. Hons. II, 1962)

4. Coaxial circular cylinders of radii $a, b\ (b>a)$, are placed with axis vertical, and a viscous incompressible fluid occupies the annular region between them. The outer cylinder is fixed and the inner cylinder is constrained to move vertically downwards with constant velocity W. If it is assumed that the fluid is in steady motion under gravity with the pressure everywhere constant, and if the streamlines are parallel to the vertical

generators, show that w the speed downwards of the fluid satisfies the equation

$$\frac{d^2w}{dr^2}+\frac{1}{r}\frac{dw}{dr}+\frac{g}{\nu}=0$$

where r is the radial distance measured from the axis, g is the acceleration due to gravity, and ν is the kinematic viscosity.

Show that

$$w=W\frac{\log(r/b)}{\log(a/b)}-\frac{g}{4\nu}\left\{r^2-\frac{b^2\log(r/a)-a^2\log(r/b)}{\log(a/b)}\right\}.$$

Find the force on the inner cylinder due to the viscous fluid, and deduce that if the cylinder were free to fall under gravity its terminal speed would be given by

$$W=\tfrac{1}{4}g(b^2-a^2)/\nu,$$

assuming the inner cylinder is solid with the same density as the fluid.

(London Univ., B.Sc. Spec. Hons., 1965)

5. Viscous liquid flows steadily under no forces along a pipe of length l in the direction of the z-axis. The cross-section of the pipe is the ellipse $\frac{1}{4}x^2+y^2=1$ and a pressure difference P is maintained across the ends. If the velocity is such that $u=v=0$, and w does not depend on z, prove that the pressure gradient along the tube is constant. Find a solution of the equation for w in the form

$$w=(x^2+4y^2-4)f(x,y),$$

and deduce that the total mass of liquid delivered per second through the pipe is $(2\pi P\rho/5\mu l)$.

(Leeds Univ., B.Sc. Spec. Hons., 1962)

6. If $f(z,\bar{z})$ is a function of the variables $z(=x+iy)$ and $\bar{z}(=x-iy)$ which is continuous and differentiable in the area S enclosed by $\mathscr{C}$, prove by using Stokes' theorem or otherwise that

$$\oint_{\mathscr{C}} f(z,\bar{z})\,dz=2i\int_S\frac{\partial f}{\partial\bar{z}}\,dS.$$

Liquid of viscosity μ flows steadily through a straight pipe of uniform cross-section S, whose boundary is $\mathscr{C}$ with speed $q(x,y)$ parallel to the generators of the pipe which are perpendicular to the (x,y)-plane. If the external forces are neglected, show that the pressure gradient is constant. If this constant is $-P$ show that

$$\frac{\partial^2 q}{\partial z\partial\bar{z}}=-\frac{1}{4}\frac{P}{\mu}.$$

Deduce that q may be expressed in the form

$$4\mu q = P\{w(z) + \bar{w}(\bar{z}) - z\bar{z}\},$$

where $w(z) + \bar{w}(\bar{z}) = z\bar{z}$ on $\mathscr{C}$.

Show that the rate of discharge R may be considered as the real part of

$$\frac{iP}{4\mu}\oint_{\mathscr{C}} \{\tfrac{1}{4}z\bar{z}^2 - \bar{z}w(z)\}\, dz.$$

Find R if the boundary $\mathscr{C}$ is the circle $z\bar{z} = a^2$.

(Nottingham Univ., Maths. Hons. II, 1963)

7 Establish the usual relation between the stress and rate of strain tensors in an isotropic inviscid compressible fluid.

Show that the normal stress on a stationary rigid boundary with which the fluid is in contact is $p + \frac{4}{3}\nu(\partial\rho/\partial t)$, where p is the pressure, ρ is the density, and ν is the kinematic viscosity.

(Cambridge Univ., M.T. II, 1960)

8. Find the velocity distribution in the steady flow of a viscous incompressible fluid along an infinitely long circular pipe due to an applied pressure gradient.

A large cylindrical tank, the horizontal circular end of which has radius a, is filled with water to a depth h which is of the same order of magnitude as a, the free surface being open to the atmosphere. The water drains from the bottom of the tank through a long and narrow vertical pipe, of uniform circular section, which is open to the atmosphere at its lower end, and, simultaneously, the level of water in the tank is maintained by an external supply. Show that, if the external supply is cut off, the tank will empty itself in a time $(8a^2\nu l/b^4 g)\log(1 + h/l)$, approximately, where l and b are the length and radius of the pipe, and ν is the kinematic viscosity of the water.

Discuss any approximations you make, and explain how their validity is related to the restrictions on the various lengths occurring in the problem.

(Cambridge Univ., M.T. II, 1958)

9. Establish the formula $\frac{1}{8}(\pi a^4/\mu l)(p_1 - p_2)$ for the rate of steady flow of an incompressible liquid through a uniform circular pipe of radius a, p_1 and p_2 being the pressures of two sections of the pipe distant l apart.

A viscous gas, having equation of state $p = \rho RT$, flows isothermally through such a pipe and ρ_1, ρ_2 are its densities at the two sections. Show that if the velocity gradient in the direction of flow is negligible in comparison with that in a radial direction, then the mass of gas flux per unit time across any section is

$$\tfrac{1}{16}(\pi a^4/\mu l)(p_1 - p_2)(\rho_1 + \rho_2).$$

(*Hint:* Let p, ρ be the pressure and density at distance x from the inlet and $(p + \delta p), (\rho + \delta\rho)$ their values at distance $(x + \delta x)$. Treating the flow in the section δx as incompressible, show the mass flux m per unit time is given by

$$m=-\frac{1}{8}\frac{\rho\pi a^4}{\mu}\frac{dp}{dx}.$$

Then use $p=k\rho$ to show

$$m=-\frac{1}{8}\frac{k\rho\pi a^4}{\mu}\frac{d\rho}{dx},$$

$$\rho\, d\rho/dx=\text{const.},$$

so that

$$\rho^2=\rho_1^2+(\rho_2^2-\rho_1^2)\,(x/l),$$

$$\rho\, d\rho/dx=\tfrac{1}{2}(\rho_2^2-\rho_1^2)/l, \text{ etc.})$$

10. Infinite fluid streams steadily in two dimensions with speed V past a thin flat plate of finite length fixed in the direction of the stream. Taking axes Ox along the plate and Oy perpendicular to the plate at the leading edge, show that the velocity $[u, v]$ in the boundary layer satisfies the equation

$$u\frac{\partial u}{\partial x}+v\frac{\partial u}{\partial y}=\frac{\mu}{\rho}\frac{\partial^2 u}{\partial y^2}$$

subject to certain assumptions to be stated.

If $h(x)$ is the thickness of the boundary layer at a distance x from the leading edge, show that

$$\frac{\partial}{\partial x}\int_0^h (u-V)u\, dy=-\frac{\mu}{\rho}\left(\frac{\partial u}{\partial y}\right)_{y=0}$$

and find h in terms of x if $u=V\sin(\pi y/2h)$.

(London Univ., B.Sc. Spec. Hons., 1962)

9

MAGNETOHYDRODYNAMICS

9.1 Nature of Magnetohydrodynamics

When a conductor carrying an electric current moves in a magnetic field it experiences a force tending to move it at right-angles to the electric field. Conversely, when a conductor moves in a magnetic field a current is induced in the conductor in a direction mutually at right-angles to both the field and the direction of motion. These two statements, first enunciated by Faraday, constitute the *laws of electromagnetism*. The first is the principle of the electric motor: the second that of the dynamo. There is nothing in them to suggest that the conductor must be a solid. In fact suggestions have been made that the motion of the sea may produce perturbations in the earth's magnetic field. Further, tidal waves sweeping up the estuary of a river will cut the terrestrial lines of force and generate a current which can be detected in a cable connecting two electrodes placed in the river on opposite banks.

In the case when the conductor is either a liquid or a gas, electromagnetic forces will be generated which may be of the same order of magnitude as the hydrodynamical and inertial forces. Thus the equations of motion will have to take these electromagnetic forces into account as well as the other forces. The science which treats these phenomena is called *magnetohydrodynamics* (MHD). Other variants of nomenclature are: hydromagnetics, magneto-fluid dynamics, magneto-gas dynamics, etc.

Most liquids and gases are poor conductors of electricity. As a consequence their motions can normally be treated by the principles of fluid dynamics which have so far been studied in this book. However, it is possible to make some gases very highly conducting by ionizing them. For ionization to take effect, the gas must be very hot—at temperatures upwards of 5000°K or so. Such ionized gases are called *plasmas*. The material within a star is a plasma of very high conductivity and it exists within a strong magnetic field. Consequently we expect MHD effects to be realized in stars. Further, at the engineering level, experiments have been made for electric power generation by passing an ionized gas between the poles of a strong electromagnet so that an electric current would be generated at right-angles to the magnetic field and to the direction of flow of the plasma, the current being collected by two spaced electrodes at right-angles to the direction of the current flow. At the present time MHD generators are not a practical possibility owing to the difficulties of producing suitably efficient and stable plasmas and sufficiently refractory materials to withstand the high temperatures of the plasmas.

MHD effects in conducting liquids have been studied in the Laboratory by Hartmann and Williams. We shall describe in Section 9.14 how the viscosity of mercury seems to be enhanced when flow takes place across a strong magnetic field. Also, the effects have been exploited in the case of molten sodium moving in a magnetic field and in the design of electromagnetic pumps and flowmeters.

9.2 Maxwell's Electromagnetic Field Equations: Medium at Rest

Before embarking on the mathematical theory of magnetohydrodynamics we summarize Maxwell's electromagnetic field equations. Their derivations will be found in any standard treatise on electromagnetic theory. We work throughout in electromagnetic units. Some knowledge will be presumed of fundamental electromagnetic theory.*

The electromagnetic field equations for a uniform conducting medium at rest are

$$\nabla \cdot \mathbf{E} = (4\pi/\varepsilon)q, \tag{1}$$

$$\nabla \cdot \mathbf{H} = 0, \tag{2}$$

$$\nabla \wedge \mathbf{E} = -\mu(\partial \mathbf{H}/\partial t), \tag{3}$$

$$\nabla \wedge \mathbf{H} = 4\pi \mathbf{j} + \varepsilon(\partial \mathbf{E}/\partial t), \tag{4}$$

where $\mathbf{E}$ is the electric field intensity, ε the dielectric constant of the medium, $\mathbf{H}$ the magnetic field intensity, μ the permeability, $\mathbf{j}$ the conduction current density vector, q the charge density of the medium per unit volume, and t signifies time. In addition Ohm's law for the conducting medium is formulated mathematically as

$$\mathbf{j} = \sigma \mathbf{E} \tag{5}$$

where σ is the conductivity of the medium. The magnetic induction vector $\mathbf{B}$ and the electric displacement vector $\mathbf{D}$ are given by the constitutive equations

$$\mathbf{B} = \mu \mathbf{H}, \tag{6}$$

$$\mathbf{D} = \varepsilon \mathbf{E}. \tag{7}$$

* See, for example, V. C. A. FERRARO, *Electromagnetic Theory*, Athlone Press (1962).

Also, since μ, ε are both in e.m.u.,

$$c=(\mu\varepsilon)^{-1/2} \tag{8}$$

where c is the velocity of light in the medium. For a vacuum,

$$c \simeq 3\times10^{10}\ \text{cm sec}^{-1}.$$

Equations (1), (2) are derived by applying Gauss's theorem to a closed surface. Equation (3) is the mathematical formulation of Faraday's law of induction which states that when the flux of electric induction through a closed circuit varies, the induced electromotive force is proportional to the rate of decay of the flux. Equation (4) describes how the magnetic field **H** depends on the conduction current **j** and on the 'displacement current' $(\varepsilon/4\pi)\partial\mathbf{E}/\partial t$ due to change of displacement vector **D** in the dielectric medium. It is known from theory that electromagnetic disturbances are propagated as waves in the dielectric medium with the velocity of light c. For this reason light waves are regarded as electromagnetic in kind.

Another result which we borrow from electromagnetic theory is the expression for the Lorentz force on a moving charge. Such a force **F** on a charge e moving with velocity **v** is given by

$$\mathbf{F}=e(\mathbf{E}+\mu\mathbf{v}\wedge\mathbf{H}) \tag{9}$$

where all units are e.m.u., as in the previous equations. We next see how the above equations are modified when the conducting dielectric medium is in motion.

9.3 Maxwell's Electromagnetic Field Equations: Medium in Motion

We now suppose that the dielectric medium is a moving fluid, such as an ionized gas. As in the development of hydrodynamic theory we suppose the medium to be continuous and to possess isotropic properties at all points. Such an assumption is reasonable when the mean free paths of the particles are very much smaller than the typical lengths associated with the system considered. It should, however, be pointed out that there are plasmas for which this assumption is invalid. In these circumstances the medium cannot be treated as a continuous entity; analysis has to proceed by considering discrete particle motions. The branch of magnetohydrodynamics in which the medium cannot be regarded as continuous is often called *plasma dynamics*. In astrophysics one is dealing with variations over extremely large distances and since particle paths are much smaller it is regarded as sufficiently accurate to treat the analysis macroscopically.

Let **v** be the local velocity of a plasma. Then, in addition to the conduction current, there is a current due to convection. We first estimate the convection current density.

Figure 9.1 shows that in time δt the local movement of the medium is

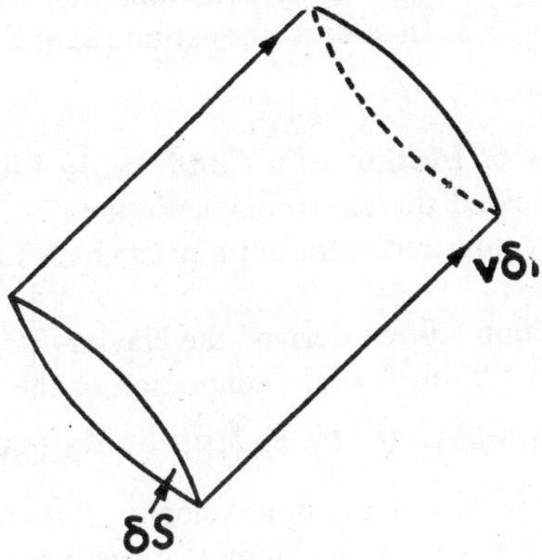

FIG. 9.1

$\mathbf{v}\delta t$. Hence the volume generated in the motion of an element of cross-section δS (normal to the motion) is $|\mathbf{v}|\delta t\delta S$. The charge within this volume is $q|\mathbf{v}|\delta t\delta S$. Thus the rate of flow of charge over the section δS is $q|\mathbf{v}|\delta S$ in the direction of $\mathbf{v}$. Hence the current density per unit area due to convection is $q|\mathbf{v}|$. Since the direction of current flow due to convection is that of $\mathbf{v}$ it follows that the current density vector due to convection is $q\mathbf{v}$.

The current density vector due to conduction is $\mathbf{j}$. Hence the total current density vector due to both conduction and convection is $\mathbf{j}+q\mathbf{v}$. Thus we modify Maxwell's equation (4) of Section 9.2 to the form

$$\nabla\wedge\mathbf{H}=4\pi(\mathbf{j}+q\mathbf{v})+\varepsilon(\partial\mathbf{E}/\partial t). \tag{1}$$

In (1) three types of current feature:

(i) The conduction current $\mathbf{j}$;
(ii) The convection current $q\mathbf{v}$ due to the motion of the medium;
(iii) The displacement current $(\varepsilon/4\pi)\partial\mathbf{E}/\partial t$ or $(1/4\pi)\partial\mathbf{D}/\partial t$ due to time variation of the electric induction $\mathbf{D}$ in the medium.

We shall show later that in many cases only (i) is important and that (ii) and (iii) are often negligible.

The other three Maxwell equations ((1), (2), (3) of Section 9.2) remain unaltered for a moving medium because they do not involve electric currents of any kind.

In addition, the form for Ohm's Law, equation (5) of Section 9.2 must be modified for a moving medium. For a unit charge moving locally with the medium with velocity $\mathbf{v}$, the Lorentz force is $\mathbf{E}+\mu\mathbf{v}\wedge\mathbf{H}$. Hence Ohm's law now assumes the form

$$\mathbf{j}=\sigma(\mathbf{E}+\mu\mathbf{v}\wedge\mathbf{H}). \tag{2}$$

To sum up, we may say that Maxwell's electromagnetic field equations for a moving medium are given by equations (1), (2), (3) of Section 9.2 and by (1) of Section 9.3. In addition equation (2) of Section 9.3 expresses Ohm's law.

9.4 The Equations of Motion of a Conducting Fluid

So far we have treated the electromagnetic field equations of a moving conducting fluid. We now derive the appropriate hydrodynamical equations of motion.

In Chapter 8, Section 8.9, we derived the Navier–Stokes vector equation of motion of a viscous fluid. A slight adaptation of the results gives

$$\rho(d\mathbf{v}/dt)=\mathbf{F}-\nabla p+\tfrac{1}{3}\rho\nu\nabla\nabla\cdot\mathbf{v}+\rho\nu\nabla^2\mathbf{v} \tag{1}$$

where ρ is the fluid density, $\mathbf{v}$ the fluid velocity, $\mathbf{F}$ the body force per unit volume, p the fluid pressure, ν the kinematic coefficient of viscosity. (The coefficient of viscosity will be denoted by $\rho\nu$ and not by μ since the latter symbol is required for the magnetic permeability of the fluid.)

When the fluid moves in an electric and magnetic field, the body force $\mathbf{F}$ per unit volume consists of three parts: gravitational, electric and magnetic. The gravitational body force per unit volume is $\rho\mathbf{g}$, where $\mathbf{g}$ is the acceleration due to gravity. An elemental volume δv of the fluid would contain a charge of amount $q\delta v$ so that the force on it due to an electric field of intensity $\mathbf{E}$ would be $(q\delta v)\mathbf{E}$. Hence the electric body force per unit volume is $q\mathbf{E}$.

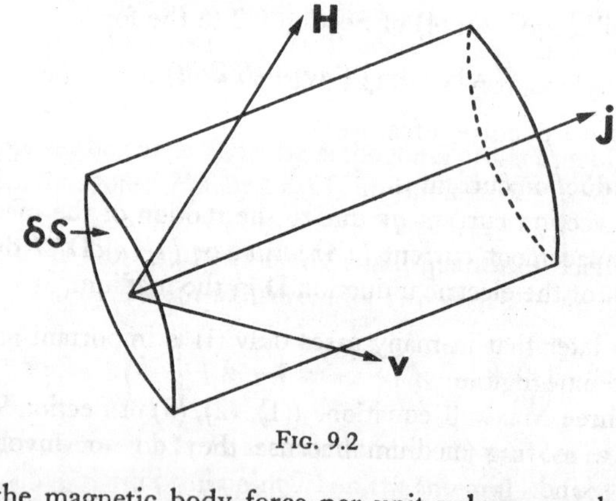

FIG. 9.2

To find the magnetic body force per unit volume, we know that the total current density vector is $\mathbf{j}+q\mathbf{v}$, but in constructing the Navier–Stokes equations the fluid element chosen was a mobile element moving along with the local velocity $\mathbf{v}$. Hence the conductive component $\mathbf{j}$ makes an effective contribution to the magnetic body force but not the convective part $q\mathbf{v}$. In Fig. 9.2 δS is the normal cross-section of a fluid element whose length δs is in the direction of $\mathbf{j}$. This element moves along with the local fluid velocity $\mathbf{v}$ in a magnetic field of intensity $\mathbf{H}$. Thus the current flowing through the element is $i=|\mathbf{j}|\,\delta S$. From electromagnetic theory, the Biot–Savart law gives as the magnetic force in the element

$$\begin{aligned}\delta\mathbf{F} &= i\delta\mathbf{s}\wedge\mathbf{B}\\ &= |\mathbf{j}|\,\delta\mathbf{s}\delta S\wedge\mathbf{B}\\ &= (\mathbf{j}\wedge\mathbf{B})\delta s\delta S,\end{aligned}$$

giving a magnetic body force per unit volume

$$\mathbf{j}\wedge\mathbf{B}=\mu\mathbf{j}\wedge\mathbf{H}.$$

In summary, then, the total body force per unit volume is

$$\mathbf{F}=\rho\mathbf{g}+\mu\mathbf{j}\wedge\mathbf{H}+q\mathbf{E} \tag{2}$$

and equation (1) may be rewritten

$$\rho(d\mathbf{v}/dt)=-\nabla p+\rho\mathbf{g}+\mu\mathbf{j}\wedge\mathbf{H}+q\mathbf{E}+\tfrac{1}{3}\rho\nu\nabla\nabla\cdot\mathbf{v}+\rho\nu\nabla^2\mathbf{v}. \tag{3}$$

In addition the equation of continuity is

$$(\partial\rho/\partial t)+\nabla\cdot(\rho\mathbf{v})=0. \tag{4}$$

9.5 Rate of Flow of Charge

Taking the divergence of both sides of equation (1), Section 9.3,

$$0=4\pi\nabla\cdot\mathbf{j}+4\pi(q\nabla\cdot\mathbf{v}+\mathbf{v}\cdot\nabla q)+\varepsilon(\partial/\partial t)\nabla\cdot\mathbf{E}.$$

Using $d/dt\equiv\partial/\partial t+\mathbf{v}\cdot\nabla$, equation (1) of Section 9.2 and equation (2) of Section 9.3, this becomes

$$\frac{dq}{dt}+q\nabla\cdot\mathbf{v}+\frac{4\pi\sigma}{\varepsilon}q+\mu\sigma\nabla\cdot(\mathbf{v}\wedge\mathbf{H})=0, \tag{1}$$

an equation expressing the rate of flow of charge, moving along with the fluid. Two cases arise for discussion.

Case I: Fluid at Rest

Taking $\mathbf{v}=\mathbf{0}$, equation (1) becomes

$$\frac{\partial q}{\partial t}+\frac{q}{\tau}=0, \tag{2}$$

where $\tau=\varepsilon/(4\pi\sigma)$, a quantity having the dimensions of time and called the *relaxation time*. τ is usually very small, since $\tau=1/(4\pi\mu\sigma c^2)$ where $\mu\sim1$ and σ, c^2 are both very large. The solution of (2) is

$$q=q_0\exp(-t/\tau). \tag{3}$$

Equation (3) shows that at any point within a conducting fluid at rest, the charge decays very rapidly in an exponential manner. Electrostatic theory asserts that when static conditions prevail the electric charge resides exclusively on the outer boundaries of a conductor. Thus the charge within the conducting fluid at rest decays exponentially at interior points, distributing itself on the outer boundaries.

Case ii: Fluid in Motion

In general (1) is not integrable. We can show, however, using dimensional considerations, that in some important cases the term $q\nabla \cdot \mathbf{v}$ is negligible. To this end, suppose L is a characteristic length in the sense of being of the same order of magnitude as a distance in which the variables in the equation change by appreciable proportions. Then

$$0\left(\frac{4\pi\sigma q}{\varepsilon}\right)=\frac{\sigma q}{\varepsilon},$$

$$0\{\mu\sigma\nabla \cdot (\mathbf{v}\wedge\mathbf{H})\}=\mu\sigma vH/L.$$

If these two are of comparable magnitude, then

$$q\sim\frac{\mu\varepsilon vH}{L}=\frac{vH}{Lc^2}. \tag{4}$$

When (4) is true,

$$q\nabla \cdot \mathbf{v}\sim\left(\frac{vH}{Lc^2}\right)\times\left(\frac{v}{L}\right)=\frac{v^2H}{L^2c^2}.$$

Thus

$$\frac{0\{q\nabla \cdot \mathbf{v}\}}{0\{\mu\sigma\nabla \cdot (\mathbf{v}\wedge\mathbf{H})\}}=\frac{v^2HL}{L^2c^2\mu\sigma vH}=\frac{v}{\sigma L\mu c^2}=\frac{\varepsilon v}{\sigma L}\sim\frac{\tau}{(L/v)},$$

since $\tau=\varepsilon/(4\pi\sigma)$. The relaxation time τ is very small and in general it is found that $(L/v)\gg\tau$. This shows that the term $q\nabla \cdot \mathbf{v}$ may be neglected when (4) holds.

9.6 Simplification of the Electromagnetic Field Equations

The result $q\sim vH/(Lc^2)$ of the last section enables the field equation

$$\nabla\wedge\mathbf{H}=4\pi\mathbf{j}+4\pi q\mathbf{v}+\varepsilon(\partial\mathbf{E}/\partial t)$$

to be simplified. For

$$0(|\nabla\wedge\mathbf{H}|)=H/L,$$
$$0(|q\mathbf{v}|)=v^2H/(Lc^2),$$

and so

$$\frac{0(|q\mathbf{v}|)}{0|\nabla\wedge\mathbf{H}|}=\left(\frac{v}{c}\right)^2.$$

For non-relativistic velocities, $(v/c)^2 \ll 1$, so that the term $4\pi q\mathbf{v}$ is negligible.

Further,

$$0\left(\left|\varepsilon\frac{\partial \mathbf{E}}{\partial t}\right|\right)=\frac{\varepsilon E}{L/v}=\frac{\varepsilon v E}{L}=\frac{vE}{\mu c^2 L},$$

and so

$$\frac{0(|\varepsilon\partial \mathbf{E}/\partial t|)}{0(|q\mathbf{v}|)}=\frac{vE}{\mu c^2 L}\times\frac{Lc^2}{v^2 H}=\frac{E}{\mu H v}.$$

This ratio is of finite size. Hence the term $\varepsilon\partial\mathbf{E}/\partial t$ is also negligible.

Thus we may ignore convective and displacement currents and take

$$\nabla\wedge\mathbf{H}=4\pi\mathbf{j}=4\pi\sigma(\mathbf{E}+\mu\mathbf{v}\wedge\mathbf{H})$$

whenever $q\sim vH/(Lc^2)$ and velocities are of non-relativistic order.

9.7 The Magnetic Reynolds Number

Taking the curl of both sides of the equation

$$\nabla\wedge\mathbf{H}=4\pi\sigma(\mathbf{E}+\mu\mathbf{v}\wedge\mathbf{H})$$

gives $\nabla(\nabla\cdot\mathbf{H})-\nabla^2\mathbf{H}=4\pi\sigma\{\nabla\wedge\mathbf{E}+\mu\nabla\wedge(\mathbf{v}\wedge\mathbf{H})\}$, or, since $\nabla\cdot\mathbf{H}=0$, $\nabla\wedge\mathbf{E}=-\mu\partial\mathbf{H}/\partial t$,

$$\partial\mathbf{H}/\partial t=\nabla\wedge(\mathbf{v}\wedge\mathbf{H})+\eta\nabla^2\mathbf{H}, \tag{1}$$

where

$$\eta=1/(4\pi\mu\sigma). \tag{2}$$

Let L denote a characteristic length, v a characteristic velocity. Then

$$0(|\nabla\wedge(\mathbf{v}\wedge\mathbf{H})|)=vH/L,$$
$$0(|\eta\nabla^2\mathbf{H}|)=\eta H/L^2,$$

so that

$$\frac{0(|\nabla\wedge(\mathbf{v}\wedge\mathbf{H})|)}{0(|\eta\nabla^2\mathbf{H}|)}=\frac{vL}{\eta}=R_m.$$

The dimensionless quantity

$$R_m=vL/\eta=4\pi\mu\sigma vL \tag{3}$$

is called the *Magnetic Reynolds number* on account of its close similarity to the Reynolds number in viscous flow theory. The equation (1) does not permit of precise analytic solution, but one may discuss its cardinal features for low and high values of R_m.

CASE I

$R_m \ll 1$, corresponding to low conductivity.

Equation (1) approximates to the *diffusion equation*

$$\partial \mathbf{H}/\partial t = \eta \nabla^2 \mathbf{H}. \tag{4}$$

For an equation of this type the rate of decay of the magnetic field is very rapid. In a poor conductor such as a liquid this phenomenon is evident; it is one reason why experiments with liquids on the laboratory scale are difficult to carry out.

CASE II

$R_m \ll 1$, corresponding to high conductivity.

Equation (1) now approximates to

$$\partial \mathbf{H}/\partial t = \nabla \wedge (\mathbf{v} \wedge \mathbf{H}). \tag{5}$$

This condition prevails within the material of a star which is a plasma of almost perfect conductivity. Such a case leads to the following theorem due to Alfvén and first formulated in 1942. It forms the theoretical basis of most of the modern developments in MHD which have subsequently been made.

9.8 Alfvén's Theorem

For a perfectly conducting fluid the flux of magnetic field intensity through any closed circuit of fluid particles moving along with the fluid remains constant.

Proof:

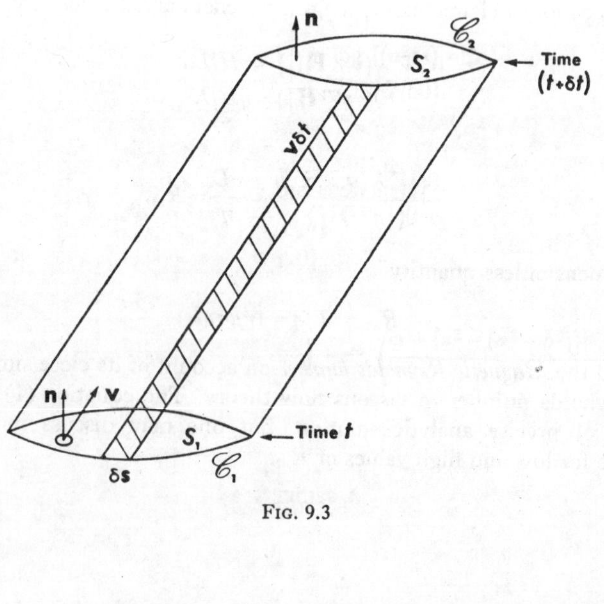

FIG. 9.3

Figure 9.3 shows the positions $\mathscr{C}_1, \mathscr{C}_2$ of a moving circuit of fluid particles at time t, $(t+\delta t)$, S_1, S_2 being the areas they contain. At time t, the normal flux of $\mathbf{H}$ over S_1 is

$$N_1 = \int_{S_1} \mathbf{n} \cdot \mathbf{H}(t)\, dS, \quad (\mathbf{n} = \text{unit vector})$$

and at time $(t+\delta t)$ the normal flux across S_2 is

$$N_2 = \int_{S_2} \mathbf{n} \cdot \mathbf{H}(t+\delta t)\, dS.$$

Thus at time t, the rate of change of flux through $\mathscr{C}_1$ is

$$\frac{dN}{dt} = \lim_{\delta t \to 0} \left(\frac{N_2 - N_1}{\delta t} \right)$$

$$= \lim_{\delta t \to 0} \left[\frac{1}{\delta t} \left\{ \int_{S_2} \mathbf{n} \cdot \mathbf{H}(t+\delta t)\, dS - \int_{S_1} \mathbf{n} \cdot \mathbf{H}(t)\, dS \right\} \right].$$

Now at time t, by Gauss's divergence theorem,

$$\int_{S_2} \mathbf{n} \cdot \mathbf{H}(t)\, dS - \int_{S_1} \mathbf{n} \cdot \mathbf{H}(t)\, dS + \oint_{\mathscr{C}_1} \mathbf{H} \cdot (d\mathbf{s} \wedge \mathbf{v}\delta t) = \int_{\Delta v} \nabla \cdot \mathbf{H}(t)\, dv,$$

where Δv is the volume generated by the moving circuit in time δt, the L.H.S. being the total outward normal flux of $\mathbf{H}$ over the surface of Δv. But $\nabla \cdot \mathbf{H} = 0$, so that

$$\int_{S_1} \mathbf{n} \cdot \mathbf{H}(t)\, dS = \int_{S_2} \mathbf{n} \cdot \mathbf{H}(t)\, dS + \oint_{\mathscr{C}_1} \mathbf{H} \cdot (d\mathbf{s} \wedge \mathbf{v}\delta t).$$

$$\therefore \quad \frac{dN}{dt} = \lim_{\delta t \to 0} \left[\frac{1}{\delta t} \int_{S_2} \mathbf{n} \cdot \{\mathbf{H}(t+\delta t) - \mathbf{H}(t)\}\, dS - \oint_{\mathscr{C}_1} \mathbf{H} \cdot (d\mathbf{s} \wedge \mathbf{v}) \right]$$

$$= \lim_{\delta t \to 0} \left[\int_{S_2} \mathbf{n} \cdot \left\{ \frac{\mathbf{H}(t+\delta t) - \mathbf{H}(t)}{\delta t} \right\} dS \right] - \oint_{\mathscr{C}_1} d\mathbf{s} \cdot (\mathbf{v} \wedge \mathbf{H})$$

$$= \int_{S_1} \mathbf{n} \cdot \frac{\partial \mathbf{H}}{\partial t}\, dS - \int_{S_1} \mathbf{n} \cdot \operatorname{curl} (\mathbf{v} \wedge \mathbf{H})\, dS$$

$$= \int_{S_1} \mathbf{n} \cdot \left\{ \frac{\partial \mathbf{H}}{\partial t} - \operatorname{curl} (\mathbf{v} \wedge \mathbf{H}) \right\} dS$$

$$= 0, \quad \text{since } \partial \mathbf{H}/\partial t = \operatorname{curl} (\mathbf{v} \wedge \mathbf{H}).$$

Thus $N = \text{const.}$ through the moving circuit. This completes the proof.

Alfvén's theorem permits of important interpretations. The above result shows that lines of force move with the fluid. In this respect they are like vortex lines in inviscid flow. Alfvén spoke of the lines of magnetic force as being 'frozen' into the fluid. This means that the fluid can readily slip along the lines of force, but that any motion at right-angles to the lines of force will carry them along with the fluid, i.e. the lines of force are convected by the fluid in its motion. Looked at in another way, we can say that motion of the fluid across the lines of force will induce currents which will be very large for a perfectly conducting fluid. Thus for the induced currents to be finite, the relative velocity component perpendicular to the field must be small, showing the binding effect that the field has upon the fluid. We shall discuss several aspects of Alfvén's theorem, but we first return to the Navier–Stokes equation to give an important interpretation of the magnetic body force term.

9.9 The Magnetic Body Force

The equation of motion of the fluid (Section 9.4, equation (3)) becomes

$$\rho\frac{d\mathbf{v}}{dt} = -\nabla p + \rho\mathbf{g} + q\mathbf{E} + \frac{\mu}{4\pi}(\nabla\wedge\mathbf{H})\wedge\mathbf{H} + \tfrac{1}{3}\rho\nu\nabla\nabla\cdot\mathbf{v} + \rho\nu\nabla^2\mathbf{v}$$

when we use the approximation curl $\mathbf{H} = 4\pi\mathbf{j}$ obtained in Section 9.6. Thus the magnetic body force per unit volume is $(\mu/4\pi)\,(\nabla\wedge\mathbf{H})\wedge\mathbf{H}$.

Now

$$\begin{aligned}(\nabla\wedge\mathbf{H})\wedge\mathbf{H} &= -\mathbf{H}\wedge\textstyle\sum\{\mathbf{i}\wedge(\partial\mathbf{H}/\partial x)\}\\ &= -\textstyle\sum\{(\mathbf{H}\cdot\partial\mathbf{H}/\partial x)\mathbf{i}\} + \sum\{(\mathbf{H}\cdot\mathbf{i})\partial\mathbf{H}/\partial x\}\\ &= -\textstyle\sum\{\mathbf{i}(\partial/\partial x)\,(\tfrac{1}{2}\mathbf{H}^2)\} + \mathbf{H}\cdot(\sum\mathbf{i}\partial/\partial x)\mathbf{H}\\ &= -\nabla(\tfrac{1}{2}\mathbf{H}^2) + (\mathbf{H}\cdot\nabla)\mathbf{H}.\end{aligned}$$

To interpret the magnetic body force, we integrate $(\mu/4\pi)\,(\nabla\wedge\mathbf{H})\wedge\mathbf{H}$ throughout a volume Δv bounded by a closed surface ΔS. Then from the above identity,

$$\int_{\Delta v}(\nabla\wedge\mathbf{H})\wedge\mathbf{H}\,dv = \int_{\Delta v}(\mathbf{H}\cdot\nabla)\mathbf{H}\,dv + \int_{\Delta S}\mathbf{n}(-\tfrac{1}{2}\mathbf{H}^2)\,dS. \qquad (1)$$

We next show that

$$\int_{\Delta v}(\mathbf{H}\cdot\nabla)\mathbf{H}\,dv = \int_{\Delta S}(H\cdot\mathbf{n})\mathbf{H}\,dS.$$

Let $\mathbf{a}$ be an arbitrary, constant, non-zero vector. Then

$$\mathbf{a}\cdot\int_{\Delta v}(\mathbf{H}\cdot\nabla)\mathbf{H}\,dv = \int_{\Delta v}\mathbf{a}\cdot\{(\mathbf{H}\cdot\nabla)\mathbf{H}\}\,dv.$$

Now

$$
\begin{aligned}
\mathbf{a}\cdot\{(\mathbf{H}\cdot\nabla)\mathbf{H}\} &= \mathbf{a}\cdot\sum\{(\mathbf{H}\cdot\mathbf{i})\,(\partial\mathbf{H}/\partial x)\}\\
&= \sum\{(\mathbf{H}\cdot\mathbf{i})\,(\partial/\partial x)\,(\mathbf{H}\cdot\mathbf{a})\}\\
&= \mathbf{H}\cdot\{\sum\mathbf{i}(\partial/\partial x)\}(\mathbf{H}\cdot\mathbf{a})\\
&= \mathbf{H}\cdot\nabla(\mathbf{H}\cdot\mathbf{a})\\
&= \mathbf{H}\cdot\nabla(\mathbf{H}\cdot\mathbf{a})+(\nabla\cdot\mathbf{H})\,(\mathbf{H}\cdot\mathbf{a})\\
&= \nabla\cdot\{\mathbf{H}(\mathbf{H}\cdot\mathbf{a})\}.
\end{aligned}
$$

$$
\begin{aligned}
\therefore\ \mathbf{a}\cdot\int_{\Delta v}(\mathbf{H}\cdot\nabla)\mathbf{H}\,dv &= \int_{\Delta v}\nabla\cdot\{\mathbf{H}(\mathbf{H}\cdot\mathbf{a})\}\,dv\\
&= \int_{\Delta S}\mathbf{n}\cdot\{\mathbf{H}(\mathbf{H}\cdot\mathbf{a})\}\,dS\\
&= \int_{\Delta S}(\mathbf{a}\cdot\mathbf{H})\,(\mathbf{n}\cdot\mathbf{H})\,dS\\
&= \mathbf{a}\cdot\int_{\Delta S}\mathbf{H}(\mathbf{n}\cdot\mathbf{H})\,dS.
\end{aligned}
$$

Since $\mathbf{a}$ is arbitrary and non-zero, this shows that

$$\int_{\Delta v}(\mathbf{H}\cdot\nabla)\mathbf{H}\,dv=\int_{\Delta S}(\mathbf{n}\cdot\mathbf{H})\mathbf{H}\,dS. \tag{2}$$

From (1), (2) the magnetic body force on Δv is

$$\frac{\mu}{4\pi}\int_{\Delta v}(\nabla\wedge\mathbf{H})\wedge\mathbf{H}\,dv=\int_{\Delta S}\frac{\mu\mathbf{H}}{4\pi}(\mathbf{n}\cdot\mathbf{H})\,dS+\int_{\Delta S}\left(-\frac{\mu}{8\pi}\mathbf{H}^2\mathbf{n}\right)dS. \tag{3}$$

Equation (3) shows that the original body force is equivalent to two kinds of surface force on each surface element δS, viz.

$$\frac{\mu\mathbf{H}}{4\pi}\mathbf{n}\cdot\mathbf{H}\delta S;\qquad \frac{-\mu\mathbf{H}^2}{8\pi}\mathbf{n}\delta S.$$

To interpret the first surface force, let θ be the angle between $\mathbf{n}, \mathbf{H}$ so that $\mathbf{n}\cdot\mathbf{H}=H\cos\theta$. Also let $\delta S'=\delta S\cos\theta$, the projection of δS normal to $\mathbf{n}$. Then, if $\mathbf{H}=H\hat{\mathbf{H}}$,

$$\frac{\mu\mathbf{H}}{4\pi}\mathbf{n}\cdot\mathbf{H}\delta S=\frac{\mu\mathbf{H}^2}{4\pi}\delta S'\hat{\mathbf{H}},$$

which shows that the force per unit area is $(\mu H^2/4\pi)$ in the direction of $\mathbf{H}$. This represents a tensile force per unit area of amount $(\mu H^2/4\pi)$ in the direction of the magnetic field.

The surface force $(-\mu\mathbf{H}^2/8\pi)\mathbf{n}\delta S$ gives a force $(\mu\mathbf{H}^2/8\pi)$ per unit area in the direction $-\mathbf{n}$. This is a hydrostatic pressure $(\mu\mathbf{H}^2/8\pi)$.

In summary, then, we may say that *for a conducting fluid in a magnetic field, the magnetic body force per unit volume, viz.* $(\mu/4\pi)(\nabla\wedge\mathbf{H})\wedge\mathbf{H}$, *is equivalent to a tension* $(\mu/4\pi)\mathbf{H}^2$ *per unit area along the lines of force, together with a hydrostatic pressure* $(\mu/8\pi)\mathbf{H}^2$. This is equivalent to *a tension* $(\mu/8\pi)\mathbf{H}^2$ *per unit area along the lines of force together with a pressure* $(\mu/8\pi)\mathbf{H}^2$ *transverse to them.* The second result is the same as the Maxwell stress in a dielectric.

9.10 Ferraro's Law of Isorotation

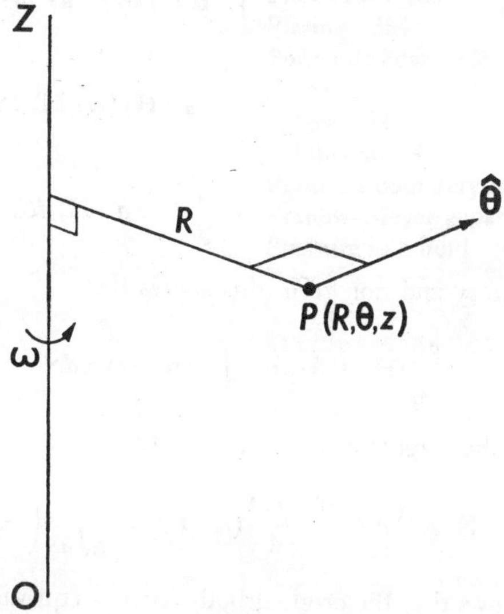

FIG. 9.4

In Fig. 9.4, OZ is the axis of symmetry of a star of high electrical conductivity, rotating about this axis with angular velocity ω. Then in accordance with Alfvén's theorem the lines of magnetic force are frozen into the material. P has cylindrical polar coordinates (R, θ, z). All variables are independent of θ and when the state is steady they are also independent of t. We shall assume steady rotation so that $\partial/\partial\theta \equiv 0$, $\partial/\partial t \equiv 0$.

We take the cylindrical polar components of $\mathbf{H}$ to be in the form

$$\mathbf{H} = \{H_R(R, z), \quad 0, \quad H_z(R, z)\}.$$

Since $\nabla \cdot \mathbf{H} = 0$,

$$\frac{1}{R}\left\{\frac{\partial}{\partial R}(RH_R) + \frac{\partial}{\partial z}(RH_z)\right\} = 0.$$

This result implies that there exists a scalar function $\Psi(R, z)$ such that

$$H_R = -\frac{1}{R}\frac{\partial \Psi}{\partial z}, \quad H_z = \frac{1}{R}\frac{\partial \Psi}{\partial R}. \tag{1}$$

We call Ψ the *magnetic stream function* on account of its close similarity with the stream function ψ of hydrodynamics.

We assume the fluid velocity at $P(R, \theta, z)$ is

$$\mathbf{v} = R\omega\hat{\boldsymbol{\theta}}, \tag{2}$$

where $\omega = \omega(R, z)$. Since $\mathbf{H}$ is independent of time, the equation $\partial\mathbf{H}/\partial t = \nabla \wedge (\mathbf{v} \wedge \mathbf{H})$ gives

$$\nabla \wedge (\mathbf{v} \wedge \mathbf{H}) = \mathbf{0}.$$

Now

$$\mathbf{v} \wedge \mathbf{H} = R\omega\hat{\boldsymbol{\theta}} \wedge \left(-\frac{1}{R}\frac{\partial \Psi}{\partial z}\hat{\mathbf{R}} + \frac{1}{R}\frac{\partial \Psi}{\partial R}\mathbf{k}\right)$$

$$= \omega\frac{\partial \Psi}{\partial R}\hat{\mathbf{R}} + \omega\frac{\partial \Psi}{\partial z}\mathbf{k}.$$

Hence

$$\nabla \wedge (\mathbf{v} \wedge \mathbf{H}) = \frac{1}{1 \cdot R \cdot 1}\begin{vmatrix} \hat{\mathbf{R}} & R\hat{\boldsymbol{\theta}} & \mathbf{k} \\ \dfrac{\partial}{\partial R} & \dfrac{\partial}{\partial \theta} & \dfrac{\partial}{\partial z} \\ \dfrac{\omega\partial \Psi}{\partial R} & 0 & \dfrac{\omega\partial \Psi}{\partial z} \end{vmatrix}$$

$$= \left(-\frac{\partial}{\partial R}\left(\frac{\omega\partial \Psi}{\partial z}\right) + \frac{\partial}{\partial z}\left(\frac{\omega\partial \Psi}{\partial R}\right)\right)\hat{\boldsymbol{\theta}},$$

giving

$$\frac{\partial}{\partial R}\left(\frac{\omega\partial \Psi}{\partial z}\right) - \frac{\partial}{\partial z}\left(\frac{\omega\partial \Psi}{\partial R}\right) = 0,$$

or

$$\frac{\partial \omega}{\partial R}\frac{\partial \Psi}{\partial z} - \frac{\partial \omega}{\partial z}\frac{\partial \Psi}{\partial R} = 0,$$

i.e.

$$\frac{\partial(\omega, \Psi)}{\partial(R, z)} = 0. \tag{3}$$

Equation (3) shows that

$$\omega = f(\Psi), \tag{4}$$

or that ω is constant on the surface $\Psi = \text{const}$.

Now the equations of the lines of force are

$$\frac{dR}{H_R} = R\frac{d\theta}{H_\theta} = \frac{dz}{H_z},$$

i.e.

$$\frac{dR}{-(1/R)\,(\partial\Psi/\partial z)} = \frac{R\,d\theta}{0} = \frac{dz}{-(1/R)\,(\partial\Psi/\partial R)},$$

or

$$\begin{cases} d\theta = 0, \\ \dfrac{\partial\Psi}{\partial R}\,dR + \dfrac{\partial\Psi}{\partial z}\,dz = 0. \end{cases}$$

These integrate to give

$$\begin{cases} \theta = \text{const.} \\ \Psi = \text{const.} \end{cases}$$

as the lines of magnetic force. Thus the surface $\Psi = \text{const.}$ is the surface generated by revolving a magnetic line of force once completely round the axis of symmetry. Such a surface may be termed a 'magnetic stream surface'. Equation (4) shows that $\omega = \text{const.}$ over such a surface.

Ferraro's law of isorotation, embodied in equation (4), states that the *angular velocity is constant over a magnetic stream surface when the star is rotating steadily about its axis of symmetry*. It follows that the angular velocity is constant along a line of force. One can also see this result rather naïvely as follows. If the angular velocity were not constant along a line of force, different parts of the line of force would take different times to describe a complete revolution. There would be shearing effects set up at points along the line and since the lines of force move with the fluid, this would create a time-dependent component of the magnetic field in the azimuthal direction, thereby violating the assumption of steady state conditions. Actually, Ferraro discovered this law in 1937, before Alfvén published his theory of frozen-in fields.

9.11 Magnetohydrostatics

When the fluid is at rest and the volume density of charge is zero, equation (3) of Section 9.4 becomes

$$\nabla p - \rho\mathbf{g} = \mu\mathbf{j} \wedge \mathbf{H}, \tag{1}$$

where $\mathbf{j} \equiv (1/4\pi)\nabla \wedge \mathbf{H}$, and equation (1) of Section 9.7 becomes

$$\partial\mathbf{H}/\partial t = \eta\nabla^2\mathbf{H}, \tag{2}$$

which is the diffusion equation that has been seen to hold for small magnetic Reynolds numbers.

For a liquid, ρ is constant, and since the field of $\mathbf{g}$ is conservative, we can find a scalar potential function Ω such that $\mathbf{g}=-\nabla\Omega$. Then (1) gives

$$\nabla(p+\rho\Omega)=\mu\mathbf{j}\wedge\mathbf{H}. \tag{3}$$

Equation (3) shows that the vectors $\mathbf{j}, \mathbf{H}$ both lie in the family of surfaces

$$p+\rho\Omega=\text{const.} \tag{4}$$

a result which bears close resemblance to the case of steady motion of a fluid under conservative body forces studied in Section 3.7, where we showed that the streamlines and vortex lines lie within certain surfaces.

The magnetic body force per unit volume of the conducting fluid is $(\mu/4\pi)(\nabla\wedge\mathbf{H})\wedge\mathbf{H}$. If this is zero, the corresponding magnetic field is said to be *force-free* and

$$(\nabla\wedge\mathbf{H})\wedge\mathbf{H}=0. \tag{5}$$

Equation (5) has the special solutions

$$\mathbf{H}=0, \quad \nabla\wedge\mathbf{H}=0, \quad \nabla\wedge\mathbf{H}=\alpha\mathbf{H}.$$

We next consider some simple problems in magnetohydrostatics.

9.11.1 Pinch Confinement of a Plasma

The confinement of a plasma is of great importance in nuclear engineering. Here we describe a process based on the 'pinch effect'.

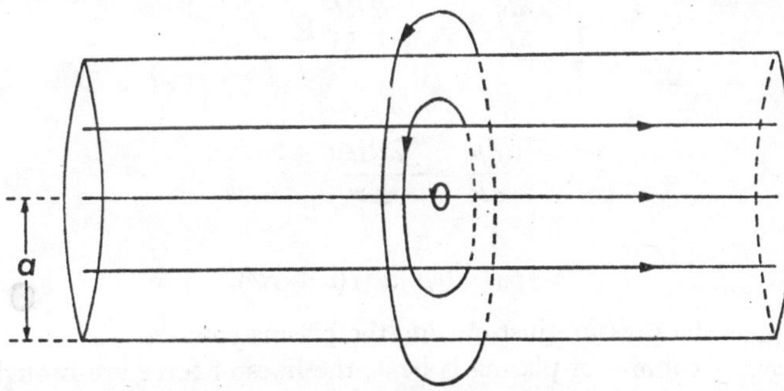

Fig. 9.5

The plasma is in the form of a long cylinder and carries a current in the longitudinal direction (Fig. 9.5). It is supposed to be uniform so that the current density vector $\mathbf{j}$ is the same at all internal points. Such a condition is realized when the conductivity is finite so that the current can quickly diffuse through the material. Then the magnetic field $\mathbf{H}$ both inside and outside the plasma consists of circular loops whose centres are on the axis of the cylinder and whose planes are at right-angles to it. It is a well-known property of electrodynamics that parallel wires carrying currents flowing in the same direction tend to attract one another. In the same way, the longitudinal filaments of the plasma tend to attract one another, thereby causing a tendency towards constriction. This is the *pinch effect*.

If we neglect gravitational forces, equation (3) gives

$$\nabla p = \mu \mathbf{j} \wedge \mathbf{H}. \tag{6}$$

Let (R, θ, z) be the cylindrical polar coordinates of any point, the z-axis being the axis of the cylinder. Then clearly

$$p = p(R), \quad \mathbf{H} = H(R)\hat{\boldsymbol{\theta}}.$$

If the total current flowing is I and if a is the radius of the section, then $\mathbf{j} = \mathbf{k}I/(\pi a^2)$. The current flowing over a section of radius $R \leqslant a$ is $\pi R^2 j = (R/a)^2 I$. Let $\mathscr{C}$ denote a circle having radius R, its centre being on the z-axis. Then, from electromagnetic theory,

$$\oint_{\mathscr{C}} \mathbf{H} \cdot d\mathbf{s} = 4\pi (R/a)^2 I,$$

since $\mathscr{C}$ links a current of $(R/a)^2 I$. Thus

$$2\pi R H(R) = 4\pi (R/a)^2 I$$

or

$$\mathbf{H} = H(R)\hat{\boldsymbol{\theta}} = (2R/a^2) I \hat{\boldsymbol{\theta}}.$$

The equation (6) now gives

$$\frac{dp}{dR}\hat{\mathbf{R}} = -\mu \frac{2RI^2}{\pi a^4}\hat{\mathbf{R}}$$

or

$$\frac{dp}{dR} = -\frac{2RI^2\mu}{\pi a^4}.$$

Hence

$$p - p_0 = (I^2\mu/\pi a^4)\,(a^2 - R^2), \tag{7}$$

where p_0 is the pressure just outside the plasma column.

When the column of plasma is bent, the lines of force are brought together on the concave side and drawn apart on the convex side. This means magnetic pressure is increased on the concave side and decreased on the convex. Thus the tendency is to produce greater distortion. This shows that the configuration is unstable.

9.11.2 Equilibrium of Sunspots

Sunspots are dark umbrae surrounded by less dark penumbrae in the photosphere, i.e. the visible layers of the sun. They extend into the photospheric material in the form of long cylindrical columns. Sunspots occur periodically, may endure for a month or so, have temperatures of about 4500°K when in full maturity and may have magnetic fields of intensity

2000 gauss associated with them. The temperature of the photosphere is about 5500°K so that the temperature in a sunspot is considerably less than that in the photosphere.

When a sunspot is in equilibrium, if we assume that the plasma densities inside and outside the spot are equal, and also that the equation of state $p=R\rho T$ holds (p=pressure, ρ=density, T=temperature, R=const.), then the decreased temperature within the spot implies a decreased pressure. One would, therefore, expect the hotter surrounding plasma to flow into the spot from outside, thereby submerging it. This probably happens when the spot decays, but when it is in equilibrium it is thought that the powerful magnetic field intensity within the sunspot material produces a sufficiently great pressure to prevent this happening. Thus if p_i is the pressure within the sunspot column, p_0 that in the material outside, then when the spot is in equilibrium,

$$p_i+(\mu\mathbf{H}^2/8\pi)=p_0,$$

if we assume a uniform magnetic field of intensity $\mathbf{H}$ within the column. Here we use the result that $\mathbf{H}$ gives rise to a Maxwellian stress system equivalent to a hydrostatic pressure $(\mu\mathbf{H}^2/8\pi)$ and a tension per unit area of $(\mu\mathbf{H}^2/4\pi)$ along the lines of force.

9.12 Magnetohydrodynamic Waves

In Chapter 7 we saw that a non-conducting compressible fluid could transmit sound waves which are longitudinal in kind. No form of transverse propagation is possible for such a fluid. Moreover, an incompressible fluid which is non-conducting cannot support any kind of wave motion at all. When the fluid is conducting, however, transverse wave motion through it is possible whenever a magnetic field is present. We first illustrate this distinctive feature of MHD by a simple model.

We suppose the fluid to be infinitely conducting so that, in accordance with Alfvén's theorem, the magnetic lines of force are glued to the fluid particles. Suppose $\mathbf{H}_0$ is the undisturbed field intensity in a tube of magnetic force of section δA. Then the mass per unit length of such a tube is m where $m=\rho\delta A$. The magnetic forces acting on the tube are equivalent to a hydrostatic pressure $\mu\mathbf{H}_0^2/8\pi$ and a tension per unit area $\mu\mathbf{H}_0/4\pi$ along the lines of force. The former is balanced by a decrease in fluid pressure leaving the tubes in tension. Thus the tension T along the line of force is $T=\mu\mathbf{H}_0^2\delta A/4\pi$. If the fluid is incompressible, the lines of force are like stretched strings in tension T and mass m per unit length. This suggests that when they are disturbed, by giving the fluid particles a small velocity component normal to them, they will execute transverse vibrations, the waves having velocity of propagation

$$\left(\frac{T}{m}\right)^{1/2}=\left(\frac{\mu H_0^2}{4\pi\rho}\right)^{1/2}=V_A.$$

Such waves are known as *Alfvén waves* after Alfvén who first pointed out the string analogy in 1942. V_A is called the ***Alfvén velocity***.

Now suppose the perfectly conducting fluid is incompressible. Then longitudinal propagation is also feasible. The nature of the wave motion produced depends on the direction of **H** relative to the particle velocity **v**.

If the particle velocity and direction of wave propagation are both parallel to $\mathbf{H}_0$, then no magnetic effects are called into play since the fluid moves along the lines of force. Hence small propagations are transmitted as ordinary plane sound waves, the velocity of propagation being a, the speed of sound in the fluid.

If the particle velocity is parallel to the direction of propagation and each is perpendicular to $\mathbf{H}_0$, the undisturbed magnetic field intensity, we can show that a new type of wave is excited. For a perfectly conducting fluid, if **H** is the field intensity at time t and **v** the particle velocity, then

$$\partial \mathbf{H}/\partial t = \nabla \wedge (\mathbf{v} \wedge \mathbf{H}).$$

Taking $\mathbf{v} = u(x)\mathbf{i}$, $\mathbf{H} = H(x)\mathbf{j}$, we obtain $\mathbf{v} \wedge \mathbf{H} = uH\mathbf{k}$ and

$$\nabla \wedge (\mathbf{v} \wedge \mathbf{H}) = \mathbf{i} \wedge \left(\frac{d}{dx}\right)(uH\mathbf{k}) = -\left(\frac{d}{dx}\right)(uH)\mathbf{j}.$$

Thus for steady motion, $\partial \mathbf{H}/\partial t = \mathbf{0}$ and so

$$(d/dx)(uH) = 0$$

or

$$uH = \text{const.}$$

But the one-dimensional equation of continuity $\nabla \cdot (\rho \mathbf{v}) = 0$ gives

$$\rho u = \text{const.}$$

From the last two equations,

$$\frac{H}{\rho} = \text{const.} = \frac{H_0}{\rho_0},$$

the suffix 0 referring to undisturbed conditions. Since the magnetic pressure is $\mu \mathbf{H}^2/8\pi$, the effective pressure is $\bar{p}$ where

$$\bar{p} = p + (\mu H^2/8\pi),$$

so that

$$\frac{d\bar{p}}{d\rho} = \frac{dp}{d\rho} + \frac{\mu H}{4\pi}\frac{dH}{d\rho}.$$

Now

$$\frac{dH}{d\rho} = \frac{H_0}{\rho_0} = \frac{H}{\rho},$$

and

$$a^2 = \frac{dp}{d\rho}.$$

Hence

$$\frac{d\bar{p}}{d\rho}=a^2+\frac{\mu H^2}{4\pi\rho}=a^2+V_A^2$$

where V_A is the Alfvén velocity. Thus the speed of propagation is $\sqrt{(a^2+V^2{}_A)}$. Such a wave is called a *magnetohydrodynamic wave.* Alfvén waves are transverse and are propagated in conducting incompressible fluids, but MHD waves are longitudinal in kind and their propagation requires a conducting incompressible fluid.

9.12.1 A More Detailed Investigation of Alfvén Waves

The string analogy considered above is a powerful model for investigating Alfvén waves in an elementary manner. Here we treat them as waves in their own right and give a more detailed theoretical investigation.

For an Alfvén wave, the velocity of propagation is along the lines of magnetic force and the particle velocity at right-angles to them. Suppose that the undisturbed field is uniform and of intensity $\mathbf{H}_0=H_0\mathbf{k}$, $\mathbf{k}$ being the unit vector in a fixed direction $\overline{OZ}$. Suppose $\mathbf{h}$ is the perturbation produced in the field due to a small disturbance. Then the resultant field is

$$\mathbf{H}=\mathbf{H}_0+\mathbf{h}. \tag{1}$$

For a conducting incompressible fluid for which there is no charge accumulation at internal points, the equation of motion, neglecting viscous effects, is

$$\rho\frac{d\mathbf{v}}{dt}=-\nabla p+\frac{\mu}{4\pi}(\nabla\wedge\mathbf{h})\wedge\mathbf{H}. \tag{2}$$

Also, the equation $\nabla\cdot\mathbf{H}=0$ becomes

$$\nabla\cdot\mathbf{h}=0. \tag{3}$$

The equation of continuity is simply

$$\nabla\cdot\mathbf{v}=0. \tag{4}$$

Suppose the disturbances are so small that squares and products of $\mathbf{h}$ and $\mathbf{v}$ are of negligible magnitude. Then

$$\frac{d\mathbf{v}}{dt}=\frac{\partial\mathbf{v}}{\partial t}+\mathbf{v}\cdot\nabla\mathbf{v}\simeq\frac{\partial\mathbf{v}}{\partial t}$$

$$(\nabla\wedge\mathbf{h})\wedge\mathbf{H}=(\nabla\wedge\mathbf{h})\wedge(\mathbf{H}_0+\mathbf{h})\simeq(\nabla\wedge\mathbf{h})\wedge\mathbf{H}_0.$$

But

$$\begin{aligned}(\nabla\wedge\mathbf{h})\wedge\mathbf{H}_0&=-\mathbf{H}_0\wedge\sum(\mathbf{i}\wedge\partial\mathbf{h}/\partial x)\\&=-\sum\{\mathbf{i}(\mathbf{H}_0\cdot\partial\mathbf{h}/\partial x)\}+\sum\{(\mathbf{i}\cdot\mathbf{H}_0)\partial\mathbf{h}/\partial x\}\\&=-\sum\{\mathbf{i}(\partial/\partial x)\,(\mathbf{H}_0\cdot\mathbf{h})\}+H_0\partial\mathbf{h}/\partial z.\end{aligned}$$

Hence

$$(\nabla\wedge\mathbf{h})\wedge\mathbf{H}\simeq-\nabla(\mathbf{H}_0\cdot\mathbf{h})+H_0\partial\mathbf{h}/\partial z.$$

Thus (2) approximates to

$$\nabla\left(p+\frac{\mu}{4\pi}\mathbf{H}_0\cdot\mathbf{h}\right)=\frac{\mu H_0}{4\pi}\frac{\partial\mathbf{h}}{\partial z}-\rho\frac{\partial\mathbf{v}}{\partial t}. \tag{5}$$

Taking the divergence of both sides of (5) and using (3), (4),

$$\nabla^2\{p+(\mu/4\pi)\mathbf{H}_0\cdot\mathbf{h}\}=0. \tag{6}$$

Thus $p+(\mu/4\pi)\mathbf{H}_0\cdot\mathbf{h}$ is a harmonic function. Two cases arise, according to whether the liquid is of infinite or finite extent.

Case I

Liquid of infinite extent.

The solution to (6) which is regular at all points including infinity is

$$p+(\mu/4\pi)\mathbf{H}_0\cdot\mathbf{h}=\text{const.} \tag{7}$$

Then (5) reduces to

$$\rho\frac{\partial\mathbf{v}}{\partial t}=\frac{\mu H_0}{4\pi}\frac{\partial\mathbf{h}}{\partial z}. \tag{8}$$

The equation governing magnetic field variations, viz.

$$\partial\mathbf{H}/\partial t=\nabla\wedge(\mathbf{v}\wedge\mathbf{H})$$

becomes, to the first order,

$$\partial\mathbf{h}/\partial t=\nabla\wedge(\mathbf{v}\wedge\mathbf{H}_0). \tag{9}$$

Now $\mathbf{v}\wedge\mathbf{H}_0=H_0\mathbf{v}\wedge\mathbf{k}$, and so $\nabla\wedge(\mathbf{v}\wedge\mathbf{H}_0)=H_0\partial\mathbf{v}/\partial z$, using $\nabla\cdot\mathbf{v}=0$. Hence (9) gives

$$\frac{\partial\mathbf{h}}{\partial t}=H_0\frac{\partial\mathbf{v}}{\partial z}. \tag{10}$$

From (8), (10) we readily obtain the wave equations

$$\left(\frac{\partial^2}{\partial t^2}-V_A^2\frac{\partial^2}{\partial z^2}\right)(\mathbf{h},\mathbf{v})=0, \tag{11}$$

for $\mathbf{h}$ and $\mathbf{v}$, where

$$V_A^2=(\mu\mathbf{H}_0^2/4\pi\rho). \tag{12}$$

Thus **h** *and* **v** *are propagated as transverse waves along the lines of force with the Alfvén velocity* V_A.

For a wave travelling in the positive direction of the z-axis, (11) gives the solutions

$$\left.\begin{aligned}\mathbf{h}&=\mathbf{h}(z-V_A t),\\ \mathbf{v}&=\mathbf{v}(z-V_A t)\end{aligned}\right\} \tag{13}$$

which are both vector functions of $z-V_A t$. From (13),

$$\frac{\partial \mathbf{h}}{\partial t}=-V_A\frac{\partial \mathbf{h}}{\partial z}$$

and substitution into (10) gives

$$\frac{\partial \mathbf{h}}{\partial z}=-\frac{H_0}{V_A}\frac{\partial \mathbf{v}}{\partial z}=-\left(\frac{4\pi\rho}{\mu}\right)^{1/2}\frac{\partial \mathbf{v}}{\partial z},$$

which is satisfied by

$$\mathbf{h}=-(4\pi\rho/\mu)^{1/2}\mathbf{v}. \tag{14}$$

Similarly for a wave travelling in the negative direction of the z-axis we have

$$\mathbf{h}=(4\pi\rho/\mu)^{1/2}\mathbf{v}. \tag{15}$$

The relations (14), (15) are due to Walén. From them we see that in both cases $(\mu/8\pi)\mathbf{h}^2=\frac{1}{2}\rho\mathbf{v}^2$ which shows that the magnetic energy of the perturbed field is equal to the kinetic energy of the motion.

Case II

Liquid of finite extent.

In this case $p+(\mu/4\pi)\mathbf{H}_0\cdot\mathbf{h}$ may not be constant. Moreover, boundary reflections and transmissions would occur so that the solution of (6) would be much more complicated. If, however, we take the curl of both sides of (5), (10) and use $\nabla\wedge\mathbf{h}=4\pi\mathbf{j}$, $\boldsymbol{\zeta}=\nabla\wedge\mathbf{v}$, where $\boldsymbol{\zeta}$ is the vorticity vector, then

$$\mu H_0\frac{\partial \mathbf{j}}{\partial z}=\frac{\rho\,\partial\boldsymbol{\zeta}}{\partial t},$$

$$4\pi\frac{\partial \mathbf{j}}{\partial t}=H_0\frac{\partial \boldsymbol{\zeta}}{\partial z}$$

from which it readily follows that

$$\left(\frac{\partial^2}{\partial t^2}-V_A^2\frac{\partial^2}{\partial z^2}\right)(\mathbf{j},\boldsymbol{\zeta})=0. \tag{16}$$

This equation (16) shows that *the vorticity vector* $\boldsymbol{\zeta}$ *and the current density vector* $\mathbf{j}$ *both satisfy the one-dimensional wave equation and are propagated with speed* V_A *along the lines of force.*

9.12.2 Reflection and Transmission of Alfvén Waves at a Discontinuity in Density

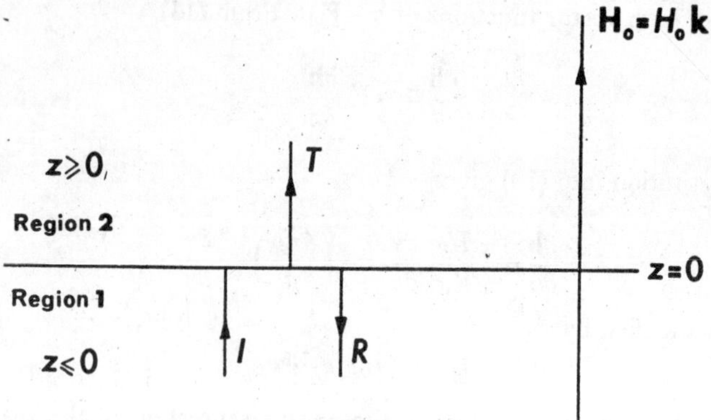

FIG. 9.6

The surface $z=0$ separates two conducting liquids of infinite extent into Regions 1, 2 which are $z<0$, $z\geqslant 0$ respectively. We consider the simplest case of an incident Alfvén wave I in Region 1 polarized in a plane at right-angles to the permanent magnetic field $\mathbf{H}_0 = H_0\mathbf{k}$ and moving in the direction $\mathbf{k}$. Then, for the incident wave, $\mathbf{v}$ and $\mathbf{h}$ are at right-angles to the z-direction. Suppose their directions are specified by the unit vector $\mathbf{i}$. We recall Walén's relations for liquids of infinite extent, viz.

$$\mathbf{h} = \mp(4\mu\rho/\mu)^{1/2}\mathbf{v},$$

the minus sign holding for propagation in the direction $\mathbf{k}$ and the plus sign for propagation in the direction $-\mathbf{k}$. On meeting the surface $z=0$, the wave suffers partial reflection back into Region 1 and partial transmission into Region 2. The regions have the same permeability μ throughout but the density ρ is discontinuous at $z=0$.

The incident, reflected and transmitted waves accordingly have the following forms:

Incident Wave I (moving parallel to $\mathbf{H}_0$)

$$\begin{cases} \mathbf{v}_I = v_I\mathbf{i} \exp\{ip(t - z/V_{A1})\}, \\ \mathbf{h}_I = -(4\pi\rho_1/\mu)v_I\mathbf{i} \exp\{ip(t - z/V_{A1})\}; \end{cases}$$

Reflected Wave R (moving parallel to $-\mathbf{H}_0$)

$$\begin{cases} \mathbf{v}_R = v_R\mathbf{i} \exp\{ip(t + z/V_{A1})\}, \\ \mathbf{h}_R = (4\pi\rho_1/\mu)v_R\mathbf{i} \exp\{ip(t + z/V_{A1})\}; \end{cases}$$

Transmitted Wave T (moving parallel to $\mathbf{H}_0$)

$$\begin{cases} \mathbf{v}_T = v_T\mathbf{i} \exp\{ip(t - z/V_{A2})\}, \\ \mathbf{h}_T = -(4\pi\rho_2/\mu)^{1/2}v_T\mathbf{i} \exp\{ip(t - z/V_{A2})\}, \end{cases}$$

where V_{A1}, V_{A2} are the Alfvén velocities in the Regions 1, 2 defined by $V_{A1}=H_0(\mu/4\pi\rho_1)^{1/2}$, $V_{A2}=H_0(\mu/4\pi\rho_2)^{1/2}$.

The total waves in Regions 1, 2 are given by:

Region 1

$$\begin{cases} \mathbf{v}_1=\mathbf{v}_I+\mathbf{v}_R, \\ \mathbf{h}_1=\mathbf{h}_I+\mathbf{h}_R; \end{cases}$$

Region 2

$$\begin{cases} \mathbf{v}_2=\mathbf{v}_T, \\ \mathbf{h}_2=\mathbf{h}_T. \end{cases}$$

At the boundary $z=0$, $\mathbf{v}_1$ and $\mathbf{v}_2$ are equal and so also are $\mathbf{h}_1$ and $\mathbf{h}_2$. Thus we obtain

$$\begin{cases} v_I+v_R=v_T, \\ -\rho_1^{1/2}v_I+\rho_1^{1/2}v_R=-\rho_2^{1/2}v_T, \end{cases}$$

which on solution yield

$$v_R=\frac{\rho_1^{1/2}-\rho_2^{1/2}}{\rho_1^{1/2}+\rho_2^{1/2}}v_I, \qquad v_T=\frac{2\rho_1^{1/2}}{\rho_1^{1/2}+\rho_2^{1/2}}v_I$$

for the velocity amplitudes of the reflected and transmitted waves.

9.13 Magnetohydrodynamic Shock Waves

In Chapter 7, shock waves in gases were studied ignoring the effects of viscosity and thermal conductivity. Here we first discuss these effects on one-dimensional steady shock formation in the case of a non-conducting gas and then introduce suitable modifications to account for a conducting gas in a transverse magnetic field. We now assume the shock zone to be a thin régime and not just a plane discontinuity as before.

9.13.1 Shock Wave in Non-Conducting Gas with Finite Viscosity and Thermal Conductivity

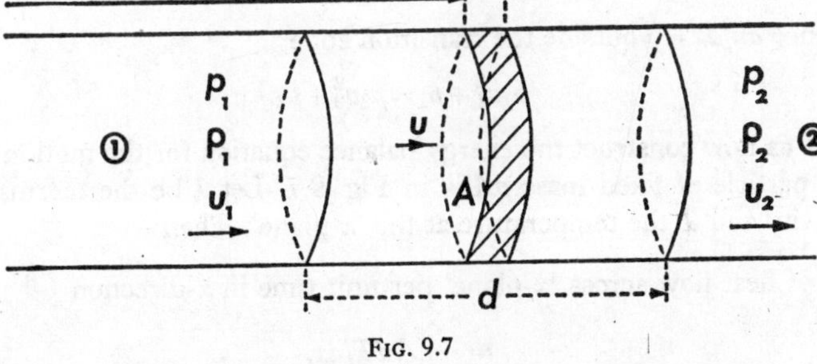

FIG. 9.7

Figure 9.7 shows the formation of a stationary shock zone of thickness d in a tube of uniform cross-section A. The flow is one-dimensional throughout and the zone is taken to separate uniform Regions 1, 2. The shock zone is a transition zone in which the pressure p, density ρ, and particle velocity u and temperature T change continuously from p_1, ρ_1, u_1, T_1 in Region 1 to p_2, ρ_2, u_2, T_2 in Region 2. The continuous change is due to the finite viscosity and thermal conductivity of the gas and the thickness d is of the order of a few mean free paths of the gas molecules. Within the shock zone we consider the motion of an element of fixed mass $\rho A\delta x$ moving along with the velocity u of the fluid. The flow is steady and one-dimensional throughout.

The equation of conservation of mass is $(d/dx)(\rho u)=0$, whence

$$\rho_1 u_1=\rho_2 u_2=m, \text{ say.} \tag{1}$$

The general Navier–Stokes equation of motion of a viscous fluid moving with velocity $\mathbf{v}$ is (Section 8.9, Equation (2))

$$\frac{d\mathbf{v}}{dt}=\mathbf{F}-\nabla\int\frac{dp}{\rho}+\frac{\nu}{3}\nabla(\nabla\cdot\mathbf{v})+\nu\nabla^2\mathbf{v}.$$

Taking $\mathbf{v}=u\mathbf{i}$, body force per unit mass $\mathbf{F}=\mathbf{0}$ and steady conditions,

$$\nabla\cdot\mathbf{v}=\frac{du}{dx},\quad \nabla(\nabla\cdot\mathbf{v})=\frac{d^2u}{dx^2}\mathbf{i},\quad \nabla^2\mathbf{v}=\frac{d^2u}{dx^2}\mathbf{i},$$

and so

$$u\frac{du}{dx}=-\frac{1}{\rho}\frac{dp}{dx}+\frac{4}{3}\nu\frac{d^2u}{dx^2},$$

whence

$$\rho u\frac{du}{dx}=-\frac{dp}{dx}+\frac{4}{3}\rho\nu\frac{d^2u}{dx^2}.$$

Remembering that ρu, $\rho\nu$ are constants this equation integrates to give

$$\left[(\rho u)u+p-\tfrac{4}{3}(\rho\nu)(du/dx)\right]_1^2=0,$$

or, since $du/dx=0$ outside the transition zone,

$$\rho_1 u_1^2+p_1=\rho_2 u_2^2+p_2. \tag{2}$$

Let us now construct the energy balance equation for the motion of the fluid particle of fixed mass $\rho A\delta x$ in Fig. 9.7. Let λ be the thermal conductivity and T the temperature at the 'x-plane'. Then

Rate of heat flow across 'x-plane' per unit time in x-direction

$$=-\lambda A\,dT/dx.$$

Rate of heat flow across '$(x+\delta x)$-plane' in x-direction

$$= -\lambda A \frac{dT}{dx} - \delta x \frac{d}{dx}\left(\frac{\lambda A\, dT}{dx}\right).$$

Net rate of heat conduction per unit time into element

$$= \lambda A \delta x (d^2T/dx^2).$$

The derivation of the equation of motion shows that the effective fluid pressure at the 'x-plane' in the x-direction is $p - \frac{4}{3}\rho v(du/dx)$.

Force on 'x-plane' due to fluid in element

$$= -(p - \tfrac{4}{3}\rho v\, du/dx)A.$$

Rate of working of this force

$$= -(pu - \tfrac{4}{3}\rho v u\, du/dx)A.$$

Rate of working of effective force on '$(x+\delta x)$-plane'

$$= \left(pu - \tfrac{4}{3}\rho v u \frac{du}{dx}\right) A + \delta x \frac{d}{dx}\left(pu - \tfrac{4}{3}\rho v u \frac{du}{dx}\right) A.$$

Total rate of working of fluid pressure on element

$$= A\delta x \frac{d}{dx}\left(pu - \tfrac{4}{3}\rho v u \frac{du}{dx}\right).$$

Internal energy of gas in element

$$= (\rho A \delta x) C_v T$$

$$= \frac{RT}{\gamma - 1}(\rho A \delta x) = \frac{p}{(\gamma-1)\rho}(A\rho\delta x).$$

Rate of gain of internal energy

$$= \frac{A\rho\delta x}{\gamma - 1}\frac{d}{dt}\left(\frac{p}{\rho}\right) = \frac{A\rho\delta x u}{\gamma-1}\frac{d}{dx}\left(\frac{p}{\rho}\right).$$

K.E. of element

$$= \tfrac{1}{2}(\rho A \delta x)u^2.$$

Rate of gain of K.E. of element

$$= (\rho A \delta x) u \frac{du}{dt} = (\rho A \delta x) u^2 \frac{du}{dx}.$$

Rate of supply of heat to element by conduction = Rate of performing external work + Rate of gain of I.E. + Rate of gain of K.E.

$$\lambda A\delta x\,\frac{d^2T}{dx^2}=A\delta x\,\frac{d}{dx}\left(pu-\tfrac{4}{3}\rho\nu u\,\frac{du}{dx}\right)+\frac{A\rho\delta x}{\gamma-1}\,u\,\frac{d}{dx}\left(\frac{p}{\rho}\right)+(\rho A\delta x)u\left(u\,\frac{du}{dx}\right).$$

Since $\rho u=$ const., dividing by $\rho uA\delta x$ gives

$$\frac{d}{dx}\left(\frac{\gamma}{\gamma-1}\,\frac{p}{\rho}-\tfrac{4}{3}\nu\,\frac{du}{dx}-\frac{\lambda}{\rho u}\,\frac{dT}{dx}+\tfrac{1}{2}u^2\right)=0.$$

Integrating through the transition zone and using $du/dx=0$, $dT/dx=0$ in Regions 1, 2,

$$\frac{\gamma}{\gamma-1}\,\frac{p_1}{\rho_1}+\tfrac{1}{2}u_1^2=\frac{\gamma}{\gamma-1}\,\frac{p_2}{\rho^2}+\tfrac{1}{2}u_2^2. \tag{3}$$

Note that equations (1), (2), (3) are just the same as those which were obtained by assuming the shock wave to be a plane discontinuity. However, the methods used in their derivation may also be used to establish conditions within the finite shock zone and to obtain an estimate of its thickness.

9.13.2 MHD Effects in Shock Formation

Now suppose the gas considered in the above section is conducting and that there is a magnetic field of intensity $\mathbf{H}=H(x)\mathbf{j}$ at right-angles to its motion where H varies continuously from H_1 in Region 1 to H_2 in Region 2. Then the presence of the magnetic field will increase the pressure by $(\mu H^2/8\pi)$. The magnetic energy per unit volume of the medium is $(\mu H^2/8\pi)$, a result well known in electromagnetic theory.

The equation of conservation of mass remains unaltered:

$$\rho_1u_1=\rho_2u_2=m. \tag{1}$$

The momentum equation $p_1+\rho_1u_1^2=p_2+\rho_2u_2^2$ obtained in Section 9.13.1 is modified to

$$p_1+\rho_1u_1^2+(\mu H_1^2/8\pi)=p_2+\rho_2u_2^2+(\mu H_2^2/8\pi). \tag{2}$$

In Section 9.13.1 we obtained the energy equation

$$\lambda A\delta x\,\frac{d^2T}{dx^2}=A\delta x\,\frac{d}{dx}\left(pu-\tfrac{4}{3}\rho\nu u\,\frac{du}{dx}\right)+\frac{A\rho\delta x}{\gamma-1}\,u\,\frac{d}{dx}\left(\frac{p}{\rho}\right)+(\rho A\delta x)u^2\frac{du}{dx}.$$

To account for the effects of the magnetic field we must modify the effective pressure from $p-\tfrac{4}{3}\rho\nu(du/dx)$ to $p+(\mu H^2/8\pi)-\tfrac{4}{3}\rho\nu(du/dx)$, and also to the R.H.S. we must add the rate of gain of magnetic energy of the elment. Now the magnetic energy of the element of mass $\rho A\delta x$ is $(A\delta x)(\mu H^2/8\pi) = (A\rho\delta x)(\mu H^2/8\pi\rho)$. Hence the rate of increase of this magnetic energy is

$$\frac{d}{dt}\left\{(A\rho\delta x)\left(\frac{\mu H^2}{8\pi\rho}\right)\right\} = (A\rho\delta x)\frac{d}{dt}\left(\frac{\mu H^2}{8\pi\rho}\right) = A\rho u\delta x\frac{d}{dx}\left(\frac{\mu H^2}{8\pi\rho}\right).$$

Hence the modified energy equation becomes

$$\lambda A\delta x\frac{d^2T}{dx^2} = A\delta x\frac{d}{dx}\left(pu + \frac{\mu H^2 u}{8\pi} - \tfrac{4}{3}\rho\nu\frac{du}{dx}\right) + \frac{A\rho\delta x}{\gamma-1}u\frac{d}{dx}\left(\frac{p}{\rho}\right)$$
$$+ (\rho A\delta x)u^2\frac{du}{dx} + (A\rho u\delta x)\frac{d}{dx}\left(\frac{\mu H^2}{8\pi\rho}\right).$$

Dividing through by $\rho uA\delta x$ gives, since $\rho u = \text{const.}$,

$$\frac{d}{dx}\left\{\frac{\gamma}{\gamma-1}\frac{p}{\rho} + \frac{\mu H^2}{4\pi\rho} + \tfrac{1}{2}u^2 - \tfrac{4}{3}\nu\frac{du}{dx} - \frac{\lambda}{\rho u}\frac{dT}{dx}\right\} = 0.$$

Integrating through the shock zone and using $du/dx = 0$, $dT/dx = 0$ in Regions 1, 2,

$$\frac{\gamma}{\gamma-1}\frac{p_1}{\rho_1} + \frac{\mu H_1^2}{4\pi\rho_1} + \tfrac{1}{2}u_1^2 = \frac{\gamma}{\gamma-1}\frac{p_2}{\rho_2} + \frac{\mu H_2^2}{4\pi\rho_2} + \tfrac{1}{2}u_2^2. \tag{3}$$

We thus have three equations for the four ratios ρ_2/ρ_1, u_2/u_1, p_2/p_1, H_2/H_1. A fourth one arises from the electromagnetic relation

$$\frac{\partial \mathbf{H}}{\partial t} = \nabla \wedge (\mathbf{v} \wedge \mathbf{H}) + \eta\nabla^2\mathbf{H}. \quad \left(\eta = \frac{1}{4\pi\mu\sigma}\right).$$

This assumes that the gas in the shock zone is not a perfect conductor. Here $\partial\mathbf{H}/\partial t = \mathbf{0}$ for a steady field, $\mathbf{H} = H(x)\mathbf{j}$, $\mathbf{v} = u(x)\mathbf{i}$, so that

$$\mathbf{v} \wedge \mathbf{H} = uH\mathbf{k},$$

and $$\nabla \wedge (\mathbf{v} \wedge \mathbf{H}) = \begin{Vmatrix} \dfrac{\partial}{\partial x} & \dfrac{\partial}{\partial y} & \dfrac{\partial}{\partial z} \\ \\ 0 & 0 & u(x)H(x) \end{Vmatrix} = -\frac{d(uH)}{dx}\mathbf{j}.$$

Thus $-d(uH)/dx + \eta\, d^2H/dx^2 = 0$. Integrating this through the shock zone,

$$u_1H_1 = u_2H_2. \tag{4}$$

The detailed solution of equations (1)–(4) will not be attempted here. It may be shown, though, that one of the effects of the magnetic field is to decrease the strength $(p_2 - p_1)/p_1$ of the shock. Also, for a non-conducting gas the velocity in Region 1 must be greater than the speed of sound a_1 of the medium, but with the transverse field present for a conducting gas shock formation requires $u_1 > \sqrt{(a_1^2 + V_{A1}^2)}$.

9.14 Laminar Flow of a Viscous Conducting Liquid between Parallel Walls in a Transverse Magnetic Field

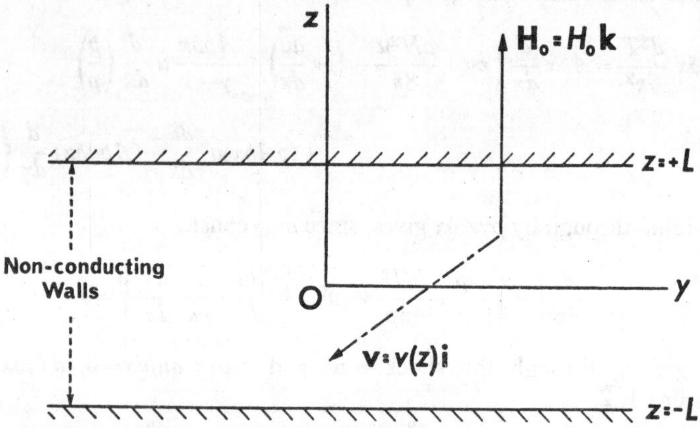

FIG. 9.8

We have seen (Section 8.10.1) that the velocity profile of a viscous fluid flowing between two parallel plates is parabolic. Here we treat the case of a highly conducting viscous liquid, such as mercury, flowing between two parallel non-conducting planes in a uniform transverse magnetic field perpendicular to the planes. As the fluid particles tend to bind themselves to the magnetic field it is obvious that the field will in some way inhibit the motion of the liquid. The motion will produce tension in the lines of force, but because of the finite conductivity they can revert to their initial positions.

We take the planes to be $z = \pm L$ and the magnetic field across them $\mathbf{H}_0 = H_0\mathbf{k}$ (Fig. 9.8). The motion of the fluid across the field will induce electric current at right angles to the velocity $\mathbf{v} = v(z)\mathbf{i}$ of the liquid and to the uniform magnetic field $\mathbf{H}_0 = H_0\mathbf{k}$, and the Lorentz force on the moving stream will oppose the motion together with the viscous forces. The velocity satisfies the continuity equation $\nabla \cdot \mathbf{v} = 0$.

The motion of the liquid will produce a perturbation field intensity $\mathbf{h} = h(z)\mathbf{i}$ as the tendency is for the liquid to drag the lines of force along in its direction of motion. Thus the total field is

$$\mathbf{H} = \mathbf{H}_0 + \mathbf{h}.$$

This satisfies $\nabla \cdot \mathbf{H} = 0$.

We assume the pressure in the liquid to be of the form

$$p = p_0(x) + p_1(z).$$

The term $p_0(x)$ gives rise to a pressure gradient $-dp_0/dx$ in the direction of motion: $p_1(z)$ is ascribable to hydrostatic causes.

For steady flow $\partial\mathbf{H}/\partial t=\mathbf{0}$, and the magnetic induction equation becomes

$$\nabla\wedge(\mathbf{v}\wedge\mathbf{H})+\eta\nabla^2\mathbf{H}=\mathbf{0}\quad(\eta=1/4\pi\mu\sigma). \tag{1}$$

The general equation of motion

$$\rho\frac{d\mathbf{v}}{dt}=-\nabla p+\rho\mathbf{g}+\frac{\mu}{4\pi}(\nabla\wedge\mathbf{H})\wedge\mathbf{H}+\tfrac{1}{3}\rho\nu\nabla\nabla\cdot\mathbf{v}+\rho\nu\nabla^2\mathbf{v}$$

gives, for steady conditions,

$$\rho(\mathbf{v}\cdot\nabla)\mathbf{v}=-\nabla(p_0+p_1)-\rho g\mathbf{k}+(\mu/4\pi)(\nabla\wedge\mathbf{h})\wedge\mathbf{H}+\rho\nu\nabla^2\mathbf{v}. \tag{2}$$

Since

$$\mathbf{v}\wedge\mathbf{H}=-v(z)H_0\mathbf{j},$$

$$\nabla\wedge(\mathbf{v}\wedge\mathbf{H})=\begin{Vmatrix}\dfrac{\partial}{\partial x} & \dfrac{\partial}{\partial y} & \dfrac{\partial}{\partial z}\\ 0 & -H_0v(z) & 0\end{Vmatrix}=H_0v'(z)\mathbf{j}.$$

Thus (1) gives

$$\frac{d^2h}{dz^2}+4\pi\sigma\mu H_0\frac{dv}{dz}=0. \tag{3}$$

Equation (2) gives

$$\mathbf{0}=-\frac{dp_0}{dx}\mathbf{i}-\frac{dp_1}{dz}\mathbf{k}-\rho g\mathbf{k}+\frac{\mu}{4\pi}h'(z)\mathbf{j}\wedge(h\mathbf{i}+H_0\mathbf{k})+\rho\nu\frac{d^2v}{dz^2}\mathbf{i},$$

whence

$$-\frac{dp_0}{dx}+\frac{\mu H_0}{4\pi}\frac{dh}{dz}+\rho\nu\frac{d^2v}{dz^2}=0, \tag{4}$$

$$\frac{dp_1}{dz}+\rho g+\frac{\mu}{4\pi}h(z)\frac{dh}{dz}=0. \tag{5}$$

From (4),

$$\frac{\mu H_0}{4\pi}\frac{dh}{dz}+\rho\nu\frac{d^2v}{dz^2}=\frac{dp_0}{dx}=-P,\text{ a const.} \tag{6}$$

Thus for steady laminar flow, *the pressure gradient in the direction of motion is constant throughout the liquid.* From (5),

$$p_1(z)=C_1-(\mu h^2/8\pi)-\rho gz. \tag{7}$$

Integrating (3),

$$(dh/dz)+4\pi\mu\sigma H_0v=C_2. \tag{8}$$

From the equations $\nabla \wedge \mathbf{H} = 4\pi \mathbf{j}$, $\mathbf{j} = \sigma(\mathbf{E} + \mu \mathbf{v} \wedge \mathbf{H})$, $\nabla \wedge \mathbf{H} = h'(z)\mathbf{j}$, $\mathbf{v} \wedge \mathbf{H} = -H_0 v(z)\mathbf{j}$ and, taking $\mathbf{j} = [j_1, j_2, j_3]$,

$$j_1 = \sigma E_1 = 0, \tag{9}$$

$$j_2 = \sigma(E_2 - \mu H_0 v) = (1/4\pi)(dh/dz), \tag{10}$$

$$j_3 = \sigma E_3 = 0. \tag{11}$$

Thus

$$\mathbf{E} = [0, E_2, 0], \tag{12}$$

where

$$E_2 = \mu H_0 v + (1/4\pi\sigma)(dh/dz) = (C_2/4\pi\sigma),$$

from (8). Thus (8) gives

$$(dh/dz) + 4\pi\mu H_0 v = 4\pi\sigma E_2. \tag{13}$$

Equation (4) now gives

$$-\frac{dp_0}{dx} + \frac{\mu H_0}{4\pi}(4\pi\sigma E_2 - 4\pi\mu\sigma H_0 v) + \rho v \frac{d^2 v}{dz^2} = 0,$$

or

$$\rho v(d^2 v/dz^2) - \sigma\mu^2 H^2 v = -(P + \mu\sigma H_0 E_2) = \text{const.} \tag{14}$$

As there is no slipping on the boundaries, $v = 0$ when $z = \pm L$. Also, assuming the walls to be non-conducting we require $j_3 = 0$ at $z = \pm L$. These latter conditions are identically satisfied by (11). The required solution to (14) is now

$$v(z) = \frac{(P + \mu\sigma H_0 E_2)\{\operatorname{ch} M - \operatorname{ch}(Mz/L)\}}{\sigma\mu^2 H_0^2 \operatorname{ch} M}, \tag{15}$$

where M is a dimensionless quantity called the *Hartmann number* defined by

$$M = \mu H_0 L \sqrt{(\sigma/\rho v)}. \tag{16}$$

If there is no externally applied current, then

$$\int_{-L}^{L} j_2 \, dz = 0.$$

From (10),

$$j_2 = \sigma(E_2 - \mu H_0 v)$$

$$= \sigma\left[E_2 - \frac{\mu H_0 (P + \mu\sigma H_0 E_2)\{\operatorname{ch} M - \operatorname{ch}(Mz/L)\}}{\sigma\mu^2 H_0^2 \operatorname{ch} M}\right].$$

Integrating this from $-L$ to L and equating to zero, we find

$$E_2=(P/\mu\sigma H_0)(M\coth M-1), \tag{17}$$

whence (15) gives

$$v(z)=(PM/\sigma\mu^2H_0^2\,\mathrm{sh}\,M)\{\mathrm{ch}\,M-\mathrm{ch}\,(Mz/L)\}. \tag{18}$$

Equation (13) integrates to give

$$h(z)=\frac{4\pi PL}{\mu H_0}\left\{\frac{\mathrm{sh}\,(Mz/L)}{\mathrm{sh}\,M}-\frac{z}{L}\right\}, \tag{19}$$

choosing $h=0$ at $z=\pm L$ to satisfy continuity requirements of $\mathbf{H}$ at the boundaries.

From (18), the mean velocity over the section is

$$\bar{v}=\frac{1}{2L}\int_{-L}^{L}v(z)\,dz=\frac{P(M\,\mathrm{ch}\,M-\mathrm{sh}\,M)}{\sigma\mu^2H_0^2\,\mathrm{sh}\,M},$$

and so

$$v(z)=\frac{\bar{v}M\{\mathrm{ch}\,M-\mathrm{ch}\,(Mz/L)\}}{M\,\mathrm{ch}\,M-\mathrm{sh}\,M}. \tag{20}$$

For a weak magnetic field, $M\sim 0$. Let us find $v(z)$ when $M=0$. We have

$$\lim_{M\to 0}\frac{M\{\mathrm{ch}\,M-\mathrm{ch}\,(Mz/L)\}}{M\,\mathrm{ch}\,M-\mathrm{sh}\,M}$$

$$=\lim_{M\to 0}\frac{\mathrm{ch}\,M-\mathrm{ch}\,(Mz/L)+M\{\mathrm{sh}\,M-(z/L)\,\mathrm{sh}\,(Mz/L)\}}{M\,\mathrm{sh}\,M}$$

$$=\lim_{M\to 0}\frac{2\{\mathrm{sh}\,M-(z/L)\,\mathrm{sh}\,(Mz/L)\}+M\{\mathrm{ch}\,M-(z/L)^2\,\mathrm{ch}\,(Mz/L)\}}{\mathrm{sh}\,M+M\,\mathrm{ch}\,M}$$

$$=\lim_{M\to 0}\frac{3\{\mathrm{ch}\,M-(z/L)^2\,\mathrm{ch}\,(Mz/L)\}+M\{\mathrm{sh}\,M-(z/L)^3\,\mathrm{sh}\,(Mz/L)\}}{2\,\mathrm{ch}\,M+M\,\mathrm{sh}\,M}$$

$$=\tfrac{3}{2}\{1-(z/L)^2\}.$$

Hence $v(z)=\frac{3}{2}\bar{v}\{1-(z/L)^2\}$, when $M=0$. This recovers the parabolic profile obtained for viscous laminar flow in the absence of a magnetic field. Figure 9.9 gives a sketch of the velocity profiles for various values of Hartmann number.

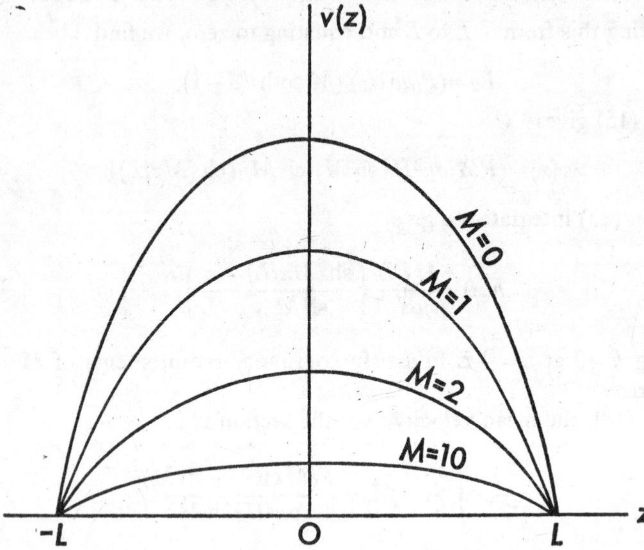

FIG. 9.9

EXERCISE 9

1. (a) Starting from the induction equation

$$\partial \mathbf{H}/\partial t = \text{curl}(\mathbf{v} \wedge \mathbf{H})$$

for an infinitely conducting fluid show that the magnetic flux across any closed contour moving with the fluid remains constant. Interpret this result in terms of the motion of the lines of force.

(b) An infinitely conducting fluid moves with velocity $\mathbf{v}$ in two dimensions relative to rectangular Cartesian axes x, y, z so that $\mathbf{v}=[v_x, v_y, 0]$. If the vector potential $\mathbf{A}$ of the magnetic field $\mathbf{H}$ ($\mathbf{H}=\text{curl } \mathbf{A}$) is such that $\mathbf{A}=[0, 0, A]$ and all variables are independent of z, show that

(i) $\partial \mathbf{A}/\partial t = \mathbf{v} \wedge \mathbf{H}$,

(ii) $\partial A/\partial t = -(\mathbf{v} \cdot \nabla)A$,

and (iii) $dA/dt = 0$.

Show that this last result is a special case of the theorem contained in part (a).

(London Univ., B.Sc. Spec. Hons., 1963)

2. Show that the vector potential of the magnetic field $\mathbf{H}$ in an infinitely conducting fluid moving with velocity $\mathbf{v}$ may be defined so that

$$(\partial \mathbf{A}/\partial t) \cdot \text{curl } \mathbf{A} = 0.$$

Deduce that if the vector potential is maintained constant ($\partial \mathbf{A}/\partial t = 0$) over a fixed surface S enclosed in a volume of infinitely conducting liquid τ,

$$I=\int_\tau \mathbf{A}\cdot \text{curl}\,\mathbf{A}\,d\tau=\text{constant}.$$

Show also that the stationary values of the *magnetic energy* subject to the constraint $I=$constant with $\delta\mathbf{A}=0$ on S correspond to the force-free fields curl $\mathbf{H}=\alpha\mathbf{H}$, where α is constant.

(London Univ., B.Sc. Spec. Hons., 1963)

3. *Steady parallel flow of a conducting liquid in a conservative field of force.* In the notation of Section 9.4, show that if a conducting liquid of uniform density ρ moves in a magnetic field of intensity $\mathbf{H}$ and in a field of conservative body forces having potential Ω per unit volume (so that $\mathbf{F}=-\nabla\Omega$) then when displacement currents are negligible and there is no volume distribution of electric charge, the equation of steady motion is

$$\rho\mathbf{v}\cdot\nabla\mathbf{v}=-\nabla(p+\Omega)+\frac{\mu}{4\pi}(\nabla\wedge\mathbf{H})\wedge\mathbf{H}+\rho\nu\nabla^2\mathbf{v}.$$

Flow for which $\mathbf{H}$ and $\mathbf{v}$ are always parallel so that $\mathbf{H}=\lambda\mathbf{v}$ is called *parallel flow.* Show that for such flow the above equation becomes

$$\left(\rho-\frac{\lambda^2\mu}{4\pi}\right)\mathbf{v}\cdot\nabla\mathbf{v}=-\nabla\left(p+\Omega+\frac{\mu\lambda^2\mathbf{v}^2}{8\pi}\right)+\rho\nu\nabla^2\mathbf{v}.$$

(First show that $\nabla\cdot\mathbf{H}=0$ and $\nabla\cdot\mathbf{v}=0$ lead to

$$(\nabla\wedge\mathbf{H})\wedge\mathbf{H}=-\nabla(\tfrac{1}{2}\lambda^2\mathbf{v}^2)+\lambda^2\mathbf{v}\cdot\nabla\mathbf{v}.)$$

Defining the 'effective pressure' $\bar{p}$ and the 'effective density' $\bar{\rho}$ by

$$\bar{p}=p+\Omega+\frac{\mu\lambda^2\mathbf{v}^2}{8\pi},$$

$$\bar{\rho}=\rho-\frac{\lambda^2\mu}{4\pi},$$

show that this equation reduces to the 'Navier–Stokes form'

$$\bar{\rho}\mathbf{v}\cdot\nabla\mathbf{v}=-\nabla\bar{p}+\rho\nu\nabla^2\mathbf{v}$$

subject to $\nabla\cdot\mathbf{v}=0$ and $\mathbf{v}\cdot\nabla\bar{\rho}=0$. The last relation shows that $\bar{\rho}$ is constant along the streamlines.

In the special case when the flow is inviscid, show further that

$$\mathbf{v}\cdot\nabla(\bar{p}+\tfrac{1}{2}\bar{\rho}\mathbf{v}^2)=0$$

and hence deduce 'Bernoulli's equation for steady incompressible parallel inviscid flow' in the form

$$\bar{p}+\tfrac{1}{2}\bar{\rho}\mathbf{v}^2=\text{constant along a streamline.}$$

4. If ψ represents the velocity stream function in an incompressible

infinitely conducting fluid and Ψ represents the magnetic stream function, and the flow is two-dimensional so that the Cartesian components of **v** and **H** are given by

$$\mathbf{v}=[-\partial\psi/\partial y, \partial\psi/\partial x, 0], \quad \mathbf{H}=[-\partial\Psi/\partial y, \partial\Psi/\partial x, 0],$$

show that the induction equation reduces to

$$\frac{\partial\Psi}{\partial t}=\frac{\partial(\Psi,\psi)}{\partial(x,y)}.$$

Show further that $d\Psi/dt=0$ and deduce that the lines of force move with the fluid. Show further that when conditions are steady, $\Psi=f(\psi)$, so that the streamlines and the lines of magnetic force coincide.

5. *Couette Flow.* A viscous incompressible conducting liquid of uniform density ρ is confined between the horizontal conducting planes $z=0$ (lower) and $z=L$ (upper). A uniform magnetic field $H_0\mathbf{k}$ acts vertically upwards. The lower plane is held at rest and the upper one is moved horizontally with uniform velocity $V\mathbf{j}$, there being no pressure gradient in the liquid. If, when conditions are steady, the velocity of the liquid at (x, y, z) is $v(z)\mathbf{j}$, and the new field is $\mathbf{H}=H_0\mathbf{k}+h(z)\mathbf{j}$, prove that

$$v(z)=\frac{V\,\mathrm{sh}\,(Mz/L)}{\mathrm{sh}\,M} \quad \left[M=L\left(\frac{\mu^2H_0^2\sigma}{\rho\nu}\right)^{1/2}\right],$$

and show that the pressure $p(z)$ at this point satisfies

$$p(z)+\rho gz+\frac{\mu}{8\pi}h^2(z)=\text{constant}$$

where the symbols have their usual meanings.

APPENDIX

CHARACTERISTICS OF SECOND ORDER PARTIAL DIFFERENTIAL EQUATIONS

1. Cauchy's Problem for the Solution of Second Order Partial Differential Equations

Suppose it is desired to find the solution of the partial differential equation

$$Ar+2Bs+Ct=D, \tag{1}$$

where x and y are the independent variables, z the dependent variable, $p=\partial z/\partial x$, $q=\partial z/\partial y$, $r=\partial^2 z/\partial x^2$, $s=\partial^2 z/\partial x\partial y$, $t=\partial^2 z/\partial y^2$, and A, B, C, D are each general functions of x, y, z, p, q. Suppose further that the solution to (1) is to pass through a given curve $\mathscr{C}$ in space and having the parametric equations

$$x=x(T), \quad y=y(T), \quad z=z(T). \tag{2}$$

We also suppose that at each point (x, y, z) of $\mathscr{C}$, p and q are known. Then the solution will be some surface—'integral surface'—passing through $\mathscr{C}$. The solution will be of the form $z=z(x, y)$. Hence at each point of $\mathscr{C}$, in virtue of the relations (2),

$$\dot{z}=p\dot{x}+q\dot{y}, \tag{3}$$

the dots here signifying total derivatives of x, y, z w.r.t. T. Furthermore, since $p=p(x, y)$, $q=q(x, y)$, at each such point of $\mathscr{C}$,

$$\dot{p}=r\dot{x}+s\dot{y}, \tag{4}$$

$$\dot{q}=s\dot{x}+t\dot{y}. \tag{5}$$

Knowing A, B, C, D, $\dot{x}, \dot{y}$, p, q, $\dot{p}, \dot{q}$ at each point of $\mathscr{C}$, we may regard equations (1), (4), (5) as linear simultaneous equations for the unknowns r, s, t at each point of $\mathscr{C}$. In fact their solution is given by Cramer's rule:

$$\frac{r}{\Delta_1}=\frac{s}{\Delta_2}=\frac{t}{\Delta_3}=\frac{1}{\Delta} \tag{6}$$

provided $\Delta \neq 0$, where

$$\Delta = \begin{vmatrix} A & 2B & C \\ \dot{x} & \dot{y} & 0 \\ 0 & \dot{x} & \dot{y} \end{vmatrix},$$

and $\Delta_1, \Delta_2, \Delta_3$ are generated from Δ by replacing the elements in its first, second, and third columns by $D, \dot{p}, \dot{q}$ respectively. Thus the equations (1), (4), (5) enable us to evaluate the second derivatives r, s, t at each considered point of (6).

We now show that all higher order partial derivatives of z at each point of $\mathscr{C}$ may be evaluated and hence, using a Taylor expansion, that an elemental integral strip may be constructed through each point of $\mathscr{C}$.

Differentiating (1) partially w.r.t. x gives the form

$$Az_{xxx} + 2Bz_{xxy} + Cz_{xyy} = D_1 \tag{7}$$

and differentiating (1) partially w.r.t. y gives

$$Az_{xxy} + 2Bz_{xyy} + Cz_{yyy} = D_2, \tag{8}$$

where $z_{xxx} = \partial^3 z/\partial x^3$, etc., and D_1, D_2 are each functions of x, y, z, p, q, r, s, t only. Furthermore, since $r = r(x, y)$, etc.,

$$\dot{r} = z_{xxx}\dot{x} + z_{xxy}\dot{y}, \tag{9}$$

$$\dot{s} = z_{xyx}\dot{x} + z_{xyy}\dot{y}, \tag{10}$$

$$\dot{t} = z_{yyx}\dot{x} + z_{yyy}\dot{y}, \tag{11}$$

so that the third order derivatives $z_{xxx}, z_{xyx}, z_{xxy}, z_{yyy}$ may be found from any four of the equations (7)–(11). Proceeding in this way it is clear *that partial derivatives of all orders of $z(x, y)$ may be evaluated at each point of $\mathscr{C}$ if they are known to exist.*

Now suppose the curve $\mathscr{C}_0$ is the orthogonal projection of $\mathscr{C}$ on the plane $z=0$. Further suppose $P(x_0, y_0, z_0)$ is any point on $\mathscr{C}$ and $P_0'(x_0, y_0, 0)$ its projection on $z=0$ so that P_0 lies on $\mathscr{C}_0$. Let $P_0'(x_0+h, y_0+k, 0)$ be a point on $\mathscr{C}_0$ near to P_0, $|h|$ and $|k|$ being small quantities. Let us evaluate the ordinate z at P_0'. By Taylor's theorem for two variables we have

$$z = z_0 + \left(h\frac{\partial}{\partial x} + k\frac{\partial}{\partial y}\right) z_0 + \frac{1}{2!}\left(h\frac{\partial}{\partial x} + k\frac{\partial}{\partial y}\right)^2 z_0 + \dots \tag{12}$$

Since all derivatives of z at P_0 are calculable, (12) enables z to be evaluated. Hence we can construct the point P' on the integral surface corresponding to P_0. If $|h|, |k| < \varepsilon$, the radius of convergence of (12), then at each point of P_0 of $\mathscr{C}_0$ we can draw a small circle, centre P_0 and radius ε. These circles overlap to form a strip in the plane $z=0$. Within each point $(x, y, 0)$ of a circle we can find the ordinate z which enables the corresponding point (x, y, z) on the integral surface through $\mathscr{C}$ to be obtained: the aggregate of all such points gives an integral strip through $\mathscr{C}$. This is illustrated in Fig. A1.

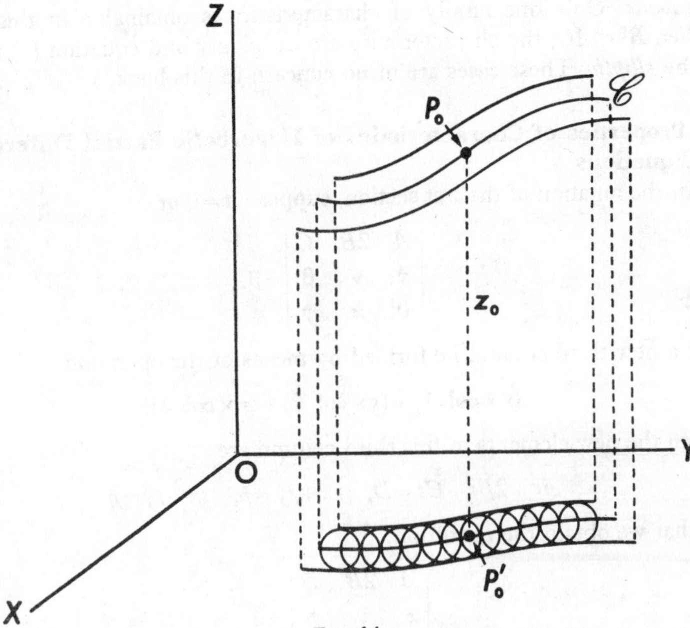

FIG. A1.

The above argument presupposes that $\Delta \neq 0$. When $\Delta = 0$, successive derivatives are not calculable in the manner described and Cauchy's method of solution breaks down. This critical case leads to the condition

$$A\dot{y}^2 - 2B\dot{x}\dot{y} + C\dot{x}^2 = 0, \tag{13}$$

or

$$A\left(\frac{dy}{dx}\right)^2 - 2B\left(\frac{dy}{dx}\right) + C = 0. \tag{13'}$$

At each point $(x, y, 0)$ of $\mathscr{C}_0$, (13) or (13′) would give a pair of directions along which $\Delta = 0$. These directions are said to be *characteristic*. They are real and distinct, coincident or imaginary according as the discriminant $B^2 - AC$ is positive, zero or negative.

When $B^2 - AC > 0$, a pair of real characteristic directions can be drawn through each point $(x, y, 0)$ of $\mathscr{C}_0$, and the original partial differential equation (1) is said to be *hyperbolic*. In this case a pair of real characteristic curves can be drawn through every point $(x, y, 0)$ of the base curve $\mathscr{C}_0$, simply by approximating them by straight-line segments whose directions are taken to coincide with those of the tangents given by the roots for dy/dx obtained from (13′). This case is realized in supersonic flow and it forms the basis of the theory developed in Chapter 7. It affords a process of numerical integration alternative to the Cauchy construction.

When $B^2 = AC$, the roots of (13') coincide and equation (1) is said to be *parabolic*. Only one family of characteristics is obtainable in this case. When $B^2 < AC$, the characteristics are imaginary and equation (1) is said to be *elliptic*. These cases are of no concern in this book.

2. Properties of Characteristics of Hyperbolic Partial Differential Equations

In the notation of the last section, suppose $\Delta = 0$ or

$$\begin{vmatrix} A & 2B & C \\ \dot{x} & \dot{y} & 0 \\ 0 & \dot{x} & \dot{y} \end{vmatrix} = 0.$$

Let a new third column be formed by means of the operation

$$(r \times \text{col. } 1) + (s \times \text{col. } 2) + (r \times \text{col. } 3).$$

Then the new elements in this third column are

$$Ar + 2Bs + Ct = D, \quad r\dot{x} + s\dot{y} = \dot{p}, \quad s\dot{x} + t\dot{y} = \dot{q}$$

so that we obtain the condition

$$\begin{vmatrix} A & 2B & D \\ \dot{x} & \dot{y} & \dot{p} \\ 0 & \dot{x} & \dot{q} \end{vmatrix} = 0. \tag{14}$$

Equation (14) expresses the conditions which hold along the characteristics $\Delta = 0$ when these are real. In a similar way it is easy to show that all such third order determinants vanish in the following 3×4 array:

$$\begin{Vmatrix} A & 2B & C & D \\ \dot{x} & \dot{y} & 0 & \dot{p} \\ 0 & \dot{x} & \dot{y} & \dot{q} \end{Vmatrix}.$$

In Chapter 7 the corresponding results in supersonic flow theory yield important physical results which hold along the characteristics.

A further important result is the following: *if there is a discontinuity at any point of a characteristic then there must be one at all points of the characteristic.* In effect this means physically that discontinuities in supersonic flow are propagated along the characteristics. This result is proved in *Elements of Partial Differential Equations* by I. N. Sneddon, pp. 111–14 (McGraw-Hill).

SOLUTIONS

EXERCISE 1

3. $(\lambda x_2-\mu x_1)/(\lambda-\mu)$, $(\lambda y_2-\mu y_1)/(\lambda-\mu)$, $(\lambda z_2-\mu z_1)/(\lambda-\mu)$.
6. (i) $n|\mathbf{r}|^{n-2}\mathbf{r}\cdot\mathbf{a}$; (ii) $(n+2)r^n\mathbf{a}-nr^{n-2}(\mathbf{r}\cdot\mathbf{a})\mathbf{r}$; (iii) 1; (iv) 2.
7. curl $\mathbf{u}=0$. Let $\mathbf{a}=[a_1, a_2, a_3]$.
   (i) No; (ii) Yes: $\varphi=c-\frac{1}{2}(a_1^2x^2+a_2^2y^2+a_3^2z^2+2a_2a_3yz+2a_3a_1zx+2a_1a_2xy)$; (iii) No; (iv) No.
10. $6\pi a^3$.
16. $V=5-r^2$ [Solve $(1/r^2)\,d/dr\,(r^2\,dV/dr)=-6$, where $dV/dr=-2$, $V=4$ when $r=1$].
17. $U=\frac{1}{4}r^2(1-3\cos^2\theta)$.
18. $\dfrac{a^2-b^2}{a}\displaystyle\int_S\frac{V\,ds}{R^3}-4\pi V=0.$
20. $\dfrac{\partial}{\partial\rho_1}\left[\dfrac{\rho_1^3}{\rho_1^2+\rho_2^2}\cdot\dfrac{\partial V}{\partial\rho_1}\right]+\dfrac{\partial}{\partial\rho_2}\left[\dfrac{\rho_2^4}{\rho_1(\rho_1^2+\rho_2^2)}\dfrac{\partial V}{\partial\rho_2}\right]+\dfrac{\rho_1}{\rho_1^2+\rho_2^2}\dfrac{\partial^2V}{\partial\varphi^2}=0.$

EXERCISE 2

1. $\zeta=0$.
6. (i) Hyperbolic cylinders; (ii) Paraboloids of revolution.

EXERCISE 3

7. (i) $p=P_0+\rho g(h-z)-\frac{1}{2}\rho\omega^2a^2$; (ii) $p=P_0+\rho gh-\frac{1}{2}\rho\omega^2a^4/r^2$.
16. Impulse $=\frac{2}{3}\pi a^3\rho U\{(2\sigma/\rho)+(2a^3+b^3)/(b^3-a^3)\}=2T/U$.
18. First show velocity potential due to motion of inner sphere alone is of

form $(A+B/r^3)x$. Then show that due to motion of outer sphere alone is of form $(C+D/r^3)y$. Take $\varphi=(A+B/r^3)x+(C+D/r^3)y$ to give stated form.

$$\mathbf{q}=\{-a^3(U+\tfrac{1}{2}b^3Ur^{-3}+\tfrac{3}{2}b^3Vxyr^{-5}-\tfrac{3}{2}b^3Ux^2r^{-5})\mathbf{i}+b^3(V+\tfrac{1}{2}a^3Vr^{-3}-\tfrac{3}{2}a^3Vy^2r^{-5}+\tfrac{3}{2}a^3Uxyr^{-5})\mathbf{j}+\tfrac{3}{2}a^3b^3r^{-5}(Ux-Vy)z\mathbf{k}\}/(b^3-a^3).$$

Exercise 4

2. $r(dr/d\theta)+r^2\cot\theta=-m/(U\sin\theta)$.
5. (i) $\psi=\frac{1}{2}Ur^2\sin^2\theta$, for unif. stream parallel to $\overline{OX}$;
   (ii) $\psi=(m/a)\{r-\sqrt{(r^2-2ra\cos\theta+a^2)}\}$ for uniform line source of strength m from $(0,0,0)$ to $(a,0,0)$. Thickness $=2\sqrt{(m/U)}$.
6. $p=p_\infty+\frac{1}{2}\rho U^2\cos^2(\theta/2)\{1-3\sin^2(\theta/2)\}$.
7. When $2m>3Ua^2$, the particle drains away through the sink or is repelled from the source farther away from the plane $x=0$.

Exercise 5

1. Impulse components per unit length $=\pi\rho a^2[-u^2(a^2+b^2), 2b^2v^2]/(b^2-a^2)$.
3. Take $\varphi=\{g(t)\cos\theta\}/r$ and find $g(t)$.
7. Pressure at $(0,y)=-2\rho m^2y^2(y^2+c^2)^2$. Along positive x-axis.
9. Modify formula to $T=-\dfrac{\rho}{2}\left(\iint_{S_1}+\iint_{S_2}\right)\psi\dfrac{\partial\psi}{\partial n}dS$; $\psi=\frac{1}{2}(K/\pi)\log r$.
10. K.E. per unit length $=\pi\rho U^2a(3+5/\sqrt{3})$.
14. $w=U\{z-(a^2/z)\}-m\log(z-2a)-m\log\{(a^2/z)-2a\}$.
16. $w=\log\{z(z^2+1)/(z-1)^4\}$; vel. $=[-2/5, 7/60]$.
19. Put $\psi=k(1-x^2/a^2-y^2/b^2)$ and find $k=\frac{1}{2}\omega a^2b^2/(a^2+b^2)$. Press. max. at 0.
22. C is circle centre $z=i(b-a^2/b)$ and radius c/b.
25. Force per unit length $=[0, -\frac{1}{2}\rho k^2/a]$.
26. Force per unit length $=[\frac{1}{3}\rho\pi\kappa^2/a, 0]$.
29. $w=V(\zeta e^{-i\alpha}+e^{i\alpha}/\zeta)$

$$=V\left\{e^{-i\alpha}\left(\frac{z+\{z^2-(a^2-b^2)\}^{1/2}}{a+b}\right)+e^{i\alpha}\left(\frac{z-\{z^2-(a^2-b^2)\}^{1/2}}{a-b}\right)\right\};$$

$M=\pi\rho(b^2-a^2)\sin\alpha\cos\alpha$.

Exercise 6

6. $\Delta U=(p_2V_2-p_1V_1)/(\gamma-1)$.

Exercise 7

8. (i) $q > a$.

(ii)
$$\begin{vmatrix} (a^2-u^2) & u(a^2+v^2)/r & (a^2-v^2)/r^2 \\ dr & -du & 0 \\ 0 & -r\,dv - v\,dr & d\theta \end{vmatrix} = 0.$$

Exercise 8

4. Viscous drag per unit length $= \dfrac{2\pi W\mu}{\log(a/b)} - g\pi a\rho\left\{a - \dfrac{b^2-a^2}{2a\log(b/a)}\right\}$.

10. $h^2 = \{2\pi^2\mu x/\rho U(4-\pi)\}$.

INDEX